THE DISSECTION OF VERTEBRATES

A Laboratory Manual

THIRD EDITION

THE DISSECTION OF VERTEBRATES

A Laboratory Manual

THIRD EDITION

GERARDO DE IULIIS, PhD

*Department of Ecology and Evolutionary Biology, University of Toronto;
Royal Ontario Museum; and George Brown College of Art and Technology*

DINO PULERÀ, MScBMC, CMI

ELSEVIER

ACADEMIC PRESS
An imprint of Elsevier

Academic Press is an imprint of Elsevier
125 London Wall, London EC2Y 5AS, United Kingdom
525 B Street, Suite 1650, San Diego, CA 92101, United States
50 Hampshire Street, 5th Floor, Cambridge, MA 02139, United States
The Boulevard, Langford Lane, Kidlington, Oxford OX5 1GB, United Kingdom

Library of Congress Cataloging-in-Publication Data
A catalog record for this book is available from the Library of Congress

British Library Cataloguing-in-Publication Data
A catalogue record for this book is available from the British Library

ISBN: 978-0-12-410460-0

For information on all Academic Press publications visit our website at
https://www.elsevier.com/books-and-journals

Publisher: Andre Gerhard Wolff
Acquisition Editor: Andre Gerhard Wolff
Editorial Project Manager: Sandra Harron
Production Project Manager: Kiruthika Govindaraju
Cover Designer: Dino Pulerà

Typeset by TNQ Technologies

Working together
to grow libraries in
developing countries

www.elsevier.com • www.bookaid.org

This book is for our spouses,
Virginia and Cinzia,
and children,
Daniel, Theodore, and Jacob,
whose love, support, encouragement, and infinite patience
were crucial for seeing this project to fruition,
and our parents,
Italo, Maria, Vittorio, and Pina,
for their love, support, and sacrifice
in providing us the opportunity to follow our dreams.

With deep respect, admiration, and gratitude, we dedicate this book to three teachers at the University of Toronto, who early on in our academic paths instilled in us a passion for anatomy, paleontology, and art as well as the intellectual discipline required to make these passions our careers. Their contributions are evident throughout the following pages.

Charles S. "Rufus" Churcher
Stephen G. Gilbert
Thomas S. Parsons

To our readers:

Despite our best efforts, some errors are bound to have escaped our notice, and we would appreciate being informed of these. We encourage you to contact us directly with your comments, suggestions, and possible ideas for text and illustrations in future editions of this book.
We look forward to hearing from you.

Sincerely,

Dr. Gerry De Iuliis
gerry.deiuliis@utoronto.ca

Dino Pulerà
dino.pulera@utoronto.ca

Contents

CONTENTS GUIDE

Preface

The resurgence of comparative vertebrate anatomy and the recognition of its broad value to many aspects of biological inquiry, including the history of vertebrates, have continued since the publication of the second edition of this manual—Budd and Olsson's (2007:1) assertion that there "has never been a better time to study morphology" rings truer than ever. Not only does vertebrate anatomy provide the only direct evidence for the steps, sequence, and timing of the changes that occurred during earlier parts of vertebrate evolution, but its value in helping elucidate the relationships among more recent members of the vertebrate clade is becoming increasingly apparent among researchers, particularly those that have often been dismissive of morphological data. This development, which might almost—somewhat facetiously—be described as a radical change has, resulted in several recent analyses that employ combined molecular and morphological data sets in trying to determine vertebrate relationships. Unfortunately, however, the study of anatomy at the introductory and intermediate college levels continues to suffer, and its stature and perceived importance have yet to regain their luster. The reasons remain the same as those noted in earlier editions of this manual—regrettably, the trend at most major academic institutions has followed a path away from whole organismal biology, and positions related to vertebrate anatomy seem to be fewer and farther between. Despite the intense (and certainly justified) focus on genetics over the last several decades, it remains "clear that morphology still stands at the heart of comparative biology. From functional morphology to ultrastructure, morphological topics link into ecology and physiology, developmental genetics and the fossil record; they connect them all" (Budd and Olsson, 2007:1).

Increased resistance from some quarters to the use of animals in various scientific endeavors has also hindered introductory study in vertebrate anatomy. Further, easily accessible computer software has been developed that allows convenient visual journeys through vertebrate bodies without the effort, expense, and "mess" of actual dissection.

The study of anatomy and morphology has much to offer the student wishing to pursue a career in biological or medical fields. Proper training in vertebrate anatomy must include a practical component that involves dissection, in addition to lectures. No other method, regardless of how intricate in presentation and scope, can replace the actual hands-on experience. It is only through a careful, patient, and repeated practical approach that we gain the expertise and practice required for understanding the spatial relationships that are essential to learning how a vertebrate body is constructed, how its component structures are related to each other, and how form and function interact.

There are those who would suggest that such a course of study is unnecessary and that anatomy can be learned through texts or software. While such materials (this text among them) may prove invaluable as aids or tools for learning, we ought not to substitute these adjuncts for the means through which we *must* come to know the vertebrate body. To do so would be akin to preparing for an acting career by watching films, rather than through rehearsing and attending acting workshops. Few of us would feel comfortable with mechanics trained solely through the Internet, trust a surgeon who has learned the craft strictly through instructional videos, or fly with a pilot who has only flown missions on a flight simulator. We would not be suspicious because we believe that such instructional methods are not useful. Rather, we recognize that for fields whose subject matter includes components arranged in complex spatial relationships, these media are meant to be used as tools that supplement and guide the trainee through a methodical, firsthand experience with the subject matter itself. And science, of course, is an empirical endeavor. When practiced, science is based on observation and experience. If we are to train students to become scientists, it is not sufficient to merely present them with the products of science; we must engage students in actual scientific practice.

The debate on the value of dissection is particularly lively for human medical anatomy (see, for example, Sugand et al., 2010; Roseth and Saltarelli, 2014; Ross, 2015; Van Wyk and Rennie, 2015). Many researchers are clearly in favor of dissection but also see the need to incorporate the advanced imaging technologies currently available. Indeed, a combination of the two techniques apparently leads to superior results (see Biasutto et al., 2006). The same logic should apply for any vertebrate, but similarly advanced technologies are unlikely to be applied to a broad range of vertebrates in the foreseeable future.

The central theme of most previous dissection manuals has been structural changes in vertebrates through their evolution from fish to mammals, with the ultimate goal being to place mammalian anatomy in context. This is certainly a necessary prerequisite for one interested principally in mammalian systematics or medicine. However, not all students or instructors are interested primarily in mammals. Two of the important lessons emphasized by phylogenetics are that all living vertebrates have as long an independent evolutionary history as mammals have, and that their anatomy has just as much to tell us about evolution, function, and morphology. Indeed, a common complaint among academic faculty is that comparative vertebrate anatomy courses have become courses on the anatomy of the cat. Be that as it may, it is important to remember that negative perceptions can often be detrimental to the well-being of a field of study and may sway departmental decisions on whether the continuation of some courses is worth the effort and expense. It is up to those of us who teach comparative anatomy to push forward and maintain its vigor and centrality—in part by relating its wide applicability to related fields, such as systematics, evolutionary biology, paleontology, paleobiology, and functional morphology—and as a prerequisite for higher level zoology courses, such as mammalogy, herpetology, ornithology, ichthyology, developmental biology, and vertebrate paleontology.

Our format and coverage is aimed at striking a balance between presenting an evolutionary sequence to "higher" vertebrates and the anatomy of each representative vertebrate as inherently important. The sequence of vertebrates is similar to those presented by other authors, but we must remember that living vertebrates are not and cannot be used as intermediates. For this reason, we provide discussions of the important features of each group based on the derived features that diagnose a particular phylogenetic grouping. We thus do not treat vertebrates by traditional grouping methods; we would rather, from the beginning, present the student with information that reflects our formal thinking and classification.

The main goal of this text is to provide today's visually oriented student population with a manual that links succinct and pedagogically effective textual direction with relevant, accurate, and attractive visual references to promote efficient learning of the complex, spatially abstract subject matter in the limited time available in a laboratory setting. Thus, a critical feature of *The Dissection of Vertebrates* is the inclusion of numerous high-quality, didactic, color illustrations. Each depicts the vertebrate approximately as it would appear in a particular stage of dissection, rather than presenting an idealized figure or photographs, as is the case for most other manuals. This in itself facilitates the use of these illustrations, both in learning and later during recall for studying purposes. Photographs are used sparingly. We have chosen illustration over photography

in the vast majority of cases because illustration is the method that affords the most control in communicating the pertinent features of a particular dissection. Photos are objective and indiscriminately record what the camera sees, whereas illustrations (or rather the illustrator) can be subjective, selective, and interpretative with the ability to omit unnecessary details and emphasize anatomy important to the purpose of the dissection. Indeed, we have taken great care to ensure that the background anatomy in the illustrations is accurate. This is important because it gives the users (instructor and student) a context for the anatomical structures under study.

Although students aspiring to careers in systematics, vertebrate paleontology or functional morphology are the primary intended audience of this manual, *The Dissection of Vertebrates* is sufficiently flexible in scope and organization that it may be used in any course on vertebrate anatomy. We present a wide-ranging and encompassing reference manual that will both help students learn the basic anatomy of vertebrates and function as a guide once they are ready to venture into the primary literature.

The Dissection of Vertebrates presents dissection instructions on more vertebrates than is normally the case, and this third edition also includes sections on "prevertebrates"—a cephalochordate and a urochordate. The primary focus is on the shark, mudpuppy, and cat, as is usual, but it also provides detailed information on vertebrates either not usually considered or treated very superficially by most other manuals that include multiple vertebrates. It is ironic that the two most speciose groups of vertebrates, birds and ray-finned fishes, are not adequately covered (if covered at all) in other dissection manuals. We hope that by providing reasonably detailed guides for these vertebrates, instructors will feel more inclined to include these readily available and inexpensive vertebrates in their courses.

This manual is organized by vertebrate. The anatomy of each is then presented systemically. This approach allows all the information on a particular vertebrate to be studied at one time and in sequence. We believe, based on years of instruction, that this method provides a more straightforward integration of the systems. The inclusion of many vertebrates and the organization by vertebrate makes *The Dissection of Vertebrates* more flexible for use in a broad-based full or half-year course at the introductory college level, and allows more convenient organization of course content, depending on time and availability of specimens and the instructor's preferences.

At the same time, we omit many topics that are often covered in most other manuals. Sections on vertebrates or structures that students are unlikely ever to dissect at the intended level of study are not included. Instead, we have focused the material on examples that are likely to be encountered in an introductory lab course, leaving those topics best presented in texts that accompany the lecture portion of a course.

Much of the required background information is presented in the Introduction and Chapter 1. This includes sections on planes of dissection and orientational terminology (see later), as well as an introduction to vertebrates and their relatives (Chapter 1). We suggest that these sections be included as part of the assigned readings for a particular laboratory for each vertebrate. This method will expose students repeatedly to the broad evolutionary development of each system. Terms that are required learning are placed in boldface print throughout the manual. Boldfaced terms are listed in a Key Terms section (which also provides common synonyms in parentheses) following each major component. Students will know at a glance the structures for which they are responsible. We suggest that students use this section as a key to learning the structures by writing a short description for each. The Key Terms sections also allow instructors to adapt this manual to their personal preferences in running their course. Structures that are not required can be identified and crossed out so students know they are not responsible for them. This method effectively allows an instructor to limit the detail of the dissection.

We believe that the concise presentation of dissection instructions combined with minimal background information results in a straightforward text that will facilitate and focus the student's learning of anatomy in laboratory. In contrast to most other manuals, much of the background material presented in lecture is omitted here, so *The Dissection of Vertebrates* is less cumbersome to use even though it covers more vertebrates than do other manuals. All the information is relevant for laboratory purposes. This should facilitate matters for the instructor as well. Among other things, it will allow a clear answer to the often-asked question, "What am I responsible for reading?" The response can be, without too much exaggeration, "All of it."

Acknowledgments

Many colleagues, students, friends, and members of our families have contributed to the publication of the first, second, and now third editions of this book, from carrying out simple tasks, to proofreading, to providing emotional encouragement and support. We are grateful to them all, although we can directly acknowledge only a few of them here: Drs. Matthew F. Bonnan (Western Illinois University), Thomas Carr (Carthage College), Randy Lauff (St. Francis Xavier University), Hans-Dieter Sues (Smithsonian Institution), Jeff Thomason (University of Guelph), and Sergio F. Vizcaíno (Museo de La Plata) for their thorough, methodical, and thoughtful reviews of earlier versions of this book and providing valuable input on text and illustrations. We are also grateful to several anonymous reviewers for their helpful comments, suggestions, and corrections.

We are indebted to our intellectual forbears for imparting not only their knowledge but for impressing upon us the need for rigor and discipline, and to our many colleagues for the resources they have produced—it is clear that we have leaned heavily on their efforts in producing this manual. Many of their works are indicated explicitly in the following pages, but several others, still relevant and useful, are noted here. Prominent among them is the work of the late Thomas S. Parsons and the late Stephen G. Gilbert (e.g., Gans and Parsons, 1964; Romer and Parsons, 1986; and Gilbert 1973a,b, 1974, 1989, 1991, and 2000); both have had a strong influence on our lives and careers. In addition, we acknowledge the work of Chiasson (1962, 1966, 1984), Chiasson and Radke (1993), Cooper and Schiller (1975), Ecker (1971), Frank (1934), George and Berger (1966), King and McLelland (1985), Marinelli and Stregner (1954), Oldham and Smith (1975), Romer (1997), Stuart (1947), Wake (1979), White and Folkens (2005), Wyneken (2001), Wischnitzer and Wischnitzer (2006), and Zug et al. (2001).

We appreciate the efforts of Corey Goldman (University of Toronto) and Dr. Cory Ross (George Brown College of Applied Arts and Technology) for academic and institutional support; Stephen Mader (Artery Studios Inc.) for his encouragement and support; Peter von Bitter, Kathy David, Brian Iwama, and Peter Reali for help with photography; Theodore and Jacob De Iuliis for their help in preparing specimens and photography; Daniel Pulerà for thoughtful suggestions on improving some figures; Celestino De Iuliis for reading earlier drafts of the manuscript; Dr. Marco Zimmer-De Iuliis for preparation of specimens; Dr. Kevin Seymour for access to the collections of the Royal Ontario Museum; Dr. Hans-Rainer Duncker (Institut für Anatomie und Zellbiologie der Justus-Liebig-Universität Giessen) for providing specimens of the avian air sac system and reviewing our illustrations (Figure 9.13); Dr. Walter Joyce (Universität Tübingen) for help with turtle skulls; Dr. Steven Huskey (Western Kentucky University) for providing skeletal specimens; Dr. Mathew J. Wedel (University of California Berkeley) for photographs of the pigeon air sac system and reprint of a rare paper; Dr. Philip J. Motta (University of South Florida) and students for their suggestions for the second and third editions, and to Dr. Motta for kindly allowing us to use the photo of the tunicate larva (Figure 1.28a); Dr. Barry Bruce (CSIRO Division of Marine Research, Tasmania), Mark McGrouther and Elizabeth Cameron (Australian Museum), and Andrew and Silvy Fox (Rodney Fox Shark Museum, Australia) for providing shark dissection photos; a special thank you to Dr. Steven E. Campana (University of Iceland) for also providing labeled shark dissection photos, and very kindly allowing us to reprint the SEM photo of the spiny dogfish skin (Figure 3.12f). We are grateful to Henry Hong and Audrey Chong of the Imaging Facility (Department of Cell and Systems Biology, University of Toronto) for the use of several microscopes and help with microscopy. We also appreciate the assistance of Tim McGill (Leica Microsystems Canada) in the latter regard. We are also thankful to the many colleagues who took the time to write and point out errors in previous editions of this manual, in particular Avery D. Franzen (University of South Dakota), Dr. Rebecca Hartley (Seattle University), Dr. Randy Lauff (St. Francis Xavier University), Dr. Mark S. Mills (Missouri Western State University), and Dr. Philip J. Motta (University of South Florida).

Sincere thanks to Drs. Rosana Moreira da Rocha (Universidade Federal do Paraná) and Michael W. Konrad (Science is Art) for assistance with aspects of tunicate anatomy, and the former for also allowing us to cite the Smithsonian Tropical Research Institute instructional video in Chapter 1; and Dr. Denis G. Baskin (University of Washington School of Medicine) for assistance on aspects of cephalochordate anatomy.

We appreciate the efforts of the following Elsevier staff who have participated in this and previous editions of the book: André Wolff, Mica Haley (Publishing Director - Life Sciences), Sandra Harron, Kiruthika Govindaraju, Lisa Tickner, our former Elsevier partners Katie Fawkes, Janice Audet, Sarah Binns, Mary Preap, and Andrea Cowan for patience, skill, guidance, and commitment to this project. We are indebted to David Cella (former Publishing Director of Elsevier) for initial consideration of our proposal and recognizing the potential for this book, and to Stephen G. Gilbert for showing us how to get started on creating our own book, and for his encouragement, support, advice, and continued inspiration. Lastly, we thank Virginia and Cinzia for being there beside us every step of the way in seeing the third edition of this book through to the end—it has been yet another long and challenging journey.

Introduction

The study of vertebrate anatomy is an interesting and valid field of study for gaining insight into the structure and function of vertebrates. But why should this be important? Of the numerous reasons, we shall mention just a few:

- It provides us with knowledge of the structures of different organisms and the great variety of form among vertebrates.
- It allows us to examine how the form of these structures is related to their function and thus how morphology is suited to a particular mode of life.
- The characteristics or features of vertebrates preserve information on their ancestry; the features are modified and passed on through the course of generations, and we may use such knowledge to discover the genealogical relationships among vertebrates.
- Comparative anatomical studies help us understand how the major transitions in vertebrate design might have occurred. Soft tissues do not fossilize, meaning that (with rare exceptions) only transformations of the hard parts of the vertebrate body are preserved in the fossil record. For other parts of the body, we must rely on a sequence of living forms. There are problems with this approach, but if we begin with a robust phylogenetic hypothesis and keep in mind that the living members of some groups are highly derived, we may be confident in this method as a reasonable approach for deducing the major steps in the evolution of different vertebrate groups.

We will consider all these aspects in the following course on comparative vertebrate anatomy. Before beginning this study, however, we review several important terms that unambiguously describe position and direction. These indispensable terms greatly facilitate navigating through the complex three-dimensional structure of vertebrate bodies.

DIRECTIONAL TERMINOLOGY AND PLANES OF SECTION

As with all advanced fields of research, anatomical study requires the use of specialized terminology. Such terminology includes not only special words for the anatomical structures themselves and concepts or processes

(such as homology, for example), but also terms to designate unambiguously the orientation and direction of structures of the vertebrate body. These terms may at first seem superfluous, but that is because most people have never dealt with anatomy in a comprehensive and detailed manner. It is perfectly adequate in everyday life to say that the stomach is lower than the heart or the appendix is in the lower right part of the belly. But this is not anatomy. You will quickly come to realize the importance of the terms presented in this section, and you are urged to learn, become familiar with, and use them.

There are two main sets of terms. One is used in medicine and by some anthropologists, the other by comparative anatomists, paleontologists, and veterinarians. To compound the problem, various synonyms exist for some terms in each set. These circumstances may be cause for confusion, but we may simplify matters by adhering to one set of terms. As we are studying comparative anatomy, we will use the system commonly used for nonhuman vertebrates.

Unlike humans, the vast majority of vertebrates go through life with the long axis of the body oriented horizontally, parallel to the substrate. It is with reference to this position that the main directional terms are defined; of course, these positions apply to the body, even though it may not be in the position shown in Figure 1 (see later). Most of these terms are coupled; that is, there are two terms that describe opposite directions along a single axis. Refer to Figure 1 while reading the following explanations. **Anterior** and **posterior** refer to the horizontal longitudinal axis and respectively designate the directions toward the head and tail. Synonyms for these terms that you may encounter are *cranial* or *rostral* for anterior, and *caudal* for posterior. The vertical direction toward the belly or the ground is **ventral**; toward the back or up is **dorsal**. **Medial** refers to the horizontal direction toward the sagittal midline (see later) of the body, whereas **lateral** refers to the direction away from the midline. These are the main terms, but another set also is useful. **Proximal** and **distal** are terms often used with a particular reference. At times this reference may be the trunk of the body; at other times a particular structure, such as the heart, may be the reference point. Proximal designates a position closer to the trunk or structure of reference; and distal farther from the trunk or structure of reference. Thus, for example, the fingers (phalanges) are distal to the upper arm (brachium); and the proximal

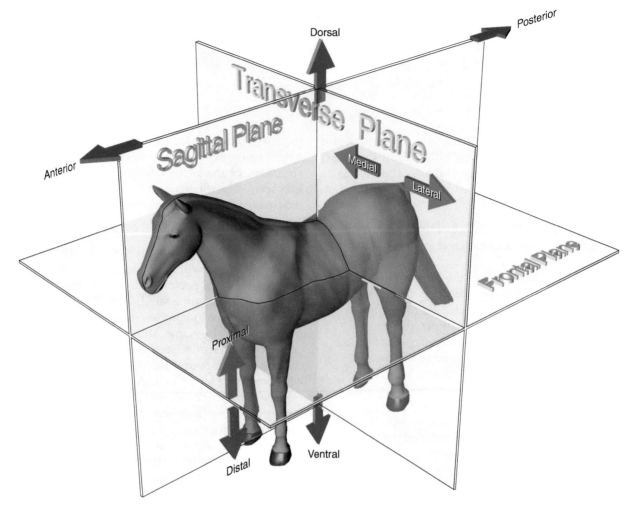

FIGURE 1   Directional terms and main planes or sections through the body shown on a horse.

end of the brachium is that end closest to the trunk. If the reference point is another structure, say the heart, then the proximal part of a blood vessel is the part closer to the heart, and the distal end that part farther away. Two other useful directional terms are **deep** and **superficial**. Deep refers to a position farther from the body surface, and superficial refers to a position closer to the body surface. For example, the skin is superficial to the muscles, and the bones are deep to the muscles.

Combinations of these terms may be used, and indeed are used often in this manual, to describe directions that are oblique to the main axes. For example, anterolateral combines anterior and lateral, and indicates a simultaneous direction toward the head and to the side. Thus, taking the umbilicus (navel or belly button) as a reference, we may describe the shoulder as anterolateral to the umbilicus. Figure 2 provides examples of these terms.

Dissection often involves cutting the body in various planes to obtain internal or sectional views, which are extremely useful for comprehending the spatial arrangement of structures. There are three main sections or planes that pass through the body (Figure 1). The **sagittal** section is vertical and lies along the longitudinal axis of the body (e.g., Figures 2.8, 7.5 and 7.40). It separates the body into right and left halves. We often consider this the **midsagittal** section or plane. Sections that are parallel to and on one side of the sagittal plane are termed **parasagittal**. A second major section is in the transverse plane, which is also vertical but is perpendicular to the sagittal plane. A **transverse** section cuts across the main longitudinal axis and subdivides the body into anterior and posterior parts (e.g., Figures 2.9 and 7.70). The last major section is in the frontal plane, which is horizontal and perpendicular to the sagittal and transverse planes. A **frontal** section separates the body into dorsal and ventral parts (e.g., Figure 3.41).

It is important to keep in mind that the directional and sectional terms are used with reference to the body

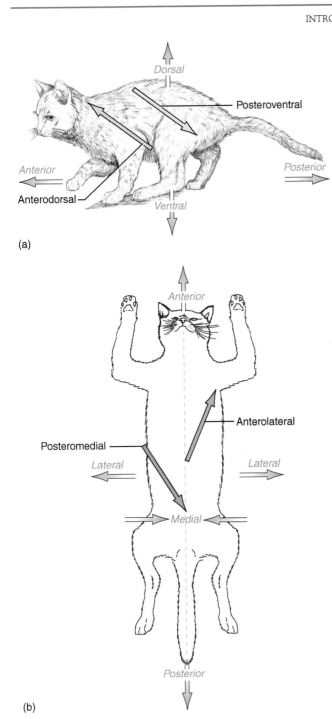

(a)

(b)

FIGURE 2　Combined directional terms shown on a cat in (a) left lateral and (b) ventral views.

of the animal, and not with reference to the observer or space. Thus, left and right refer to the animal's left and right. If the animal, say the horse in Figure 1, is lying on its side, the sagittal section (or plane) still passes through the midline of the horse and divides it into left and right halves, and ventral is still toward the horse's belly, rather than the ground.

Key Terms: Introduction

anterior (cranial, rostral)
deep
distal
dorsal
frontal
lateral
medial
midsagittal
parasagittal
posterior (caudal)
proximal
sagittal
superficial
transverse
ventral

1

Vertebrates and Their Kin

The vertebrates, or Vertebrata, form an ancient group with a history spanning some 545 million years. On the one hand, they include the organisms most familiar to us, such as fish, birds, cats, and dogs, as well as humans; on the other, few people are aware of the great diversity in their form, structure, and habits. Indeed, they include some of the largest and most complex organisms ever evolved. But vertebrates are part of a larger grouping of animals, and to understand their history and the development of their structure, they must be placed in phylogenetic context.

In discussing vertebrates, several other groups of organisms are usually considered. A *monophyletic* or natural group (see later) of organisms is referred to as a *taxon* (pl., *taxa*) or *clade*. The vertebrates are part of a larger clade, Deuterostomia (from the Ancient Greek *deuteros*, second, and *stoma*, mouth). This clade plus Protostomia (from the Ancient Greek, *protos*, first, and *stoma*, mouth) are included in Bilateria (Figure 1.1). The organisms included as deuterostomes are considered to have a secondary mouth during embryonic development (deuterostomy); that is, the primary opening, the blastopore, develops into the anus, whereas a secondary opening develops into the mouth. By contrast, in most Protostomia, the blastopore develops into the mouth (see Lowe et al., 2015; Soukup and Kozmik, 2016; this is a somewhat simplified overview of the development of these structures—see Nielsen, 2015, for further details). However, it is largely unknown whether the mouth among all deuterostomes is homologous (Kaji et al., 2016), and deuterostomy is known to occur in some non-deuterostomes (Nielsen, 2015).

Among deuterostomes (Figure 1.1), in order of increased remoteness of common ancestry to vertebrates, are Urochordata (tunicates, or sea squirts), Cephalochordata (amphioxus), and Ambulacraria, which consists of Echinodermata (sand dollars, sea lilies, star fish, sea cucumbers, and sea urchins) and Hemichordata (acorn worms and pterobranchs). These are the typical nonvertebrate (or "invertebrate") relatives of the group in which we are mainly interested. These taxa and the evolutionary relationships among them are briefly outlined in the following pages to provide an organizational framework for undertaking the dissection of the vertebrates discussed in this manual. The phylogenetic scheme presented in Figure 1.1 excludes the xenoturbellid worms; some authors have considered these as possibly deuterostomes, as related to or possibly ambulacrarians, as protostomes, or as falling outside the bilaterian clade (see Lowe et al., 2015 and Rouse et al., 2016 for further details).

Before this, however, it is necessary to present an explanation of several important terms used in discussions of *phylogeny*. Phylogeny refers to the pattern of descent among taxa. It describes, in other words, the evolutionary or genealogical relationships among them. Evolution is a historical (and ongoing) process. Therefore, the evolution of organisms has occurred, and occurs, in only one way—for example, humans and chimpanzees are, among living creatures, either each other's closest relatives (they share a common ancestor not shared by other organisms) or they are not. Only one of these possibilities is correct. Evolutionary biologists try to recover the pattern of descent based on the data available to them. Phylogenies are therefore hypotheses that approximate the true pattern of descent. As hypotheses, they are testable and thus open to falsification when new data become available. If a hypothesis is falsified, then another one may be proposed—for example, new evidence might show that humans share a recent common ancestor with a different great ape than chimpanzees, such as orangutans.

Many researchers are involved in the task of reconstructing phylogeny. The vast majority are specialists in one field or another, so that each scientist (or more commonly, team of researchers) presents phylogenies based on data that he or she is most familiar with and most competently trained in. Thus, molecular biologists almost exclusively use molecular sequence data, and morphologists usually rely on morphological data.

Ideally, one would think, the resulting phylogenies would converge on each other and result in cladograms that have consistent branching patterns. However, the cladograms, to a greater or lesser degree, are rarely entirely congruent with one another. The discrepancies occur not only among mainly molecular or mainly morphological analyses, but between molecular and morphological analyses as well. That differences should exist is understandable, given that scientists use different data and methods in their analyses. For example, Zhong et al. (2009) noted that selection of the analytical model and the type of data analyzed (mitochondrial vs. nuclear DNA sequences) both strongly influence results and can produce strikingly different tree topologies. Thomson et al. (2014) evaluated the validity of recent attempts based on microRNA data to resolve relationships among several groups that have been particularly difficult to understand, such as the relationships among hagfishes, lampreys, and vertebrates. These authors found several types of problems with such analyses and urged caution in the application and interpretation of microRNA studies.

Over much of the past 150 or so years, essentially since Darwin (1859) convinced scientists that the patterns of similarity among organisms are due to descent with modification, phylogeny reconstruction has been based almost exclusively on morphological data. During the 1990s, however, techniques were developed for discovering and analyzing molecular information, and phylogenies based on such data have become very popular. The impression that many scientists seem to have is that the molecular-based analyses are more reliable than those based mainly or exclusively on morphological data. Indeed, the trend among scientists has been to view molecular-based studies as authoritative and rely on them as a base for their own further research. For example, one study often cited as support for the idea that urochordates (and not cephalochordates) are the closest relatives of vertebrates (the Olfactores hypothesis; see later) is that by Delsuc et al. (2006), even though this study suggested that cephalochordates are not even chordates—i.e., that Chordata is not monophyletic—an idea for which there is neither much support among morphological or other molecular analyses nor acceptance among scientists. A later study by Delsuc et al. (2008) recovered a monophyletic Chordata, but not a monophyletic Deuterostomia, the latter result in contrast to the earlier study. In part this impression of the superiority of molecular-based analyses is almost certainly due to the (somewhat mistaken) idea that molecular data sets contain considerably larger quantities or more informative kinds of data, but reliance on such molecular analyses to the exclusion (and even dismissal) of morphology-based

analyses is troublesome, and the superiority of molecular-based analyses is far from established. For example, Ruppert (2005) had already conducted an in-depth study of characters commonly used to diagnose chordates (see later), and based chiefly on morphological characters, presented the phylogenetic hypothesis depicted in Figure 1.1 that has since been generally supported by mainly molecular-based, morphology-based, and combined molecular–morphology-based studies (i.e., that echinoderms and hemichordates are sister taxa, a group that is in turn sister taxon to the chordates, within which urochordates and vertebrates are sister taxa; in other words, that urochordates and not cephalochordates are the closest relatives of vertebrates). Yet the significance and contribution of Ruppert's (2005) work seem to have been largely underappreciated. Stach's (2014) analysis incorporated some of Ruppert's (2005) data and resulted in support for the traditional view, presented in Figure 1.2, that cephalochordates, rather than urochordates, are more closely related to vertebrates, and that hemichordates are more closely related to chordates than to echinoderms. Stach (2008, 2014, 2015) presented cogent and convincing arguments on the importance of the morphology of both extant and extinct organisms, as well as of molecular and evolutionary development data, in helping elucidate our views on the relationships among organisms, sentiments echoed by, among others, Jenner (2015)—indeed the exchange between Stach (2015) and Jenner (2015) makes for interesting reading. The just-noted series of papers by Stach, as well as Thomson et al. (2014), provided examples of the possible reasons why molecular-based analyses are often in conflict with each other.

The comments in the preceding paragraphs are not intended to promote the value of one kind of study over another. Rather, it is important to realize that several lines of evidence are informative, valuable, and necessary. Indeed, several researchers have recently begun to combine a wider range of data in their analyses. The more information available, the more confident we may be in our results, as reflected in the statement by Lowe et al. (2015: 456) that our "understanding of vertebrate origins is powerfully informed by comparative morphology, embryology and genomics of chordates, hemichordates and echinoderms, which together make up the deuterostome clade." With respect to the hypothesis that views a closer affinity between urochordates and vertebrates than between cephalochordates and vertebrates, it has become widely accepted because a variety of data from different disciplines supports it (Satoh, 2016). Gee (2018) provides an insightful and very readable account of the main ideas relevant to the evolutionary origin of vertebrates.

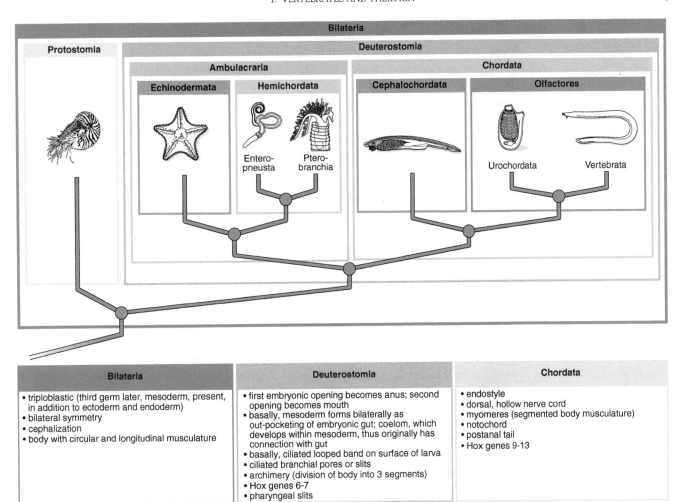

Bilateria	Deuterostomia	Chordata
• triploblastic (third germ later, mesoderm, present, in addition to ectoderm and endoderm) • bilateral symmetry • cephalization • body with circular and longitudinal musculature	• first embryonic opening becomes anus; second opening becomes mouth • basally, mesoderm forms bilaterally as out-pocketing of embryonic gut; coelom, which develops within mesoderm, thus originally has connection with gut • basally, ciliated looped band on surface of larva • ciliated branchial pores or slits • archimery (division of body into 3 segments) • Hox genes 6-7 • pharyngeal slits	• endostyle • dorsal, hollow nerve cord • myomeres (segmented body musculature) • notochord • postanal tail • Hox genes 9-13

FIGURE 1.1 Cladogram showing phylogeny of Bilateria. Several synapomorphies of the main groups are provided in the boxes below the cladogram. *Adapted from Benton, 2015; Brusca et al., 2016; Satoh, 2016.*

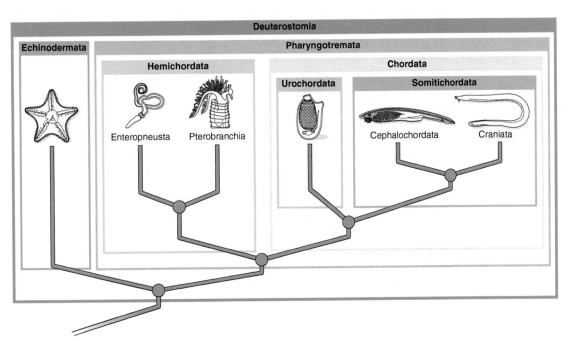

FIGURE 1.2 Cladogram showing the traditional phylogeny of Deuterostomia; as noted in the text, most biologists no longer tend to accept this hypothesis.

PHYLOGENY AND CLASSIFICATION

For most of the past 250 years, the classification of organisms has followed the Linnean system, which uses ranks to designate levels of organization for organisms being classified. Most readers will be familiar with the main formal Linnean ranks, ordered hierarchically from most to least inclusive: kingdom, phylum, class, order, family, genus, and species. Researchers have differed in assigning rank to the vertebrates and their relatives. For example, some authors recognize three phyla: Echinodermata, Hemichordata, and Chordata. Others consider Urochordata and Cephalochordata as phyla on their own, separate from Chordata. Still others view Urochordata as a separate phylum, but Cephalochordata as a subphylum of the phylum Chordata. If you find this confusing, you're not alone! The different designations did—or at any rate were meant to—have some grounding in biological reality. They reflected a particular researcher's perception of the magnitude of the difference in the levels of organization (a quality that may be referred to as a *grade*) among the taxa under consideration. Thus, if a taxon was considered a phylum, it mainly implied that its members had a fundamentally more distinct basic body plan than if it was considered only a subphylum of a larger taxon. As you have probably already realized, researchers' perceptions along these lines are subjective.

In recent years, however, the formal Linnean ranking system has fallen increasingly into disuse as systematists have become aware that no particular taxon has an intrinsic or special value that justifies its recognition at a higher or lower rank compared with other taxa (Laurin, 2010). In other words, there is no special reason for "elevating" birds, or Aves, to the rank of *class*, equal to and thereby excluded from the class Reptilia. In fact, it is improper to do so, because birds are properly part of the taxon named Reptilia. Here, formal ranks are not used, and taxa are referred to simply by their name (except for this and the preceding paragraphs, in which ranks are used to reflect the historical understanding of the groups).

Formal names are applied to natural or *monophyletic* groups. A monophyletic group includes an ancestor and all of its descendants (provided that the phylogeny has been carefully reconstructed). Such groups are termed *clades*. Clades are recognized based on common ancestry. If two taxa are in a clade, it is because they are linked by a common ancestor. Biologists infer such ancestral relationships through the presence of shared derived characters or synapomorphies (see later). If two (or more) taxa share a character that is exclusive to them, then we assume that they share this feature because they have

inherited it from a common ancestor, rather than each having evolved the character independently, and so infer that the taxa are descendants of the same ancestor (which we are not able to actually recognize, and thus refer to it as hypothetical). Biologists use many characters in trying to reconstruct phylogeny. The practice is complicated by the fact that organisms can and do evolve very similar characters independently of each other, an occurrence referred to as *homoplasy*. In reconstructing phylogeny, researchers consider (ideally, at least) the totality of evidence. It is rare that a single character alone can be used to reconstruct phylogeny.

The pattern of relationships among taxa is depicted visually by a *cladogram*, which is essentially a diagram of nodes and branches, with the nodes representing ancestors and the branches that diverge from a node representing the descendant taxa of the ancestor.[1] The node, then, may be thought of as representing the hypothetical ancestor of the two taxa that diverge from it. The pattern of branching represents the pattern of relationship. Examine the cladogram in Figure 1.1. Note the node from which Chordata and Ambulacraria diverge. This node represents the ancestral species that split to produce two lineages, one that evolved into Chordata and the other into Ambulacraria. The two branches that diverge from this ancestor represent the evolutionary paths to the divergent taxa. Only the branching pattern is of concern. The length of the branches is immaterial in terms of absolute time, but relative time is implied by branching sequence. Clearly, the divergence of Urochordata and Vertebrata occurred after the divergence of Cephalochordata and Olfactores.

Informal names, set between quotation marks, are used to designate a group of organisms that do not descend from the same common ancestor but do possess (or lack) some features of the taxon in which we are interested. Many such terms were considered formal names in earlier classifications. For example, the terms "protochordates" or "prevertebrates" are commonly used to refer to the hemichordates, cephalochordates, and urochordates. Grouping them together is a shorthand way of referring to them as close relatives of vertebrates (no quotation marks here, so this is the vernacular form of the formal name Vertebrata), and that they lack various characters that vertebrates possess. We must be clear that informal groups, though often convenient, do not reflect phylogeny; they are not monophyletic.

In discussing how biologists reconstruct phylogeny, the nature of the similarity among organisms must be considered, because it is necessary to differentiate between those similarities that are useful in

---

[1] Technically, nodes are speciation events of one ancestor into two or more descendant or daughter species. The root and internodes (segments between the nodes) represent the common ancestors and loci of character transformations.

reconstructing phylogeny and those that are not. One kind, termed *plesiomorphic*, refers to similarity based on the presence of ancestral conditions or states. Consider Vertebrata, in which the presence of vertebrae (or their precursors) is an ancestral feature—in other words, this condition was present in the common ancestor of all vertebrates. These structures may inform us that all vertebrates share a common ancestor, but their presence per se cannot be used to decipher the relationships *among* vertebrates. As a practical example, let us consider a turtle, a bird, and a mammal. All possess vertebrae, and may therefore be considered vertebrates, but the presence of these structures does not allow us to say which two of these forms are more closely related to each other than either would be to the third. This similarity, therefore, is due to the retention of a trait that is ancestral for vertebrates. When an ancestral character is shared among taxa, it is described as *symplesiomorphic*.

A second kind of similarity is due to the inheritance of a *modified* character state. Such modification is considered derived or *apomorphic*. When organisms share a derived trait, it is described as *synapomorphic*. Synapomorphies do indicate phylogenetic relationships. In the most basic sense, sharing a derived trait is a shorthand way of saying that the organisms under consideration possess a modified trait because it was inherited from an ancestor that first acquired or evolved the modification. An assortment of organisms united by synapomorphies forms a natural group or clade; that is, the clade is a real entity in evolutionary terms. It means that all the organisms included within the clade were ultimately derived from the same ancestor. For example, all vertebrates that possess jaws do so because this character was inherited from an ancestor that had evolved them. If we wish to understand the relationships among a lamprey (which lacks jaws), a fish, and a dog, the presence of jaws is a character state that indicates that the dog and fish are more closely related to each other than either is to a lamprey. When two groups are each other's closest relatives, they are said to be *sister groups*.

A natural or monophyletic group may be recognized formally by a name. The only restriction imposed is that a monophyletic group include the ancestor and all descendants of the ancestor, even though the identity of the ancestor cannot be determined. A monophyletic group may also be termed a clade, from which is derived the term *cladistics*. An alternate name for cladistics is *phylogenetic systematics*. Cladistics is the methodology that recognizes shared derived traits as the only valid indicators for inferring phylogenetic relationships.

The third type of similarity is termed *homoplasy* and results from morphologically similar solutions to particular selection pressures. For example, the fusiform body shape of fishes and porpoises, which are mammals, is not due to inheritance from a common ancestor, but to

selection pressure to adopt a form suitable for moving efficiently through water. Such similarity does not indicate phylogenetic relationship, although in some cases the similarity may be so profound that it leads us to inaccurately attribute its cause to phylogenetic proximity (i.e., recent common ancestry). The reliable method of recognizing homoplasy is to identify it as similarity in different monophyletic groups following, of course, a phylogenetic analysis.

In addition to clades or monophyletic groups, we may speak of *grades*, which are not natural groups. A grade recognizes a group of organisms based on a comparable level of organization or complexity. A new grade may be achieved through the accumulation of a number of derived characters so that a new mode of living is made possible. In the past, some such groups were formally recognized, but they were united essentially because their members shared a particular grade of evolution. We now recognize such groups as artificial rather than natural. An artificial group is one that does not include an ancestor and all of its descendants or combines two or more groups while excluding one or more ancestors (e.g., "Haemothermia" = mammals + birds).

Probably the most familiar example is the case of Reptilia. Formerly, Reptilia included living and fossil crocodiles, turtles, snakes, and lizards, as well as their extinct relatives, such as dinosaurs, pterosaurs, and plesiosaurs. Reptilia was given a rank (class) equivalent to that of Aves (birds) and Mammalia (mammals), even though the ancestors (and early relatives) of these two groups were considered reptiles. As so defined, however, Reptilia is not a natural group because it does not include all of its descendants, as the birds and mammals are excluded (and each placed in a group of equal rank). Current usage of Reptilia varies (examples are noted again later). As its traditional concept is so embedded in our thinking, some authors have preferred to abandon it entirely for formal purposes (and prefer to use Sauropsida) but retain it in its colloquial sense. In this latter meaning, "reptile" represents a grade that includes cold-blooded amniote tetrapods, with scales (lacking hair or feathers); that is, the features we usually associate with living reptiles such as turtles, crocodiles, snakes, and lizards. Other authors redefine Reptilia as a formal group that includes the typical reptiles and birds. The more basal fossil allies of mammals, termed mammal-like reptiles, are excluded from Reptilia and properly united with their mammalian descendants in Synapsida.

The discussion given here provides the basic background information required to interpret cladograms and how they are constructed. For more detailed discussions on cladistics and classification, consult a text in comparative anatomy that provides more detailed explanations of these concepts. Liem et al. (2001) provide a particularly thorough discussion.

VERTEBRATE RELATIVES

Nearly all the taxa mentioned so far belong to Deuterostomia, a major clade of coelomate triploblastic metazoans, multicellular animals that have a true body cavity that houses the viscera and that possess three primary body layers (ectoderm, mesoderm, and endoderm). The other major clade is Protostomia, which includes nematodes, annelids, arthropods, mollusks, brachiopods, and several other groups (see Benton, 2015).

The synapomorphies (shared derived characters) of deuterostomes that indicate they are a clade are mainly long-recognized similarities of early embryonic development and, as more recently elucidated, molecular similarities as well (see Benton, 2015, Figure 1.1). They include type of cleavage of the fertilized egg, pattern of mouth and anus formation, and formation of the mesoderm and coelom (body cavity). Among further shared morphological features are ciliated pharyngeal pores (or openings) and an endostyle, as well as genetic features, such as several Hox genes (see Lowe et al., 2015; Nielsen, 2015). Although these are shared among deuterostomes, some have become modified in several derived members. For example, echinoderms seem radically different given their adult pentaradial symmetry and lack of pharyngeal slits. These characters are modifications in more recent, including extant, echinoderms, but their larvae are still bilaterally symmetric, and fossil evidence provides compelling evidence that pharyngeal slits were present in stem (i.e., those that branch along the stem leading to) echinoderms (Lowe et al., 2015).

Next, we must consider the pattern of relationships, or phylogeny, among deuterostomes. The phylogenetic scheme presented here follows the one that has become most commonly cited among researchers over the past dozen or so years and is based on morphology, genetics, and molecular evolutionary development (or evo-devo; see Stach, 2014). It is generally termed the Olfactores hypothesis. This scheme is quite in contrast to the more traditional view (the Notochordata hypothesis) that considers cephalochordates, urochordates, and hemichordates as successive outgroups to vertebrates (Figure 1.2). Although the Olfactores hypothesis is more commonly recognized, it should be mentioned that the Notochordata hypothesis does have recent support in the work of, among others, Stach (2008; 2014), Wang et al. (2010) and Zhong et al. (2009). Yet another alternative was presented by Nielsen (2015), who considered ambulacrarians to consist of echinoderms + enteropneusts, and deuterostomes as including hemichordates + chordates.

One group traditionally recognized in vertebrate history is Chordata, which includes Cephalochordata, Urochordata, and Vertebrata. One reason Chordata has been considered particularly important is because several easily recognizable characters are clearly shared by chordates. Without belaboring the point, such distinctions as "important" or "major" often imply a status that may not be justified. There is no real reason the chordates should be considered more "important" than the next most inclusive group, for example. It is more a matter of convenience and tradition, and, perhaps, because we have only recently begun to fully comprehend that all branches in the tree of life may be considered equally important.

In any event, beginning with Chordata is convenient. The chordates are united by the presence of the following morphological synapomorphies: an endostyle; a dorsal, hollow nerve cord; a notochord; myomeres; and a postanal tail. Another feature important in chordates is the presence of pharyngeal slits, although as noted earlier this feature is not exclusive to them (it arose in the deuterostome ancestor but was subsequently lost in echinoderms). These features are present at some point during the lives of all chordates, although they may be expressed to varying degrees and restricted to part of the life cycle in different vertebrate groups, or modified in derived members. Humans, for example, do not possess a tail (an obvious one, at any rate), notochord, and pharyngeal slits; but pharyngeal pouches, a notochord, and a tail are transient features present during embryonic development. The endostyle is represented by its homologue, the thyroid gland.

Pharyngeal slits are bilateral apertures that connect the pharynx (essentially the "neck" of the animal), which is the anterior part of the gut, with the outside. In forms that are familiar to us, such as fish, the slits are part of the respiratory, or gas exchange, system: The gills reside on the walls of the slits and perform gas exchange as water passes over them. In some fishes, like sharks, the slits open individually onto the surface of the body; in most other fishes, the slits open into a common chamber that then leads out to the body surface by a common opening. Originally the slits did not function in gas exchange. Ancestral vertebrates were suspension or filter feeders, and the slits were the means for allowing water to exit the oral cavity and pharynx. This is the feeding method present still in cephalochordates and nearly all urochordates. As water passes out of the pharynx through the slits of a filter feeder, food particles are filtered out and directed toward the rest of the digestive system, as explained more thoroughly in the following sentences. The endostyle, a midventral groove (on the floor of the pharynx), has ciliated cells that secrete mucus that is spread around the inner walls of the pharynx. Food particles suspended in the water are trapped by the mucus, and the water leaves the pharynx through the slits. The mucus and trapped food particles are then passed back into the digestive system. The slits and endostyle were thus originally part of the feeding mechanism (the vessels associated with the pharynx of urochordates also contribute to gas exchange; see later).

The notochord (from the Ancient Greek *noton*, back, and Latin *chorda*, cord or rope) is a relatively thin, rod-like structure extending dorsally along the length of the trunk and tail in less derived chordates. It is an important support structure, and the name Chordata is derived from notochord. It is a hydrostatic structure consisting of a fibrous sheath that encloses a fluid-filled central core. It is flexible along its length, but as it is filled with fluid, cannot be easily compressed anteroposteriorly (or telescoped). The notochord provides support for the body and allows the side-to-side locomotory movements characteristic of less derived vertebrates. In derived vertebrates the notochord is largely replaced functionally by the bone or cartilage of the vertebral column. It is present embryologically; and in adult humans, notochordal tissue may persist as part of the intervertebral disks that lie between adjacent vertebrae.

The presence of a tubular nerve cord enclosing a fluid-filled central canal occurs only in chordates. Another distinctive feature of the chordate nerve cord is that it is formed by an embryological process called *invagination*, a rolling and sinking into the body of ectodermal tissue. Further, it is dorsal to the digestive tract, whereas in most "nonchordates" the nerve cord is solid and ventral.

Myomeres are the segmented paired muscular blocks that extend through the trunk and tail. Alternating contraction of the musculature of the right and left sides of the body exerts forces on the notochord, noted earlier as a laterally flexible rod, that allow the side-to-side locomotory movements characteristic of less derived chordates and vertebrates.

The postanal tail is a continuation past the anus of the trunk musculature and notochord. This extension is an important development that allows the locomotion particular to vertebrates. A few chordates do not possess a postanal tail as adults (e.g., frogs), but a tail is present in nearly all chordate larvae.

The main chordate characters, plus pharyngeal slits, are clearly present during the life of a cephalochordate. The name Cephalochordata (from the Ancient Greek *kephalos*, head, and Latin *chorda*, cord or rope) is derived from the presence of a notochord extending from the tail nearly to the tip of the head. Cephalochordate species commonly are referred to as amphioxus (which means sharp at both ends) or lancelet (little spear). Given the fact that these little creatures essentially lack a head and so are pointed at both ends, amphioxus is an especially appropriate designation. Amphioxus has a fish-like body, but it is not an active swimmer as an adult, although it is capable of swimming rapidly (Lacalli and Stach, 2016). Instead, it burrows into the substrate, usually just out from sandy beaches, and assumes a position with only its mouth exposed. Its filter-feeding lifestyle was described earlier. Intake of water and its movement through the pharynx is accomplished by ciliary action.

The pharynx has numerous slits that collectively empty into a surrounding chamber, the atrium, before leaving the body through a common opening. The endostyle secretes mucus, which traps food particles suspended in the water.

The basal position of cephalochordates among chordates suggests that the similarities between cephalochordates and vertebrates represent ancestral features lost in urochordates, and that the latter are therefore derived in this regard (they are also derived in their ability to produce cellulose; see later). Thus, additional features present in cephalochordates and vertebrates may be viewed as present in the common ancestor of chordates, including arrangement of the circulatory system with dorsal and ventral aortae, segmentally arranged spinal nerves (or roots), and development of mesoderm, including the hypomere (or lateral plate mesoderm) and mesodermal somites, which develop into myomeres (which in amphioxus extend through to the anterior tip of the body). The presence and similarity of such features have long conferred to amphioxus the status of being a model organism in the study of early chordate evolution.

The urochordates include sea squirts, or tunicates, which are sessile, saclike organisms as adults. In the larval stage, however, chordate characters, plus the pharyngeal slits, are present. Predictably, the chordate characters lost in adults—the tail, notochord, and nerve cord—are used in locomotion by the free-swimming larva as it searches for a suitable place to anchor itself to metamorphose into the adult form. During this transformation, the tail is absorbed, along with the nerve cord and notochord, and only small remnants remain in the adult. The name Urochordata (from the Ancient Greek *uron*, tail, and Latin *chorda*, cord or rope) is derived from the notochord being present in the tail. Conversely, the pharyngeal region expands dramatically into a barrel-shaped structure with numerous slits. Water and suspended food particles are drawn into this "barrel," which is lined with mucus from the endostyle. Food particles are trapped by the mucus, and water leaves through the slits into the atrium, the chamber surrounding the pharynx.

The phylogenetic position of Hemichordata has been recently clarified. They were traditionally grouped with the chordates, but molecular (in particular) and morphological evidence has been mounting over at least a decade that points to a monophyletic relationship of hemichordates with echinoderms, the two groups within Ambulacraria, as noted earlier. Hemichordates comprise two clades, Enteropneusta (acorn worms) and Pterobranchia, both of which are marine animals. Acorn worms are reasonably diversified and well known, but pterobranchs are not as well understood. Some but not all pterobranchs have a single pair of pharyngeal apertures, whereas all acorn worms have several such openings.

VERTEBRATES

Vertebrata include organisms differentiated by distinct feeding apparatuses: those without jaws and those with jaws. Among living vertebrates, the former include relatively few forms, the lampreys and hagfishes. These are clearly more derived than cephalochordates and urochordates, and share various characteristics with the much more abundantly diversified jawed vertebrates, such as sharks, bony fishes, and mammals. However, they are undoubtedly less derived than the latter in the lack of jaws, the feature to which they owe their designation as "agnathans" (from the Ancient Greek *a*, without; and *gnathos*, jaw). Their mouths are circular and they are grouped together as Cyclostomata (round-mouthed; from the Ancient Greek *kyklos*, circle, and *stoma*, mouth). Vertebrates with jaws are grouped together as Gnathostomata (jaw-mouthed; from the Ancient Greek *gnathos*, jaw, and *stoma*, mouth), as indicated in Figure 1.3. In addition to the living "agnathans" are several fossil groups that evolved early in vertebrate history. Many of these jawless forms were excessively bony, but most of this bone was dermal and formed shields or plates that covered and protected the body. These forms are informally termed "ostracoderms" (from the Ancient

Greek *ostrakon*, shell, and *derma*, skin). At one time, both the living and extinct jawless vertebrates were grouped together formally as "Agnatha," but it has been clear for some time that this grouping is paraphyletic and that the "ostracoderms" represent successive groups that branch along the stem (and are thus stem groups) leading to jawed vertebrates; although mentioned here only briefly, some of them are represented in Figure 1.4.

The phylogenetic scheme followed here has been stabilized within the last few years, but for the previous several decades, differences of opinion existed about precisely which of the forms discussed so far were to be regarded formally as Vertebrata. In part this is because a main character of vertebrates is, of course, the presence of vertebrae, a repeating series of articulating cartilaginous or bony elements forming the spinal column that provide support for the body, muscular attachment, and protection for the nerve or spinal cord. Vertebrae form around the notochord during embryonic development and enclose the spinal cord. However, not all chordates included in Vertebrata, as recognized here, have complete vertebrae as just described; and some have almost no trace of vertebrae at all. In large part, which chordates were to be recognized formally as vertebrates hinged largely on the relationships of the hagfishes

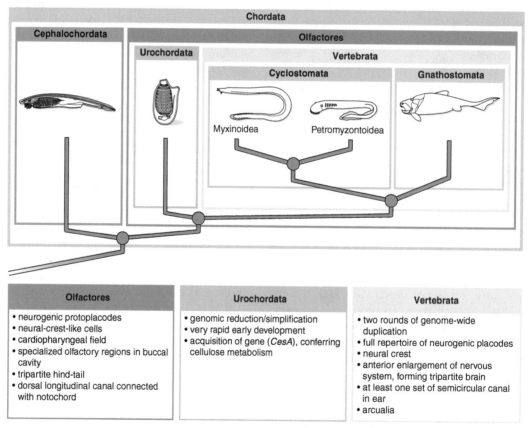

FIGURE 1.3　Cladogram showing phylogeny of Chordata. Several synapomorphies of the main groups are provided in the boxes below the cladogram. *Adapted from Benton, 2015; Brusca et al., 2016; Diogo et al., 2015; and Satoh, 2016.*

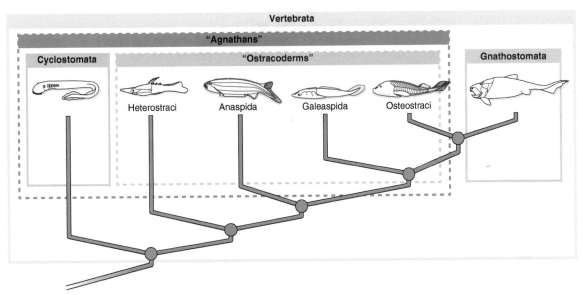

FIGURE 1.4 Cladogram showing phylogeny of Vertebrata, including representatives of early extinct members termed "ostracoderms." The cladogram also indicates the paraphyletic grouping of vertebrates that lack jaws as "agnathans." The "Ostracoderms" + Gnathostomata clade is not usually named formally, but some of its synapomorphies include: cerebellum; utricle and loop of vertical semicircular canal well separated; trunk dermal skeleton; sensory-line system with neuromasts; and dermal head covering. Dashed lines denote a paraphyletic group. *Adapted from Benton, 2015.*

(Myxinoidea) and lampreys (Petromyzontoidea), both to each other and to gnathostomes.

For most of the 20th century, the hagfishes and lampreys were considered each other's closest relatives and grouped as Cyclostomata. The cyclostomes and gnathostomes constituted Vertebrata. Beginning in the late 1970s, the idea that cyclostomes were paraphyletic, with lampreys being more closely related to gnathostomes than to hagfishes, began to gain acceptance based mainly on morphological evidence. Because lampreys possess rudimentary vertebrae (or at least precursors of true vertebrae), termed *arcualia*, which are essentially cartilaginous blocks on either side of the nerve cord, most researchers began to restrict Vertebrata to the lamprey + gnathostome assemblage, while the hagfishes, in which vertebral elements were thought to be absent, were considered the sister group to this Vertebrata. An important feature shared by both hagfishes and the lamprey + gnathostome assemblage is the development of a true head (see later), and so the name Craniata (or Craniota) was applied to the hagfish and lamprey + gnathostome clade (Janvier, 2011), as indicated in Figure 1.5. Soon after, though, molecular sequencing data and morphological reassessment began to provide convincing evidence of cyclostome monophyly, and Ota et al. (2011) reported that small cartilaginous nodules associated with the notochord in embryos of a hagfish species are homologous with portions of the vertebrae of gnathostomes (Janvier, 2011). And so we have come full circle, in a sense, with the general consensus being to recognize a monophyletic Cyclostomata as a sister group to

Gnathostomata; under this scheme, Craniata effectively becomes synonymous with Vertebrata.

It is instructive in any event to consider the development of our ideas on phylogeny by examining the changes in our thinking regarding the relationship among hagfishes, lampreys, and vertebrates. Whereas many molecular-based analyses have supported the pairing of hagfishes and lampreys as Cyclostomata, Near (2009) suggested that these findings were largely due to methodological artifacts and argued instead for a closer relationship between lampreys and gnathostomes based on a combination of morphological and molecular evidence. Near's findings were "flatly rejected" (Benton, 2015: 47) based on microRNA data by Heimberg et al. (2010) (see also Janvier, 2010), who also noted that reanalysis of morphological data could not strongly resolve the question of relationships. However, Thomson et al. (2014: 5) (without necessarily ruling out the validity and usefulness of microRNA data) critically reappraised the Heimberg et al. (2010) analysis and noted "that the miRNA data are essentially equivocal regarding the phylogenetic affinity of lampreys." Of note is not necessarily the jumping back and forth on our ideas of the relationships; rather, it furnishes another example of the seeming reluctance to consider morphological evidence as useful as molecular evidence in phylogeny reconstruction. Many researchers, including morphologists, seem too willing, as noted by Stach (2008, 2014), to defer to molecular-based analyses, even when such analyses are in conflict with each other and morphological evidence provides support for one hypothesis over another.

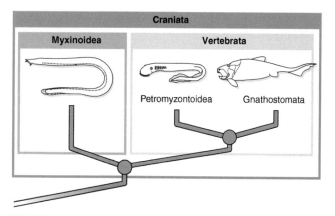

FIGURE 1.5 Cladogram showing relationships of living vertebrates that reflect the distinction between craniates and vertebrates. As noted in the text, this concept is generally considered outdated, and most biologists recognize the phylogeny presented in Figure 1.6.

As noted earlier, a major feature of vertebrates is the development of a true head. A distinct anterior enlargement of the nervous system, forming a brain, and of sensory organs occurs in vertebrates. The brain is tripartite, with three primary subdivisions; and the specialized sense organs—eyes, ears, and nose—are complex. These structures are protected and supported by a bony or cartilaginous cranium or braincase. Closely associated with a head is the neural crest, a unique feature of vertebrates. This comprises embryonic tissue formed of cells assembled near the developing neural tube that migrate through the embryo to give rise to a great variety of structures. In the head region, neural crest cells initiate and largely form the vast array of cranial structures characteristic of vertebrates. Another unique feature associated with the head are neurogenic placodes. Placodes are thickenings of the ectoderm occurring early in embryonic development that differentiate and help form a variety of the body's structures. Neurogenic placodes occur only on the head and are involved in forming sensory receptors and neurons, and contribute to the cranial nerves. Thus, the neural crest and neurogenic placodes, transitory though they may be, are distinguishing features of vertebrates (Schlosser, 2015). Associated with such innovations in head structures are cranial muscles and a chambered heart. Diogo et al. (2015) recently explored the link between cranial and cardiac musculatures in the emergence of vertebrates. Pharyngeal slits are still present, of course, but in vertebrates they are associated with gills and are thus used for gas exchange rather than feeding. Yet another innovation of vertebrates is that moving water into the pharynx and out through the pharyngeal slits is accomplished by muscular, rather than ciliary, action.

Before continuing with the phylogenetic story of vertebrates, it is useful to examine the feeding habits of the earliest vertebrates, as evolutionary innovations of the feeding apparatus reflect the major transitions in vertebrate evolution. It is generally hypothesized that the earliest step in becoming a vertebrate occurred in creatures considered to represent a "prevertebrate" stage, although such forms may have qualified as true vertebrates. Their precise phylogenetic position, in any event, is not of concern here, because we are interested in feeding mode. The "prevertebrate" was probably a suspension or filter feeder (that is, it filtered food material that was suspended in water) and used ciliary action to generate a current of water into its mouth and out of the pharyngeal slits; in other words, a creature very much like amphioxus. It did not have jaws, and its pharyngeal bars were probably collagenous. The ciliary pump imposed limits on size, as it restricted the amount and type of food the animal was capable of ingesting.

The next step involved a change from the ciliary pump with collagenous bars to a muscular pump with cartilaginous bars. This is the "agnathan" stage. The combination of these characteristics meant that the intake of water, and thus food, was controlled by active expansion and compression of the pharyngeal region, which allowed a diversification in size and type of food and thus of the vertebrates themselves. The cartilaginous, rather than collagenous, bars were instrumental in this. Musculature could be used to compress the pharynx, including the bars, but once the muscles relaxed, the cartilaginous bars could spring back into shape, expanding the pharynx.

The third level of development is the gnathostome stage, which involved the development of jaws. Jaws conferred the ability to grasp prey and close the mouth to prevent its escape; hence to seek and capture food. These features set the stage for the predaceous, active lifestyle of vertebrates, in sharp contrast to the sedentary lifestyles of "prevertebrates."

Returning to the phylogenetic story, several extinct early vertebrates have extensive fossil records and were clearly the dominant forms during early vertebrate history. Some of these groups are recognized as being more closely related to more derived vertebrates. One particularly interesting group is the conodonts, which for nearly 200 years were considered "invertebrates." Recent fossil evidence suggests that not only may they be vertebrates, they may, according to some authors, be more derived than cyclostomes and represent stem gnathostomes (see, for example, Donaghue and Rücklin, 2016; Murdock et al., 2013), although it seems more probable that they represent stem vertebrates or stem cyclostomes (Janvier, 2015). However, other scientists remain skeptical of a close relationship to vertebrates (see Turner et al., 2010). Several other extinct groups are more surely related to gnathostomes; these possessed excessive dermal bone arranged as protective broad plates or shields, particularly around the head, and are termed "ostracoderms," as noted earlier. They include Heterostraci, Anaspida, Galeaspida, and Osteostraci (among others) and are indicated in Figure 1.4. A textbook on vertebrate paleontology, such as Benton's (2015), provides further discussion of their anatomy and phylogenetic relationships.

The remaining vertebrates form the clade Gnathostomata (Figures 1.3, 1.4, and 1.6). It is worth noting that some authors restrict this name to crown group jawed vertebrates (i.e., Chondrichthyes and Osteichthyes), but here this assemblage is termed Eugnathostomata, as explained later. As their name implies, gnathostomes have jaws. Their development was a significant evolutionary advancement, perhaps the most important in vertebrate history, because jaws controlled by muscles allow animals to grasp objects firmly. The development of teeth conferred a more certain hold and further allowed the reduction of food to smaller pieces. These abilities allowed the exploitation of many feeding opportunities. A second innovation was necessary before vertebrates could fully exploit potential new food sources, because the mouth, and hence body, must be guided toward an object. The control of the body in three dimensions is allowed by the presence of paired fins with internal skeletal and muscular support that permitted control of the body in locomotion. Gnathostomes have an internal ear with three semicircular ducts, two vertically and one horizontally oriented. These structures are concerned with balance and position of the organism.

The number of inner ear ducts can and has been used to support ideas on the relationships among vertebrates, and thus merits additional remarks. Until recently, hagfishes have been recognized as having one vertical duct and lampreys as having two vertical ducts, the horizontal duct missing in both. This has been used as evidence to support a progressive sequence of complexity of the inner ear, with the hagfishes possessing the basal state (one duct), lampreys and gnathostomes being more derived in possessing at least two ducts, and gnathostomes being even further derived in evolving a third (the horizontal) duct. This sequence fit nicely with the phylogenetic hypothesis recognizing Craniata (hagfishes, lampreys + gnathostomes) and Vertebrata (lampreys + gnathostomes), as described earlier. However, other researchers pointed out that the condition in hagfishes could represent a "degenerate" condition resulting from the loss of one of the two ducts originally present, and thus that the inner ear could not be used to support the Craniata/Vertebrata hypothesis. Recent work on the inner ear of lampreys suggests another alternative—that the evolution of inner ear structures in gnathostomes and lampreys (at least) may represent parallel evolution of the two systems, rather than one having been derived from the other. Further, the system of lampreys is no less suitable than that of gnathostomes for controlling the body in three dimensions (Maklad et al., 2014). The analysis by the latter authors indicates

a more complex lamprey inner ear than was previously supposed. Among other features, Maklad et al. (2014) noted that the lamprey inner ear does possess horizontal ducts and that their topology differs from that present in gnathostomes, lying respectively medial and lateral to the surface of the labyrinth. Despite these differences, the unique horizontal semicircular duct of gnathostomes may still be considered synapomorphic.

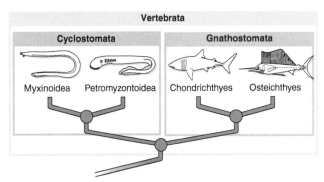

FIGURE 1.6 Cladogram showing phylogeny of Vertebrata. In this hypothesis, the jawless lamprey and hagfish are considered sister taxa forming a monophyletic Cyclostomata, whereas the scheme presented in Figure 1.5 separated them, with lampreys considered more closely related to gnathostomes (the jawed vertebrates). Although morphological evidence supporting a monophyletic Cyclostomata is equivocal (there may be valid synapomorphies, such as a mouth possessing retractable keratinous teeth, and form of the craniofacial musculature), nearly all molecular analyses strongly support this clade.

Other gnathostome synapomorphies include the presence of five pharyngeal slits and jointed visceral arches. In fishes the arches are embedded deep in the body, adjacent to the pharyngeal wall, whereas in cyclostomes they are not articulated structures and lie superficially, just beneath the skin toward the outside of the body. The traditional hypothesis on jaw origins considers jaws as arising from the modification of an anterior visceral arch (located close to the original mouth); i.e., an already existing arch, initially similar to and part of the series of skeletal structures in the branchial region, was modified to form upper and lower jaws. This arch is termed the *mandibular arch*. Other hypotheses on the development of the mandibular arch have been advanced. A recent and novel explanation for its origin, the mandibular confinement hypothesis, reverses the idea proposed in traditional hypotheses. It suggests that the jaws do not effectively represent (or are homologous with) an already existing arch in ancestral gnathostomes; rather, the region anterior to the series of arches formed skeletal structures that functioned as jaws and subsequently came to resemble those in the series of arches. Miyashita (2015) and Miyashita and Diogo (2016) provide further details of this hypothesis and summarize other hypotheses that attempt to explain jaw origins.

The familiar group of vertebrates that possess these features, at least initially, is the fishes, which are also the earliest gnathostomes. Most people know what a fish is, but few recognize that not all fishes are the same with respect to their relationships to other vertebrates. Although they were once all included as "Pisces," they do not form a natural group because some possess features that indicate a common ancestry with tetrapods. Therefore, if "Pisces" were to be retained as a formal term, then the tetrapods would have to be included in the taxon, but it would then be the equivalent of Gnathostomata.

Fish, by and large, all have a similar way of getting on in the world, and it should thus be clear that our everyday concept of fish represents a grade rather than a clade. That they are fish conveys the general idea that locomotion is accomplished essentially through lateral undulations of the trunk and tail with guidance supplied by paired pectoral and pelvic fins; gas exchange occurs primarily through gills located in the pharyngeal slits; the heart is a simple, tubular, single-barreled pump; and so on. However, as some fish are more closely related to other types of vertebrates, including birds and mammals, our classification must reflect this.

Among gnathostomes two extant lineages, Chondrichthyes (from the Ancient Greek *khondros*, cartilage, and *ikthys*, fish) and Osteichthyes (from the Ancient Greek *osteon*, bone, and *ikthys*, fish), are almost universally accepted as monophyletic sister taxa based on molecular and morphological evidence (Figure 1.6; Brazeau and Friedman, 2014). However, two fossil groups, Placodermi and Acanthodii, were long considered monophyletic and traditionally viewed as prominent assemblages in early gnathostome history. Although several authors cautioned that these groups were not unproblematic, their phylogenetic positions and the nomenclature reflecting them, illustrated in Figure 1.7, had become imbedded in our concept of this early history. Indeed, the position of its members as either placoderms or acanthodians was largely due to difficulties in accommodating them as typical chondrichthyans or osteichthyans; and, as noted by Brazeau and Friedman (2014), this hindered our ability to tease out the steps and timing of the evolutionary transition between jawless and jawed vertebrates. This "neat" arrangement, however, has recently been subjected to rigorous character and cladistic analyses, which suggest that placoderms and acanthodians (as formerly conceived) may not exist as independent groups—i.e., they are not monophyletic.

The term placoderm (from the Ancient Greek *plax*, flat, platelike, and *derma*, skin) denotes the general feature of these vertebrates—typically large, platelike bones cover the head and anterior trunk regions as in the "ostracoderms," but they share numerous features that unite them with all other gnathostomes, including jaws and paired fins. An important feature usual of placoderms is that the armored head and trunk regions were linked by a movable joint (not present in "ostracoderms") that allowed the head shield to be lifted (Benton, 2015). Some 350 placoderm genera have been described and vary considerably in size and form—they are not rare, and elucidation of their relationships to other gnathostomes should greatly enhance our understanding of early gnathostome evolution. Although they were generally small

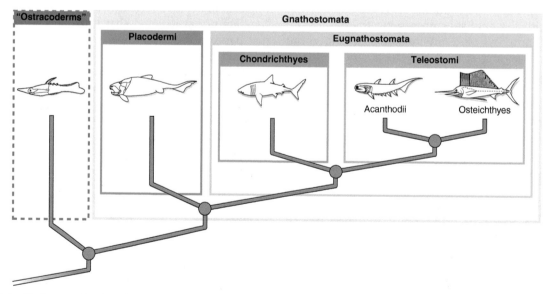

FIGURE 1.7 Cladogram showing the traditional relationships among Gnathostomata. As noted in the text, this outdated concept viewed Placodermi and Acanthodii as monophyletic groups, the former as sister group to Eugnathostomata and the latter as sister group to Osteichthyes within the clade Teleostomi. A more recent concept of gnathostome relationships is presented in Figure 1.8. Cyclostomes are omitted; dashed lines denote a paraphyletic group.

and possessed extensive dermal coverings, some were massive predators; the North American Late Devonian *Dunkleosteus*, at 6–7 m in length, was the largest vertebrate that had evolved up to that time (Benton, 2015). Although details of placoderm relationships are still being worked out, suggestions are that several groups of placoderms (e.g., Acanthothoraci, Petalichthyida, Antiarchi, and Arthrodira) may lie, phylogenetically, between "ostracoderms" and osteichthyans (acanthodians are discussed later).

Despite the similarity between placoderms and osteichthyans (i.e., bony fishes, as opposed to chondrichthyan or cartilaginous fishes) in having large bony skeletal elements, for most of the last 100 or so years this resemblance was considered to have been acquired independently. With chondrichthyans and osteichthyans firmly established in a group as sister taxa, thus excluding placoderms, the nature of the morphology of the immediate ancestor of the chondrichthyan-osteichthyan clade— was it covered by a bony or soft hide?—could not be discerned. Yet, the sharklike form was usually assumed to be the more "primitive" condition, although it turns out that we had it backwards. The reasons for this were bluntly (though truthfully) provided by Friedman and Brazeau (2013: 176): "The status of sharks as surrogate ancestors seems well established, but this is an illusion of dogmatic repetition combined with spurious portrayals of present-day cartilaginous fishes as unchanged 'living fossils'. The popular model of a shark-like ancestor is, in the end, more of a hangover of the 'great chain of being' of ancient philosophy and pre-Darwinian archetypes than a product of modern comparative biology and phylogenetic 'tree-thinking'."

The change in thinking mainly got underway with the work of Brazeau (2009), who noted that placoderms may not form a clade at all, but that the various placoderm groups lie, as noted earlier, phylogenetically between "ostracoderms" and osteichthyans. The work of Zhu et al. (2013, 2016) reinforced this idea in reporting on the remarkable fossil fish, considered maxillate placoderms, *Entelognathus* and *Qilinyu* from the Silurian of China. These authors reported on dermal elements typical of osteichthyan jaws, including the maxilla, premaxilla (upper jawbones that integrate with the cheek bones), and dentary (a lower jawbone that is part of several bony plates and cartilage), which they interpreted as homologous to the gnathal plates (simple platelike bones surrounding the mouth) that are typical of placoderms and that help elucidate the gradual transformation of jaw elements between placoderms and more derived vertebrates. However, the recent reanalysis by King et al. (2017) of a revised and expanded data set suggested that parsimony analysis indicates that the paraphyly and monophyly of placoderms (excluding maxillate forms) are essentially equally parsimonious, but that Bayesian tip-dated clock methods strongly support placoderm monophyly (again, excluding maxillate forms), although the authors noted the need for further investigation. Resolving the phylogeny of placoderms is highly important for understanding events in early vertebrate evolution, as the paraphyletic and monophyletic hypotheses are starkly different frameworks for establishing plesiomorphy and tracing morphological change. Under the paraphyletic scenario, shared placoderm features are presumed plesiomorphic (thus ancestral for all gnathostomes), with the simple gnathal plates of the placoderm jaw being the "starting" condition, and dermal bones (e.g., maxilla, dentary) being added as the gnathostome node was approached. However, if placoderms are monophyletic, then the placoderm condition may be viewed as due to unique specializations. The osteichthyan-like condition could then be viewed as the ancestral condition for gnathostomes, from which the bony elements were reduced during placoderm evolution to produce their gnathal plates. Pending resolution of their phylogenetic status, the latter are represented simply as a group in Figure 1.8, without intent regarding its monophyly or paraphyly.

In the remaining gnathostomes, Eugnathostomata (from the Ancient Greek, *eu*, good or true + *gnathos*, jaw, and *stoma*, mouth), the second visceral arch is modified into a hyoid arch, a supporting element for the jaw (Figures 1.7, 1.8, and 1.9; in the latter figure and in Figure 1.6, they are grouped as Gnathostomata, as they represent the living gnathostomes). In addition, eugnathostomes possess true teeth (but see Rücklin and Donoghue, 2015, 2016; Burow et al., 2016). They include the remaining fishes and the terrestrial vertebrates. As alluded to earlier, two clearly differentiated groups are recognized among living eugnathostomes: Chondrichthyes (the cartilaginous fishes: sharks, rays, and ratfishes) and Osteichthyes (the bony fishes, including bony fishlike forms, such as ray-finned fishes, and lobe-finned fishes and their terrestrial relatives). An obvious difference between them is that sharks and their relatives bear skin equipped with toothlike scales, and the bony fishes have platelike bones over the head and shoulder regions.

Chondrichthyes (Figure 1.9) includes the sharks and rays (Elasmobranchii) and chimaeras (Holocephalimorpha), and is united by various derived features, such as placoid scales, a cartilaginous skeleton with prismatic calcification, an endolymphatic duct connecting the inner ear with the exterior, and the presence of claspers in males. Elasmobranchii (from the Ancient Greek *elasmos*, thin plate, and *branchia*, gills) includes the living sharks and rays (Neoselachii). Among the sharklike forms are Galeomorphii, the larger assemblage of sharks, and several other smaller clades commonly grouped together as Squalea. The rays are grouped as Batoidea. Elasmobranchs have partitions between the pharyngeal slits that bear the gills. The holocephalimorphans

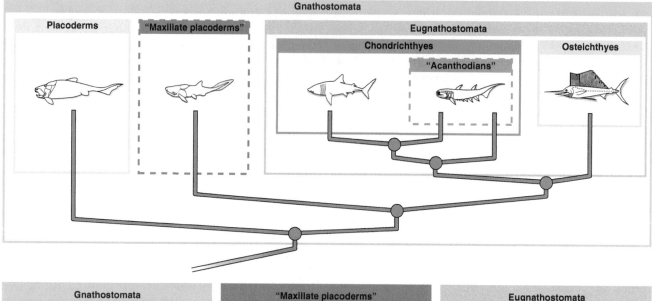

FIGURE 1.8 Cladogram showing phylogeny of Gnathostomata. Placoderms are depicted as a monophyletic group here (but this is still open to debate; see text), whereas "maxillate placoderms" are considered stem eugnathostomes (some authors, e.g., Zhu et al., 2012, who restrict Gnathostomata to crown group jawed vertebrates, would consider them stem gnathostomes; see text). "Acanthodians" are considered as chondrichthyans, whereas they were previously viewed as the sister group of osteichthyans (see Figure 1.7). Several synapomorphies of the main groups are provided in the boxes below the cladogram. Dashed lines denote a paraphyletic group. *Adapted from Benton, 2015; Miyashita, 2015; and Qiao et al., 2016.*

(chimaeras or ratfishes; from the Ancient Greek, *holos*, whole, and *kephale*, head) differ in having a fleshy operculum covering the slits. Also, the upper jaw is fused to the braincase, a feature from which the group gets its name, whereas in elasmobranchs the upper jaw is separate from the braincase. Despite these and other specializations, chondrichthyans retain plesiomorphic features in their basic anatomy. The near absence of bone, however, is not ancestral, but a secondary loss. The combination of this condition and of ancestral features is a main reason the shark is used so extensively for dissection: The ancestral features allow an understanding of the basic vertebrate systems, and the specialized absence of bone facilitates the dissection of these systems.

A third radiation of fishes, the fossil acanthodians (from the Latin, *acanthodes*, spiny, thorny) noted earlier, was long considered to form a clade, Acanthodii, more closely related to the bony fishes, and traditionally grouped with them (and their terrestrial descendants, of course) as Teleostomi (Figure 1.7; it was this arrangement that rendered chondrichthyans the basal-most eugnathostomes, and thus their characteristics were largely considered ancestral; see also Brazeau and Friedman, 2015). The bony fishes, as their name implies, were considered

fishes that retained and improved on a bony skeleton, in contrast to the general trend of reduction of bone in sharks and their close relatives. The acanthodians were viewed as a relatively minor group of very early, extinct bony fishes, characterized by long stout spines associated with paired fins, of which more than two were often present (see Maisey et al., 2017 for further discussion). More recently, their monophyly and their phylogenetic position among eugnathostomes have been questioned. Their precise relationships are still being worked out, but it seems sufficiently clear that "acanthodians" are not monophyletic (hence we have added the quotation marks). Some authors, such as Brazeau (2009) and Davis et al. (2012), considered some basal gnathostomes, most basal chondrichthyans, and others basal osteichthyans. Zhu et al. (2013), on the other hand, viewed them all as more closely related to chondrichthyans and moved them to the chondrichthyan stem, as indicated in Figure 1.8. Brazeau and Friedman (2014), however, were more reserved. These authors suggested that the two most plausible alternatives (that the acanthodians are all stem chondrichthyans, or that most are stem chondrichthyans but some are stem osteichthyans) differ very little in terms of support.

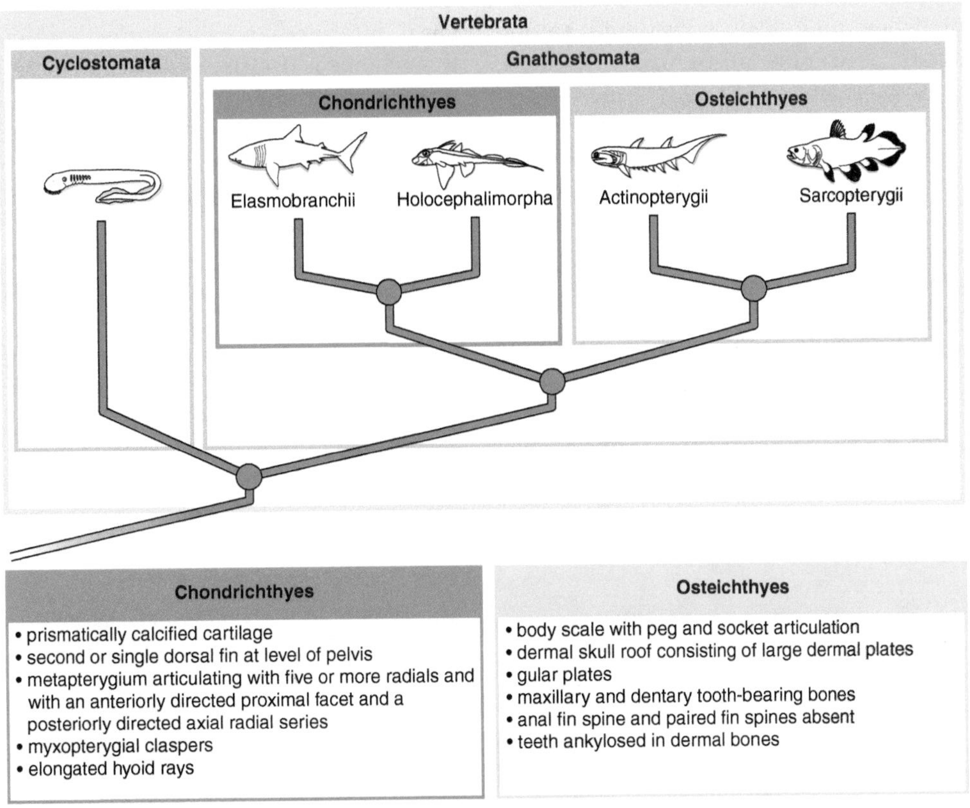

FIGURE 1.9 Cladogram showing phylogeny of living vertebrates. Several synapomorphies of the main groups are provided in the boxes below the cladogram. *From Benton, 2015.*

With the repositioning of "acanthodians" as in Figure 1.8, the remaining eugnathostomes may all be considered Osteichthyes (Figures 1.8 and 1.9), which includes two major radiations of bony fishes (one of which includes the tetrapods), Actinopterygii (from the Ancient Greek *aktis*, ray, and *pterux*, wing) and Sarcopterygii (from the Ancient Greek, *sarx*, flesh, and *pterux*, wing). A lung or air sac is considered an ancestral trait for this group. From it a swim bladder evolved in some derived bony fishes.

Actinopterygii (Figure 1.10) or ray-finned fishes are the most diverse and numerous vertebrates (about half of all living vertebrates are actinopterygians; Pough and Janis, 2018) and inhabit nearly all aquatic habitats. Their fins are supported internally mainly by fibrous rays (rather than an endoskeletal support) and are controlled by muscles that lie mainly within the body wall. Actinopterygians include a staggering diversity of fossil and living forms (see Benton, 2015, for further detail), and only the most general of evolutionary outlines of living actinopterygians is possible here (although additional detail on their characteristics is provided in Figure 1.10). Several groupings may be recognized. The most basal is Polypteridae, a clade that includes the extant *Polypterus* (bichir) and *Erpetoichthys* (ropefish or reedfish), as well as several fossil forms; these fishes are also

grouped together as Cladistia. Polypterids retain several of the early features of early actinopterygians, such as ganoid scales, a well-ossified skeleton, and paired ventral lungs (air sacs) connected to the pharynx for aerial respiration. Many students are surprised that lungs would be important in fishes, but aerial respiration is so important that bichirs drown if deprived of it. For many years, not only were polypterids considered the most basal of living actinopterygians, but close to the origin of the entire actinopterygian clade; and many very early fossil actinopterygians, informally considered "palaeoniscoids," were considered more closely related to the remaining actinopterygians than were the polypterids. Thus, many of the highly unusual (for an actinopterygian) features, such as their muscular pectoral fins, were considered primitive (Coates, 2017). The recent effort of Giles et al. (2017) strongly suggests, however, that polypterids are not as basal as previously assumed, and indeed, that they are much more closely related to living actinopterygians than are the "palaeoniscoids." This in turn suggests that many apparently plesiomorphic features of polypterids are actually reversals, and some (such as the architecture of the pectoral girdle) are probably autapomorphic (autapomorphy refers to a derived character unique to a clade). The analysis by Giles et al. (2017) is partly based on a group of extinct fishes known as scanilepiform fishes, previously considered among the "palaeoniscoids" but now recognized by these authors as stem polypterids.

Actinopteri includes more derived actinopterygians, with Chondrostei (sturgeons and paddlefishes, as well as extinct relatives) and its sister clade Neopterygii (Figure 1.10). The latter includes Holostei and Teleostei. Of these, teleosteans (or, more simply, teleosts) are by far more diversified and numerous—indeed, this clade includes nearly 30,000 species, more than all living amphibians, reptiles (including birds), and mammals combined, whereas holosteans include only eight species distributed in Lepisosteiformes (gars) and Amiidae (bowfins) (Clarke et al., 2016). These three groups of actinopterygians—the chondrosteans, holosteans, and teleosteans—may be considered as reflecting a sequence of

basal, intermediate, and derived actinopterygians, and this scheme is useful in following, in broad outline, some of the main trends in actinopterygian evolution. These include changes in the feeding apparatus, fin form and position, and body shape.

The feeding apparatus of bony fishes is structured so that the lower jaw was ancestrally capable only of simple opening and closing movements. In this system, the upper jaw was fused to the braincase. The upper and lower jaws were long, with the articulation far back under the skull, permitting a wide gape. These features are reflected by the orientation of the hyomandibular, which sloped posteroventrally. The feeding apparatus underwent modifications, resulting in a complex kinetic system in which the jaws are protruded and allow inertial suction feeding. The main anatomical changes are

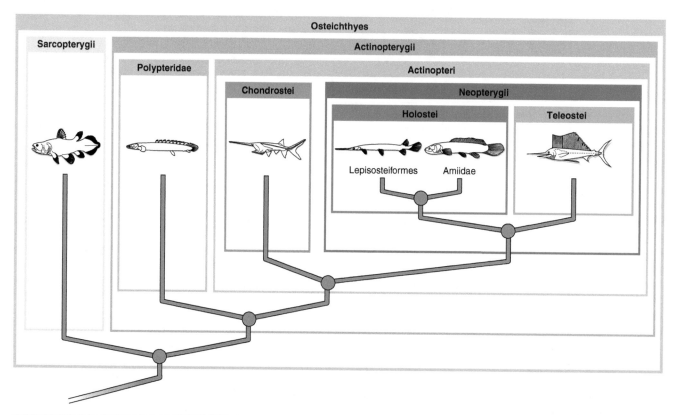

Sarcopterygii	Actinopterygii	Actinopteri	Neopterygii
• muscular pectoral and pelvic limbs, limb bones substantial • true prismatic enamel on teeth • more than four plates forming sclerotic ring • triradiate scapulocoracoid • splenial present in lower jaw • intracranial joint between otico-occipital and ethmosphenoid • one or more squamosals present	• dermosphenotic T-shaped, contacts nasals • postorbital absent • squamosal absent • denatary with enclosed mandibular canal • one or two pairs of extrascapulars • single dorsal fin	• accessory vomerine tooth plate • all fins possess branching rays	• maxilla and preopercular do not contact palatoquadrate • mobile maxilla • maxilla has peg-like anterior head • quadratojugal forms brace for quadrate • symplectic present • coronoid process present

FIGURE 1.10 Cladogram showing phylogeny of Osteichthyes, with emphasis on Actinopterygii; some early fossil groups are omitted. Several synapomorphies of the main groups are provided in the boxes below the cladogram. *Adapted from Benton, 2015, and Schultze, 2016.*

that the jaws shortened, so the hyomandibular swung forward to assume, in derived teleosteans, an antero-ventral orientation. The maxilla, a bone of the upper jaw, was freed from the jaw margin to function as a lever participating in movements of the premaxilla, the most anterior element of the upper jaw. Inertial suction feeding opened up numerous opportunities and is one of the main features cited in the success of actinopterygians (but see later). Associated changes occurred in the position and form of the fins and of the body. Ancestrally, the tail was heterocercal and the paired fins were in positions similar to those in the sharks: relatively ventral, with the pectoral fins lying anteriorly and the pelvic fins lying posteriorly. Also, the body is, again as in sharks, fusiform, or torpedo-shaped. These features make for fast swimming. In teleosteans the pectoral fins are moved dorsally and the pelvic fins anteriorly. The tail is homocercal (superficially symmetrical) and the body laterally compressed. These changes allowed for different swimming styles, with considerably more precision control (for example, the dorsal position of the pectoral fins allows them to function as "brakes").

The changes discussed in the preceding paragraph provide a sweeping overview of the broad changes that occurred during actinopterygian evolution. It is worth noting, though, that they are among the several reasons (or key innovations) that have been advanced to explain the extraordinary radiation of teleosteans. Others include reproductive biology and, more recently, genomic architecture—that is, that the teleostean ancestor underwent a duplication of its genome approximately 160 million years ago, with the biological assumption that a copy of extra genes was available for evolving novel functions and morphologies, thereby accelerating evolutionary change (Pennisi, 2016). The work of Clarke et al. (2016: 11531), however, challenges these ideas; these authors noted that arguments for enhanced phenotypic evolution in teleosteans "draws heavily on the snapshot of taxonomic and phenotypic imbalance apparent between living holosteans [which did not undergo genomic duplication] and teleosts." The paleontological evidence reveals an extraordinary taxonomic richness and morphological diversity for extinct holosteans, including periods during which this group exceeded teleosteans in taxonomic richness; and that teleosteans do not exhibit enhanced phenotypic diversification throughout much of their history. In other words, despite genomic duplication, for example, the teleosteans were not particularly more diverse or abundant than holosteans for much of the first 150 million years of their history (see Clarke et al., 2016, for further details).

The sister group to Actinopterygii is Sarcopterygii (the second group of bony fishes that also includes tetrapods), which possess paired fins with internal skeletal support and muscles, and so are known as the lobe- or fleshy-finned fishes (Figures 1.9, 1.10, and 1.11). In actinopterygians the proximal part of the fin consists of a series of parallel bones called radials. These radials articulate with the girdle proximally and with bony fin rays, or lepidotrichia, distally. By contrast, the proximal part of the fin of sarcopterygians consists of a single radial, with more distal radials extending along a main axis (Clack, 2012). As fish, sarcopterygians are not as diverse or successful as the actinopterygians (although this was not always the case, as they were quite diverse earlier in their history; see Benton, 2015 and Clack, 2012). As a clade, however, sarcopterygians are extremely successful, owing to the radiation of tetrapods. The fleshy fins of sarcopterygian fishes were not used for walking on terrestrial environments, but for maneuvering in shallow waters. Interestingly, a group of living sarcopterygian fishes, coelacanths (see later), swim by moving their fins the same way a terrestrial vertebrate uses its limbs to move on land.

Other sarcopterygian synapomorphies are provided in Figure 1.10. The most basal clade is Actinista, the coelacanths, represented only by two living species; its sister group, generally termed Rhipidistia (see, for example, Mondéjar-Fernández, 2018; Pough and Janis, 2018), includes Dipnomorpha and Tetrapodomorpha (Figure 1.11). The former includes Dipnoi, the living lungfishes, of which only three genera survive. Among the characteristics that allow recognition of tetrapodomorphs is the unique presence of an internal nostril within the oral cavity. The vertebrates that belong to this clade include those with a well-formed limb with digits (fingers or toes) and those more closely related to them than to Dipnomorpha. Within Tetrapodomorpha is Tetrapoda (which some authors prefer to call Neotetrapoda, and use Tetrapoda as a more inclusive group; see, for example, Benton, 2015; Shubin et al., 2014), a crown clade that includes the living limbed vertebrates (with the understanding that some members have secondarily lost limbs, such as snakes and whales) and their most recent common ancestor. This grouping leaves out some extinct, limbed forms, such as *Acanthostega* and *Ichthyostega* (see later). Several authors (e.g., Buffrénil et al., 2015; Konietzko-Meier et al., 2016; Laurin and Soler-Gijòn, 2006, 2010) use Stegocephalia for the clade that also includes these last two, but this usage has apparently not gained widespread acceptance (e.g., see Benton, 2015; Clack, 2012; Schoch, 2014). In any event, the concept of Tetrapodomorpha views several extinct close relatives of Tetrapoda, some certainly with fishlike fins and others possessing limbs with digits, as successive stem taxa (i.e., occupying positions on the stem leading to Tetrapoda). The divergent usage of Tetrapoda (as with many other names) is related to the method preferred among different researchers: some prefer character-based definitions, whereas others prefer phylogeny-based definition. The

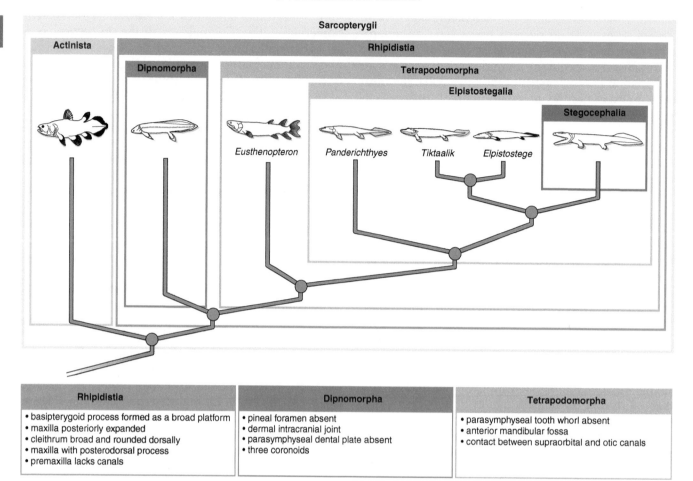

FIGURE 1.11 Cladogram showing phylogeny of Sarcopterygii; some early fossil groups are omitted. Several synapomorphies of the main groups are provided in the boxes below the cladogram. *Adapted from Benton, 2015.*

latter tends to restrict common and long-used names (such as Tetrapoda, Reptilia, and Mammalia) to crown groups, as is done here. The interested reader is referred to Laurin and Anderson (2004) and Schoch (2014) for further explanation and discussions of the advantages and disadvantages of these two methods.

We may note several of the more familiar and better-known fossil fish–like forms, such as *Eusthenopteron, Panderichthyes, Elpistostege,* and *Tiktaalik* (Figure 1.11). Most of these fishlike relatives of tetrapods are included with their more derived tetrapodomorph relatives in Elpistostegalia (see, for example, Shubin et al., 2014). Among the notable features of such fish are an elongated humerus and loss of the dorsal and anal fins. Until recently, the gap between the more basal fishlike forms and the earliest stegocephalians such as *Acanthostega* and *Ichthyostega* was rather pronounced—on the one hand, we had forms that were clearly fish and, on the other hand, forms with fully developed limbs with digits. The discovery of *Tiktaalik,* closely related to stegocephalians, has filled in a good deal of this gap (Ahlberg and Clack, 2006; Clack, 2012; Daeschler et al., 2006; Shubin et al.,

2006, 2014). Its pectoral fin is morphologically and functionally transitional between a fin and a limb, bearing expanded skeletal elements with mobile joints and an array of distal segments similar to the distal limb pattern of basal tetrapods. The fin could assume a range of postures, including a position in which the shoulder and elbow were flexed and the distal elements extended to provide a limb-like supporting stance (Shubin et al., 2006). The pelvic girdle is transitional in size and robustness between those of limbed and other finned tetrapodomorphs (Shubin et al., 2014). Our most recent understanding of these groups indicates that limbs first evolved in vertebrates that lived almost entirely in an aquatic environment. It is thought that the limbs were useful particularly in shallow areas, helping maintain the animal's position so that it could wait for prey and allow it easy access to air. Perhaps these vertebrates could also clamber out of the water to escape predators. After all, the terrestrial environment would at that time have been relatively free of predators and competition.

Tetrapoda includes Amphibia and Reptilomorpha (Figure 1.12). Tetrapods were and have remained mainly

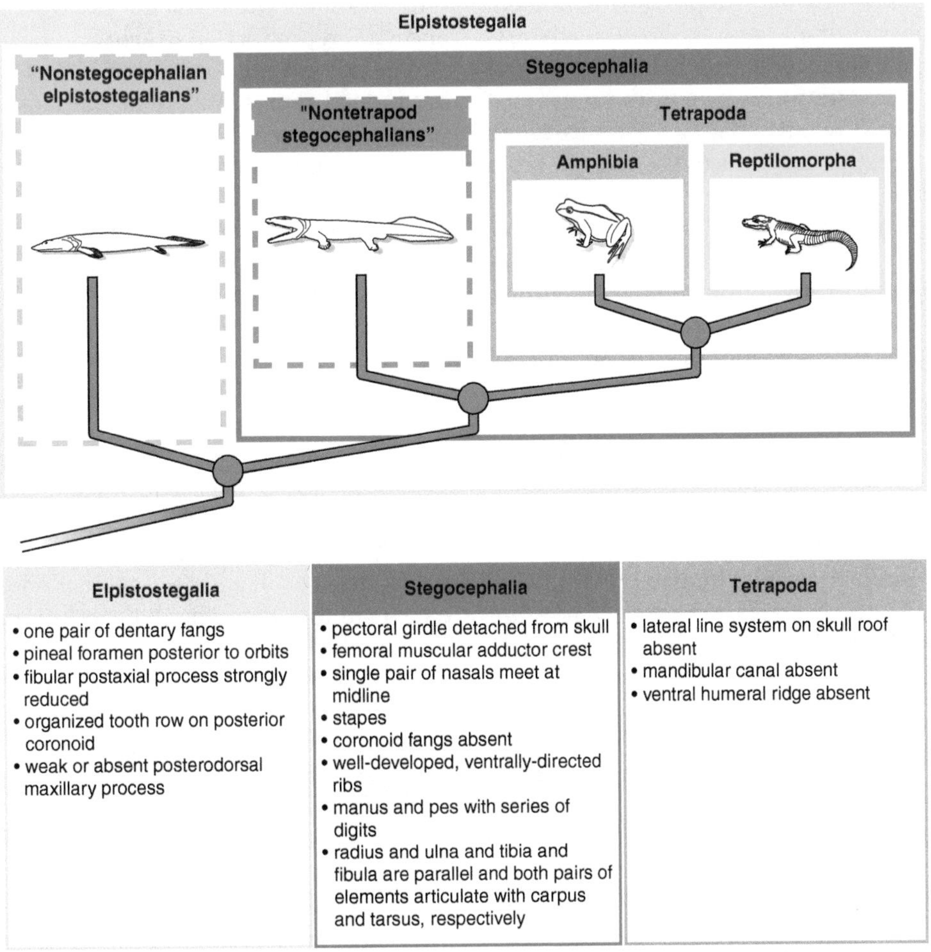

FIGURE 1.12 Cladogram showing phylogeny of Elpistostegalia; some fossil groups are omitted. Several synapomorphies of the main groups are provided in the boxes below the cladogram. *Adapted from Benton, 2015.*

amphibious or terrestrial, although derived members of several lineages readopted a mainly or entirely aquatic existence (e.g., the extinct ichthyosaurs among reptiles and the whales among mammals), and others are (or were) capable of flight (e.g., birds and the extinct pterosaurs among reptiles, and bats among mammals). They share (together with several groups omitted here) five or fewer digits. There is, as usual, a lack of consensus on these names—for example, Schoch (2014) followed the usage of Amphibia (as is done here), whereas Benton (2015) preferred Batrachomorpha—as well as on the position of several of the fossil groups; Schoch (2014) and Marjanović and Laurin (2013) provide useful discussions on these topics.

Lissamphibia and Amniota (Figure 1.13) represent, respectively, the living amphibians and reptilomorphs (although, of course, they also include extinct members— Lissamphibia includes fossil and living frogs, for example), and thus are the living tetrapods. The three lissamphibian (a term derived from the nature of their skin; from the Ancient Greek *lissos*, smooth) groups, quite distinct from one another, are Anura (frogs and toads), Caudata (newts and salamanders), and Gymnophiona (caecilians). Frogs and salamanders are reasonably familiar vertebrates, and are generally considered sister groups, forming the clade Batrachia (Benton, 2015; Schoch, 2014). Frogs are highly specialized for saltatory locomotion, whereas salamanders possess a more general body form and locomotion. Caecilians are specialized in being legless burrowers or swimmers.

The term amphibian is derived from the Ancient Greek *amphi*, both, and *bios*, life, and reflects the duality in the lifestyle of many living members of the group— often a larval aquatic stage and a terrestrial adult stage are present. This term was traditionally applied to living amphibians, formally Lissamphibia (including frogs, salamanders, and caecilians) as well as to the many fossil groups between fishes and amniotes (i.e., not only the stem groups leading to Lissamphibia), and in this sense, the living amphibians were generally viewed as representatives of the stage between fishes and amniotes. This was due largely to the fact that their reproductive strategy is still mainly tied to an aquatic environment (though this is not true of all lissamphibians), whereas reproduction in amniotes is more nearly independent of water. This general impression is true in the sense that amphibians do tend to retain what has been assumed to represent an ancestral reproductive strategy. From this it is but a small step to the view that all amphibians, including the living forms, are therefore primitive tetrapods. However, this is both misleading and incorrect: it is wrong to think of any living organism as primitive, because the word primitive is well entrenched in everyday language as meaning "not as good as" or "not good enough." A creature may retain ancestral (basal is the appropriate alternative) features, but that does not make the creature itself primitive. Each living organism is the product of a long evolutionary history and is a mosaic of both basal and derived (often "advanced" is used, but this term is also inappropriate to describe organisms) features. For example, humans retain bone, an ancestral vertebrate character, whereas sharks are derived ("advanced") in the loss of bone. The presence of this ancestral feature does not make humans "more primitive" or "less advanced" than sharks. Using the same logic, frogs are not more "primitive" than humans just because they "may" (the reason for italicizing "may" will become evident in the next paragraph) retain an ancestral reproductive strategy.

The second misconception is that living amphibians are indeed representative of the lifestyle of the earliest terrestrial vertebrates. In some ways these early forms were intermediate between fishes and more derived terrestrial vertebrates, and in the past we have lumped these forms together with living amphibians. But we must be careful. Living amphibians, while retaining an *apparently* ancestral reproductive mode, are clearly very specialized. They are not like the early terrestrial vertebrates, and in fact are highly derived vertebrates. As noted by Schoch (2014), modern salamanders were long considered as the model living organisms to understand the extinct amphibians and early tetrapods—both their apparently primitive mode of terrestrial locomotion and biphasic life cycle (with a water-living larva that metamorphoses into a terrestrial adult) were seen as essential features of all early tetrapods. Indeed, more recently recovered fossils and analytic techniques that have shed light on the lives of early amphibians indicate that metamorphosis was not present in most of them and that the salamander model is inappropriate as a model for understanding early terrestrial vertebrates (see Schoch, 2014, for further details).

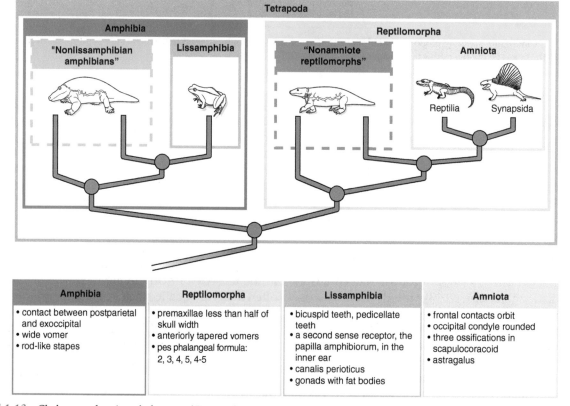

FIGURE 1.13   Cladogram showing phylogeny of Tetrapoda. Several synapomorphies of the main groups are provided in the boxes below the cladogram. *Adapted from Benton, 2015, and Schoch, 2014.*

As noted earlier, the sister group of Amphibia is Reptilomorpha, which includes the living Amniota and fossil kin, as well as several fossil groups as stem taxa (Figures 1.12 and 1.13). The main innovation of amniotes is the amniotic egg. Amniote embryos develop within extraembryonic membranes that are usually encased in a calcareous or leathery egg (the term "anamniotes" refers to vertebrates whose eggs do not have such extraembryonic membranes). The membranes provide the embryo with a "watery" environment that is protected against desiccation, and thus amniote reproduction has become essentially independent of an aqueous environment in which vertebrates ancestrally reproduced (however, a relatively moist environment is still essential).

The amniotes (Figures 1.13 and 1.14) include two great lineages, Reptilia and Synapsida, which have followed independent evolutionary paths since the early history of amniotes. The synapsids include "mammal-like reptiles" and mammals. The latter, Mammalia, are the living synapsids. One mammalian group is Monotremata, a relatively small clade, including the echidnas and platypus, that retains the ancestral reproductive strategy of laying eggs. The other group, Theria, includes the marsupials (Marsupialia) and the placental mammals (Eutheria). These mammals have evolved reproductive

modes where embryos are retained in and nourished by the mother's body.

Reptilia (effectively equivalent to Sauropsida of some authors; see Benton, 2015) consists of the typical living and fossil reptiles, such as turtles, lizards, snakes, and crocodiles, along with other familiar and mainly extinct groups, such as dinosaurs (including birds, which of course are not extinct), pterosaurs, plesiosaurs, and ichthyosaurs (Figures 1.14, 1.15, and 1.16). As indicated in Figure 1.14, two main reptilian groups may be recognized, Parareptilia and Eureptilia (which some authors refer to as Reptilia). There is reasonable consensus, reached mainly during the late 1990s, that the former is monophyletic and includes several extinct groups such as Bolosauridae, Procolophonidae, Pariesauromorpha, Millerosauria, and Mesosauridae, among others (see Benton, 2015; Modesto et al., 2015; Tsuji and Müller, 2009). One issue that remained contentious, however, was the position of Testudines (turtles and tortoises), which is among the groups on which Parareptilia was originally erected during the 1940s; a characteristic of parareptiles was assumed to be the anapsid skull condition (see later). The position of turtles was the subject of continued and considerable debate following the analysis by Rieppel and de Braga (1996) that repositioned

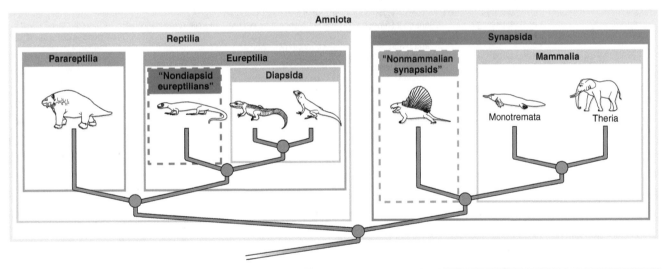

Reptilia	Parareptilia	Eureptilia	Diapsida	Synapsida	Mammalia
• jugal separates maxilla from quadratojugal • tabular small or absent • parasphenoid wings absent • single coronoid • single centrale in tarsus	• caniniform maxillary tooth absent • supraglenoid foramen absent • long and slender femoral diaphysis • posterior hemal arches attached to anterior centrum	• bilaterally embayed posterior skull margin • no contact between supratemporal and postorbital • narrow iliac blade • small supratemporal	• dorsal and ventral temporal fenestrae • suborbital fenestra • ossified sternum • complex tibio-astragalar joint	• maxilla contacts quadratojugal • caniniform maxillary teeth • ventral temporal fenestra • paroccipital process contacts tabular and squamosal	• surfaces of upper and lower molars match precisely • distinct masseteric fossa with well-defined ventral margin • elongated cochlear canal • greatly enlarged gyrencephalic cerebral hemispheres

FIGURE 1.14 Cladogram showing phylogeny of Amniota. Several synapomorphies of the main groups are provided in the boxes below the cladogram. *Adapted from Benton, 2015.*

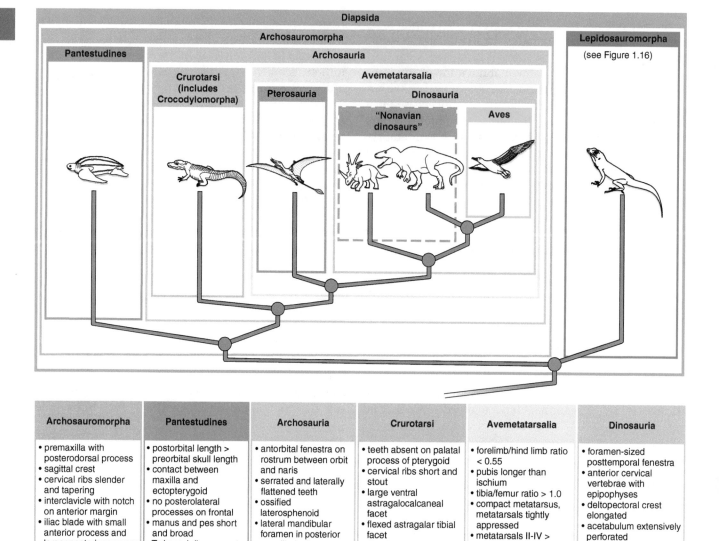

FIGURE 1.15 Cladogram showing phylogeny of Diapsida, with emphasis on Archosauromorpha (some groups omitted; see Figure 1.16 for Lepidosauromorpha). Several synapomorphies of the main groups are provided in the boxes below the cladogram. *Adapted from Benton, 2015, and Schoch and Sues, 2017.*

Testudines within Eureptilia, a hypothesis that was soon after supported by molecular evidence (see Tsuji and Müller, 2009, for further details). Evidence began to mount that supported a eureptilian relationship for Testudines, although its position among eureptilians (that is, to which eureptilian group they are most closely related) remained uncertain.

With regard to a parareptilian or eureptilian relationship for Testudines, a main point of contention revolved around whether the turtle skull is ancestrally anapsid (that is, evolved from ancestors that lacked temporal fenestrae) and thus justifiably placed among parareptiles,

or apparently anapsid, with the anapsid condition being a secondary redevelopment from originally diapsid ancestors (as would be required if they are considered eureptilians); this latter proposal is itself not particularly recent, and dates back to at least the 1920s (see Schoch and Sues, 2017). Despite the existence of several early fossil turtle relatives, until recently none preserved definitive evidence that allowed the conclusion that the diapsid condition was originally present in Testudines. The recently discovered stem turtle *Pappochelys*, however, does preserve such evidence. The recent analyses by Schoch and Sues (2015, 2017) of *Pappochelys*, which

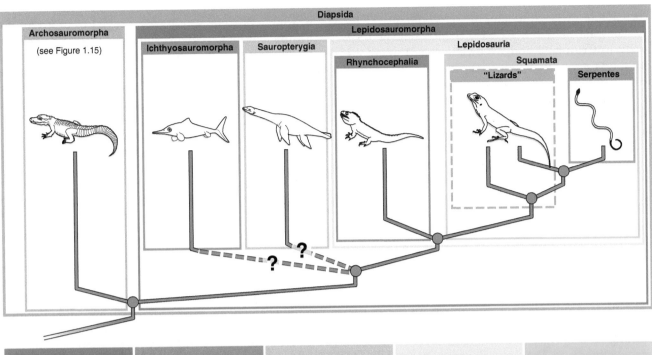

FIGURE 1.16 Cladogram showing phylogeny of Diapsida, with emphasis on Lepidosauromorpha (some groups omitted; see Figure 1.15 for Archosauromorpha). Several synapomorphies of the main groups are provided in the boxes below the cladogram. *Adapted from Benton, 2015, and Motani et al., 2015.*

preserves a small upper temporal opening and a large ventrally-open lower opening, strongly support the diapsid (and thus eureptilian) affinities of turtles. A diapsid condition in an older stem turtle, *Eunotosaurus*, was described by Bever et al. (2016). Testudines and its stem taxa may be grouped as Pantestudines (following Joyce, 2015, and Schoch and Sues, 2017). The position of turtles among eureptilians is considered in the next paragraph.

Eureptilia includes fossil stem groups and Diapsida (Figure 1.14). The diapsids consist of two main lineages, Archosauromorpha and Lepidosauromorpha (Figures 1.15 and 1.16). With these two groups now identified, we may return to the position of turtles (and their fossil kin), which have been considered recently to occupy several positions (see Joyce, 2015, for further details):

as a basal group (outside the archosauromorph + lepidosauromorph grouping or even Diapsida), or more closely related to either lepidosauromorphs or archosauromorphs. Recent works defend both basal (e.g., Goloboff et al., 2009; Lyson et al., 2010; Werneburg and Sánchez-Villagra, 2009) and more derived positions, with several studies suggesting a lepidosauromorph relationship (e.g., Lyson et al., 2012) and others, apparently more robustly, an archosauromorph relationship (e.g., Crawford et al., 2012, 2015; Shaffer et al., 2017; Thomson et al., 2014). The latter view is followed here, as represented in Figure 1.15.

In addition to the turtles and kin (Pantestudines), Archosauromorpha also includes Archosauria, a much larger and more diverse assemblage (Figure 1.15).

The living archosaurs are the crocodylians (Crocodylomorpha) and birds (Aves), which belong respectively to Crurotarsi and Avemetatarsalia (Figure 1.15), but the group also includes their many extinct kin. The extinct flying reptiles, the pterosaurs (Pterosauria), are a familiar example of archosaurs, as are the dinosaurs (Dinosauria), a clade to which the birds belong; so birds are dinosaurs. Dinosaurs other than birds are generally referred to as nonavian dinosaurs but this is an informal designation, as it is paraphyletic (because some nonavian dinosaurs are more closely related to birds than to other nonavian dinosaurs). More commonly, the term dinosaur is used to refer to nonavian dinosaurs. Though formally incorrect, this reflects our shared perception of what dinosaurs are. Such usage is fine, so long as we remember that it does not reflect formal classification, hence phylogeny. We may mention here that a debate has recently developed over the relationships among dinosaurs. This is rather a surprise, as for many years, essentially since the late 1800s, there had been great stability in our views on the higher-level relationships among dinosaurs. The classical view is that there are two main groups of dinosaurs based on the structure of their hip bones; the bird-hipped forms or Ornithischia (one of the hip bones, the pubis, points backwards, as in modern birds), and the lizard-hipped dinosaurs or Saurischia (the pubis points forward, as in modern lizards); the latter are further subdivided into Sauropodomorpha and Theropoda. An analysis by Baron et al. (2017a) upset this long-standing hypothesis, and instead linked Theropoda and Ornithischia in a clade termed Ornithoscelida, leaving sauropodomorphs as sister group to this clade. This view was almost immediately challenged by Langer et al. (2017), who reestablished the validity of the traditional arrangement, although they admitted that the difference between the two arrangements was not significantly different. In reply, Baron et al. (2017b) reinforced the preference for their Ornithoscelida hypothesis. It seems that the relationships among dinosaurs is destined for continued intensive scrutiny.

The other main diapsid lineage, Lepidosauromorpha, includes the extinct Sauropterygia (e.g., placodonts, plesiosaurs), Ichthyosauromorpha (e.g., ichthyosaurs), and Lepidosauria, following the scheme outlined by Benton (2015) and presented in Figure 1.16. Lepidosauria consists of Rhynchocephalia and Squamata. Rhynchocephalia is represented today by only two species of *Sphenodon* (tuatara). Its much more speciose sister group Squamata consists of the lizards and snakes. Within the squamates, lizards and snakes are not sister groups. This situation is rather like that of Dinosauria mentioned earlier. The term lizard denotes, colloquially, the group of squamates other than snakes, but snakes are nested within this group. They are, in other words, a group of reduced-limbed or limbless lizards. Snakes (Serpentes, which

is a monophyletic group) are derived from a group of lizards, but there remains some debate over which particular group. However, the perception attached to the colloquial meanings of lizards and snakes is a convenient shortcut, so long as we remember that the term lizard does not refer to a monophyletic clade (unless, of course, we choose to define the term so as to include snakes, but then it would be the equivalent of squamates).

Two other reasonably familiar clades of diapsid reptiles are the extinct aquatic Sauropterygia (e.g., plesiosaurs and placodonts) and Ichthyosauromorpha (most familiarly represented by ichthyosaurs). Ichthyosaurs had a fish- or porpoise-like body, whereas sauropterygians were more diverse. For example, plesiosaurs had wide bodies with paddle-like appendages to propel them through the water, whereas placodonts had wide and flattened crushing teeth and heavily built bodies, but their limbs were not modified into paddles. Although some uncertainty remains—for example, Neenan et al. (2013) included them in a clade that is sister group to a clade including Lepidosauromorpha and Archosauromorpha—they seem to be more closely allied to lepidosauromorphs within Diapsida (Benton, 2015), although their relationships to other lepidosauromorphs remain to be resolved, as indicated in Figure 1.16.

AMNIOTE SKULLS AND CLASSIFICATION

Amniotes have long been subdivided on the condition of the temporal region of the skull, that portion posterior to the orbit. This region can either be solid or have openings, termed temporal fenestrae (from the Latin, *fenestra*, window). Anapsid describes the condition where there are no temporal fenestrae and is characteristic of turtles and most basal amniotes (Figure 1.17a). Three basic patterns of fenestration are recognized. The diapsid condition has two temporal fenestrae, one above the other on either side of the skull (Figure 1.17b). These are the supratemporal (upper or dorsal) fenestra and the infratemporal (lower or ventral) fenestra. This is the condition present in most reptiles. A single fenestra occurs as two variants. The euryapsid skull, characteristic of two reptile groups (ichthyosauromorphans and sauropterygians), bears a supratemporal fenestra (Figure 1.17c), whereas the synapsid skull bears an infratemporal fenestra and is characteristic of synapsids (Figure 1.17d).

The pattern of fenestration is defined by the typical configuration of bones that border the fenestra. The typical configurations of bones for the skull types are illustrated in Figure 1.17. In the synapsid condition the dorsal border of the fenestra is formed mainly by the squamosal and postorbital bones, although the parietal may occasionally participate, whereas the ventral border is formed mainly by the squamosal and jugal bones, with the quadratojugal bone occasionally contributing. The infratemporal fenestra of the diapsid skull is bordered by the jugal, squamosal, and postorbital bones, with the quadratojugal occasionally participating. The supratemporal fenestra is bordered by the postorbital, squamosal, parietal, and, in many cases, the postfrontal bones. Two bony bars, temporal bars (or arcades), are clearly defined in the diapsid skull, a ventral bar formed mainly by the jugal and quadratojugal bones, and a dorsal bar, between the fenestrae, formed by the postorbital and squamosal bones. The fenestra of euryapsid skulls is bordered usually by the parietal, postfrontal, postorbital, and squamosal bones, with the last two meeting ventrally below the fenestra. From these basic patterns, several specializations have evolved.

For much of the past century, the classification of amniotes closely reflected fenestration; and hence Anapsida, Diapsida, Euryapsida, and Synapsida were used as formal names for amniote radiations. More recently, however, we have realized that while the pattern of fenestration does broadly reflect amniote evolution, it is not tied as strictly to phylogeny as was once presumed. At least two of these groups, Diapsida (includes archosauromorphs and lepidosauromorphs) and Synapsida (includes mammals), are still considered monophyletic, but for the former we recognize that at least the earliest basal members of the clade had anapsid skulls.

Within Diapsida a variety of specializations evolved. Among those that display a fully diapsid pattern are tuataras (*Sphenodon*) and crocodylians. Birds, lizards, and snakes, however, have tended to lose one or both temporal bars. This has decoupled the posterior part of the skull, allowing the potential for considerable flexibility among the functional regions of the skull. In general, lizards have lost the lower temporal bar and snakes both the lower and upper temporal bars. At this point, we must elaborate on the presence of the classic diapsid condition in *Sphenodon*. We noted earlier that *Sphenodon* possesses the lower temporal bar. For many years, this condition was interpreted as the retention of an ancestral diapsid condition, and so *Sphenodon* was long designated as a living fossil. However, recent work has demonstrated that all lepidosaurs (rhynchocephalians and squamates) inherited a skull without a lower temporal bar; that is, it is the ancestral condition for this clade. The presence of the lower temporal bar in *Sphenodon* is thus due to secondary redevelopment of this structure, rather than to retention of the ancestral diapsid condition (see Mo et al., 2010). Birds have lost the upper temporal bar so the infratemporal and supratemporal fenestrae have merged into a single large opening. Further, they have lost the bony bar posterior to the orbit, merging the orbit

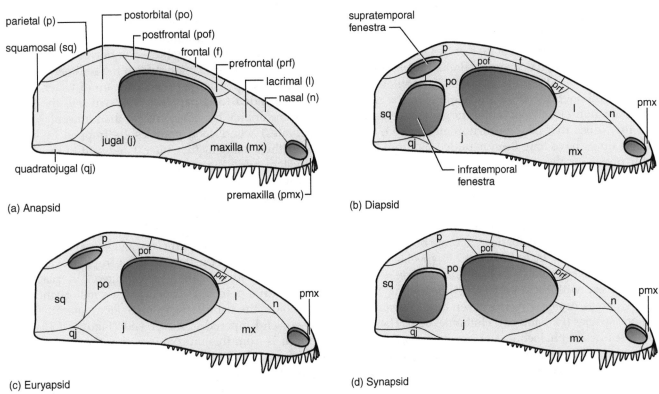

FIGURE 1.17 Diagrammatic illustrations of the four main amniote skull patterns.

and the fenestrae. Among synapsids, in mammals the fenestra increases in size and the bony bar behind the orbit generally disappeared so that orbit and fenestra merge. Although this resembles the condition in birds, the morphologies are convergent.

Euryapsida is no longer considered a valid name. It was applied to ichthyosaurs and plesiosaurs, but we now recognize that the euryapsid pattern evolved independently in these groups from the diapsid condition, and thus that these groups (and relatives, such as placodonts) are diapsids. The euryapsid condition apparently evolved from a secondary "filling in" of the lower temporal fenestra.

Anapsida was applied to the more basal amniotes, including turtles, as the anapsid skull is the basal and nearly ubiquitous type among early tetrapods. As noted earlier, however, recent phylogenetic analyses have revealed that turtles are more closely related to Diapsida than to other anapsid-skulled early amniotes. Anapsida is generally no longer used as a formal name and many early anapsid amniotes are included in Parareptilia (as noted earlier), although some members of this clade do exhibit temporal fenestration (see Modesto et al., 2009).

VERTEBRATE RELATIVES: CEPHALOCHORDATES

Following the preceding outline of vertebrates and their relatives, study of two "prevertebrates" closest to the vertebrates may be undertaken, because such consideration is useful in elucidating the evolutionary development of several key vertebrate features and functions. These include the cephalochordate *Branchiostoma* and the urochordate *Ciona*. They are presented in phylogenetic sequence, although the adult morphology of the cephalochordate more closely resembles the general vertebrate form than does that of the urochordate; indeed, these little creatures have long been regarded as providing a useful proxy for the latest invertebrate morphological ancestor of vertebrates (Garcia-Fernàndez, 2006) because cephalochordates clearly display several important chordate synapomorphies, such as a notochord and a dorsal hollow nerve cord (among others), but possess a less specialized version of the basic vertebrate body plan.

Cephalochordata includes some 32 species distributed among the genera *Asymmetron*, *Epigonichthys*, and *Branchiostoma*, with most species (24) belonging to the latter, whereas *Asymmetron* has seven and *Epigonichthys* only one species (Holland and Holland, 2010). The commonly studied cephalochordates are *Branchiostoma lanceolatum*, *B. floridae*, and *B. belcheri* (Lacalli and Stach, 2016). Two common names are usually applied to cephalochordates: amphioxus or lancelet; the derivation of these names is considered below. Amphioxus is an older term once applied formally to *Branchiostoma*. Although it is no longer considered taxonomically valid, amphioxus (but only as a common name) is pervasive in the literature. These names are used interchangeably in this manual.

Cephalochordates are small organisms (adults range from 1 to 8 cm in length; Ruppert, 1997) that mainly live in the coarse sediments (sands, gravel, or shell material) of temperate and tropical marine environments (Vargas and Harlan, 2010). Although often viewed as occurring in clean sediments, they have also been reported from anthropologically impacted sediments (Da Silva et al. 2008). Most have restricted geographical distributions, but *Asymmetron lucayanum* has a wide circumtropical distribution (Kon et al. 2006; Poss and Boschung, 1996; Sibaja-Cordero et al. 2012). They hatch as larvae at about 1 mm in size and grow to about 1 cm in size, after which metamorphosis occurs. An asymmetric pharynx and a mouth located on the left side of the body characterize the larval stage. At metamorphosis, the pharynx becomes symmetric and the mouth is repositioned anteroventrally (Lacalli and Stach, 2016).

Larval lancelets are planktonic, hovering in the water through ciliary action (Stokes and Holland, 1998). Although the embryos are symmetric, the larvae become markedly asymmetric, perhaps more so than occurs in any other animal (Ruppert, 1997). For example, the left and right side pharyngeal slits appear at different times and the mouth opens originally on the left side of the body. During metamorphosis to the adult stage, bilateral symmetry is largely reestablished (several asymmetric structures persist in the adult). As adults, lancelets spend most of their time partially buried in burrows, with only their anterior end sticking out from the sediments. They are filter or suspension feeders: a stream of water, along with suspended food particles, is drawn into the pharynx of the animal by ciliary (rather than muscular, as in vertebrates) action; food particles are trapped by sticky mucus (thus acting as the "filter") and sent to the digestive system, while the water is removed through openings that lead from the inside to the outside of the pharynx. The openings, or gill slits, are therefore part of the feeding apparatus, rather than the primary respiratory structures present in most fishes. Ruppert et al. (2000) indicated that these organisms feed on particles smaller than 100 microns and transfer this production to higher trophic levels.

Begin your investigation of amphioxus by placing a specimen in a petri dish containing a small quantity of water. Note the fishlike body, elongated, laterally compressed, and tapered at either end (Figures 1.18a and 1.19a). It is from this morphology that the term amphioxus (from the Greek *amphi*, both, and *oxus*, sharp, as noted earlier) is derived. The alternate common name lancelet is derived from lancet, a surgical instrument that usually has a sharp, double-edged blade.

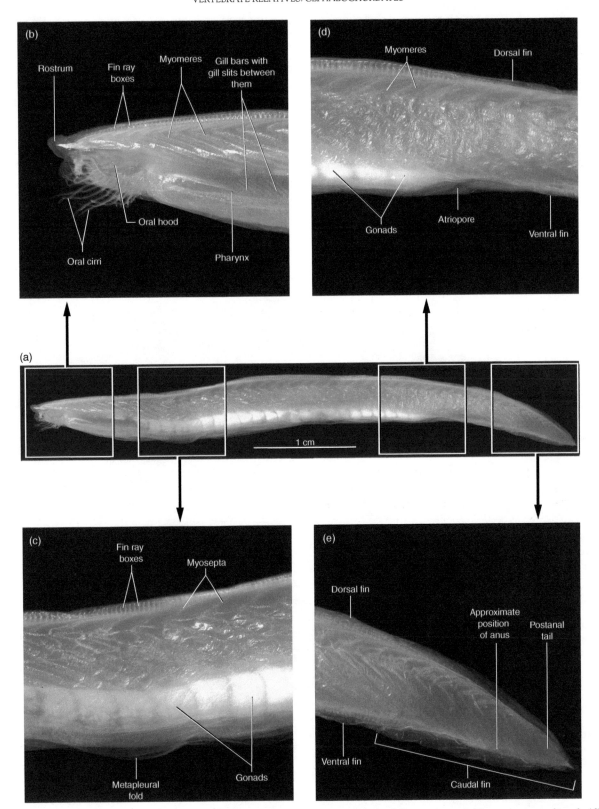

FIGURE 1.18 External features of adult amphioxus in (a) left lateral view, with details of (b) anterior end, (c) anterior part of trunk, (d) posterior part of trunk, and (e) tail region.

Consideration of characteristics beyond these general external features requires that the specimen be examined under approximately 10X magnification with a dissection microscope. Identify the segmented bilateral series of pinkish muscular blocks or **myomeres** (see page 7) extending along much of its body. The myomeres, separated by connective tissue partitions termed **myosepta** (sing., **myoseptum**) are evident through the epidermis, which is formed by a single layer of cells and is therefore thin and semitransparent (Figure 1.18b–d). The myomeres are V-shaped, with the pointy end of the V directed anteriorly. Despite this generalized overall fishlike appearance, however, cephalochordates lack several typical vertebrate features. For example, note the absence of paired eyes and paired fins. Further, although a **postanal tail** (Figure 1.18e) is present, there is no obvious anatomical distinction between a head and trunk (Ruppert, 1997). Recall that the development of a true head, one that includes paired sensory organs, cranium, and a well-developed brain, is a hallmark of vertebrates. As well, the reproductive organs or **gonads** are arranged ventrally as a series of oval, light colored structures. Their ventral portion may be observed, in mature individuals, just below the margins of the myomeres, approximately along the central third of the body (Figure 1.18a, c, and d).

Examine the anterior end of the animal. The feeding apparatus lies just posterior to the **rostrum** and ventral

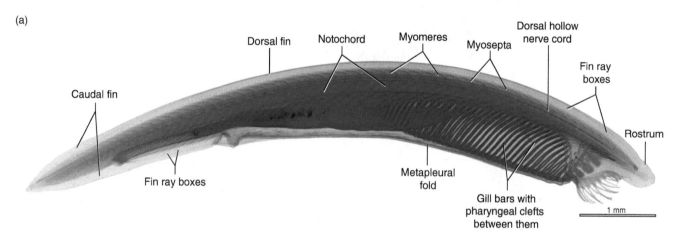

(a)

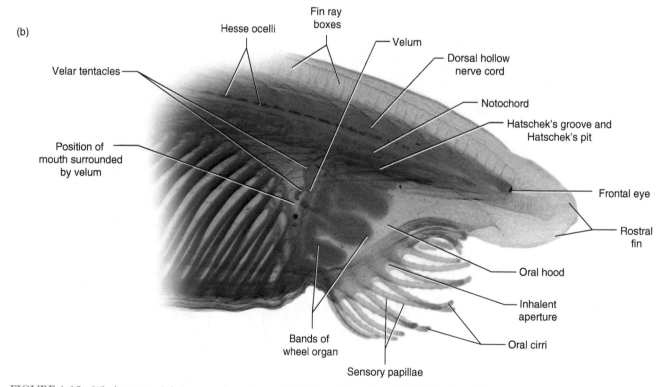

(b)

FIGURE 1.19 Whole mount slide images of amphioxus in left lateral view; juvenile individuals in (a) right lateral view; (b) detail of anterior end of same individual; (c) and (d) right lateral views of another individual; and (e) right lateral view of an adult individual.

to the first several myomeres. Externally, the apparatus includes the **oral hood** and **oral** (or buccal) **cirri**, slender tentacles that fringe the opening of the oral hood. The latter forms a "cheek-like" enclosure for the **oral** (or buccal) **cavity** (Figure 1.19b) or vestibule, the chamber into which water and suspended food particles first enter the amphioxus feeding apparatus. The entrance into the oral cavity is the **inhalant aperture**, located anteroventrally. The oral cavity and oral hood may not be initially apparent, however, as they are often partially collapsed and the oral cirri folded within or across the inhalant aperture. Gentle pressure applied to the region, with a fine probe or the side of a needle, usually renders the oral hood, oral cavity, and oral cirri observable; but they will be considered again below, so do not spend too much time on them. The **mouth** is a

smaller opening lying within the **velum** (Figure 1.19b); these structures are further described later), which demarcates the oral cavity from the **pharynx**. The latter, containing the gill bars, is clearly shown in Figures 1.19, 1.20a, 1.21, and 1.22.

A **dorsal fin** extends along the dorsal surface of the body (Figures 1.18 and 1.19a, d). It is supported by a series of **fin ray boxes** (Figures 1.18b, c and 1.20a), which are filled by a gel-like substance (Ruppert, 1997). A **rostral fin** extends anteriorly from the dorsal fin and wraps around the rostrum and a **ventral fin**, also supported by fin ray boxes, lies approximately along the posterior quarter of the ventral surface (Figures 1.18e and 1.19b, c). The dorsal and ventral fins extend into the tail and expand slightly to form the **caudal fin**, which is not supported by fin ray boxes (Figure 1.18e). The caudal fin

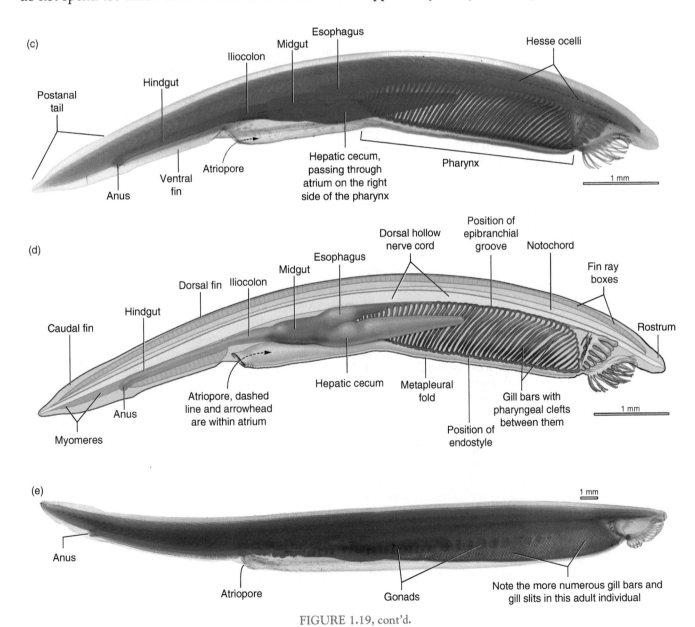

FIGURE 1.19, cont'd.

is often folded over onto the body surface, so delicate manipulation is required to extend it into its life position. However, do not spend too much time on this, as these structures will be observed presently under different conditions. The **metapleural folds** (Figures 1.18e and 1.19a) are ventrolateral fins extending anteriorly from the ventral fin almost to the anterior end of the body. The position of two openings that lead to the exterior may be noted at this time. One is the **atriopore**, which is the exit of the atrium, an internal chamber that will be examined shortly. The atriopore lies midventrally between the posterior ends of the metapleural folds, hence just anterior to the ventral fin (Figures 1.18d, e and 1.19c–e). The **anus**, the exit of the digestive tract, opens ventrally just posterior to the ventral fin. It usually lies just to the left of the midventral line, so it may be described as opening to the left of the caudal fin. Several of the features discussed in these paragraphs are also labeled in Figures 1.20–1.25.

Observation of additional features of the lancelet or amphioxus requires examination of commercially available stained slides of smaller, usually immature specimens. Slides of adult specimens may be available for you to examine (but are not usually commercially available), although these are typically unsuitable for detailed observation of many important structures. As you work your way through this description, keep in mind that the slides at your disposal, even of immature specimens, may not be of the same quality as those illustrated here, so some structures will not be as clearly apparent.

Examine first a whole mount slide of a lancelet. It is often advantageous to examine more than one because some features are better represented in some than other specimens, but most of the features described below are readily apparent in most slides. Three specimens, each displaying some structures more clearly than others, are presented in Figure 1.19a and c (immature individuals) and Figure 1.19e (an adult individual).

Notice first the much greater level of detail, particularly of internal structures, available in such slides. Begin by reidentifying the structures described above. Just posterior to the rounded rostrum is the feeding apparatus. The oral hood, oral cavity, and oral cirri are now clearly apparent. The cirri bear sensory papillae (Figure 1.19b) and are supported by L-shaped skeletal rods, the structure of which (chondrocytes embedded in a collagen-rich extracellular matrix) is very similar to vertebrate cartilage (Kaneto and Wada, 2011).

The myomeres, separated by myosepta, are apparent in all the specimens of Figure 1.19, but are more clearly seen as individual structures in Figure 1.20a. The dorsal, rostral, ventral, and caudal fins are also clearly evident. Note the anus, opening just posterior to the fin ray boxes of the ventral fin, and atriopore, opening just anterior to the ventral fin. The fin ray boxes (Figure 1.19a; see also Figure 1.20b) do not extend into the caudal fin, which may instead be stiffened by hollow, cutaneous canals (which, however, are not visible in Figure 1.19a). The metapleural folds extend bilaterally along the ventrolateral surface of the body, more or less parallel to the ventral body wall. They are more clearly observed in transverse sections (for example, Figures 1.21 and 1.22). Finally, compare the absence of gonads in the immature specimens and their presence in the adult individual (Figure 1.19a, c, and d compared with Figure 1.19e).

Consider next additional structures that have been noted as typical of chordates. Most obvious, perhaps, is the presence of a pharynx with numerous bilateral openings located between strips of pharyngeal tissue, generally termed gill bars (Baskin and Detmers, 1976). The openings are obliquely oriented, elongated and oval. This branchial region takes up nearly the anterior half of the animal. The bars, oriented posteroventrally, are darkly staining and supported by skeletal rods (Figures 1.19 and 1.20a).

The number of openings depends on both the species and growth, with the number of openings and bars increasing during growth (contrast the immature with mature individuals in Figure 1.19a, c, and d with Figure 1.19e). Understanding the development of these structures requires differentiating between the original pharyngeal openings present earlier in development with those present later. A distinct terminology may be applied (as is done here) to these structures. The original openings are termed **pharyngeal clefts**, which are separated or defined by **gill bars** (or primary gill bars). During development each cleft (except the pair at the very front; Ruppert, 1997) becomes subdivided into two **gill slits** by the down growth of a new gill bar, this one termed a **tongue bar** (or secondary gill bar). In the immature individuals of Figure 1.19 splitting of some of the gill bars appears to be occurring. However, as is clear in Figure 1.19b, this is due to the superposition of some of the bars of the right side bars on those of the left side. The differences between gill and tongue bars are difficult to note in lateral view (e.g., Figure 1.19) but are fairly clear in a transverse section (Figure 1.22). The gill bars tend to be larger than the tongue bars (Baskin and Detmers, 1976). Two skeletal rods support a gill bar, whereas one rod supports a tongue bar, and a gill bar has a coelomic cavity, located medially (Figure 1.22d). Regularly spaced bars termed **synapticulae**, oriented approximately at right angles to the gill and tongue bars, help support the tongue bar between two gill bars.

The **endostyle** is a midventral groove extending longitudinally along the floor of the pharynx, whereas the **epibranchial groove** extends longitudinally dorsally along the pharynx. Their position is indicated in

(a)

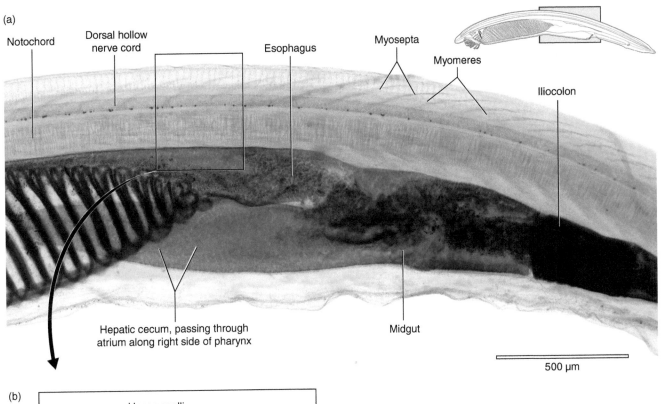

Notochord

Dorsal hollow
nerve cord

Esophagus

Myosepta

Myomeres

Iliocolon

Hepatic cecum, passing through
atrium along right side of pharynx

Midgut

500 μm

(b)

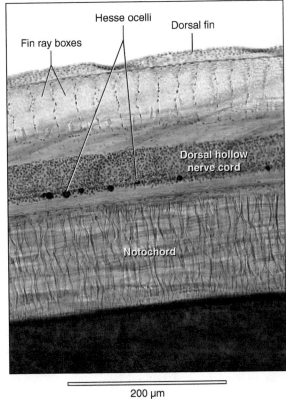

Fin ray boxes

Hesse ocelli

Dorsal fin

Dorsal hollow
nerve cord

Notochord

200 μm

FIGURE 1.20   Left lateral views of amphioxus; (a) middle region of body and (b) blowup showing detail of structures.

Figure 1.19d, but they are best appreciated in transverse views (Figures 1.21 and 1.22). The endostyle secretes mucus (as do cells lining the bars), the function of which is considered later, as is that of the epibranchial groove. The endostyle is considered homologous to the vertebrate thyroid gland because some endostylar cells concentrate iodine and produce thyroid-like hormones. Further, molecular studies have shown the expression of Thyroid Transcription Factor, which regulates thyroid-specific genes. Unlike the thyroid gland of vertebrates, however, the endostyle does not act like a true endocrine gland because its secretions are released into the pharynx (Di Fiore et al., 2012).

Two other chordate synapomorphies present in cephalochordates are the **notochord** and **dorsal hollow nerve cord**. The notochord extends from the tip of the rostrum (a feature to which, as noted earlier, cephalochordates owe their name) into the tip of the tail. In lateral view, the notochord and dorsal hollow nerve cord are labeled in Figure 1.19b and d, but are best seen in Figure 1.20. They are also clear in transverse sections, with the notochord lying ventral to the dorsal hollow nerve cord (Figures 1.22–1.25). The notochord has a striated appearance and lies just dorsal to the pharynx anteriorly and the digestive tract (described later) farther posteriorly, a position that reflects its embryologic origin from the roof of the

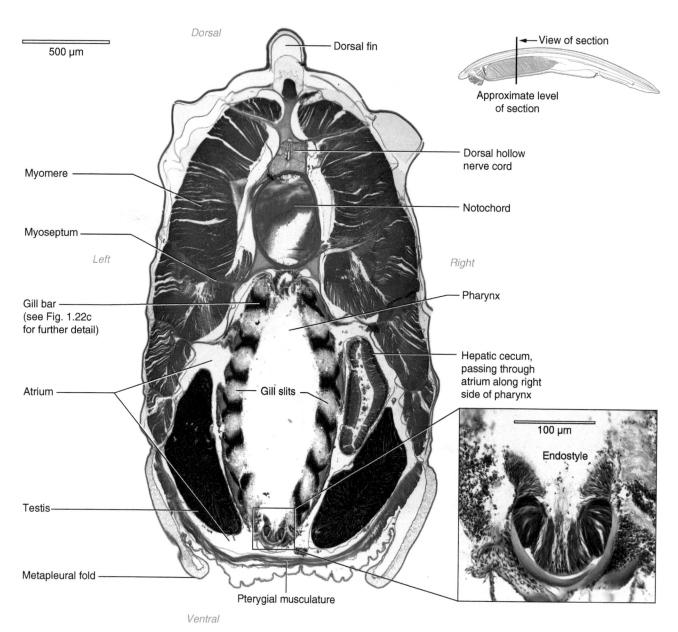

FIGURE 1.21   Transverse section through pharyngeal region of a male amphioxus in posterior view, with blowup showing detail of endostyle.

gastrocoele (or archenteron), as also occurs in most chordates (Kardong, 2018). It differs from the notochord of other chordates in that it is largely muscular with plate-like muscle cells arranged somewhat like a stack of coins along the length of the notochord (Ruppert, 1997). Intercellular, gel-filled spaces lie between the successive muscle cells. As well, there are fluid-filled intracellular spaces. The tough sheath of the notochord is composed of collagen fibers. The muscle fibers extend transversely, and when they contract, the internal pressure rises. As the notochordal sheath resists ballooning, the notochord stiffens (Kardong, 2018). The notochord thus functions as an axial stiffening structure, resisting longitudinal compression and dorsoventral flexure, but permitting lateral bending. As the left and right side myomeric series alternately contract, the notochord allows undulatory (i.e., fishlike) movements without shortening of the body (Ruppert, 1997). Amphioxus can use such movements to swim and burrow, activities that may be accomplished head-first or tail-first.

The dorsal hollow nerve cord lies dorsal to the notochord (Figures 1.20–1.25) and extends almost as far anteriorly and posteriorly (Figure 1.19b and d). For most of its length it is about a third as thick as the notochord. The anterior part of the cord does not display enlargement (although there is a slight expansion in young larvae) or obvious external subdivisions that might correspond to the main subdivisions of the vertebrate brain, such as forebrain, midbrain, and hindbrain (Kardong, 2018; Wicht and Lacalli, 2005). However, a brain is present, "based on detailed comparison between amphioxus and vertebrates of both cellular level morphology and the expression patterns of transcription factors and signaling molecules involved in late brain development," but it is not as distinct morphologically or histologically from the rest of the nerve cord compared to the condition in vertebrates (Lacalli and Stach, 2016: 722).

At the anterior end of the nerve cord, note the dark **frontal eye** (frontal ocellus) (Figure 1.19b). There is some evidence that it is homologous to the paired eyes of vertebrates (Lacalli, 2013). The frontal eye consists of a pigment cup containing pigment cells and two adjacent rows of possible photoreceptor cells that might form part of a simple retina, although the system is incapable of forming a proper image. In the larva the frontal eye orients the animal to light, but its function in the adult is not clear (Lacalli, 2013). Numerous other and smaller photoreceptors, termed **Hesse ocelli** (Figure 1.19b, c and 1.20b), appear as dark granules farther posteriorly along the ventral part of the nerve cord (Ruppert, 1997); they are more abundantly evident along the anterior half of the cord.

Return to the anterior end of the animal for a more detailed examination of the feeding apparatus (Figure 1.19b). The external features, including the oral hood, oral cirri, and the inhalant aperture leading into the oral cavity were noted earlier. Although it is not evident in the illustrations, each cirrus is supported internally by a skeletal rod that is connected at its base to those of adjacent cirri. Each cirrus also bears a series of sensory papillae. The cirri function in helping to strain out and prevent coarse particles from entering the oral cavity (Jandzik et al., 2015).

Lining the interior surface of the oral hood are a series of longitudinal fingerlike bands that are united posteriorly by a circumferential band (located immediately anterior to the velum) to form the **wheel organ** (Figure 1.19b). The bands, longer dorsally than ventrally, bear numerous cilia. These function in coordination to produce a swirling current of water that is drawn into the oral cavity, the inner walls of which are lined with a sticky mucus that is produced mainly by cells lining the grooves between the bands and, perhaps, by those along the outer margin of the bands. Just to the right of the dorsal midline is a circular depression, **Hatschek's pit**, that extends posteriorly as the narrow **Hatschek's groove**. These have an endocrinological function (Ruppert, 1997) and their approximate position is indicated in Figure 1.19b. The mucus and action of the cilia trap particles, which are then conveyed as mucus strings posteriorly toward the end of the oral cavity. In addition to its mucus-producing function, Hatschek's pit also has an endocrine function and is homologous with Rathke's pouch (a depression in the roof of the embryological stomodeum, which forms the oral cavity) and adenohypophysis (part of the pituitary gland) that are present in all vertebrates (Ruppert, 1997; Kardong, 2018).

The oral cavity ends posteriorly at the velum (mentioned earlier), a partial septum that is oriented transversely. It is a partial septum because there is an opening, the mouth, at its center. Water and the mucus strings pass posteriorly though the mouth and into the pharynx, which was noted above. The margin of the mouth is fringed with **velar tentacles** (smaller and less numerous than the oral cirri; Figure 1.19b), which extend across the mouth and therefore into the path of the water passing into the pharynx. Cilia are present on the interior walls of the pharynx, occurring on the internal (or pharyngeal) surface of the bars—these are termed frontal cilia—and on the endostyle and epipharyngeal groove. Cilia on the anterior and posterior surfaces of the bars, and thus extending across the slits, are termed lateral cilia (Ruppert, 1997; Nielsen et al., 2007).

The endostyle, as noted earlier, secretes mucus, which is carried dorsally along the walls of the pharynx, mainly through action of the frontal cilia. The mucus acts as a filter, to which food particles suspended in the water adhere. It is for this reason that amphioxus is considered a filter or suspension feeder. As the mucus, with attached particles, moves dorsally, it aggregates into longitudinal

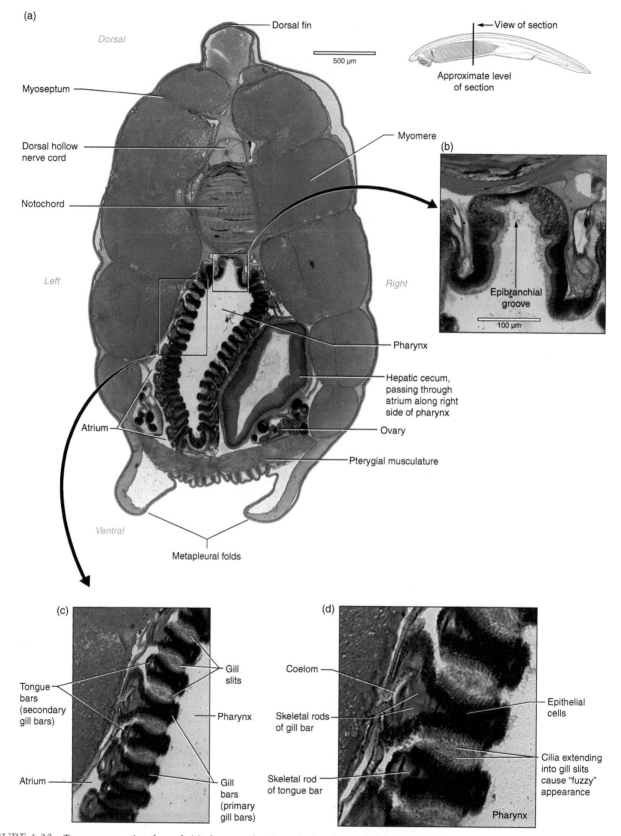

FIGURE 1.22   Transverse section through (a) pharyngeal region of a female amphioxus in posterior view, with (b) blowup showing detail of epibranchial groove, and (c) and (d) showing successive blowups of gill bars.

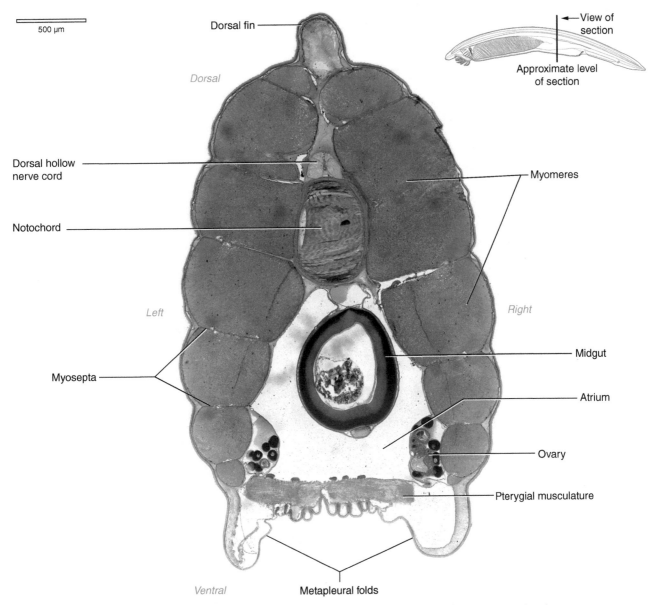

FIGURE 1.23 Transverse section through midgut region of a female amphioxus in posterior view.

strings. Each string, oriented more or less horizontally, is carried dorsally; the food particles are therefore not carried individually up the inner or pharyngeal wall of each bar. When the string reaches the epipharyngeal groove, it is passed posteriorly, mainly through action of epipharyngeal cilia, into the **esophagus** (Nielsen et al., 2007; Figure 1.19c, d and 1.20a), which is the initial part of the postpharyngeal gut (see later).

Water within the pharynx is removed by passing through the slits, mainly through action of the lateral cilia. The slits, however, do not open directly to the external environment (except in larval individuals), as does occur, for example, in the shark. Rather, the water enters a second chamber, the **atrium**, which surrounds the pharynx and extends farther posteriorly (Figures 1.19c, d and 1.21–1.24). The atrium is formed during metamorphosis from the larval stage by the down growth, on either side of the body, of the metapleural folds. These extend ventrally on either side of the pharynx and enclose the pharynx (Kardong, 2018). The floor of the atrium includes the transversely oriented **pterygial musculature** (Figures 1.21–1.23).

The atrium leads to the atriopore, which, as noted earlier, is a large posterior opening through which water leaves the body (Figures 1.19c, d and 1.24). Gametes also pass from the gonads through the atriopore. In contrast to the typical condition of vertebrates (but not cyclostomes), which possess specialized ducts to convey gametes from the gonads, the gametes of amphioxus are shed directly into the atrium and pass through the

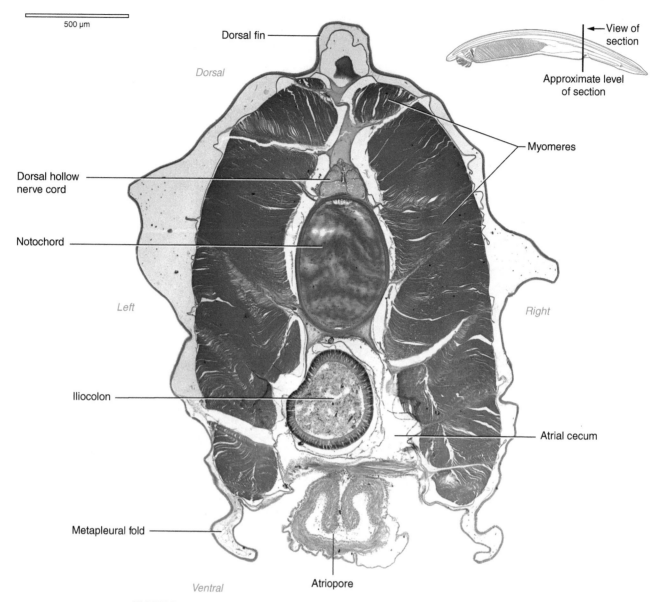

500 µm

Dorsal fin

Dorsal

View of section

Approximate level of section

Myomeres

Dorsal hollow nerve cord

Notochord

Left

Right

Iliocolon

Atrial cecum

Metapleural fold

Ventral

Atriopore

FIGURE 1.24   Transverse section through iliocolon of amphioxus in posterior view.

atriopore to leave the body. The gonads are suspended in the atrium, and the **testes** of males and **ovaries** of females are distinguishable in transverse sections, with the former having a dark, dense appearance and the latter a lighter, more granular appearance (compare Figure 1.21 with Figures 1.22a and 1.23). Amphioxus is dioecious (although hermaphroditic individuals occasionally occur) and gametes are spawned into the sea water, where fertilization occurs. The atrium continues posteriorly beyond the atriopore and into the tail as the tapering and blindly-ending **atrial cecum** (Figure 1.24), which passes along the right side of the intestine (and thereby displaces it to the left) and ends to the right of the anus (Ruppert, 1997).

The path of the water may be summarized as follows. It enters through the inhalant aperture, and then proceeds sequentially through the oral cavity, mouth, pharynx, gill slits, atrium, and, finally, atriopore, from which it leaves the body. The passage of water through the slits and into the atrium may be controlled in part by regulation of the lateral cilia. Nielsen et al. (2007: 984) noted that the entry "of distasteful water into the pharynx causes rapid inhibition of the lateral cilia." Ruppert (1997) stated that water flow may be stopped by contraction of the **atrial sphincter**, surrounding the atriopore, and strong contraction of the pterygial musculature. Contraction of the pterygial musculature results in compression of the pharynx, which causes the ventral part of the pharynx to swing posteriorly, thereby increasing the obliquity of the gill bars. This, in turn, causes the slits to narrow, interfering with the cilia and stopping their beat. The flow may also be reversed periodically. Ruppert (1997) noted that stimulation of the oral cirri or velar tentacles produces a "cough" reflex caused by a rapid peristaltic contraction of the atrial musculature that expels water from the atrium through the mouth. This action removes unwanted material.

As noted earlier, the strings of mucus collected into the epibranchial groove are passed posteriorly into the postpharyngeal gut, which is subdivided into several parts. The posterior part of the pharynx narrows markedly as it leads toward the postpharyngeal gut. The first part of the postpharyngeal gut is the esophagus, which has already been noted. It is often not clearly evident in many preparations, but is shown in Figures 1.19c, d and 1.20a. The esophagus is followed by the **midgut**, which is considerably expanded compared to the esophagus (Figures 1.19c, d, 1.20a, and 1.23). The expansion is due to the evagination of the ventral floor of the midgut to form a diverticulum, generally termed the **hepatic cecum**, which extends anteriorly through the atrium along the right side of the pharynx (Figures 1.19c, d, 1.20a, 1.21, and 1.22a). Posterior to the diverticulum the midgut narrows gradually into its posterior portion, the **iliocolon**, which is an easily recognized dark band. The gut then narrows again into the **hindgut**, or intestine, a fairly straight tube that leads to the anus (Figures 1.19c, d, 1.20a, 1.24, and 1.25).

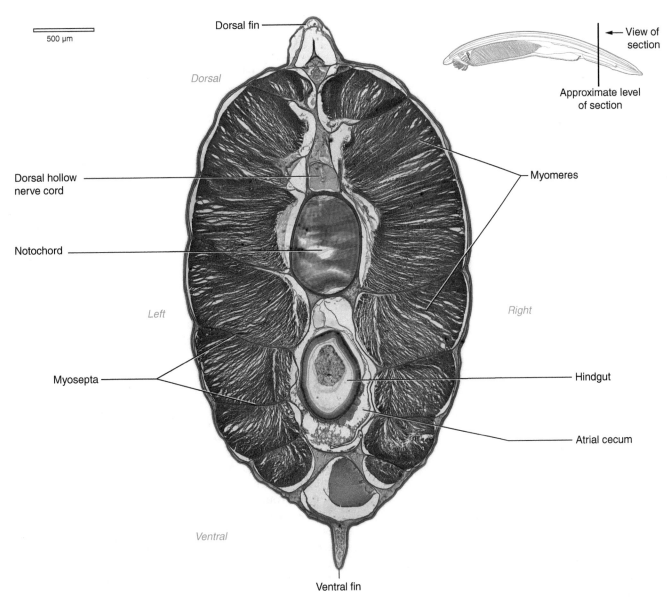

FIGURE 1.25 Transverse section through hindgut of an amphioxus individual in posterior view.

The postpharyngeal gut is lined by epithelium made of ciliated, secretory, and absorptive cells, so that all parts of the gut may contribute to the functions of moving material through the gut, as well as digestion and absorption. Its subdivisions are nevertheless specialized for specific functions. Ruppert (1997) described the process of feeding, noting that food is consolidated in a continuous cord of mucus that extends from the epipharyngeal groove to a fecal cord at the anus. The cord is in constant rotation about its long axis, through the action of the cilia of the iliocolon, which twists the cord into a braid and mixes the food with digestive enzymes. As the food cord rotates, some food particles break free of the cord, particularly in the iliocolon, and pass anteriorly to enter the hepatic cecum. The latter is apparently the main site for the production of digestive enzymes, which are distributed to the midgut, but also plays a role in absorption and storage. The cecum thus has functions that are commonly associated with the vertebrate liver (storage), pancreas (enzyme secretion) (Homberger and Walker, 2003; Liem et al., 2001), and intestine (absorption). Undigested material continues into the hindgut. Although some absorption may occur in the hindgut, its main function is the production and movement posteriorly of a fecal cord out through the anus (Figure 1.19c–e).

Key Terms: Cephalochordates

anus
atrial cecum
atrial sphincter
atriopore
atrium
caudal fin
dorsal fin
dorsal hollow nerve cord
endostyle
epibranchial groove
esophagus
fin ray boxes
frontal eye (frontal ocellus)
gill bars (primary gill bars)
gill slits
gonads
Hatschek's groove
Hatschek's pit
hepatic cecum
Hesse ocelli
hindgut
iliocolon
inhalant aperture
metapleural folds
midgut
mouth

myomeres
myosepta (sing., **myoseptum**)
notochord
oral cavity (buccal cavity)
oral cirri
oral hood
ovaries
pharyngeal clefts
postanal tail
pterygial musculature
rostral fin
rostrum
synapticulae
testes
tongue bar (secondary gill bar)
velar tentacles
velum
ventral fin
wheel organ

VERTEBRATE RELATIVES: UROCHORDATES

Urochordata is a much more diverse clade than Cephalochordata and includes some 3000 species. An alternative name for the group is Tunicata, based on the presence of an outer covering—the **tunic** (or test)—formed by the epidermis and composed mainly of cellulose (Manni and Pennati, 2016). Urochordates were traditionally classified mainly among Ascidiacea (or sea squirts), by far the most speciose group, and Thaliacea and Appendicularia (or Larvacea), with about 75 and 70 described species, respectively (Manni and Pennati, 2016). The lifestyle of urochordates is varied, and includes solitary or colonial, and benthic or pelagic forms. Three groups of ascidians were traditionally recognized, Aplousobranchiata, Phlebobranchiata, and Stolidobranchiata (Burighel and Cloney, 1997; Giribet, 2018), but molecular analyses indicate that Ascidiacea, as so recognized, is paraphyletic. The recent analyses by Delsuc et al. (2018) and Kocot et al. (2018) have to a large degree clarified the relationships among urochordates. Appendicularia is recognized as sister group to the remaining urochordates, and Ascidiacea is indeed paraphyletic, with a main division between Stolidobranchiata and a clade including Aplousobranchiata, Phlebobranchiata, and Thaliacea; the latter seems to be sister group to the Aplousobranchiata + Phlebobranchiata assemblage. "Ascidians" (given that the group is now recognized as paraphyletic, quotation marks are used) are benthic and sessile, whereas thaliaceans (salps, pyrosomes, and doliolids) are pelagic. Both groups have colonial and solitary representatives. A common theme among "ascidians" and thaliaceans is that they undergo a pelagic larval stage with a tail that includes a notochord, dorsal hollow nerve cord, and musculature, but these structures are lost during metamorphosis into the adult stage, as noted later. Conversely, appendicularians are solitary pelagic tunicates that retain the tail as adults. "Ascidians" are hermaphroditic, but self-fertilization is rare (Kardong, 2018), and may have paired, unpaired, or multiple gonads (Burighel and Cloney, 1997). A more comprehensive overview of urochordate form and diversity is provided by Holland (2016).

As noted, the adult body of the animal is invested by a thick covering, the tunic. This structure is formed mainly by a variety of cellulose, very similar to that found in plants, and the protein tunicin. Urochordates are the only animals that are able to synthesize cellulose independently, and apparently gained this ability through horizontal gene transfer of the cellulose synthase gene *CesA* from bacteria more than 520 million years ago (Satoh, 2016). The acquisition of a tunic that functions as a protective shield for the body was probably a main factor in the alteration of the urochordate lifestyle: in place of ancestral, free-living adults, they developed a sessile lifestyle by adhering strongly to objects and the substrate. Among the advantages of such a lifestyle is that it requires much less energy than active predation (see Satoh, 2016).

The urochordate considered here is the "ascidian" *Ciona* (vase tunicate), one of the most studied of urochordates and considered the closest model organism to vertebrates (Giribet, 2018). It is an "ascidian" in the broad sense, but formally a member of Phlebobranchiata. There are several species, but the most commonly studied is *C. intestinalis*. It has been determined recently from molecular studies that *C. intestinalis* includes two species, and these are formally recognized as *C. intestinalis* and *C. robusta*. Here, we avoid this distinction by referring only to *Ciona*, as the two species are nearly morphologically identical (see Brunetti et al., 2015).

Solitary urochordates such as *Ciona* are barrel shaped (Figure 1.26). They are sessile, attaching themselves to the substrate by means of rootlike extensions, the **attachment villi**, of the tunic; and are filter feeders, extracting plankton suspended in the water that enters its body. The openings through which water enters and exits the body are termed siphons, and in "ascidians" they are located close together toward the end of the body opposite the end that is attached to the substrate (by contrast, in thaliaceans they are located at opposite ends of the body). In this manual, general descriptions of the form of the adult and larva are provided. For students interested in pursuing dissections of tunicates, we recommend an excellent instructional video by Dr. R. M. da Rocha (Universidade Federal do Paraná, Brazil), available at www.youtube.com/watch?v=77mFjTYgjuA.

Examine an adult specimen of *Ciona* (Figure 1.26). The tunic is secreted mainly by the epidermis (although parts of it are also produced by tunic cells), around the body of the animal. Its form is highly variable in structure and function among tunicates; that of *Ciona* is thick, translucent, gelatinous, and composed largely of cellulose, although other polysaccharides as well as proteins, glycoproteins, and glycosaminoglycans are present. Its form, with free cells and a matrix (including ground substance and fibers), are characteristic features of connective tissue. The tunic performs several roles, including supporting and protecting the underlying tissues and organs, participating in immunology, and anchoring the animal to the substrate (Burighel and Cloney, 1997).

The siphons form the entrance into and exit out of the body and are easily distinguishable: the **oral siphon** is aligned nearly with the longitudinal axis of the animal, whereas the **atrial siphon** is oriented obliquely to this axis (Burighel and Cloney, 1997; Millar, 1953). The openings are helpful in identifying the anatomical orientations of the animal's body. They are at the animal's anterior end, with the oral siphon extending directly anteriorly. The posterior end of the body is opposite to the siphons. The attachment villi, by which the animal anchors itself, are located at this end of the body. The surface of the body from which the atrial siphon extends is the dorsal surface of the animal. This is determined by the position of the spindle-shaped **cerebral ganglion** (Figure 1.27), which lies in the body wall between the two siphons. The gently convex body surface opposite the atrial siphon is thus the animal's ventral surface, so that Figure 1.26a is a lateral view of the right side of the body and Figure 1.26c is a lateral view of the left side of the body.

The tunic extends down into and lines the inner surface of the oral siphon, but only part way into the inner surface of the atrial siphon. The free margin of the siphons is notched and a pigment spot or **ocellus** lies near a notch (Millar, 1953). A ring of **oral tentacles**, varying in length, protrude into the oral siphon from its inner surface (Figure 1.27). These apparently allow control of the size of particles that can pass into the **pharynx** or branchial basket (Burighel and Cloney, 1997).

The cerebral ganglion forms the central nervous system of the adult. Arising from the ganglion are the principal nerves of the animal, including paired anterior and posterior body wall nerves and visceral nerve. These are not considered further here and the interested reader may refer to Millar (1953), Burighel and Cloney (1997), and Mackie and Burighel (2005) for further detail. A structure associated with the cerebral ganglion is the neural gland, which lies immediately ventral to it; the two are together commonly referred to as the neural complex. Mackie and Burighel (2005) provide an instructive perspective on the commonly held (though misleading) view of the "primitiveness" or rudimentary nature of the urochordate brain.

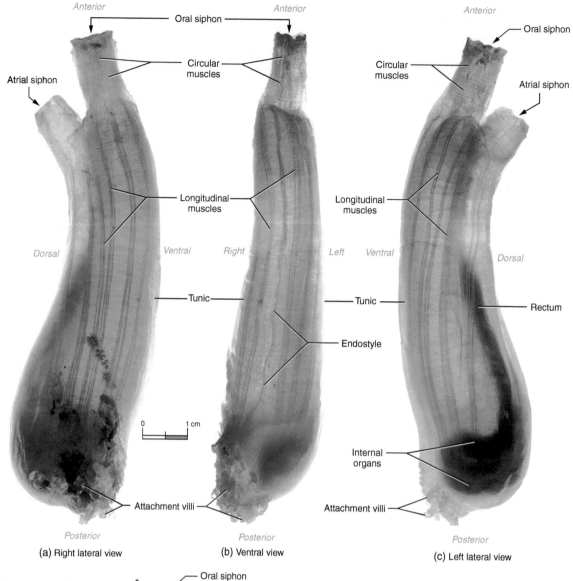

(a) Right lateral view   (b) Ventral view   (c) Left lateral view

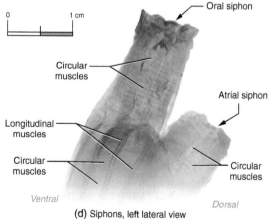

(d) Siphons, left lateral view

FIGURE 1.26   External view of an adult *Ciona* individual in (a) right lateral, (b) ventral, and (c) left lateral views; (d) detail of siphons in left lateral view.

Within the tunic is the mantle and external to it is the epidermis. The latter is a single-layered epithelium and, like the tunic, extends into the siphons. The mantle contains body wall musculature and parts of the circulatory system, connective tissue, and neural complex. The musculature is arranged into **longitudinal** and **circular muscles**. Several major longitudinal muscle bands extend from the siphons to the posterior end of the body in *Ciona* (Millar, 1953). These are easily observed in Figure 1.26 (see also Figure 1.27). The circular musculature is also perceptible, particularly on the oral siphon (Figure 1.26d). Contraction of the longitudinally-arranged musculature causes "cowering," a behavior produced by significant reduction of body length, whereas contraction of the circularly-arranged musculature can cause rapid ejection ("squirting") of water from the siphons (Burighel and Cloney, 1997), an activity from which the common designation as "sea squirts" is derived.

The viscera and gonads lie mainly toward the left side of the animal (Figure 1.26b), and so are more clearly evident in left lateral view (compare Figure 1.26a with Figure 1.26c). They do not lie in the atrium (see later), but within a separate space, the **epicardial** (or visceral) **cavity**, enclosed by epicardium. The epicardial membranes meet the posterior wall of the pharynx along the **retropharyngeal groove**. Technically, there are right and left pericardial cavities, the membranes of which surround structures within the cavity. The epicardial cavities are formed as diverticula of the posterior end of the pharynx; and in the adult the pharynx and epicardial cavities do retain a connection though two tiny pore-like openings, the left and right pharyngo-epicardiac openings, which are located along the retropharyngeal groove (see Hoshino and Tokioka, 1967; Millar, 1953), extending from the **endostyle** to the **esophagus** (Figure 1.27), as noted later. However, Millar (1953) suggested that there is probably little exchange of water between the cavities through the openings. The **heart** is surrounded by the pericardium, which encloses a pericardial cavity that has been considered a remnant of the coelom.

As indicated in Figure 1.27, an inward current of water, carrying food particles as well, is created by ciliary action and enters through the oral siphon and passes into the large barrel-shaped pharynx, the walls of which are pierced by numerous small pharyngeal slits or **stigmata** (the morphology of the pharynx makes it obvious why it is also referred to as the branchial basket). The water current is generated by cells lining the stigmata. The food particles become entangled in the mucus that lines the inner wall of the pharynx and the water exits, much as noted for amphioxus, through the stigmata into a surrounding chamber, the **atrium**. The atrium surrounds the pharynx on its dorsal, posterior, and right and left sides, and is usually subdivided into right and left peribranchial cavities. These merge dorsally into the cloacal cavity, the space near and extending into the atrial siphon. Millar (1953) noted that although there are no partitions between the peribranchial and cloacal cavities, their distinction is maintained based on early stages of development. Here, the cavity is labeled simply as atrium.

The food-entangled mucus, or alimentary cord, is moved to the **dorsal languets**, a row of projections into the pharynx (in some species these are replaced by a ridge, the dorsal lamina; see Chiba et al., 2004), along which it moves posteriorly toward the esophagus. As in amphioxus, the mucus is produced by a midventral groove, the endostyle, which extends the length of the pharynx. From the posterior end of the endostyle, the narrower retropharyngeal groove extends dorsally to the opening of the esophagus.

Water in the atrium leaves through the atrial siphon, as do gametes and fecal pellets, the remnants of the digestive process. Thus, in terms of water and food movement and gametes, the "ascidian" condition resembles that in amphioxus, but the latter differs in lacking gonadal ducts and in having its digestive tract extend separately to the body surface. The inner pharyngeal wall of *Ciona* is crisscrossed by numerous blood vessels, and the pharynx is also involved in oxygenating the blood. Besides feeding and gas exchange, the pharynx also functions in digestive processes, uptake of heavy metals, and absorbing dissolved organic matter (Burighel and Cloney, 1997).

The posterior end of the pharynx narrows markedly and leads, in its dorsal region, to the mouth of the esophagus, which receives the alimentary cord from the pharynx and moves it to the saclike **stomach** (Figure 1.27), that functions in digestion and absorption. The esophagus is relatively short and tapered, but it is not merely a connecting tube between pharynx and stomach: One of its functions, performed mainly by the ciliated cells at its mouth, is to pull the alimentary cord from the pharynx into the esophagus. At this point, therefore, the food is not moved by action of the pharyngeal cilia (Millar, 1953). The **intestine**, which continues the activities of digestion and absorption (Burighel and Cloney, 1997), begins toward the ventral end of the stomach and curves first anteriorly and then sharply dorsally. It passes anterior to the stomach and then bends anteriorly and transitions, opposite the esophagus, into the **rectum** along the dorsal surface of the pharynx. The rectum opens through the **anus** into the atrium near the atrial siphon (Figure 1.27). The intestine and rectum are nearly uniform in width and cannot be clearly distinguished externally, but they are distinct histologically (Millar, 1953).

Two other structures of the digestive system may be noted, though they are not illustrated here. The pyloric (or gastric) cecum is a diverticulum near the junction of the stomach and intestine and is more prominent in juveniles than in adults. Its histology suggests that it has no functions that are not also present in the stomach and intestine (Millar, 1953). The pyloric gland is a system of blindly-ending tubules and ampullae that form as an evagination of gastric epithelium early in development (Burighel and Cloney, 1997; Millar, 1953). These structures lie on the surface of the rectum and esophagus and communicate via a narrow duct with the stomach, near the latter's junction with the esophagus. Several functions have been postulated for the pyloric gland; its participation in sugar metabolism and glycogen storage has been demonstrated (Burighel and Cloney, 1997).

The gonads include the **ovary** and **testis** (Figure 1.27). As noted earlier, "ascidians" are hermaphroditic. Although cross-fertilization appears to be the rule, Burighel and Cloney (1997) noted that *Ciona intestinalis* (among other species) may be self-sterile or self-fertile depending on the specific population. The ovary and

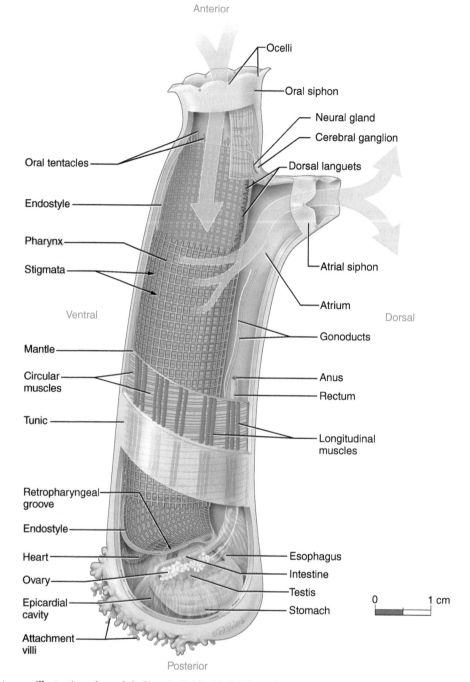

FIGURE 1.27   Cut away illustration of an adult *Ciona* individual in left lateral view, showing internal structure. Blue arrows indicate path of water flow.

testis are located approximately between the stomach and intestine, with the ovary, to the left of the stomach, forming a more discrete structure than the testis, which consists of a system of branching tubes that spread over the intestine and stomach (Millar, 1953). **Gonoducts** extend from these organs toward the atrial siphon. They are long and narrow, extending anteriorly beyond the anus.

The circulatory system consists of a heart and vessels. In the adult the heart is approximately V-shaped, with the pointed end of the V oriented anterodorsally, and located to the right of the viscera (Figure 1.27). A main ventral vessel extends anteriorly along the ventral side of the pharynx from the more ventrally positioned arm of the heart. From the ventral vessel arise paired transverse vessels that encircle the pharynx and extend into a dorsal vessel that passes along the dorsal side of the pharynx. The smaller vessels that crisscross around the stigmata arise from the transverse vessels. Posteriorly, the dorsal vessel branches into three main vessels that pass through

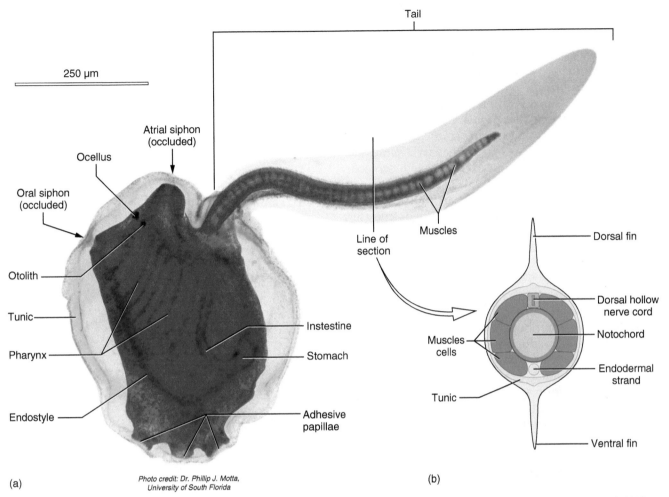

FIGURE 1.28   Larval "ascidian" tadpole; (a) whole mount and (b) transverse view through tail region. *Modified from Konno et al., 2010.*

the epicardial cavity to supply visceral structures. The blood is then recollected into the more dorsally positioned arm of the heart (Millar, 1953). The pattern of flow occurs in one direction along this route, but then reverses periodically (Konrad, 2016). Given the bidirectional flow of blood in these vessels, they cannot be distinguished as arteries and veins. This circulatory pattern is distinct from that of vertebrates, in which blood flow is unidirectional through continuous tubes: Blood is pumped away from the heart through vessels termed arteries, which branch into capillaries, the tiny vessels supplying the body's tissues, and then recollects into larger vessels termed veins that return the blood to the heart. There has been discussion as to whether the circulatory system of tunicates is an open system, as occurs in non-chordates such as insects and crustaceans, or a closed system as occurs in vertebrates, in which the vessels form a series of continuous tubes. In the open system the heart pumps blood through a short vascular tree, but then exits the vessels and flows in direct contact with cells (thereby resembling lymph) before recollecting in a large pericardial sinus and thence returning to the heart. In the

system just described for *Ciona*, however, no such sinus is present (and has not been reported for any "ascidian") and the vessels do form a system of continuous tubes. The idea of an open system seems to be based mainly on the apparent absence of endothelial cells (which line the vessels of vertebrates), rendering the vessels markedly permeable, although Konrad (2016) noted that the absence of endothelium is not well documented.

The morphology of the adult animal, just described, gives little evidence for the chordate affinities of urochordates—other than the perforated pharynx, which is more broadly characteristic of deuterostomes, the other main characters seem to be missing. As noted earlier, however, additional chordate characteristics are present in earlier life. The first stage of the life cycle includes fertilization and development into a free-swimming "tadpole" larva (Figure 1.28a), which attaches within one or two days to a solid substrate. It then begins the second stage, involving a dramatic metamorphosis, growth, and development into the sessile adult (Konrad, 2018).

The tadpole larva (Figure 1.28a) consists of a large ovoid trunk region followed by a slim tail region. In the trunk are the main organs that will ultimately develop fully into the adult structures considered earlier. At one end of the larva are the **adhesive papillae**, which serve to attach the larva to the substrate. It is this end of the animal (the larval anterior end, corresponding to the posterior end of the adult) that becomes attached, so that the larva "upends" itself to metamorphose into an adult. The pharynx, less extensive than in the adult, and intestine are clearly evident in Figure 1.28a. The oral and atrial siphons are occluded by the larval tunic and the digestive tract is not fully differentiated, so feeding cannot occur during the larval stage (Burighel and Cloney, 1997). As development into an adult proceeds, the pharynx expands and becomes modified—the slits proliferate by division of the existing slits to produce the numerous stigmata, rather than by the development of new slits (Millar, 1953).

The other chordate features are found mainly in the axial complex, which forms the bulk of the postanal tail. The tunic encloses the axial complex, and the cuticle that covers the tunic contributes to the extensions or **fins** that give the tail a paddle-like shape. The axial complex includes a central **notochord**, a **dorsal hollow nerve cord**, and striated **muscle cells** (Figure 1.28b); the latter do not, however, form segmental myomeres. The cells on one side of the body contract together to bend the tail. The notochord is a flexible, tubular rod, with a gel- or fluid-filled lumen (Kardong, 2018). It extends through much of the tail and into the posterior end of the trunk, nearly reaching the pharynx.

The nervous system includes the dorsal hollow nerve cord (Figure 1.28b), which extends along the notochord and into the trunk, and the sensory vesicle and visceral ganglion. The last two structures lie close together but are not clearly differentiable in Figure 1.28a. The position of the sensory vesicle is indicated by the two dark spots that lie between the siphons. These are the **ocellus**, a light sensitive organ, and **otolith** (or statolith), a gravity-sensitive structure. The cerebral ganglion also lies nearby, but becomes functional only after metamorphosis (Kardong, 2018).

Metamorphosis into the adult form begins soon after the larva attaches itself to the substrate. As described by Kardong (2018), the tail is withdrawn into the body as the notochord, losing its integrity and fluid, becomes limp. Over the next several days, the axial complex is resorbed, its components redistributed to support the growing juvenile. The sensory vesicle and visceral gland are also lost, whereas the pharynx enlarges, the stigmata proliferate, and the juvenile begins feeding.

Key Terms: Urochordates

adhesive papillae
anus
atrial siphon (excurrent siphon)
atrium
attachment villi
cerebral ganglion
circular muscles
dorsal hollow nerve cord
dorsal languets
endostyle
epicardial cavity (visceral cavity)
esophagus
fins
gonoducts
heart
intestine
longitudinal muscles
muscle cells
notochord
ocellus
oral siphon (incurrent siphon)
oral tentacles
otolith
ovary
pharynx (branchial basket)
rectum
retropharyngeal groove
stigmata (pharyngeal slits)
stomach
testis
tunic (test)

2

The Lamprey

INTRODUCTION

As noted in Chapter 1, the earliest vertebrates lacked jaws and so are termed "agnathans." Numerous extinct "agnathans" are known from the fossil record, and most possessed a covering armor of dermal bone from which is derived the term "ostracoderms." Living vertebrates retaining the absence of jaws are the hagfishes (Myxinoidea) and lampreys (Petromyzontoidea). A characteristic feature of these vertebrates is an anterior, rounded, sucker-like structure used to attach themselves to the body of their prey, from which the term cyclostome is derived. As noted in Chapter 1, these living vertebrates are now considered sister taxa, grouped as Cyclostomata.

Of the approximately 50 lamprey species, the marine lamprey, *Petromyzon marinus*, of the Atlantic Ocean and North American Great Lakes, is the most commonly studied cyclostome. Hagfishes are about as diverse, with 60 species generally recognized, and are exclusively marine—indeed, they are the only vertebrates with body fluids isosmotic to seawater. Though lampreys and hagfishes share several similarities, such as an eel-like body shape and a sucker-like mouth, there are important differences between them. Lampreys are generally parasitic, attaching themselves to their prey and relying mainly on a liquid diet, while hagfishes are scavengers and tear off pieces of dead or dying prey. These differences are reflected in several innovative modifications. The lamprey, for example, has a subdivided pharynx with the ventral part forming a respiratory tube that can be isolated from the mouth by a valve termed the velum. This ensures that its liquid diet neither escapes from the pharyngeal slits nor interferes with gas exchange through the gills. While the tube is isolated, the lamprey continues to ventilate its gills by pumping water in and out of the pharyngeal slits. The hagfish, on the other hand, has a more substantial diet and does not require a respiratory tube that can be isolated. It ventilates its gills by having a nasal opening that continues past the olfactory sac to communicate with the pharynx. A velum is also present in the hagfish. With the velum closed, muscular action compresses the pharynx and water moves over the gills.

SECTION I: SKELETON

The entirely cartilaginous skeleton of the lamprey is not particularly well developed (Figure 2.1). Study the skeletal elements in prepared specimens, which are usually embedded in acrylic. The head skeleton is complex and quite unlike that of other vertebrates. Endochondral elements forming a **chondrocranium** include large cartilages that partially enclose the brain and sense organs (nasal capsule, otic capsule), and others that extend anteriorly to support the **annular cartilage**, the ring-like structure that is the main skeletal element of the **oral funnel** (see later). Several of these cartilages are labeled in Figure 2.1. The median **lingual cartilage**, supporting the rasping **tongue**, extends posteriorly from the annular cartilage, ventral to the chondrocranium.

Extending posteriorly and connecting to the chondrocranium is a network of cartilages forming the **branchial basket**, which supports the pharyngeal region. The somewhat hemispherical cartilage at the posterior end of the branchial basket is the **pericardial cartilage**, which lies on the posterior wall of the pericardial cavity, the space that contains the heart. This pharyngeal skeleton has three main differences from those found in more derived fishes. One is that the structure is a connected network, rather than relatively separate and articulated arches. Another difference is the connection to the pericardial cartilage. Yet another is that the branchial basket lies superficially, lateral to the gills and thus just beneath the skin. By contrast, in other fishes the branchial arches lie more deeply, medial to the gills, and thus in the medial wall of the pharynx. Because the branchial basket lies just under the skin, it may be observed by carefully skinning one side of the head of your specimen. If you

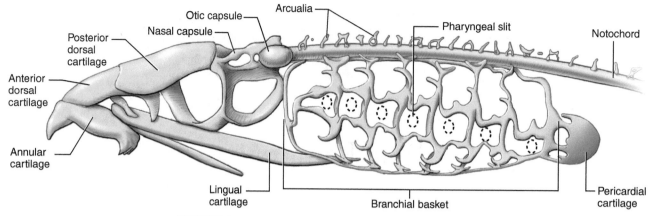

FIGURE 2.1   Skeletal elements of the lamprey in left lateral view.

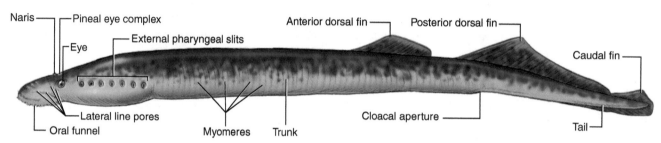

FIGURE 2.2   External features of the lamprey in left lateral view.

wish to do this, postpone your dissection until you have studied the structures described in the next section.

The large **notochord** is the axial support of the body. It is an elongated rod extending from beneath the posterior half of the brain to the tip of the tail. The vertebrae are represented by small, cartilaginous structures, termed **arcualia**, that lie dorsal to the notochord on either side of the spinal cord (see later). They are usually embedded in connective tissue or lost in prepared specimens, and are not particularly evident. Indeed, they are difficult to find, but may occasionally be observed in cross-sections.

SECTION II: EXTERNAL ANATOMY

The body of the lamprey is elongated and cylindrical and covered by smooth, scaleless skin (Figure 2.2). The **head** extends posteriorly to include the slanted row of seven rounded or oval apertures, the **external pharyngeal slits**, which lie posterior to each of the laterally placed, lidless **eyes**. The latter are of moderate size and covered by transparent skin. In preservatives, however, this skin turns opaque. Anteriorly the **oral funnel** forms a wide, sucking disk that attaches to the body of the lamprey's prey (Figures 2.2 and 2.3). The funnel's margin bears small, soft projections, the **buccal papillae**, which are primarily sensory structures. The interior surface of

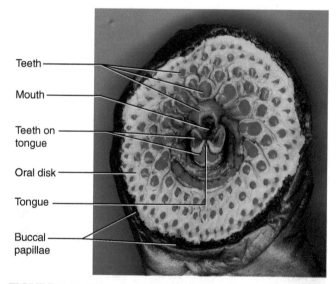

FIGURE 2.3   View of oral funnel of the lamprey showing mouth and oral disk with teeth.

the funnel is termed the **oral disk**, and it is lined with numerous horny **teeth**, which are cornified epidermal derivatives and thus not homologous with the teeth of more derived vertebrates. Remove a tooth with forceps to observe a replacement tooth immediately underneath it. At the pit of the funnel lies the rasping tongue, which also bears horny teeth and, like them, is not homologous

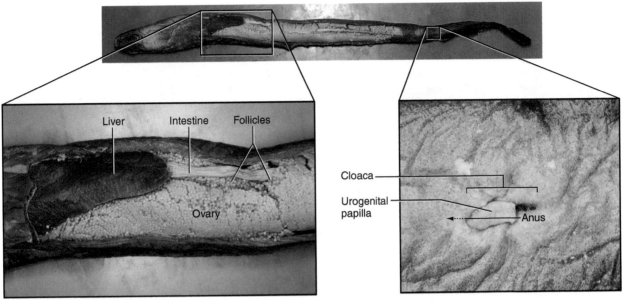

FIGURE 2.4   Ventral view of the lamprey, with pleuroperitoneal cavity exposed (top). Detail of anterior end of pleuroperitoneal cavity and contained structures (left). Detail of cloacal region (right).

with the structures of the same name in more derived vertebrates. The tongue is used to abrade the skin of prey, so that its blood and body fluids may be ingested. Dorsal to the tongue is the **mouth**. The lamprey, being an "agnathan," lacks jaws.

A single, median **naris** lies middorsally between the eyes (Figures 2.2 and 2.8). A lighter patch of skin immediately posterior to the naris denotes the position of the **pineal eye complex**, which lies just below the skin and functions as a photoreceptor that detects changes in light. A **lateral line system**, which functions in detecting vibrations in the water, is present in the lamprey, but is not typically conspicuous. Usually, however, pores for the canals are noticeable in the head. These **lateral line pores** are arranged in rows. One occurs just posterior to each eye, and several others may be found between the eye and the anterior end of the oral funnel.

The **trunk** extends from the head to the **cloaca**, a rather shallow midventral depression (Figure 2.4). The **intestine** opens into the cloaca through a slit-like **anus**. Posterior to the latter, there is a small **urogenital papilla** with a terminal **urogenital pore** through which gametes and excretory products leave the body. The **tail** extends posteriorly to the tip of the body and becomes laterally compressed. The lamprey has no trace of paired fins, but there are three median fins (Figure 2.2), the **anterior dorsal** and **posterior dorsal fins** and the **caudal fin**. The fins are supported by **fin rays**, which can be seen if the specimen is held up against light. Examine the lateral surface of the trunk or tail to observe the outline of the segmented series of muscular blocks, the **myomeres**. These are composed of longitudinal fibers extending between successive connective tissue partitions, the **myosepta**. Contraction of the myomeres on one side of the body flexes the body toward that side, and alternating contraction of the myomeres on opposite sides of the body produces the characteristic side-to-side swimming motion of fishes.

SECTION III: PLEUROPERITONEAL CAVITY AND VISCERA

Open the **pleuroperitoneal cavity** by making an incision through the midventral body wall extending from about 1 cm posterior to the last pharyngeal slit to just anterior to the cloaca. Make several vertical cuts on one side of the body from the ventral incision so the body wall may be reflected. The wall, or portions of it, may be removed to facilitate exposing the contents of the cavity, but be careful not to injure the underlying organs.

The most notable feature of the lamprey viscera, as with much of the lamprey's anatomy, is its relative simplicity. This is often manifest in the apparent absence of structures, as noted later. Keep in mind, however, that an absence or apparent lack of complexity in a particular structure does not necessarily indicate primitiveness.

On opening the cavity, the **gonad** may be the first structure visible (Figures 2.4 and 2.5). It is normally of moderate size but in breeding season it may be huge, occupying much of the pleuroperitoneal cavity. The **ovary** and **testis** are, secondarily, single median structures, supported by a mesentery from the middorsal line (Figure 2.5) and lying dorsal to the **liver** and **intestine**.

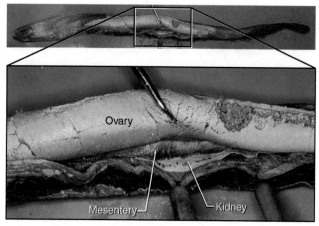

FIGURE 2.5   Ventral view of pleuroperitoneal cavity of the lamprey, showing detail of gonad, kidney, and mesentery.

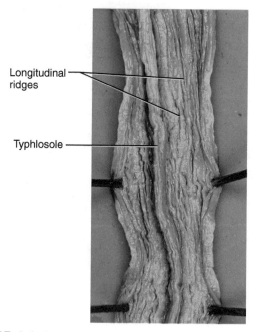

FIGURE 2.6   Intestine of the lamprey, cut to show its internal structure.

The ovary and testis are difficult to distinguish from each other. Just before breeding, however, the ovary contains many follicles, giving it a granular appearance compared with the testis. Genital ducts are absent in the lamprey, and gametes are shed into the **coelom**. They must make their way to the posterior end of the pleuroperitoneal cavity, where they enter the urogenital sinus (the single median cavity within the urogenital papilla) and exit the body through the urogenital pore.

The liver (Figure 2.4) is the large, greenish organ at the anterior end of the pleuroperitoneal cavity. If the ovary is very large, it may cover the posterior end of the liver. A gall bladder and bile duct are present in the larval stage but absent in adult lampreys. The digestive system is relatively simple and consists mainly of a long, narrow, tubular intestine (though it may be distended in some specimens) extending the length of the pleuroperitoneal cavity. There are none of the bends and folds so common in more derived vertebrates. The anterior part of the intestine extends from the "**esophagus**" (see later and Figure 2.8) and is dorsal to the liver, but more posteriorly it is easily seen in ventral view. Note the lack of any true regional specialization along the digestive tract. There is no stomach, for example. Given the mainly liquid diet of lampreys, there is no need for a separate stomach to temporarily store food and "feed" it to the intestine. Nor is there a distinct pylorus. A terminal swelling, sometimes present, may be recognized as a rectum. Make a longitudinal slit along a portion of the intestine and spread it open. Numerous longitudinal folds are present on its interior surface to increase surface area (Figure 2.6). One of these, the **typhlosole**, is notably larger than the others. It is also termed a spiral valve but its relationship to the spiral valve of other vertebrates is ambiguous.

Also note the virtual absence of either dorsal or ventral mesenteries. A few small mesenteric sheets of dorsal mesentery, carrying blood vessels to the gut, are present near the posterior end of the intestine. A spleen is absent, as is a distinct pancreas. Tissue performing pancreatic

functions is present, however, and scattered through some viscera. Exocrine pancreas is present in parts of the intestinal wall, and islet tissue (endocrine) occurs in the liver, but they are not visible grossly.

A long, thin, ribbon-like **kidney** lies on either side of the middorsal line and extends for much of the length of the pleuroperitoneal cavity (Figures 2.5 and 2.7). The **archinephric duct**, which drains the kidney, lies along its free lateral margin (Figure 2.7). Make a transverse cut through the kidney and examine the cut section to observe the duct. Posteriorly the archinephric ducts enter the urogenital sinus, but it is impractical to attempt tracing them.

SECTION IV: SAGITTAL SECTION

A sagittal section of the head and anterior part of the trunk reveals a number of interesting features, and in particular allows the observation of the extreme specializations of the lamprey's digestive and respiratory systems. Using a large, sharp scalpel, make a clean sagittal section through the head of the lamprey. Maintain your cut as close to the midsagittal plane as possible, and extend the section about 10 cm posterior to the level of the last external pharyngeal slit. Compare your specimen to Figure 2.8. If your section is not quite midsagittal, use a new scalpel blade and carefully shave tissue from the larger half until you reach the midsagittal plane.

Examine the anterior part of the head to reidentify structures already noted, such as the annular cartilage, oral funnel, horny teeth, mouth, and tongue. Trace posteriorly from the mouth, over the anterodorsal surface of the tongue, into the **oral cavity**, which has a short

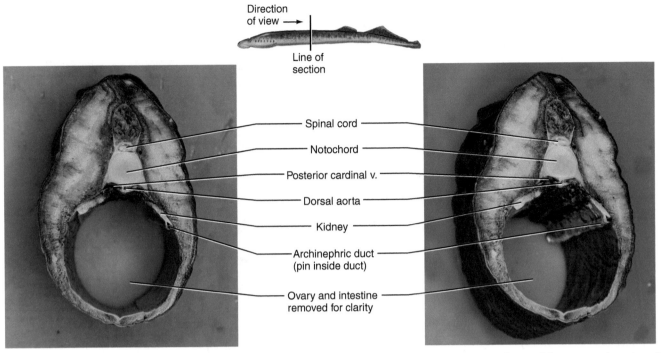

FIGURE 2.7   Cross-section through body of the lamprey, seen in anterior view. The left and right photographs are of the same section, viewed at slightly different angles.

anterodorsal extension, the **oral cecum**. Posteriorly the oral cavity passes into the "esophagus" dorsally and the **respiratory tube** or "**pharynx**" ventrally. The terms set in quotation marks denote subdivisions of the embryonic pharynx, and so are not homologous with the esophagus and pharynx of other vertebrates (although, of course, the "pharynx" is partly homologous with the pharynx). This anatomical condition in the lamprey reflects its highly specialized feeding mode. It would make little sense to feed on fluids if they could easily escape through pharyngeal slits. During metamorphosis from the larval stage, a horizontal partition develops that subdivides the originally single pharynx into dorsal and posterior portions. The dorsal subdivision, as indicated, is the "esophagus" and serves to conduct food to the intestine. It is thin-walled and usually collapsed, thus difficult to identify. The respiratory tube is the ventral subdivision. It ends blindly posteriorly. As noted earlier, the velum is a valve that can close the respiratory tube's anterior opening into the oral cavity, thereby isolating the tube. This system is necessary because the lamprey spends much of its time with the oral funnel attached to its prey, so that the mouth cannot function in ventilating the gills. With the respiratory system isolated from the ingestion mechanism, food is passed back into the intestine. The lamprey continues to ventilate its gills with respiratory water currents going both in and out through the pharyngeal slits.

The **tongue** is supported by the large **lingual cartilage**. Note the complex of muscles arranged around the lingual cartilage. This musculature is responsible for working the tongue. Protractor muscles extend anteriorly ventral to the cartilage, whereas the retractor muscles extend posteriorly from it.

The small **brain** lies dorsally. It is not particularly similar to that of more derived vertebrates, but it is tripartite, having the same major subdivisions. Trace it posteriorly as it passes into the **spinal cord**. Note the large **notochord**, which begins ventral to the posterior half of the brain and extends posteriorly, ventral to the spinal cord, to the end of the body. Also note the slender sections of the cartilaginous elements that contribute to the chondrocranium. True vertebrae are absent, but cartilaginous blocks termed arcualia are present on either side of the spinal cord. As noted earlier, however, these are difficult to find and are lost in most prepared specimens.

Locate the naris. It opens into a short tube that leads to the dark-walled **olfactory sac**, directly anterior to the brain. The sac's interior surface has numerous folds to increase surface area. The tube continues past the opening of the olfactory sac as the **hypophyseal pouch**, which ends blindly ventral to the brain and anterior end of the notochord. Dorsal to the olfactory sac and anterior end of the brain is the pineal eye complex, a region specialized primarily for light detection.

The **heart** lies in the pericardial cavity, posterior to the respiratory tube. Posteriorly it is separated from the pleuroperitoneal cavity by the **transverse septum**, which is stiffened by the pericardial cartilage. The heart has three chambers, in contrast to the four present in more derived vertebrates. These are, in order of blood flow, the **sinus venosus**, **atrium**, and **ventricle**. The spatial positions

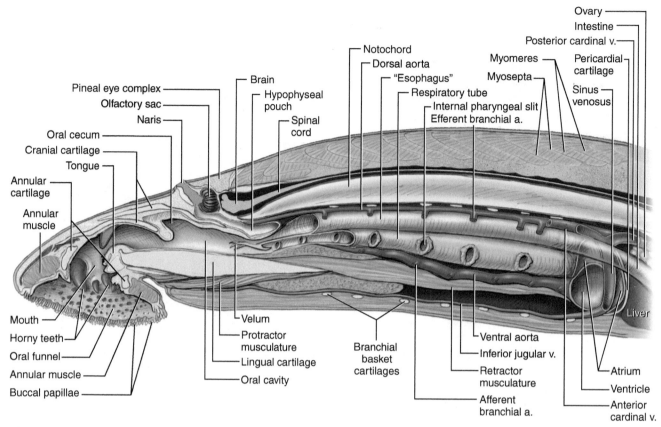

FIGURE 2.8 Schematic sagittal section through head and pharyngeal region of the lamprey. The vessels, "esophagus," respiratory tube, and parts of the heart are shown with portions cut away, rather than in sagittal section.

of these chambers do not follow the sequential posterior-to-anterior progression typical of the heart of most fishes. Instead, the sinus venosus is a tubular structure, oriented dorsoventrally, that lies between the atrium and ventricle. The atrium mainly occupies the left side of the pericardial cavity, and the ventricle mainly occupies the right side. The depiction of the heart in sagittal section (Figure 2.8) is somewhat idealized, as structures such as the intestine and inferior jugular vein are shown in their entirety, rather than as sectioned as they would be in a true midsagittal section. Compare these structures with those shown in Figure 2.9, which shows the relationships of these and other structures in transverse view.

The flow of blood passes forward from the ventricle through the **ventral aorta**, which sends out seven **afferent branchial arteries** to the capillaries in the septa between the pharyngeal slits. The blood is recollected by the **efferent branchial arteries**, which channel it into the **dorsal aorta** for distribution to the body. It is easiest to observe the dorsal aorta in a transverse section. It is a median structure extending just ventral to the notochord (Figures 2.7 and 2.9). It continues in the tail as the **caudal artery**.

Blood from the head returns to the heart mainly through the paired **anterior cardinal veins** dorsally and the median **inferior jugular vein** ventrally. Much of the posterior part of the body is drained by the paired **posterior cardinal veins** as well as the **hepatic portal vein**. The latter is difficult to identify and not considered further here except to note that is comprises a system of veins that drains blood from the intestine and conducts it to the heart. Dorsal to the heart, the right anterior and posterior cardinal veins join to form a right **common cardinal vein**, which then enters the dorsal end of the sinus venosus (Figure 2.9). The left anterior and posterior cardinal veins join the right common cardinal vein. Note that this condition is distinct from that present in most jawed fishes, such as the shark, where the anterior and posterior cardinal veins of each side of the body unite to form their own common cardinal vein that enters the sinus venosus separately. The inferior jugular vein is a median structure and enters the ventral end of the sinus venosus. For most of its length, it is a median structure, but near the sinus venosus it veers slightly to the left.

In the tail, the **caudal vein** accompanies the caudal artery. The lamprey does not have a renal portal system, as do jawed fishes, so the caudal vein does not bifurcate into renal portal veins that then enter the kidneys. Instead the caudal vein bifurcates into the right and left posterior cardinal veins (Figure 2.7), which receive blood from the kidneys.

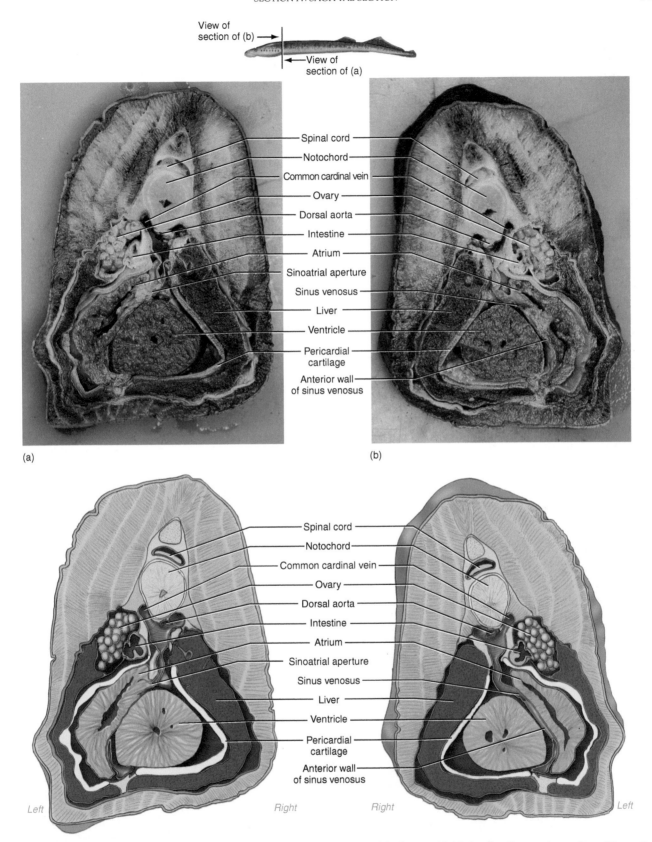

FIGURE 2.9 Cross-section through body of the lamprey to show the structure of the heart, with (a) showing the anterior portion of the section in posterior view and (b) showing the posterior portion of the section in anterior view. Interpretative illustrations are below each photograph.

Key Terms: Lamprey

afferent branchial arteries
annular cartilage
anterior cardinal veins
anterior dorsal fin
anus
archinephric duct
arcualia
atrium
brain
branchial basket
buccal papillae
caudal artery
caudal fin
caudal vein
chondrocranium
cloaca
coelom
common cardinal vein
dorsal aorta
efferent branchial arteries
"esophagus"
external pharyngeal slits
eyes
fin rays
gonad
head
heart
hepatic portal vein
hypophyseal pouch (nasohypophyseal pouch)
inferior jugular vein
intestine
kidney

lateral line system
lingual cartilage (piston cartilage)
liver
mouth
myomere
myoseptum (plur. myosepta)
naris (nostril)
notochord
olfactory sac
oral cavity
oral cecum
oral disk
oral funnel
ovary
pericardial cartilage
pineal eye complex
pleuroperitoneal cavity
posterior cardinal veins
posterior dorsal fin
respiratory tube ("pharynx")
sinus venosus
spinal cord
tail
teeth
testis
tongue (piston)
transverse septum
trunk
typhlosole
urogenital papilla
urogenital pore
velum
ventral aorta
ventricle

3

The Shark

INTRODUCTION

The spiny dogfish shark, *Squalus acanthias*, belongs to Chondrichthyes, which first appeared in the Silurian Period, and along with Osteichthyes is a main group of living gnathostomes (jawed vertebrates). Living chondrichthyans comprise the sharks and rays (Elasmobranchii, meaning plate-gilled) and the chimaeras (Holocephalimorpha, meaning whole or entire head). Among the specialized features that unite these groups are unique perichondral and endochondral mineralization, distinctive placoid scales, an inner ear that opens externally through the endolymphatic duct, pelvic claspers in males (an adaptation for the internal fertilization practiced by modern chondrichthyans), and a cartilaginous skeleton. As the earliest "agnathans" were bony, the cartilaginous condition of the chondrichthyan skeleton is apparently derived. In addition to these specialized features, they retain numerous ancestral characteristics, which is one reason why they are ideally suited as subjects for the study of a basic vertebrate. Chondrichthyans lack the swim bladder or lung that evolved early among the bony fishes, and modern sharks all possess a large, oil-filled liver that dramatically reduces their specific density.

Modern chondrichthyans are a fairly large and diverse group that includes sharks, skates, rays, sawfishes, and holocephalimorphans. Sharks have fusiform bodies with well-developed paired fins and a powerful heterocercal tail. Five to seven pharyngeal slits and usually a spiracle are present. A row of serrated, triangular, and pointed teeth commonly line the upper and lower jaws and are followed by rows of teeth that can rapidly replace those that become broken or worn. An efficient swimmer with a well-armed mouth, the shark is a formidable slashing predator. However, not all sharks fit this mold. Indeed, the largest of them, the basking and whale sharks, feed by straining food from the water.

The skates and rays, included within Batoidea, are a subgroup of Neoselachii (which also includes the squalomorph sharks such as *Squalus acanthias*). Batoids are dorsoventrally flattened, largely due to greatly expanded pectoral fins. The pharyngeal slits open ventrally, and the prominent spiracle opens dorsally. The tail and caudal fin are generally reduced and often whip-like. Locomotion is accomplished through wave-like flapping of the fins rather than lateral undulations of the trunk and tail. Some rays can generate electric shocks, produced by modified muscles, to repel an attack or to capture prey. The flattened condition of the body is typical of bottom-dwelling forms. The teeth are modified into flattened plates for crushing bottom-dwelling small vertebrates, mollusks, and crustaceans. Like the largest sharks, however, the largest rays, such as the manta ray, tend to strain food from the water.

Holocephalimorpha, the other main chondrichthyan group, includes the chimaeras. Holocephalimorphans (from the Ancient Greek *holos*=whole, and *kephale*=head) are so called because their upper jaw is fused to the cranium (unlike the condition in sharks). They possess an operculum that covers the gills (thus there is a single opening on either side of the head, as occurs in bony fishes). The peculiar globular head with dental plates as well as the long, thin tail have inspired their alternate designation as ratfish. Additional features of chimaeras are a spiracle that is present only as a transitory structure during ontogeny, anterior vertebrae forming a synarcual that articulates with an anterior dorsal fin, and the male having, in addition to pelvic claspers, a cephalic clasper.

SECTION I: SKELETON

Study a prepared specimen of the dogfish skeleton. It may be immersed in fluid, in a sealed glass jar, or set in clear acrylic blocks and available only for visual inspection. Separate specimens of the head skeleton may be available for closer inspection. Place such specimens in a tray and cover them with preservative fluid. As they are delicate and easily broken, handle them cautiously.

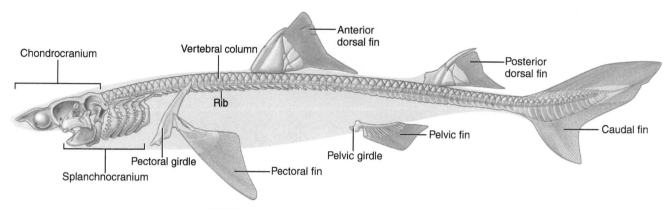

FIGURE 3.1   Skeleton of the shark in left lateral view.

Segments of the vertebral column and the fins are often set in acrylic blocks. Although these are not easily broken, handle them carefully. Do not, for example, slide them across a table or poke them with needles or pens.

Examine a specimen of a dogfish skeleton (Figure 3.1). Anteriorly, the skeleton of the head includes the **chondrocranium** and **splanchnocranium**. The **vertebral column** extends from the chondrocranium to within the tail and supports the **anterior dorsal fin**, **posterior dorsal fin**, and **caudal fin**. The **pectoral girdle**, supporting the **pectoral fins**, lies just posterior to the splanchnocranium; the **pelvic girdle**, supporting the **pelvic fins**, lies farther posteriorly. Note that the pectoral and pelvic girdles are isolated components without direct attachment to the rest of the skeleton.

Chondrocranium

The chondrocranium is the large single element of the head skeleton (Figures 3.2 and 3.3). It surrounds and provides support for the brain and sense organs. The scoop-like **rostrum** projects anteriorly and contains the **precerebral cavity**, which communicates posteriorly with the cranial cavity by way of the **precerebral fenestra**. The rostrum is supported ventrally by the **rostral carina**. Paired **rostral fenestrae** lie on either side of the carina. The paired **nasal capsules** lie lateral to the base of the rostrum. These are delicate structures and may be lost or damaged in some specimens. Paired openings, the **nares**, may be preserved on the surface of each capsule.

Posterior to the capsules lies the optic region of the chondrocranium. An **orbit**, the space occupied in life by the eyeball, lies on either side of this region. The **antorbital shelf** forms the anterior orbital wall. The orbit is defined dorsally by the **antorbital process**, **supraorbital crest**, and **postorbital process**. Posteriorly and ventrally the orbit expands into paired **basitrabecular processes**. Various openings into the chondrocranium are present in this region. On the dorsal surface between the antorbital processes is the median **epiphyseal foramen**. Several smaller

openings, the **superficial ophthalmic foramina**, pierce each supraorbital crest. The medial wall of the orbit bears numerous openings, the largest being the **optic foramen** lying anteroventrally in the orbit. Another large opening, the **trigeminofacial foramen**, is located posteriorly in the orbit. The **optic pedicle**, which supports the eyeball, takes root from the orbit just anterior to the trigeminofacial foramen. Other openings such as the **trochlear, abducens,** and **oculomotor foramina** may be identified.

Behind the orbits is the squared otic region, which contains the **otic capsules** that house the inner ears. The **basal plate** is the wide, flattened ventral part. The **carotid foramen** lies at the anterior end of the basal plate. On the dorsal surface, near the junction of the postorbital processes and otic regions, the chondrocranium bears a large, median depression, the **endolymphatic fossa**. Within the fossa are two pairs of openings. The smaller, anterior two are the **endolymphatic foramina**, and the posterior two are the **perilymphatic foramina**. The **hyomandibular foramen** pierces the chondrocranium at the anteroventral part of the otic region.

The occipital region forms the posterior part of the chondrocranium. The large median opening is the **foramen magnum**. An **occipital condyle** lies on either side and just below it. The paired condyles articulate with the first vertebra. A **vagus foramen** lies laterally to each condyle. A **glossopharyngeal foramen** lies at each ventrolateral corner of the occipital region.

Splanchnocranium

The splanchnocranium includes the seven visceral arches: the **mandibular arch, hyoid arch,** and five **branchial arches** (Figures 3.3 and 3.4). Each arch includes several segments. The mandibular and hyoid arches are highly modified for their use as jaw elements and jaw supports. The branchial arches support the interbranchial septa (see Figure 3.20) and lie between successive pharyngeal slits.

The mandibular arch, the largest of the arches, forms the jaws. Paired **palatoquadrate cartilages** fuse anteriorly to

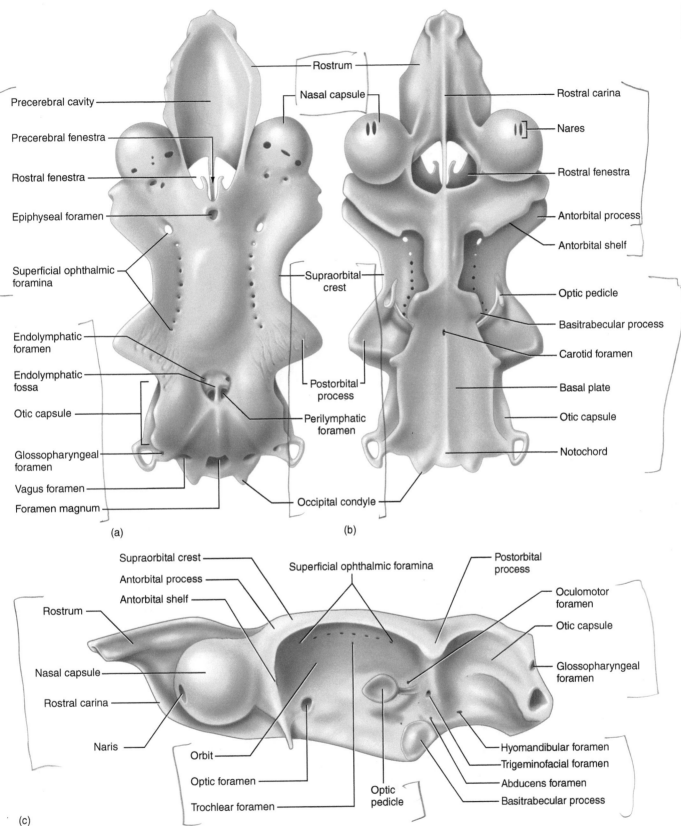

FIGURE 3.2   Chondrocranium of the shark in (a) dorsal view, (b) ventral view, and (c) left lateral view.

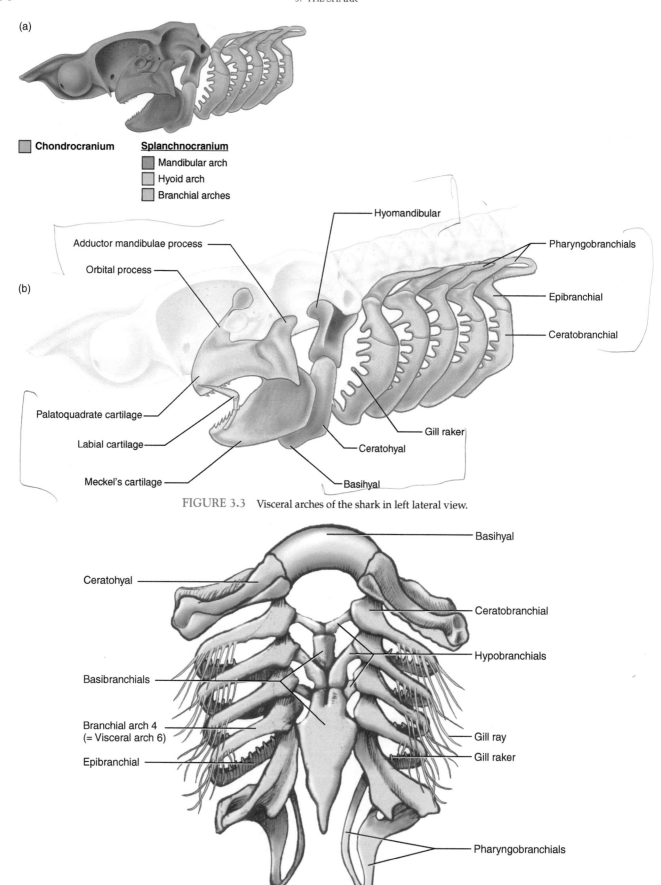

(a)

Chondrocranium

Splanchnocranium
  Mandibular arch
  Hyoid arch
  Branchial arches

(b)

Hyomandibular

Adductor mandibulae process

Orbital process

Pharyngobranchials

Epibranchial

Ceratobranchial

Palatoquadrate cartilage

Labial cartilage

Meckel's cartilage

Gill raker

Ceratohyal

Basihyal

FIGURE 3.3   Visceral arches of the shark in left lateral view.

Basihyal

Ceratohyal

Ceratobranchial

Hypobranchials

Basibranchials

Branchial arch 4
(= Visceral arch 6)

Epibranchial

Gill ray

Gill raker

Pharyngobranchials

FIGURE 3.4   Hyoid arch and branchial arches of the shark in ventral view. Note that the mandibular arch, illustrated in Figure 3.3, is absent here.

form the upper jaw, which articulates with the chondrocranium. The palatoquadrate cartilage bears two prominent dorsal projections. Anteriorly, the **orbital process** contacts the medial wall of the orbit. Posteriorly, the **adductor mandibulae process** serves for the attachment of the jaw-closing musculature, the adductor mandibulae, which is considered later. The ventral half of the mandibular arch forms the lower jaw, which consists of paired **Meckel's cartilages** fused anteriorly. The slender **labial cartilage**, supporting the labial fold (see page 61), is attached to Meckel's cartilage but may not be preserved in your specimen.

The hyoid arch is modified to support the jaws (or mandibular arch). The dorsal segment of the hyoid arch, on each side, is the **hyomandibular** (technically, this is the hyomandibular cartilage, thus hyomandibular is an adjective; the noun is hyomandibula, but hyomandibular is often used alone, with the understanding that it means hyomandibular cartilage), which abuts the otic capsule. A **ceratohyal** articulates with the other end of the hyomandibular. The ceratohyals from either side articulate ventrally with the median **basihyal**. Ligamentous attachments exist between the mandibular arch and hyoid arch, which through articulation with the otic regions acts as a support or suspensor of the jaw.

The remaining arches are branchial. Each arch is formed, in dorsal to ventral order, from paired **pharyngobranchials**, **epibranchials**, and **ceratobranchials**. Ventrally the arches are completed by three paired **hypobranchials** and two median, unpaired **basibranchials**. The pharyngobranchials of visceral arches six and seven are fused.

Vertebrae and Fins

There are two vertebral types, **trunk vertebrae** and **caudal vertebrae** (Figures 3.5 and 3.6). Sections of these types are usually set in acrylic blocks for examination. Although the vertebral column has a dorsal passageway in both the trunk and the tail, a ventral passageway is present only in the tail.

Each segment of the vertebral column is formed largely from an hourglass-shaped vertebral body or **centrum**. The **notochord** can be observed within the centrum in a sagittal section. Dorsal to each centrum is a triangular **neural plate**, the base of which sits on the centrum. The tips of the neural plates bear a low ridge termed the **neural spine**. Between successive neural plates are **intercalary plates**, also triangular but inverted so that neural and intercalary plates together form the **neural arch**. The arch helps form a passageway, the **neural canal** (for passage of the spinal cord), above the centra. Each plate is pierced by a foramen for the root of a spinal nerve. A **basapophysis** projects on either side from the ventrolateral surfaces of the centra of the trunk vertebrae. A slender **rib** (Figures 3.1 and 3.7) projects from the basapophysis.

Caudal vertebrae also bear a neural arch. In addition, the ventral surface of each centrum bears, on either side, a plate of cartilage that forms with its opposite member the **hemal arch**. The **hemal canal**, for the caudal artery and vein, passes within the arch. A thin horizontal partition, the **hemal plate**, may separate the passage for the artery dorsally and the vein ventrally. A **hemal spine** extends ventrally from the hemal arch.

The vertebral column helps support the anterior and posterior dorsal fins as well as the caudal fin. The anterior (Figures 3.1 and 3.7) and posterior (Figure 3.1) dorsal fins are similar in structure. Each has a large proximal **basal pterygiophore** to which the **fin spine** is anchored anteriorly. Spines are clipped by some suppliers prior to shipping. More distally are a series of **radial**

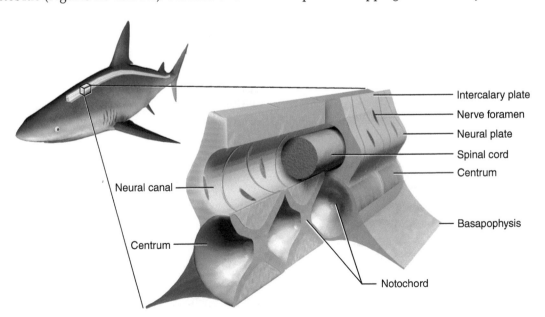

FIGURE 3.5 Anterolateral view of sagittally sectioned and whole articulated trunk vertebrae of the shark.

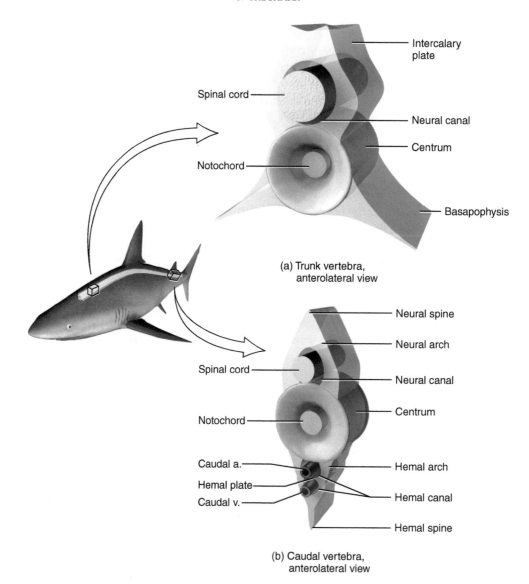

(a) Trunk vertebra,
anterolateral view

(b) Caudal vertebra,
anterolateral view

FIGURE 3.6   Vertebrae of the shark in anterolateral view. (a) Trunk vertebra, anterolateral view, (b) Caudal vertebra, anterolateral view.

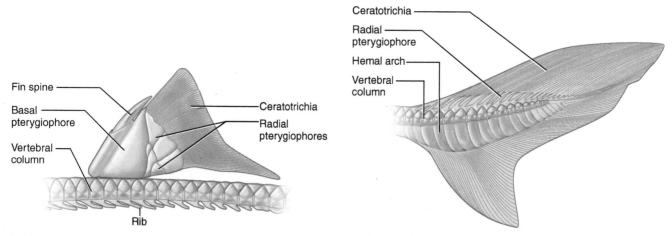

FIGURE 3.7   Skeletal elements of the anterior dorsal fin of the shark in left lateral view.

FIGURE 3.8   Skeletal elements of the caudal fin of the shark in left lateral view.

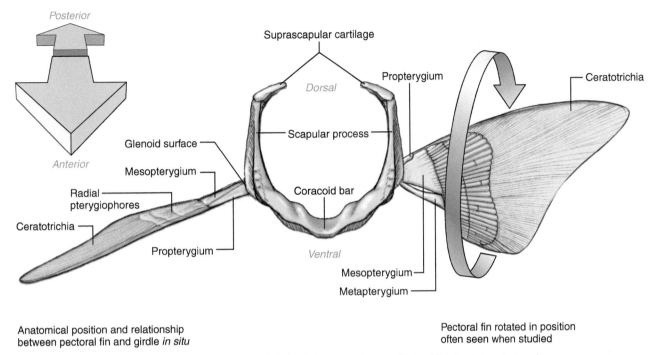

Posterior

Anterior

Suprascapular cartilage

Dorsal

Propterygium

Ceratotrichia

Glenoid surface

Scapular process

Mesopterygium

Radial pterygiophores

Coracoid bar

Ceratotrichia

Propterygium

Ventral

Mesopterygium

Metapterygium

Anatomical position and relationship
between pectoral fin and girdle *in situ*

Pectoral fin rotated in position
often seen when studied

FIGURE 3.9   Pectoral girdle and skeletal elements of pectoral fins of the shark in anterior view.

pterygiophores. and finally, the **ceratotrichia**, which are fibrous dermal rays rather than cartilaginous elements. The caudal fin (Figure 3.8) is of the heterocercal type—asymmetric, with the vertebral axis curving into the dorsal lobe of the fin. It is supported proximally by the hemal arches and radial pterygiophores.

The paired appendages, the pectoral and pelvic fins, are supported by cartilaginous girdles. The pectoral girdle (Figure 3.9) includes a stout, U-shaped cartilage to which the fins attach on either side. The portion between the fins is the **coracoid bar**, and the part extending dorsally past the attachment of each fin is the **scapular process**. A separate slender element, the **suprascapular cartilage**, attaches dorsally to each scapular process. The area that articulates with the fin is the **glenoid surface**. The pectoral fin (Figures 3.9 and 3.10) has basal and radial pterygiophores, followed by ceratotrichia. Three basals are recognized: the **propterygium**, **mesopterygium**, and **metapterygium**, in anterior to posterior order. The glenoid surface of the fin is borne mainly by the mesopterygium. The radials of the pectoral fin are rod-like structures, more regularly shaped than in the median dorsal fins, and are arranged in rows.

The pelvic girdle (Figure 3.11) consists of a single element, the **puboischiadic bar**. The pelvic fins articulate on either end of the bar at the **acetabular surfaces**. An **iliac process** lies near each acetabular surface. The pelvic fin of the female and male are similar in

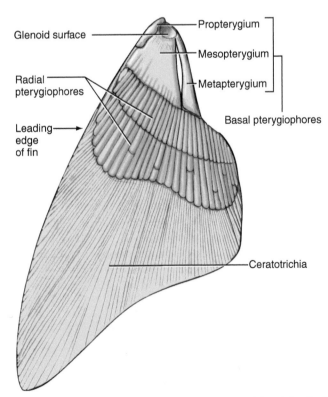

Glenoid surface

Propterygium

Mesopterygium

Radial pterygiophores

Metapterygium

Leading edge of fin

Basal pterygiophores

Ceratotrichia

FIGURE 3.10   Skeletal elements of the left pectoral fin of the shark in dorsal view.

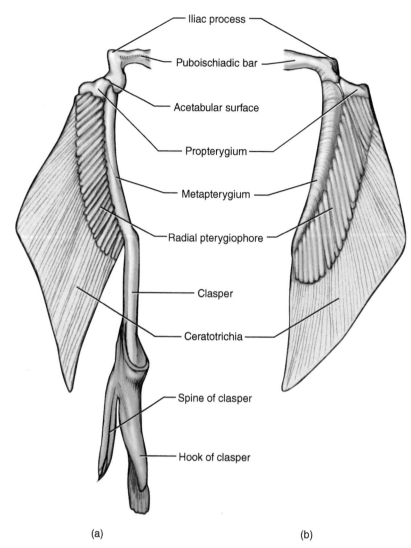

Iliac process

Puboischiadic bar

Acetabular surface

Propterygium

Metapterygium

Radial pterygiophore

Clasper

Ceratotrichia

Spine of clasper

Hook of clasper

(a)   (b)

FIGURE 3.11   Skeletal elements of the pelvic fin of the shark in ventral view. (a) right fin of the male and (b) left fin of the female.

bearing a short propterygium and a stout, elongated metapterygium that extends posteriorly (Figure 3.11). A row of cylindrical radial pterygiophores articulates mainly with the metapterygium and supports the ceratotrichia. The pelvic fin of the male differs in bearing a **clasper**, formed from modified radials, that functions in the intromission of sperm into the cloaca of the female. The clasper extends posteriorly from the metapterygium and bears a **hook** and **spine** distally (Figure 3.11a).

Key Terms: Skeleton

abducens foramen
acetabular surfaces
adductor mandibulae process
anterior dorsal fin
antorbital process

antorbital shelf
basal plate
basal pterygiophore
basapophysis
basibranchials
basihyal
basitrabecular processes
branchial arches
carotid foramen
caudal fin
caudal vertebrae
centrum (vertebral body)
ceratobranchials
ceratohyal
ceratotrichia
chondrocranium
clasper
coracoid bar

endolymphatic foramina (sing., **foramen**)
endolymphatic fossa (pl., **fossae**)
epibranchials
epiphyseal foramen
fin spine
foramen magnum
glenoid surface
glossopharyngeal foramen
hemal arch
hemal canal
hemal plate
hemal spine
hook of clasper
hyoid arch
hyomandibular
hyomandibular foramen
hypobranchials
iliac process
intercalary plates (interneural arch)
labial cartilage
mandibular arch
Meckel's cartilage (mandibular cartilage)
mesopterygium
metapterygium
nares (sing., **naris**)
nasal capsules
neural arch (vertebral arch)
neural canal (vertebral canal)
neural plate
neural spine
notochord
occipital condyle
oculomotor foramina (sing., **foramen**)
optic foramen
optic pedicle
orbit
orbital process
otic capsules
palatoquadrate cartilages
pectoral fins
pectoral girdle
pelvic fins
pelvic girdle
perilymphatic foramina
pharyngobranchials
posterior dorsal fin
postorbital process
precerebral cavity
precerebral fenestra
propterygium
puboischiadic bar
radial pterygiophores
rib
rostral carina
rostral fenestra (pl., **fenestrae**)

rostrum
scapular process
spine of clasper
splanchnocranium
superficial ophthalmic foramina
supraorbital crest
suprascapular cartilage
trigeminofacial foramen
trochlear foramen
trunk vertebrae
vagus foramen
vertebral column

SECTION II: EXTERNAL ANATOMY

The integument of sharks is relatively thin. It is subdivided into **dermis** and **epidermis**, but the latter is not keratinized and comprises a layer of live cells, which, in life, is covered by mucus. As noted earlier, bone is almost entirely absent in sharks, and so they lack extensive bony coverings. However, **placoid scales** or denticles (Figure 3.12a–c) are embedded in the skin (Figure 3.12d). A scale is built like a tooth, formed from **dentine**, covered by **enamel**, and containing a **pulp cavity** (Figure 3.12e). Such scales, which reduce the drag of water passing over the skin during swimming, are uniformly distributed over the skin (Figure 3.12b,c,f) and can be seen under low magnification, but their presence can be felt by running your hand over the skin. The integument also contains various specialized cells, such as **melanophores** that control pigmentation of the skin and secretory cells, but these cannot be seen grossly.

The body regions (Figure 3.13) of the shark, and fishes in general, are not as neatly differentiated as in tetrapods. This is due to the elongated and fusiform body, which facilitates movement through the water. The **head** extends posteriorly to the end of the row of **external pharyngeal slits** and includes the laterally placed **eyes**. The **nares** (sing. **naris**) lie ventrolaterally on the tapered **snout**. Examine a naris closely, and note that it is incompletely subdivided by a flap of skin into incurrent and excurrent apertures (see Figure 3.18). The naris leads into the blind-ended **olfactory sac** (see Figure 3.36). The ventrally located **mouth** is supported by upper and lower jaws that have rows of sharp **teeth**. A **labial pocket** lies on either side of the mouth (Figures 3.13 and 3.18). Mouth and pocket are separated by a flap, the **labial fold**, which is supported by the labial cartilage.

The **spiracle** is a relatively large opening into the pharynx and lies posterior to each eye. Its anterior wall contains a fold of tissue, the **spiracular valve**, that can be folded over the opening to close the spiracle. The valve bears a **pseudobranch**, a reduced gill, on its posterior surface. A row of five external pharyngeal slits,

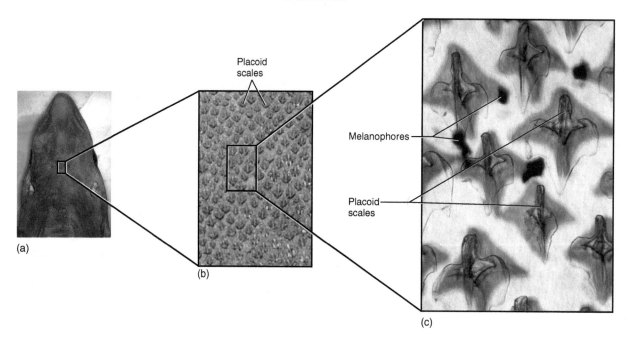

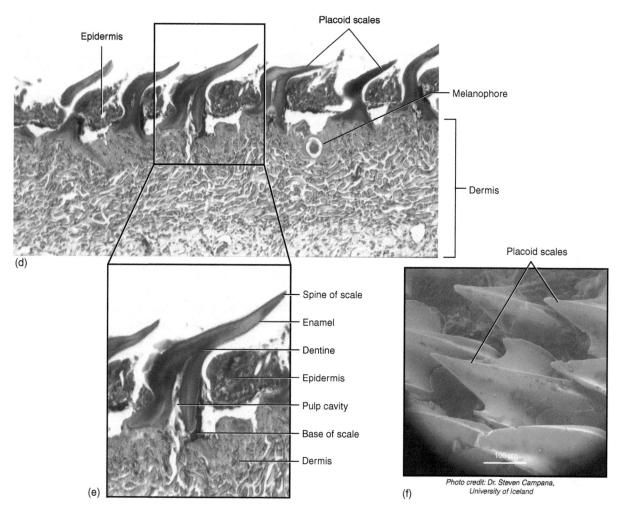

FIGURE 3.12   Skin of the shark; (a–c) successive blowups of the surface of the skin to show structure of scales; (d,e) vertical section through skin and scales, and (f) SEM image of skin showing scales.

separated by four **interbranchial septa** (Figure 3.13), lies on either side of the head; each slit leads into the pharynx. The slits are the means by which water leaves the pharynx as the water passes over the respiratory structures or **gills**. Gills are made up of **gill lamellae**, which are injected with red latex. Manipulate the interbranchial septa and note that each septum carries gill lamellae on its anterior and posterior surfaces (see Figure 3.23). Note that lamellae are present on the anterior wall of the first slit, but not on the posterior wall of the last slit.

The **trunk** follows the head and extends posteriorly to the midventrally located **cloaca** (Figure 3.14),

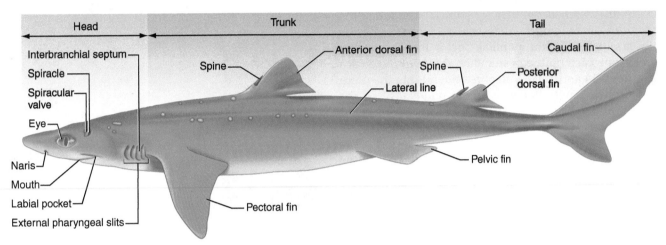

FIGURE 3.13   External features of the shark in left lateral view.

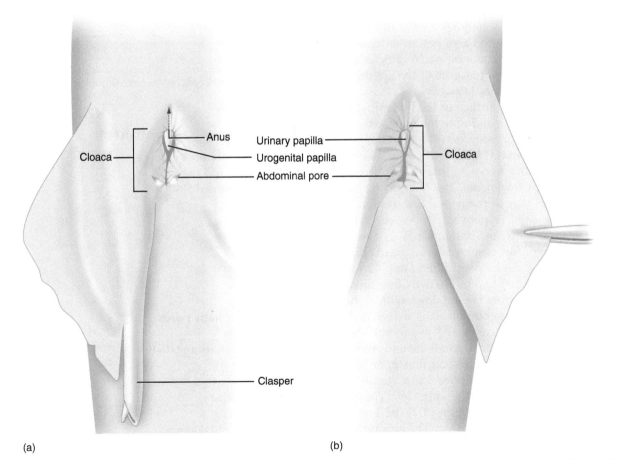

FIGURE 3.14   Ventral view of posterior portion of the shark to show cloaca and pelvic fin: (a) right side of the male and (b) left side of the female.

a chamber into which the urinary, reproductive, and digestive tracts exit. Pull the pelvic fins laterally to expose the cloaca more clearly. In the male (Figure 3.14a) the genital ducts and urinary ducts open at the tip of the **urogenital papilla**, a cone-like structure lying in the cloaca. A similar structure occurs in females (Figure 3.14b), but as only the urinary ducts open at its tip, it is termed the **urinary papilla**. The **anus**, the posterior opening of the digestive tract, opens into the cloaca anterior to the papilla. Locate the **abdominal pores**, one each on the posterolateral side of the cloaca (Figures 3.14 and 3.33). The pores extend into the pleuroperitoneal cavity and may allow removal of excess fluid from the cavity.

The **anterior dorsal fin** (Figure 3.13) lies middorsally on the trunk and carries a sharp spine anteriorly. The large, paired **pectoral fins** lie anteriorly, just behind the external pharyngeal slits, and the paired **pelvic fins** lie at the end of the trunk, one on either side of the cloaca. In the male an elongated, cylindrical, and dorsally grooved **clasper**, a copulatory organ, is a posterior extension of the medial aspect of the pelvic fin (Figure 3.14a).

The **tail** (Figure 3.13) extends posteriorly from the cloaca and carries the well-developed heterocercal **caudal fin**. Lateral undulations of the tail and caudal fin produce the propulsive force that moves the shark forward. The **posterior dorsal fin**, another medial fin, lies at the anterior part of the tail and also has an anterior spine. The spiny dogfish, the common name of *Squalus acanthias*, is derived from the presence of the spines on the anterior and posterior dorsal fins.

Examine the lateral surface of the body, just dorsal to the midlateral plane, and note the faint, pale **lateral line** extending anteroposteriorly. It marks the position of the **lateral line canal** (Figure 3.13), a cutaneous tube that contains sensory nerve endings and opens to the surface by way of tiny pores. The lateral line canal is the predominant part of the lateral line system, a system of sensory cells responsive to pressure changes caused by vibrations and movements in the water. The distribution of this system into the head and other parts of the body will be considered later. The lateral line system is modified to provide two other functions. The balancing apparatus is contained within the otic capsule of the chondrocranium. It is mentioned here because this system is connected to the surface by a pair of endolymphatic ducts (recall the endolymphatic foramina, page 54 and Figure 3.2) that open through the **endolymphatic pores** (Figure 3.15). These lie dorsally on the head between the spiracles. The second modification consists of the **ampullae of Lorenzini**, which function in electroreception. The ampullae open in numerous pores all around the head and may be noted by gently squeezing these regions to extrude a gel-like substance.

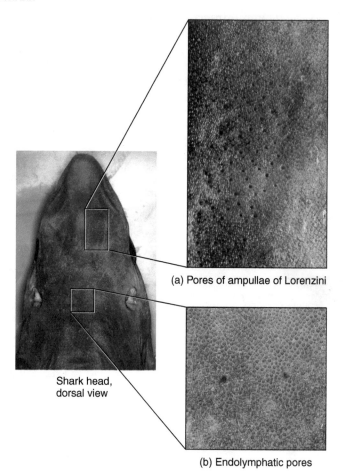

(a) Pores of ampullae of Lorenzini

Shark head, dorsal view

(b) Endolymphatic pores

FIGURE 3.15 Enlarged views of the surface of the skin of the shark: (a) pores of ampullae of Lorenzini and (b) endolymphatic pores.

Key Terms: External Anatomy

abdominal pores
ampullae of Lorenzini
anterior dorsal fin
anus
caudal fin
clasper
cloaca
dentine
dermis
enamel
endolymphatic pores
epidermis
external pharyngeal slits
eyes
gill
gill lamellae
head
interbranchial septa
labial fold
labial pocket
lateral line

lateral line canal
melanophores
mouth
naris (pl., **nares**)
olfactory sac
pectoral fins
pelvic fins
placoid scales
posterior dorsal fin
pseudobranch
pulp cavity
snout
spiracle
spiracular valve
tail
teeth
trunk
urinary papilla
urogenital papilla

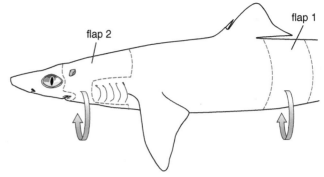

FIGURE 3.16 Diagram of the shark in left lateral view showing guidelines for cutting the skin.

SECTION III: MUSCULAR SYSTEM

Study of the musculature requires that the skin be removed. There are various techniques that allow efficient skinning. One of these is that a scalpel, in most cases, be used as little as possible. Remember that dissection does not have as its goal the frequent use of a scalpel, as many students would like to believe. Instead, we may recognize skill in dissection by selection of the proper implement for the task at hand. It may be used to cut the skin and then to help reflect it. In reflecting the skin, a blunt probe is also effective and easier. A scalpel may be used for areas where the skin adheres more strongly to the musculature. In such cases, the scalpel should be used in a sideways scraping motion. Turning the blade axially 180 degree and using its point as a pick can also be very effective. Its cutting edge should rarely be used in separating skin from underlying body structures.

Trunk and Appendicular Muscles

Table 3.1 lists the musculature of the dogfish and summarizes their origins, insertions, and main actions. The trunk musculature is considered first for several reasons: It is relatively easy to examine as an introduction to the musculature; it comprises the bulk of the musculature of the body; and the procedure will give you practice in skinning a region where mistakes and inexperience will not produce serious damage.

Peel back a strip of skin, about 10 cm in width, from one side of the trunk region just posterior to the anterior dorsal fin (Figure 3.16, flap 1). Cut with a scalpel three sides of a rectangular section of skin. Begin the incisions middorsally and continue ventrally down the body to the midventral plane. The skin is thin, so cut carefully to avoid damaging underlying structures. Do not cut the fourth side. This will produce a flap of skin that can be wrapped around exposed tissues at the end of a lab period to help protect them. After cutting the edges of the flap, lift one corner and begin to pull the skin away from the trunk musculature. As the skin begins to pull away, scrape the connective tissue between skin and muscles. Use the non-cutting edge of your scalpel, as described earlier. Skin will come off easily in some spots but will stick in others because muscle fibers attach directly to the dermis. Patiently scrape the muscle fibers from the inside of the skin toward the body with a scalpel. In some areas a probe or the ends of narrow forceps will suffice.

Once flap 1 is reflected notice that the trunk musculature is divided into the dorsal **epaxial musculature** and the ventral **hypaxial musculature** by the **horizontal skeletogenous septum**, a connective tissue sheet lying in the frontal plane (Figure 3.17). Each of these primary divisions consists of segmented, Z-shaped **myomeres**, separated by connective tissue sheets termed **myosepta**. The **linea alba** is connective tissue separating left and right myomeres midventrally (Figure 3.18). The myomeres are complex internally, extending farther anteriorly and posteriorly than their margins at the surface of the body.

To examine the appendicular muscles, skin portions from the dorsal and ventral surfaces of a pectoral fin and note the musculature revealed. It is subdivided into dorsal and ventral divisions (Table 3.1). The dorsal division includes the **pectoral abductor,** which lies dorsally and pulls the pectoral fin dorsally (Figure 3.17). The ventral division includes the **pectoral protractor** (Figure 3.17) and the **pectoral adductor** (Figure 3.18). The former lies anteriorly and, being a protractor, pulls the pectoral fin anteriorly. The pectoral adductor lies ventrally and pulls the pectoral fin ventrally. *Abductor* and *adductor* are terms that describe the actions of muscles. Abductor muscles pull a structure away from the midventral line; adductor muscles pull toward the midventral line. The pelvic fin musculature (also subdivided into dorsal and ventral divisions; Table 3.1) is comprised of the **pelvic**

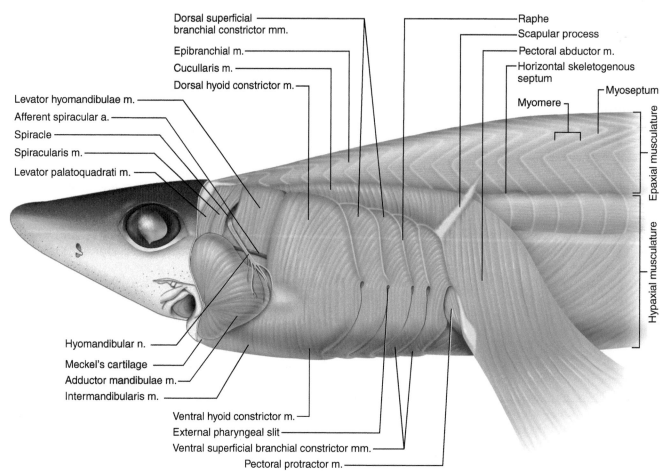

FIGURE 3.17   Head, branchial region, and anterior part of the trunk of the shark in left lateral view, skinned to reveal musculature.

TABLE 3.1   Skeletal Musculature of the Shark.

Muscle	Origin	Insertion	Main Actions
APPENDICULAR MUSCULATURE			
Pectoral Fin Musculature			
Dorsal Division			
Pectoral abductor	Hypaxial musculature and posterior surface of suprascapular process	Dorsal surface of pectoral fin	Abducts and flattens pectoral fin.
Ventral Division			
Pectoral adductor	Posterior surface of scapulocoracoid	Ventral surface of pectoral fin	Adducts pectoral fin and renders it ventrally concave.
Pectoral protractor	Lateral surface of scapulocoracoid	Propterygium	Protracts pectoral fin.
Pelvic Fin Musculature			
Pelvic abductor	Trunk musculature, iliac process of puboischiadic bar, metapterygium	Radial cartilages of pelvic fin; in males, extends onto clasper	Abducts and flattens pelvic fin.
Pelvic adductor	Proximal portion: Puboischiadic bar; distal portion: Metapterygium	Proximal portion: Metapterygium; distal portion: Radial cartilages; in males, extends onto clasper	Adducts pelvic fin.

TABLE 3.1   Skeletal Musculature of the Shark.—cont'd

Muscle	Origin	Insertion	Main Actions
AXIAL MUSCULATURE			
Branchiomeric Musculature			
Branchial Musculature			
Branchial adductor	Epibranchial cartilage	Ceratobranchial cartilage	Adducts gill arch.
Cucullaris	Epibranchial musculature	Pectoral girdle, 5th branchial arch	Elevates pectoral girdle, elevates last branchial arch.
Dorsal interarcual	Pharyngobranchial	Pharyngobranchial	Adducts gill arch.
Dorsal superficial branchial constrictors	Cucullaris	Tendinous sheath of pharyngeal slits	Compress pharyngeal pouches; those portions that close pharyngeal slits 2 to 5 are considered by some authors (see text) as branchial trematic constrictors.
Interbranchial	Gill rays	Gill rays	Adducts pharyngeal arches, move interbranchial septa.
Lateral interarcual	Pharyngobranchial	Epibranchial cartilage	Adducts pharyngeal arch.
Ventral superficial branchial constrictors	Coracoarcual	Tendinous sheath of pharyngeal slits	Compress pharyngeal pouches.
Hyoid Musculature			
Dorsal hyoid constrictor	Chondrocranium, cucullaris	Tendinous sheath, hyomandibular muscle	Compresses first pharyngeal pouch; that portion of it, considered by some authors (see text) as a hyoid trematic constrictor, closes 1st pharyngeal slit.
Interhyoideus	Ceratohyal cartilages (deep to and not anteriorly as extensive as the intermandibularis)	Midventral raphe	Left and right side interhyoideus muscles pull ceratohyals together toward midline during opening of jaws and hyoid arch; help elevate floor of oral cavity when mouth closed.
Levator hyomandibulae	Region of otic capsule of chondrocranium, epibranchial muscles	Hyomandibular cartilage	Elevates hyomandibular cartilage.
Ventral hyoid constrictor	Raphe of first pharyngeal slit	Tendinous sheath of hyomandibular cartilage	Compresses first pharyngeal pouch.
Mandibular Musculature			
Adductor mandibulae	Palatoquadrate cartilage	Meckel's (or mandibular) cartilage	Adducts lower jaw.
Intermandibularis	Meckel's (or mandibular) cartilage, medial surface; posterior portion from surface of adductor mandibulae	Midventral raphe: The left and right intermandibular muscles meet along this raphe.	Left and right side intermandibularis muscles pull lower jaws together toward midline during opening of jaws and hyoid arch; help elevate floor of oral cavity when mouth closed.
Levator palatoquadrati	Chondrocranium	Palatoquadrate cartilage	Elevates palatoquadrate (upper jaw).
Preorbitalis	Region of nasal capsule of chondrocranium	Adductor mandibulae	Adducts lower jaw, protrudes palatoquadrate (upper jaw).
Spiracularis	Chondrocranium	Spiracle	Closes spiracle.

Continued

TABLE 3.1 Skeletal Musculature of the Shark.—cont'd

Muscle	Origin	Insertion	Main Actions
EPAXIAL MUSCULATURE			
Dorsal myomeres of trunk and tail, arranged as longitudinal bundles	Myosepta	Myosepta	Left side myomeres flex trunk and tail to left; right side myomeres flex trunk and tail to right
EPIBRANCHIAL MUSCULATURE			
Dorsal myomeres; anterior extension of epaxial musculature dorsal to branchial region	Myosepta Deeper portions attach to posterior surface of chondrocranium and anterior vertebrae	Myosepta	Elevate chondrocranium during feeding.
EXTRINSIC EYE MUSCULATURE			
Oblique Muscles			
Dorsal oblique	Anterior surface of orbit	Dorsally on medial surface of eyeball, just anterior to insertion of dorsal rectus	Moves eyeball.
Ventral oblique	Anterior surface of orbit	Ventrally on medial surface of eyeball, just anterior to insertion of ventral rectus	Moves eyeball.
Rectus Muscles			
Dorsal rectus	Posterior surface of orbit	Dorsally on medial surface of eyeball, just posterior to insertion of dorsal oblique	Moves eyeball.
Lateral rectus	Posterior surface of orbit	Posterior surface of eyeball	Moves eyeball.
Medial rectus	Posterior surface of orbit	Medial surface of eyeball	Moves eyeball.
Ventral rectus	Posterior surface of orbit	Ventrally on medial surface of eyeball, just posterior to insertion of dorsal oblique	Moves eyeball.
HYPAXIAL MUSCULATURE			
Ventral myomeres of trunk and tail, arranged as longitudinal bundles	Myosepta	Myosepta	Left side myomeres flex trunk and tail to left; right side myomeres flex trunk and tail to right.
HYPOBRANCHIAL MUSCULATURE			
Posthyoid Musculature			
Coracoarcuals	Scapulocoracoid	Coracomandibular, coracohyoids	Helps depress basihyal and ceratohyal cartilages, expanding oral and pharyngeal cavities.
Coracobranchials	Scapulocoracoid and dorsal surface of coracoarcual	1st coracobranchial: Basihyal cartilage; 2nd –4th coracobranchials: Hypobranchial cartilages; 5th coracobranchial: Posterior basibranchial cartilage	Abduct hyoid and branchial arches, helping expand branchial region.
Coracohyoids	Coracoarcual	Basihyal	Abduct hyoid arch.
Prehyoid Musculature			
Coracomandibular	Coracoarcual	Symphysis of lower jaw (Meckel's or mandibular cartilages)	Abducts lower jaw.

Sources: Huber, Soares, & Carvalho, 2011; Motta & Huber, 2012; Motta & Wilga, 2001; Homberger & Walker, 2003; Wilga & Motta, 1998.

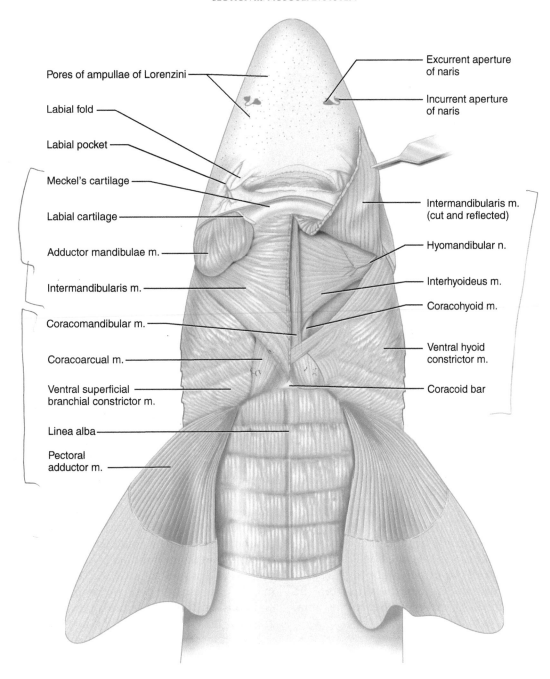

Pores of ampullae of Lorenzini

Labial fold

Labial pocket

Meckel's cartilage

Labial cartilage

Adductor mandibulae m.

Intermandibularis m.

Coracomandibular m.

Coracoarcual m.

Ventral superficial
branchial constrictor m.

Linea alba

Pectoral
adductor m.

Excurrent aperture
of naris

Incurrent aperture
of naris

Intermandibularis m.
(cut and reflected)

Hyomandibular n.

Interhyoideus m.

Coracohyoid m.

Ventral hyoid
constrictor m.

Coracoid bar

FIGURE 3.18 Branchial region and anterior part of the trunk of the shark in ventral view, skinned to reveal musculature.

adductor (including proximal and distal portions) and **pelvic abductor**; these muscles may be found associated with the pelvic fin. However, be careful if you decide to remove the skin from the ventral surface of the pelvic fin of a male. The **siphon** (Figure 3.33; see page 89 for further description), a structure associated with the reproductive system, lies on the ventral surface of the fin musculature.

Muscles of the Head and Branchial Region

Examine next the musculature of the head and branchial region. The head may be skinned from snout to the pectoral girdle, but this is not necessary. Instead, skin only half the head (as in Figure 3.16, flap 2). Make an incision through the skin middorsally between the eyes that extends posteriorly to the level of the pectoral fin. Continue the incision ventrally to the dorsal edge of the fifth pharyngeal slit. The skin may be removed from the around the pharyngeal slits, but this is difficult and time-consuming, as the skin adheres tightly here. This effort can be avoided without missing much anatomical detail by leaving a rectangular flap around the slits, as follows (Figure 3.16, flap 2). From the fifth slit, cut anteriorly to the first slit, ventrally

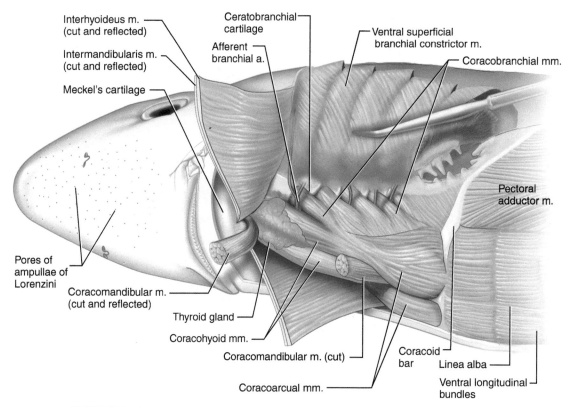

Interhyoideus m.
(cut and reflected)

Intermandibularis m.
(cut and reflected)

Meckel's cartilage

Ceratobranchial
cartilage

Afferent
branchial a.

Ventral superficial
branchial constrictor m.

Coracobranchial mm.

Pectoral
adductor m.

Pores of
ampullae of
Lorenzini

Coracomandibular m.
(cut and reflected)

Thyroid gland

Coracohyoid mm.

Coracomandibular m. (cut)

Coracoarcual mm.

Coracoid
bar

Linea alba

Ventral longitudinal
bundles

FIGURE 3.19   Branchial and hypobranchial musculature of the shark in ventrolateral view.

along the anterior margin of this slit, and then posteriorly to the ventral end of the fifth slit. Continue
the incision ventrally around the body to about 2 cm
past the midventral line. Do not cut midventrally.
Return to the beginning of the incision between the
eyes and cut posteroventrally around the eye to the
angle of the mouth. Proceed along the edge of the jaw
until you reach ventrally approximately to the same
level as the posterior cut. Separate the skin from the
body using the methods described earlier for the axial
musculature.

Branchiomeric Musculature

On exposing the musculature, view the shark in lateral view (Figures 3.17 and 3.39) and locate the spiracle.
There are two muscles anterior it, but careful dissection
is required to separate them, so proceed cautiously. The
small **spiracularis** is on the anterior wall of the spiracular valve. The larger **levator palatoquadrati** lies anterior
to it. The muscles pass from the otic capsule to the upper
jaw and serve to lift or stabilize it. The **levator hyomandibulae** lies posterior to the spiracle (Figures 3.17 and
3.39). It extends between the otic capsule and the hyomandibular, and raises the hyomandibular during jaw
closing. Ventral to these muscles and just posterior to the
angle of the mouth is the large **adductor mandibulae**,

which extends between upper and lower jaws and closes
the mouth (Figures 3.17 and 3.18). A modification of this
muscle, the **preorbitalis**, originates on the nasal capsule
and extends posteriorly to merge with the adductor
mandibulae (Figures 3.39 and 3.43); the preorbitalis (and
other muscles) will be noted later when structures in the
orbit are examined. It should be noted that some authors,
such as Wilga and Motta (1998), refer to the adductor
mandibulae, as used here, as the quadratomandibularis,
and consider the quadratomandibularis and the preorbitalis as the adductor mandibulae complex.

Two other structures may be noted briefly at this time.
The conspicuous **hyomandibular nerve** passes across
the ventral part of the levator hyomandibulae from the
spiracle. The **afferent spiracular artery** passes deep to
the nerve, and should be injected with red latex.

Posterior to the muscles just discussed are the five
superficial constrictors, comprising a block of muscles
that surrounds the pharyngeal slits (Figures 3.17–3.19).
The constrictors are subdivided into dorsal and ventral
portions. They are arranged sequentially from front to
back and separated by vertical connective tissue partitions termed **raphes**, which look like white lines on the
muscle surface. The first constrictor, more complex than
the others, includes the **dorsal** and **ventral hyoid constrictors**, which lie between the adductor mandibulae
and the first pharyngeal slit. Raphes extending dorsally

and ventrally from the first slit separate the hyoid constrictors from the first of four superficial constrictors. The **dorsal** and **ventral superficial branchial constrictors,** properly constrictors three to six, are similar in form and extend between the raphes associated with each slit. The branchial constrictors compress the branchial region. Note that the fibers of the dorsal hyoid and dorsal superficial branchial constrictors are oriented posterodorsally from the edge of the pharyngeal slits, whereas those of the ventral hyoid and ventral superficial branchial constrictors are oriented posteroventrally. Along the edge of each slit, however, the fibers are nearly dorsoventrally oriented. These function as a flap valve to close the slit. Some authors (e.g., Homberger & Walker, 2003; Huber, et al., 2011) refer to this musculature, effectively between the dorsal and ventral hyoid constrictors and the dorsal and ventral superficial branchial constrictors, as the hyoid trematic and branchial trematic constrictors, respectively.

The triangular **cucullaris** lies dorsal to the constrictors. Dorsal to the cucullaris are the dorsal longitudinal bundles of the **epibranchial musculature,** which attach anteriorly to the back of the chondrocranium and represent the anterior continuation of the epaxial musculature into the head. Indeed, some authors, such as Wilga and Motta (1998) and Motta and Wilga (1999), refer to the epibranchial musculature as expaxial musculature. The cucullaris arises from fascia covering the longitudinal bundles and inserts on the epibranchial cartilage of each branchial arch and the scapular process of the pectoral girdle.

Examine the ventral surface of the shark (Figure 3.18). The **intermandibularis** lies between Meckel's cartilages. Composed of left and right halves separated midventrally by a raphe, the fibers of the intermandibularis radiate fan-like from Meckel's cartilage toward the midline. The muscle is some 2–3mm thick, but posteriorly it may be lifted from the underlying muscle. Carefully cut through the intermandibularis parallel to and about 2–3mm to one side of the raphe. A second, thin, and less extensive muscular sheet lies deep to it. This is the **interhyoideus,** which, as its name implies, extends between the cartilages of the hyoid arch, specifically the ceratohyals. The muscle adheres tightly to the intermandibularis. It is difficult to separate them, but attempt to do so from the anterior end of the interhyoideus. The fibers of the interhyoideus are less obliquely oriented and do not extend as far anteriorly as the intermandibularis. The hyomandibular nerve may be seen lying ventrolaterally on the interhyoideus (Figure 3.18). If it becomes too difficult and time-consuming to separate the muscles, you may instead view the interhyoideus by cutting through it as well and reflecting the sheets (Figure 3.19).

The muscles discussed so far in this section are part of the branchiomeric musculature (there are others, but they lie more deeply and will be studied shortly). Now

that their distribution has been noted, their relationships may be summarized. The muscles are each associated with particular visceral arches (see Figures 3.17 and 3.18). The muscles of the mandibular arch are the levator palatoquadrati, spiracularis, adductor mandibulae, and intermandibularis. The levator hyomandibulae, hyoid constrictors, and interhyoideus belong to the hyoid arch. Each of the branchial arches, except for the last, has the typical constrictor setup that has just been described. The constrictors are differentiated into various parts; only the more superficial portions have been examined so far.

The deeper portions require further dissection. Insert scissors into the third slit and cut vertically through the dorsal and ventral superficial branchial constrictors so the slit may be spread open. You will thus be able to view the interbranchial septum of the third and of the fourth branchial arches. Compare your specimen with Figures 3.20 and 3.23. Note that most of each septum is covered by gill lamellae. This portion of the septum is formed by the **interbranchial** muscle. Its fibers are curved. Carefully remove, by scraping, the lamellae from one septum to see them. The interbranchial is a deep portion of the constrictor musculature, and the superficial branchial constrictor lies lateral to it. Next snip frontally and completely through the middle of a septum and locate the cartilaginous arch. The small circular section of muscle that has been cut is the **branchial adductor,** a short muscle extending between the ceratobranchial and epibranchial that represents a deep derivative of the constrictor sheet (Figure 3.20).

Using a probe, separate the epibranchial musculature from the cucullaris and levator hyomandibulae and push these regions apart to expose the anterior cardinal sinus, which is part of the cardiovascular system (Figure 3.21; see also Figure 3.43). On the medial side of the sinus lie the dorsal surfaces of the branchial arches. Pick away connective tissue to expose the muscles, but do not injure the nerves that lie in this region. Find two sets of small, thin muscles. The strap-like and elongated **dorsal interarcuals** lie between successive pharyngobranchial cartilages of branchial arches 1–4. Lateroventral to these lie the shorter and wider **lateral interarcuals,** which extend between the epibranchial and pharyngobranchial cartilages of each of branchial arches 1–4.

Hypobranchial Musculature

Return to the ventral surface (Figures 3.18 and 3.19). If you have not already cut through the interhyoideus, do so now and reflect it. The muscles visible between the hypaxial muscles and the lower jaw constitute nearly all the **hypobranchial musculature**. They may be divided into a prehyoid group, including the **coracomandibular,**

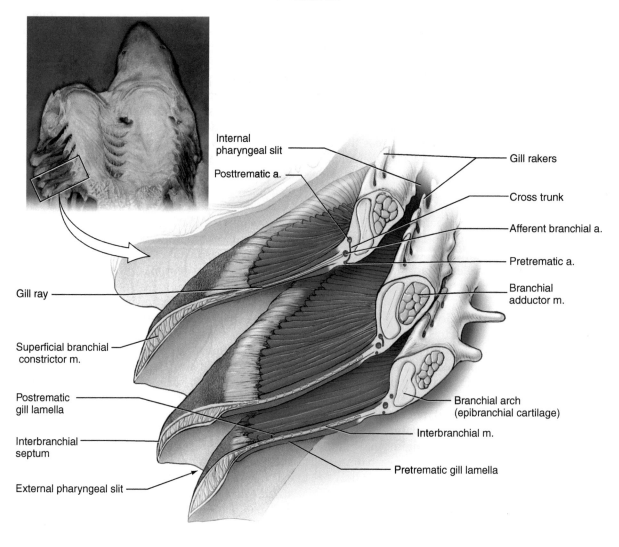

FIGURE 3.20   Section through interbranchial septa of the shark.

and posthyoid group (see later). The coracomandibular is the long, midventral, and nearly cylindrical muscle exposed on reflection of the intermandibularis and interhyoideus. It extends between the lower jaw and the muscles posterior to it, which attach to the coracoid bar. Cut the coracomandibular near its posterior end, reflect it, and note the dark, pinkish **thyroid gland** deep to the muscle's anterior end (Figure 3.19).

The remaining muscles belong to the posthyoid group. Deep to the coracomandibular lie the elongated, paired **coracohyoids** (Figures 3.18 and 3.19), which insert anteriorly on the basihyal. Posteriorly the coracohyoids are continuous with the broader, nearly triangular **coracoarcuals**, which lie medially between the ventral superficial branchial constrictors and anterior to the hypaxial musculature. Cut the ventral superficial branchial constrictor by snipping, on the same side that is skinned, from the

last branchial slit to (but *not* through) the posterolateral end of the coracoarcual. Then dissect forward between the coracoarcual and ventral superficial constrictor to separate these muscular regions. Spread them apart to reveal the **coracobranchials** (Figure 3.19), a series of five muscles that fan out from the coracoid bar, coracoarcuals and the walls of the pericardial cavity to the ceratobranchial and basibranchial cartilages. Do not attempt to further dissect these muscles at this point; doing so would require the removal and destruction of structures that have yet to be seen.

The names of the hypobranchial muscles contain the common component "coraco-" that refers to the coracoid bar, but only the coracoarcuals and parts of the coracobranchials attach directly to this structure. The others gain an indirect attachment by way of the coracoarcuals.

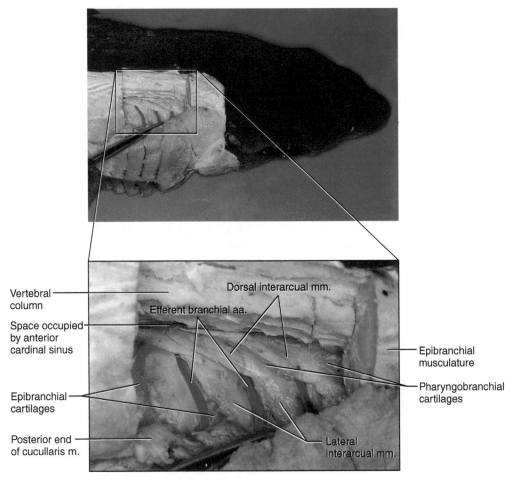

Vertebral column

Efferent branchial aa.

Dorsal interarcual mm.

Space occupied by anterior cardinal sinus

Epibranchial musculature

Pharyngobranchial cartilages

Epibranchial cartilages

Posterior end of cucullaris m.

Lateral interarcual mm.

FIGURE 3.21   Blowup of branchiomeric and epibranchial musculature of the shark in right lateral view. The epibranchial musculature has been dissected to reveal the interarcual muscles.

Key Terms: Muscular System

adductor mandibulae
afferent spiracular artery
branchial adductor
coracoarcuals
coracobranchials
coracohyoids
coracomandibular
cucullaris
dorsal hyoid constrictor
dorsal interarcuals
dorsal superficial branchial constrictors
epaxial musculature
epibranchial musculature
horizontal skeletogenous septum
hyomandibular nerve
hypaxial musculature
hypobranchial musculature
interbranchial

interhyoideus
intermandibularis
lateral interarcuals
levator hyomandibulae
levator palatoquadrati
linea alba
myomeres
myosepta
pectoral abductor
pectoral adductor
pectoral protractor (cranial muscle of ventral division)
pelvic abductor
pelvic adductor
preorbitalis
raphe
siphon
spiracularis
thyroid gland
ventral hyoid constrictor
ventral superficial branchial constrictors

SECTION IV: DIGESTIVE AND RESPIRATORY SYSTEMS

Anteriorly, the digestive system includes the **mouth**, **oral cavity**, and **pharynx**, but these structures also function in the respiratory system. To observe them, make a frontal cut, using scissors, through the gills on the side of the head that was skinned for the muscles. Insert the scissors into the angle of the mouth and cut posteriorly through the gills. Stop once you have cut into the last branchial slit. Swing open the floor of the mouth as far as you can. Refer to Figure 3.22 but do not make a transverse cut through the ventral body wall at this time.

The mouth is the opening into the oral cavity. The oral cavity is bounded anteriorly by the **teeth**. Its floor is formed by the **primary tongue**, which is not a true tongue as occurs in tetrapods. The boundary between the oral cavity and pharynx is not clearly defined in the adult, though the oral cavity is often considered to end at the posterior margin of the teeth. The pharynx is considered the region into which the pharyngeal slits lead. Posteriorly it narrows into the **esophagus**, which leads to the **stomach**. Observe the five **internal pharyngeal slits**. **Gill rakers** are the finger-like structures projecting across the slit that help keep food from escaping through the slits or damaging the **gills**, which are the organs of gas exchange. These gills are internal gills, present in adult fishes. The larvae of many fishes possess external gills, highly vascularized and filamentous processes attached to the lateral surface of the head between certain

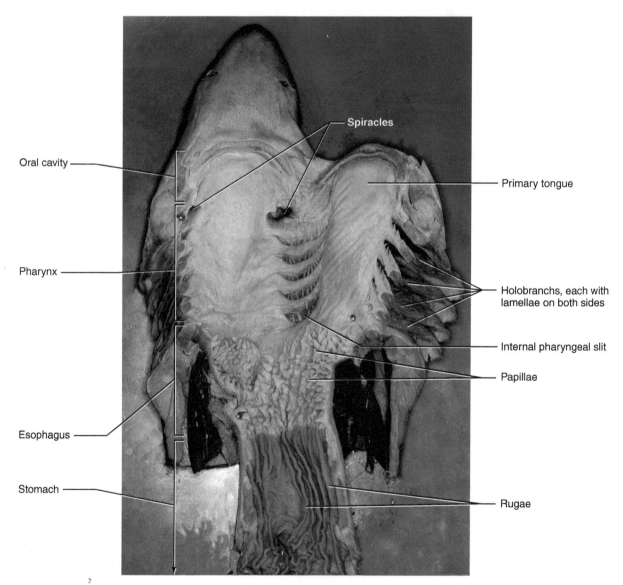

FIGURE 3.22 Anterior portion of the shark in ventral view. The right side visceral arches have been cut through to reflect the floor of the oral cavity and pharynx. The esophagus and stomach have also been cut and reflected.

pharyngeal slits (Liem, Bemis, Walker, & Grande, 2001). The **spiracle**, a smaller and dorsally placed opening, lies in front of the first pharyngeal slit. The spiracle, along with the mouth, allows water to enter the pharynx, from which it passes out through the pharyngeal slits. The spiracle is larger and more important as a passageway for water in bottom-dwelling skates and rays.

Examine the branchial region along the cut you made to open the mouth. Note the four **interbranchial septa**, the partitions that separate the five pharyngeal slits. Each slit is a passageway, often termed a pouch or chamber, between the pharynx and the external environment. The chamber is subdivided into a medial **branchial chamber** (the portion associated with gills), and lateral to it a **parabranchial chamber** (the portion not associated with gills; Figure 3.28). The anterior and posterior surfaces of each septum bear a series of nearly parallel, ridge-like **primary gill lamellae** (Figures 3.20 and 3.22). Gas exchange occurs across **secondary gill lamellae**, which are very small, tightly packed, and attached perpendicularly and transversely to the primary gill lamellae (Liem et al., 2001). They cannot be made out with the unaided eye (although

their presence can be appreciated by the ridged or striated appearance of the surface of a primary gill lamella), but are readily observable under a dissecting microscope.

A septum (with its associated supporting structures) that has lamellae on both anterior and posterior surfaces is a complete gill or **holobranch**. A **hemibranch** has lamellae on only one side of the septum, as occurs with the hyoid arch, which bears lamellae only on its posterior wall. Keep in mind that holobranch and hemibranch are used with reference to septa and not to pharyngeal slits.

Reexamine a septum in section (Figures 3.20 and 3.23). It consists of a medial cartilaginous support, the branchial arch. Review the cartilages of the arch if necessary. A **branchial adductor muscle** (noted earlier; see page 71 and Figure 3.20) lies medial to the arch. Extending laterally is a muscular wall, the **interbranchial muscle**, supporting the interbranchial septum. Cartilaginous **gill rays** (see Figure 3.20) extend from the arch into, and help support, the interbranchial muscle. Farther laterally, the septum is completed by the superficial branchial constrictor muscle. A **pretrematic artery** lies at the base of the lamellae on the posterior surface of the septum, and

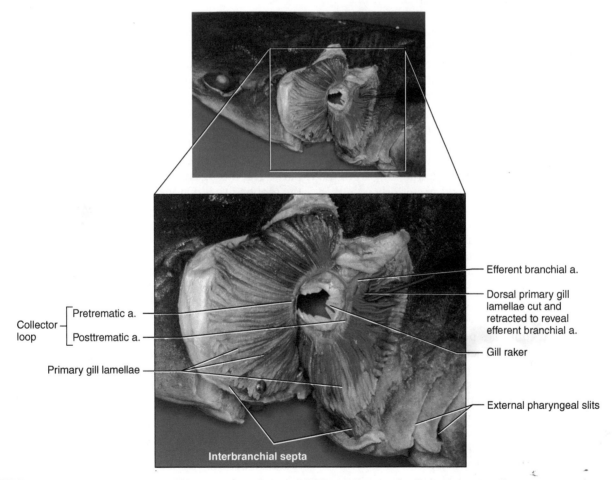

FIGURE 3.23 Branchial region of the shark in dorsolateral view, with blowup showing detail of pretrematic and posttrematic surfaces of successive interbranchial septa.

a **posttrematic artery** lies anteriorly. These names are used with reference to a pharyngeal slit, *trema* being the Ancient Greek work for hole or slit. Probe the septum for **cross trunks** linking these arteries. An **afferent branchial artery** lies near the middle of the septum between the pretrematic and posttrematic arteries.

The **coelom** or body cavity includes the **pericardial cavity** anteriorly and the **pleuroperitoneal cavity** posteriorly. They are separated by the **transverse septum**, lying very near the pectoral girdle. The pericardial cavity, which contains the heart, will be considered later together with the cardiovascular system. The pleuroperitoneal cavity contains the viscera, mainly digestive organs, and other structures.

To gain access to the pleuroperitoneal cavity, make a longitudinal cut, slightly to one side of the midline, from the posterior aspect of the pectoral girdle to the base of the tail. In doing so, cut through the puboischiadic bar and to one side of the cloaca. Then make a transverse cut on either side to produce four flaps as shown in Figure 3.24.

Note that you have cut through several layers to open the pleuroperitoneal cavity: skin, musculature, and **parietal peritoneum**, the epithelium that lines the body cavity (Figure 3.24). The epithelium covering the organs within the cavity is the **visceral peritoneum**. The organs are suspended from the body wall or connected to each other via membranes, formed by two adjacent sheets of visceral peritoneum, termed **mesenteries**. Any membrane in the peritoneal cavity that supports an organ may be called a mesentery. One mesentery is simply called *the* **mesentery**, and you will meet that one later.

First, examine the viscera. The most conspicuous is the **liver**, which occupies most of the anterior part of the cavity. **Right** and **left lobes** extend posteriorly on either side. Do not cut off these lobes (otherwise, you will have to contend with a continuous leakage of the oil present in the liver that reduces the shark's specific gravity). A small **median lobe** extends for a short distance between them and contains an elongated **gall bladder** extending along the right margin.

Spread the lobes of the liver to expose the viscera completely. Entering the cavity dorsal to the liver is the esophagus. It continues posteriorly as the stomach, a large J-shaped organ. There is no external distinction between the stomach and esophagus. Internally, however, the esophagus bears finger-like projections or **papillae**, whereas the stomach bears longitudinal ridges termed **rugae** (Figures 3.22 and 3.25). Slit open a portion of the esophagus and stomach to observe these structures. Note the possible presence of stomach contents, which could range from krill to octopus to fish.

The main part of the stomach is the **body**. The smaller, narrower, posterior part is the **pyloric region**.

It constricts at the **pylorus**, which marks the separation between the stomach and intestine. Two organs lie near the junction of the stomach and intestine. The triangular **spleen** is the large, dark organ at the posterior end of the stomach. The spleen is a lymphoid organ, not a digestive organ, and functions in the production and storage of blood cells. The **pancreas** consists of two parts linked by a narrow **isthmus**. A flattened, oval **ventral lobe** (Figure 3.24) adheres to and almost completely hides the ventral surface of the **duodenum**, the anterior segment of the intestine. The narrow, elongated **dorsal lobe** of the pancreas extends posteriorly.

The intestine of the shark is subdivided into a short duodenum, a **valvular intestine**, and a narrow **colon** (Figures 3.24 and 3.25). The valvular intestine bears a **spiral valve**, an internal subdivision that increases the effective length of the intestine. Slit open the intestine to see the valve (Figure 3.25). The colon, lacking a spiral valve, extends from the valvular intestine. It is joined by the salt-excreting **digitiform gland** (Figures 3.24 and 3.25), before continuing into the cloaca as the **rectum** (Figure 3.25).

Other structures in the pleuroperitoneal cavity should be noted only briefly at this time. The **gonads**, either paired **testes** or **ovaries**, lie anteriorly, dorsal to the liver (see Figures 3.33 and 3.34). The **kidneys** extend longitudinally along the dorsal wall of the cavity on either side of the midsagittal plane as two narrow strips. These organs, as well as various ducts associated with the urogenital system, will be considered later.

Next, consider the mesenteries. Pull the digestive tract ventrally and to the animal's left. Note the dorsal mesentery, the thin, translucent sheet that suspends the gut middorsally (Figure 3.26). It is subdivided into several parts. The **mesogaster** (or greater omentum) extends to the esophagus and stomach. The mesentery (in the limited sense of this word; as noted earlier, it can also refer to any membrane that supports viscera) supports the anterior part of the intestine. Gently probe the mesentery and note that it arises from the mesogaster. Pull the digitiform gland ventrally to observe the **mesorectum**. Finally, gently pull the spleen posteriorly and note that it is attached to the stomach via the **gastrosplenic ligament**.

In the embryo a complete ventral mesentery is also present, but only portions of this membrane remain in the adult. Spread the median and right lobes of the liver. A ribbon-like strand, the **gastrohepatoduodenal ligament** (or lesser omentum) will be seen extending from the liver toward the gut. Near the liver it is a single bundle, carrying the **bile duct** and supporting blood vessels between the liver and gut. As it approaches the pyloric region, it divides into a **hepatoduodenal ligament** that carries the bile duct to the duodenum, and a **hepatogastric ligament** that passes into the adjacent portions of the body and pyloric region of the stomach.

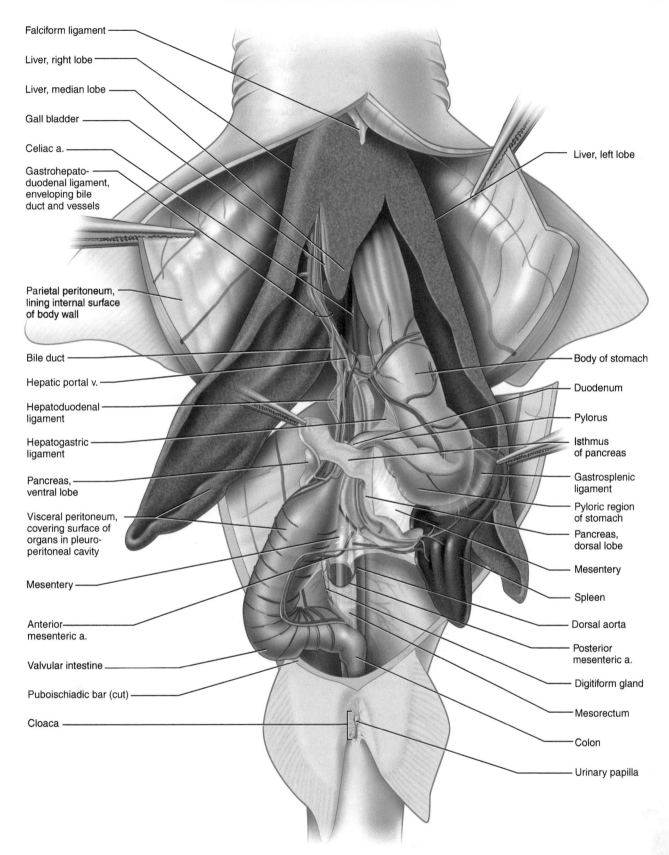

Falciform ligament

Liver, right lobe

Liver, median lobe

Gall bladder

Celiac a.

Gastrohepato-
duodenal ligament,
enveloping bile
duct and vessels

Parietal peritoneum,
lining internal surface
of body wall

Bile duct

Hepatic portal v.

Hepatoduodenal
ligament

Hepatogastric
ligament

Pancreas,
ventral lobe

Visceral peritoneum,
covering surface of
organs in pleuro-
peritoneal cavity

Mesentery

Anterior
mesenteric a.

Valvular intestine

Puboischiadic bar (cut)

Cloaca

Liver, left lobe

Body of stomach

Duodenum

Pylorus

Isthmus
of pancreas

Gastrosplenic
ligament

Pyloric region
of stomach

Pancreas,
dorsal lobe

Mesentery

Spleen

Dorsal aorta

Posterior
mesenteric a.

Digitiform gland

Mesorectum

Colon

Urinary papilla

FIGURE 3.24 Pleuroperitoneal cavity of the shark in ventral view, showing viscera and vessels.

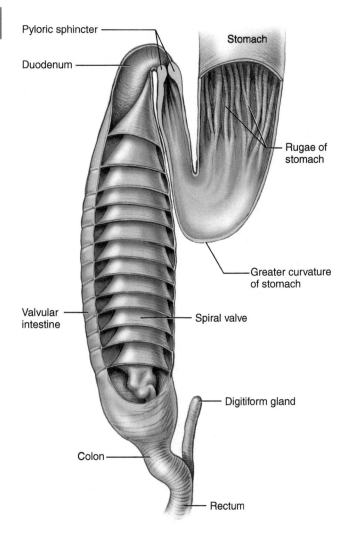

Pyloric sphincter

Duodenum

Stomach

Rugae of stomach

Greater curvature of stomach

Valvular intestine

Spiral valve

Digitiform gland

Colon

Rectum

FIGURE 3.25 Cut away view of the valvular intestine and posterior end of the stomach of the shark in ventral view, revealing the structure of the spiral valve.

Look between these ligaments and note that the pancreas is supported by a part of the dorsal mesentery. The **falciform ligament** extends between the anteroventral surface of the liver and the midventral body wall. This short mesentery also supports the common opening of the oviducts, described later. Other mesenteries supporting the reproductive tract are discussed with the urogenital system.

Key Terms: Digestive and Respiratory Systems

afferent branchial artery
bile duct
body (of **stomach)**
branchial adductor muscle
branchial chamber
coelom

colon
cross trunks between **pretrematic** and **posttrematic** **arteries**
digitiform gland (rectal gland)
duodenum
esophagus
falciform ligament
gall bladder
gastrohepatoduodenal ligament (lesser omentum)
gastrosplenic ligament (lienogastric ligament)
gill rakers
gill rays
gills
gonads
hemibranch
hepatoduodenal ligament
hepatogastric ligament
holobranch
interbranchial muscle
interbranchial septa
internal pharyngeal slits
isthmus of **pancreas**
kidneys
liver, left, median, and **right lobes**
mesenteries—general sense
mesentery—specific sense
mesogaster (greater omentum)
mesorectum
mouth
oral cavity
ovaries
pancreas, dorsal and **ventral lobes**
papillae (sing., **papilla)**
parabranchial chamber
parietal peritoneum
pericardial cavity
pharynx
pleuroperitoneal cavity
posttrematic artery
pretrematic artery
primary gill lamellae
primary tongue
pyloric region of **stomach**
pylorus
rectum
rugae (sing., **ruga)**
secondary gill lamellae
spiracle
spiral valve
spleen
stomach
testes (sing., **testis)**
transverse septum
valvular intestine
visceral peritoneum

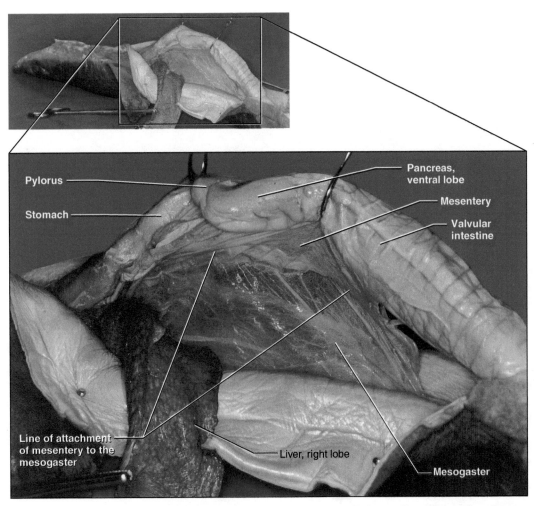

Pylorus

Stomach

Pancreas, ventral lobe

Mesentery

Valvular intestine

Line of attachment of mesentery to the mesogaster

Liver, right lobe

Mesogaster

FIGURE 3.26   Pleuroperitoneal cavity of the shark in right lateral view, with intestine pulled ventrally and liver reflected to expose mesenteries.

SECTION V: CARDIOVASCULAR SYSTEM

Heart and Arterial Circulation

Heart

The **pericardial cavity** contains the **heart**, the muscular pump that drives the blood around the cardiovascular system. To expose the pericardial cavity, continue the incision into the pleuroperitoneal cavity forward through the coracoid bar and the hypobranchial musculature. Spread the flaps to reveal the heart (Figure 3.27). Note the **parietal pericardium**, the shiny epithelium lining the cavity, and the **visceral pericardium**, covering the heart. The **transverse septum** separating the pleuroperitoneal and pericardial cavities is incomplete; the two cavities communicate through a small opening, the pericardioperitoneal canal, that lies ventral to the esophagus.

The **ventricle** is the most conspicuous structure of the heart. It is a large muscular chamber; coronary arteries, supplying the heart, will be seen on its surface. Lift the ventricle's posterior end to expose the **sinus venosus**, the

thin, triangular, posterior, sac-like chamber attached to the transverse septum. It receives venous blood from the body and passes it anteriorly into the **atrium**. The atrium is large and appears to envelop the ventricle, often giving the impression that there are two. This single atrium passes its blood to the ventricle. Anteriorly the ventricle narrows into the fourth chamber of the heart, the muscular, tube-like **conus arteriosus**, through which blood leaves the heart (but see Durán et al., 2008) and enters the **ventral aorta**. Note that the ventral aorta and its branches are not typically injected with colored latex.

Arteries of the Branchial Region

Once blood leaves the heart, it passes through the ventral aorta and the **afferent branchial arteries** (noted earlier; see page 76 and Figure 3.20) on its way to being oxygenated in the gills. To trace this arterial route, continue dissecting forward from the conus arteriosus. The ventral aorta gives off five pairs of afferent branchial arteries. The most posterior are the fourth and fifth afferent branchial arteries, which may arise separately or by a

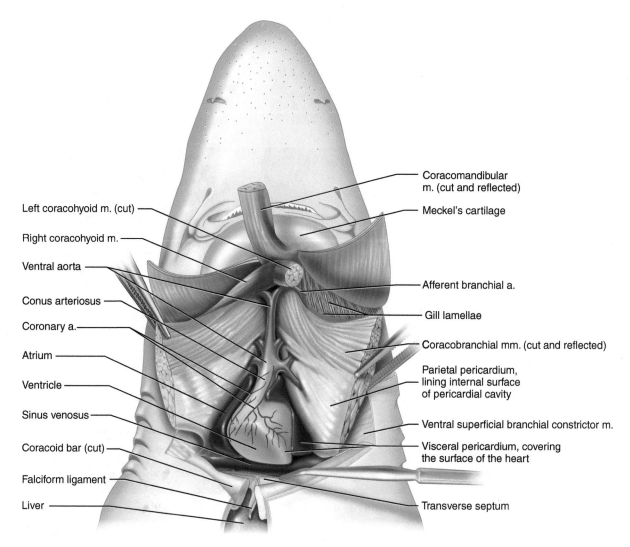

Coracomandibular
m. (cut and reflected)

Left coracohyoid m. (cut)

Right coracohyoid m.

Ventral aorta

Conus arteriosus

Coronary a.

Atrium

Ventricle

Sinus venosus

Coracoid bar (cut)

Falciform ligament

Liver

Meckel's cartilage

Afferent branchial a.

Gill lamellae

Coracobranchial mm. (cut and reflected)

Parietal pericardium,
lining internal surface
of pericardial cavity

Ventral superficial branchial constrictor m.

Visceral pericardium, covering
the surface of the heart

Transverse septum

FIGURE 3.27   Head and branchial region of the shark in ventral view, with the hypobranchial musculature dissected and reflected to expose the heart and ventral aorta.

short common trunk. The third afferent branchials arise farther anteriorly. Trace the ventral aorta anteriorly to its bifurcation. Each of the divisions divides again into the first and second afferent branchial arteries. Follow the arteries as far as possible without causing damage into the interbranchial septa. The arteries pass between bundles of the coracobranchial muscles (Figures 3.19 and 3.27).

Blood passing through the afferent branchial arteries enters the gill lamellae, where gas exchange occurs. The oxygenated blood is then collected again so it may be sent to the rest of the body. To view the system that recollects the blood, delicately remove the membrane lining the roof of the oral cavity and pharynx and the area around the uncut internal branchial slits (Figures 3.28–3.30). Oxygenated blood is initially collected from the lamellae into the **pretrematic** and **posttrematic** arteries, noted on pages 75 and 76. These arteries join dorsally and ventrally to form a complete **collector loop** around

a pharyngeal slit (Figures 3.23 and 3.30). Note that there are only four collector loops, as the last slit has only a pretrematic artery. From the dorsal end of each loop, an **efferent branchial artery** carries blood away from a gill (Figure 3.28). The efferent branchial arteries were likely observed earlier during the dissection to expose the anterior cardinal sinus and interarcual muscles (see page 71 and Figure 3.21). The four pairs of arteries pass posteromedially and empty into the **dorsal aorta** on the roof of the pharynx. The dorsal aorta passes posteriorly to supply most of the body with oxygenated blood. Its branches are considered later.

Examine first the vessels that supply the head. A **hyoidean artery** arises from the anterodorsal part of the first collector loop, just anterior to the first efferent branchial artery, and extends anteriorly (Figures 3.28 and 3.30). **Paired dorsal aortae** arise from the first efferent branchial arteries. They pass anteriorly and then veer

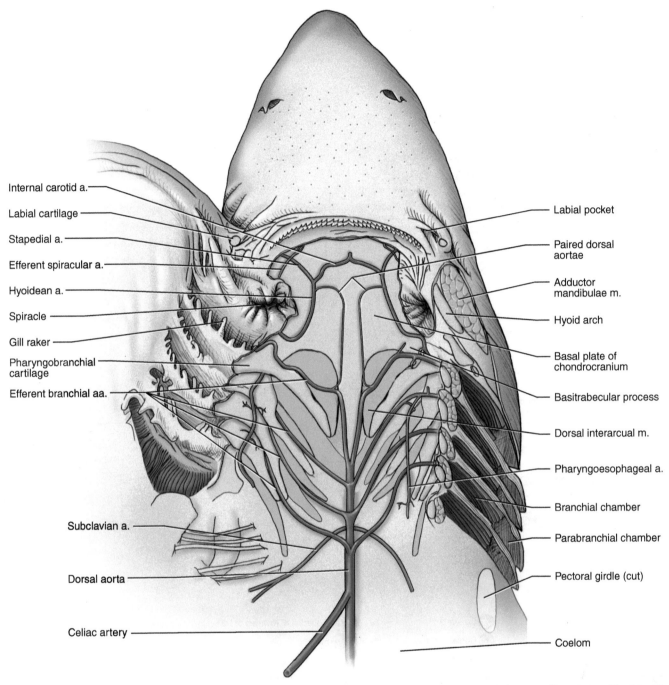

Internal carotid a.

Labial cartilage

Stapedial a.

Efferent spiracular a.

Hyoidean a.

Spiracle

Gill raker

Pharyngobranchial cartilage

Efferent branchial aa.

Subclavian a.

Dorsal aorta

Celiac artery

Labial pocket

Paired dorsal aortae

Adductor mandibulae m.

Hyoid arch

Basal plate of chondrocranium

Basitrabecular process

Dorsal interarcual m.

Pharyngoesophageal a.

Branchial chamber

Parabranchial chamber

Pectoral girdle (cut)

Coelom

FIGURE 3.28 Roof of the oral cavity and pharynx of the shark in ventral view, showing the pattern of the arterial circulation. The left side visceral arches have been cut. The floor of the oral cavity and of the pharynx have been swung open.

laterally and join a hyoidean artery. The union of the hyoidean and a paired dorsal aorta forms the **internal carotid artery**. The internal carotid extends anteriorly a short distance, and near the level of the spiracle gives rise to the **stapedial artery**, which extends anterolaterally a short distance before passing into the chondrocranium. Past the origin of the stapedial artery, the internal carotids from either side extend anteromedially and meet middorsally to form a single vessel that enters the chondrocranium.

The **afferent spiracular artery** arises from near the middle of the pretrematic artery of the first loop. Do not dissect for it here. It was noted earlier (page 70; Figure 3.17) as it passed deep to the hyomandibular nerve on the levator hyomandibulae muscle. Return to the dissection of the muscles, and trace the afferent spiracular artery anteriorly into the pseudobranch of the spiracle and posteriorly into the collector loop. The **efferent spiracular artery** collects blood from the pseudobranch. Find it on the roof of the oral cavity. It extends

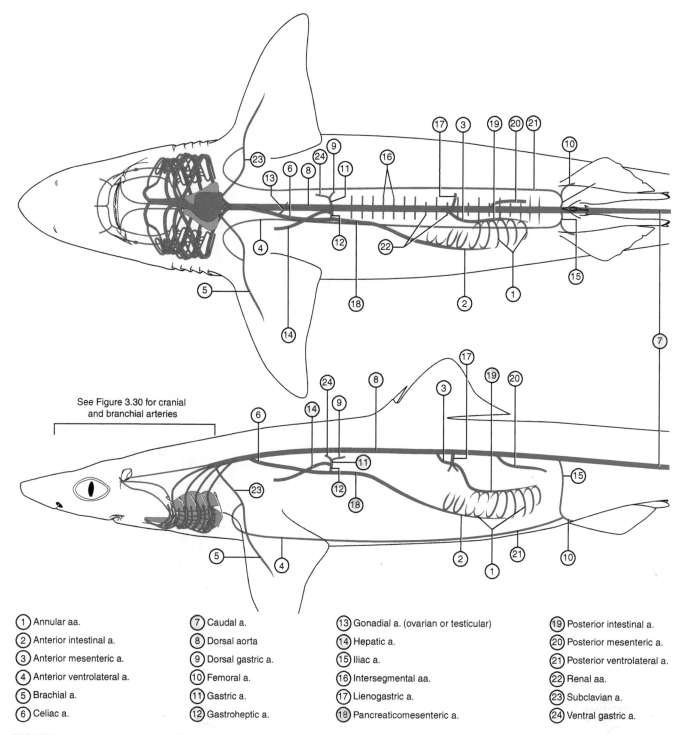

FIGURE 3.29 Schematic illustration showing the pattern of the arterial system of the shark superimposed on ventral (top figure) and left lateral (bottom figure) views of the body outline.

① Annular aa.

② Anterior intestinal a.

③ Anterior mesenteric a.

④ Anterior ventrolateral a.

⑤ Brachial a.

⑥ Celiac a.

⑦ Caudal a.

⑧ Dorsal aorta

⑨ Dorsal gastric a.

⑩ Femoral a.

⑪ Gastric a.

⑫ Gastroheptic a.

⑬ Gonadial a. (ovarian or testicular)

⑭ Hepatic a.

⑮ Iliac a.

⑯ Intersegmental aa.

⑰ Lienogastric a.

⑱ Pancreaticomesenteric a.

⑲ Posterior intestinal a.

⑳ Posterior mesenteric a.

㉑ Posterior ventrolateral a.

㉒ Renal aa.

㉓ Subclavian a.

㉔ Ventral gastric a.

anteromedially, passing ventral to the stapedial artery, and enters the chondrocranium. The lower jaw is supplied by the **external carotid artery**, which arises from the anteromedial corner of the first collector loop (Figure 3.30a). The external carotid may easily be viewed as it emerges externally onto the lower jaw by skinning the lower jaw near the angle of the mouth, just medial to the adductor mandibulae muscle (Figure 3.30b).

The **hypobranchial artery** (Figure 3.30) usually arises from the ventral end of the second collector loop, but branches from the other loops may contribute to it. The hypobranchial passes posteriorly to the conus arteriosus, where it divides into the **coronary** and the **pericardial arteries**. The former is clearly observed on the conus arteriosus and ventricle (Figures 3.27 and 3.30), the latter on the wall of the pericardial cavity. The narrow, sinuous

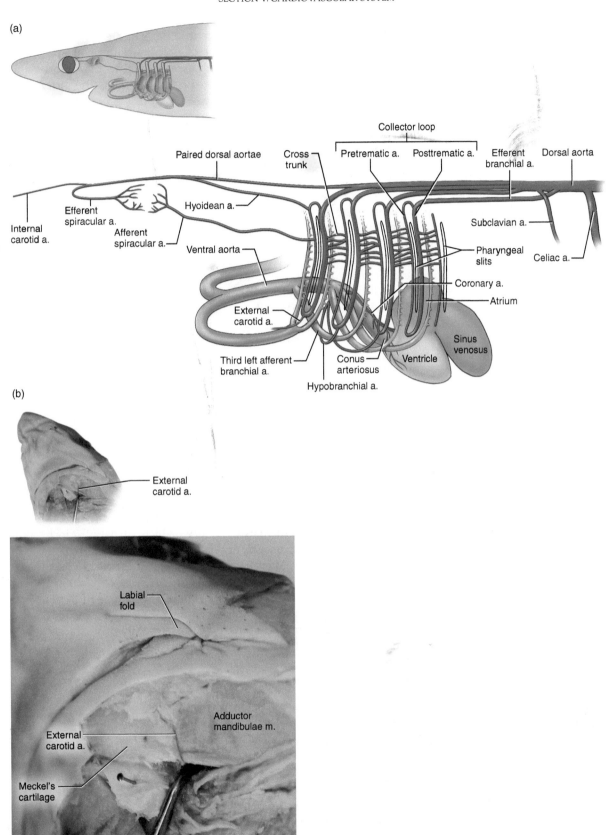

FIGURE 3.30 Arterial system in the head and branchial region of the shark: (a) schematic illustration showing the pattern in left lateral view; (b) external exposure of the external carotid artery onto the surface of the lower jaw in ventral view.

pharyngoesophageal artery arises from the second efferent branchial artery and extends posteriorly to give off branches to the pharynx and esophagus (Figure 3.28). The rest of the blood flow passing back to the body goes through the dorsal aorta, which gives off several large branches.

Branches of the Dorsal Aorta

The paired **subclavian arteries** are the first major branches of the dorsal aorta (Figures 3.28–3.30). These arise usually between the third and fourth efferent branchial arteries. Initially they pass posterolaterally. At the pectoral girdle each subclavian veers latero-ventrally, passing deep to the posterior cardinal sinus (described later), along the scapular process. Follow the artery on the side for which the sinus was dissected. It gives rise to two main branches, the **brachial** and **anterior ventrolateral arteries**, but it is difficult to neatly dissect the origins of these vessels. The brachial supplies the pectoral fin (do not confuse its spelling with *branchial*: brachial = arm, branchial = gill). To find it, pull the pectoral fin away from the body and cut through the skin between the body and posteromedial surface of the fin. This will free the fin from the body. Picking away the connective tissue from the medial surface of the fin will soon reveal the brachial artery as it passes along the medial cartilaginous fin support. Trace the artery back toward the subclavian to note the position of its origin. After the origin of the brachial, the anterior ventrolateral artery continues, on the inside of the body wall, passing first slightly anteriorly and then curving markedly posteriorly, about midway between the lateral and midventral lines. Follow it as it continues back, giving off branches that supply the myomeres, and eventually anastomoses with the posterior ventrolateral artery (a branch of the iliac artery, noted later on this page).

After giving off the subclavian arteries, the dorsal aorta continues posteriorly into the pleuroperitoneal cavity. Return to this cavity to examine the following vessels (Figures 3.29 and 3.31). The first branch of the dorsal aorta in the cavity is the **celiac artery**, a large, unpaired vessel that continues posteriorly along the right side of the stomach. Near its origin, it gives rise to a pair of **testicular** (in males) or **ovarian** (in females) **arteries** (Figure 3.29) to supply the gonads. The celiac artery continues to the anterior tip of the dorsal lobe of the pancreas, where it divides into the **pancreaticomesenteric** and **gastrohepatic arteries**. The latter is a very short branch (sometimes altogether missing) that subdivides almost immediately into a long, narrow **hepatic artery** and a short, wider **gastric artery**. The hepatic artery turns anteriorly toward the liver, accompanied by the hepatic portal vein (considered later) and the anterior part of the bile duct. The gastric artery passes to the stomach, where

it branches into the **dorsal** and **ventral gastric arteries** to the dorsal and ventral parts of the stomach, respectively. Return to the pancreaticomesenteric artery and trace it posteriorly as it passes dorsal to the pylorus and onto the ventral side of the intestine as the **anterior intestinal artery**. Before doing so it gives off several smaller branches near the pylorus (which you do not need to name). On the intestine the anterior intestinal artery gives rise to **annular arteries** that encircle the intestine.

Trace the dorsal aorta farther posteriorly. At about the level of the spleen, two arteries arise close together. The anterior one is the **anterior mesenteric artery.** This vessel extends onto the intestine as the **posterior intestinal artery**, which, like the anterior intestinal artery, gives rise to annular arteries. The posterior artery is the **lienogastric artery**, which mainly supplies the spleen and the posterior part of the stomach. Note that the anterior mesenteric and lienogastric arteries lie in the posterior edge of the greater omentum. Sometimes these arteries come off in reverse order (i.e., the lienogastric is the more anterior); trace them to identify the arteries in your specimen. After a short distance the dorsal aorta gives off a **posterior mesenteric artery**, which passes along the anterior edge of the mesorectum and onto the digitiform gland.

The final branches of the dorsal aorta are the paired **iliac arteries** and the **caudal artery** (Figure 3.29). At about the level of the cloaca, the iliac arteries arise and pass posterolaterally dorsal to the kidneys. They emerge from under the kidneys and extend toward the pelvic fins. Before entering the fin, each iliac branches into the **femoral artery**, which enters the fin, and the **posterior ventrolateral artery**, which turns anteriorly along the body wall and eventually unites with the anterior ventrolateral artery, noted earlier. The caudal artery is the posterior continuation of the dorsal aorta into the tail. Finally, the dorsal aorta gives rise to numerous **intersegmental arteries** (Figures 3.29, 3.33, and 3.34) to the axial musculature. Free the lateral margin of a kidney and lift it to observe these arteries.

Venous Circulation

Hepatic Portal System

The hepatic portal system is the venous system that returns blood from the digestive tract and spleen to the liver (where raw nutrients in blood are processed before the blood returns to the heart). Essentially, it drains the structures supplied ultimately by the celiac (except for the gonads), anterior mesenteric, lienogastric, and posterior mesenteric arteries. Thus, the branches of the hepatic portal system closely follow many of the branches of these arteries, and it is convenient to study them following your identification of the arteries. In some specimens the hepatic portal system is injected with yellow

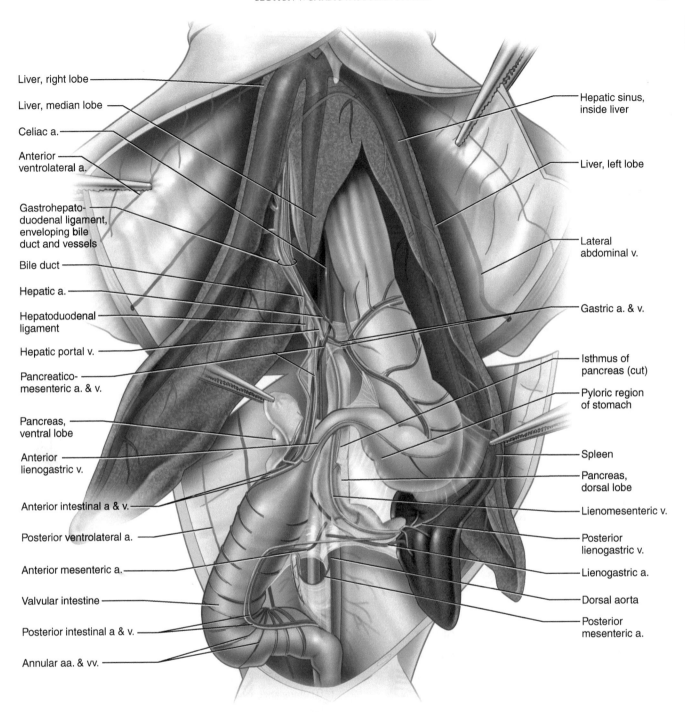

Liver, right lobe

Liver, median lobe

Celiac a.

Anterior ventrolateral a.

Gastrohepato-duodenal ligament, enveloping bile duct and vessels

Bile duct

Hepatic a.

Hepatoduodenal ligament

Hepatic portal v.

Pancreatico-mesenteric a. & v.

Pancreas, ventral lobe

Anterior lienogastric v.

Anterior intestinal a & v.

Posterior ventrolateral a.

Anterior mesenteric a.

Valvular intestine

Posterior intestinal a & v.

Annular aa. & vv.

Hepatic sinus, inside liver

Liver, left lobe

Lateral abdominal v.

Gastric a. & v.

Isthmus of pancreas (cut)

Pyloric region of stomach

Spleen

Pancreas, dorsal lobe

Lienomesenteric v.

Posterior lienogastric v.

Lienogastric a.

Dorsal aorta

Posterior mesenteric a.

FIGURE 3.31   Pleuroperitoneal cavity of the shark in ventral view, showing the pattern of the arteries and veins.

latex, which greatly facilitates its study. You may follow its branches in an uninjected specimen, but also briefly examine a shark that has the system injected.

The main vessel of the hepatic portal system is the **hepatic portal vein** (Figures 3.31 and 3.32), a large vein that lies in the gastrohepatoduodenal ligament alongside the hepatic artery and anterior part of the bile duct. The hepatic portal vein is formed by the confluence of three main vessels, the **gastric**, **pancreaticomesenteric**, and **lienomesenteric veins**. They unite to form the hepatic portal vein near the anterior tip of the dorsal lobe of the pancreas. Recall that the celiac artery splits into its branches very near this point as well. Occasionally, the gastric and lienomesenteric veins join to form a very short vessel that then unites with the pancreaticomesenteric to form the hepatic portal vein.

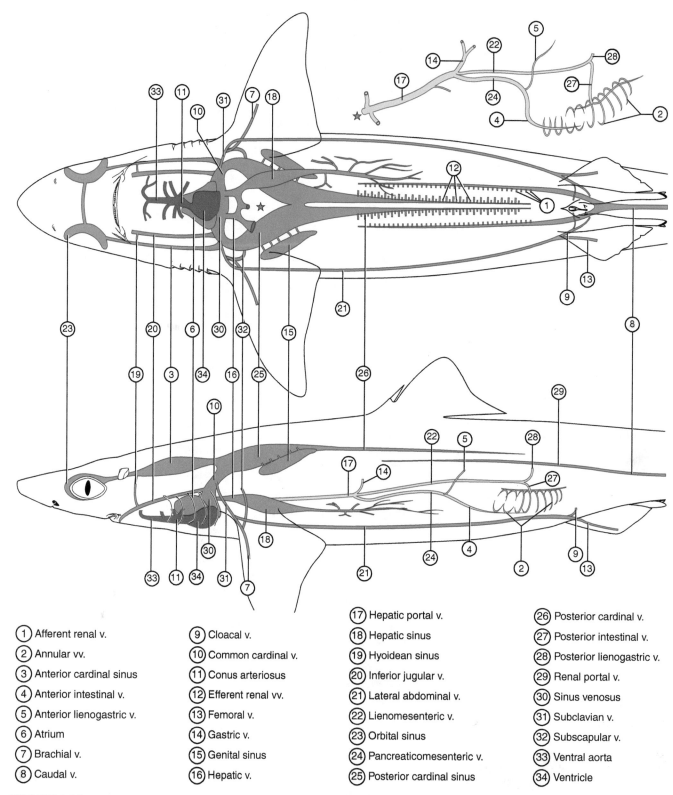

FIGURE 3.32 Schematic illustration showing the pattern of the venous system in the shark superimposed on ventral (top figure) and left lateral (bottom figure) views of the body outline. In the top figure, the hepatic portal system has been moved outside the body outline for clarity. The stars indicate the continuity between the hepatic portal system and the liver.

① Afferent renal v.
② Annular vv.
③ Anterior cardinal sinus
④ Anterior intestinal v.
⑤ Anterior lienogastric v.
⑥ Atrium
⑦ Brachial v.
⑧ Caudal v.

⑨ Cloacal v.
⑩ Common cardinal v.
⑪ Conus arteriosus
⑫ Efferent renal vv.
⑬ Femoral v.
⑭ Gastric v.
⑮ Genital sinus
⑯ Hepatic v.

⑰ Hepatic portal v.
⑱ Hepatic sinus
⑲ Hyoidean sinus
⑳ Inferior jugular v.
㉑ Lateral abdominal v.
㉒ Lienomesenteric v.
㉓ Orbital sinus
㉔ Pancreaticomesenteric v.
㉕ Posterior cardinal sinus

㉖ Posterior cardinal v.
㉗ Posterior intestinal v.
㉘ Posterior lienogastric v.
㉙ Renal portal v.
㉚ Sinus venosus
㉛ Subclavian v.
㉜ Subscapular v.
㉝ Ventral aorta
㉞ Ventricle

The gastric vein accompanies the gastric artery onto the dorsal and ventral surfaces of the stomach. Follow the lienomesenteric vein as it extends along the dorsal lobe of the pancreas. Rotate the spleen toward the left to observe that the vein is formed by the confluence, near the posterior end of the dorsal lobe, of the quite short **posterior lienogastric vein** and the **posterior intestinal vein** (Figure 3.32). The former comes from the spleen and posterior part of the stomach, the latter from the posterior part of the intestine (recall that this region is supplied by the anterior mesenteric artery, which becomes the posterior intestinal artery). The pancreaticomesenteric vein accompanies the pancreaticomesenteric artery to the beginning of the intestine. Here it is formed by several tributaries. Among these are the **anterior intestinal** and the **anterior lienogastric veins** (Figure 3.32). As you might expect, the former vein extends parallel to the anterior intestinal artery. It is larger than the anterior lienogastric vein, which comes from the spleen and adjacent regions of the pyloric region of the stomach. On the intestine **annular veins** accompany the annular arteries and enter the anterior intestinal vein and posterior intestinal vein.

Renal Portal System

Blood from the tail passes through the kidneys via the renal portal system before returning, through the posterior cardinal veins and sinuses, to the heart. The main vessels of the renal portal system are the **caudal vein** and the **renal portal veins**. The latter arise through bifurcation of the caudal vein (Figure 3.32). The caudal vein passes anteriorly through the hemal arches of the vertebrae. Make a partial transverse section—deep enough to cut through the vertebrae—of the tail just posterior to the level of the cloaca. Note the caudal artery, which will be injected, lying in the dorsal part of the hemal arch. The caudal vein, which should not be injected, lies ventral to the artery (see Figure 3.6). Make additional partial sections, spaced about 1 cm apart, anterior to the first. As these sections are anterior to the cloaca, they will be at the level of the posterior end of the body cavity. Thus, cut deeply enough to go through the caudal artery and vein, but avoid cutting the kidneys, which lie ventral to and on either side of the vertebrae. You will thus be able to observe, near the posterior end of the kidneys, the bifurcation of the caudal vein into left and right renal portal veins. The latter continue anteriorly along the dorsolateral margins of the kidneys. **Afferent renal veins**, which you will not be able to observe (but see Figure 3.32), branch from the renal portal veins and carry blood to the kidneys.

Systemic Veins

The systemic veins (Figure 3.32) are those that drain most of the body other than the viscera. Each side of the head and branchial region are drained mainly by the **anterior cardinal sinus** and **inferior jugular vein**. The anterior cardinal sinus, exposed earlier (see page 71; Figures 3.21 and 3.43), is a relatively large space (compared with the diameter of most veins) lying dorsal to the pharyngobranchial cartilages. It was exposed during the dissection of the musculature to find the dorsal and lateral interarcual muscles. The sinus receives vessels that drain the eye, brain, and head, and leads blood posteriorly toward the heart. The **orbital sinus**, for example, surrounds the eye, but it is impractical to attempt to find it. The inferior jugular vein is a thin vessel draining the floor of the branchial region. You will not see it, but will probe its course later. On each side of the head, the anterior cardinal sinus and inferior jugular vein are connected via the **hyoidean sinus**. Pass a probe into the hyoidean sinus to determine its course. It lies along the posterior surface of the hyoid arch, and so it was sectioned during the cut made to open the oral cavity and pharynx.

Return to the heart, and slit the ventral wall of the sinus venosus from side to side. Gently probe its walls and note that there are a number of openings that lead into it. Probing its posterior wall just to either side of the midsagittal plane will lead into the opening of a **hepatic vein** and then into a **hepatic sinus,** which collects blood from the liver (Figures 3.31 and 3.32). Each posterolateral corner of the sinus venosus leads to a short though large vessel, the **common cardinal vein**, which receives the rest of the main vessels leading blood back to the heart (Figure 3.32). Pass a blunt, curved probe along the posterior wall of the common cardinal vein. It should lead, without much effort, into the **posterior cardinal sinus**, which was noted earlier (page 84) and is described later. The anterior cardinal sinus enters the common cardinal vein at about the same level, but it is much more difficult to probe for by this method. Here, merely note that the tip of the probe veers anteriorly and dorsally.

The other vessels that enter the common cardinal vein are the inferior jugular vein and the subclavian vein (Figure 3.32). Gently probing the anterior wall of the common cardinal vein will lead into the inferior jugular vein. You may follow its course by palpating for the probe in the hypobranchial region. The **subclavian vein** enters the common cardinal vein just lateral to the entrance of the inferior jugular.

Return to the pleuroperitoneal cavity. Lift the gonad and anterior part of the liver so the roof of the cavity may be observed. Lying dorsolateral to the esophagus is the posterior cardinal sinus (noted earlier), a large space bounded by a thin-walled membrane. It is filled with blue latex. Note that the sinus curves toward the dorsal midline. Slit open the sinus and carefully remove the latex. Probe gently to verify that this sinus is continuous with the sinus from the other side of the body. Veins from the gonads (Figure 3.32) and esophagus empty into the sinus, but it is impractical to trace them.

Follow the sinus posteriorly, moving the viscera to one side as you do so (Figure 3.32). The sinus narrows into a **posterior cardinal vein**, which lies lateral to the dorsal aorta. The posterior cardinal veins appear as thin, translucent vessels because they are probably not injected, and may be difficult to discern. Do not confuse them with the oviducts of a female or the archinephric ducts of immature males; these are thin, flattened and straight structures that are easily identified because they continue anteriorly to lie on the posterior cardinal sinus. Note that the veins lie along the medial margins of the kidneys. They receive **efferent renal veins**, which collect blood from the kidneys and the segmentally arranged **intersegmental veins** (Figures 3.33 and 3.34) that drain blood from the body wall. These veins, however, may not be easily observable.

A **lateral abdominal vein** extends along the inside of the ventrolateral body wall on each side (Figures 3.31 and 3.32). Trace one of these veins posteriorly. It is formed, at about the level of the cloaca, by the confluence of the **cloacal vein** (from the cloaca) and the **femoral vein** (from the pelvic fin). Trace the lateral abdominal forward to the pectoral girdle. Here, you will observe a number of veins coming together. You will probably see a conspicuous **subscapular vein** (Figure 3.32), which essentially extends parallel to the subclavian artery (but does not share its name). Find the **brachial vein** on the medial surface of the fin, in company with the brachial artery. It will probably be uninjected. Trace the brachial vein toward the body. It unites with the lateral abdominal to form the subclavian vein, already noted as entering the common cardinal vein. The subscapular vein actually joins the brachial vein, but the brachial, subscapular, and lateral abdominal veins all come together very near each other in forming the subclavian vein, the vessel that continues the blood's journey toward the heart. Follow the subclavian vein as it arches dorsomedially.

Key Terms: Cardiovascular System

afferent branchial arteries
afferent renal veins
afferent spiracular artery
annular arteries
annular veins
anterior cardinal sinus
anterior intestinal artery
anterior intestinal vein
anterior lienogastric vein
anterior mesenteric artery
anterior ventrolateral artery (anterior epigastric artery)
atrium

brachial artery
brachial vein
caudal artery
caudal vein
celiac artery
cloacal vein
collector loop of gill
common cardinal vein (duct of Cuvier)
conus arteriosus
coronary artery
dorsal aorta
dorsal gastric artery
efferent branchial arteries
efferent renal veins
efferent spiracular artery
external carotid artery
femoral artery
femoral vein
gastric artery
gastric vein
gastrohepatic artery
heart
hepatic artery
hepatic portal vein
hepatic sinus
hepatic vein
hyoidean artery
hyoidean sinus
hypobranchial artery (commissural artery)
iliac arteries
inferior jugular vein
internal carotid artery
intersegmental arteries
intersegmental veins
lateral abdominal vein
lienogastric artery
lienomesenteric vein
orbital sinus
ovarian artery
paired dorsal aortae
pancreaticomesenteric artery
pancreaticomesenteric vein
parietal pericardium
pericardial artery
pericardial cavity
pharyngoesophageal artery
posterior cardinal sinus
posterior cardinal vein
posterior intestinal artery
posterior intestinal vein
posterior lienogastric vein
posterior mesenteric artery
posterior ventrolateral artery (posterior epigastric artery)
renal portal veins

sinus venosus
stapedial artery
subclavian artery
subclavian vein
subscapular vein
testicular artery
transverse septum
ventral aorta
ventral gastric artery
ventricle
visceral pericardium

SECTION VI: UROGENITAL SYSTEM

As various ducts of the excretory and reproductive systems become associated during embryonic development, it is convenient to discuss these systems together. The kidneys and gonads, noted earlier (page 76), are the main organs of the excretory and reproductive systems, respectively. Manipulate the digestive tract and liver to observe them. The paired **kidneys** are long, narrow structures on either side of the dorsal aorta on the roof of the body cavity (Figures 3.33 and 3.34). They extend nearly the length of the cavity. As they lie dorsal to the parietal peritoneum lining the pleuroperitoneal cavity, the kidneys are retroperitoneal. The paired and elongated gonads, **testes** (sing., **testis**) in the male and **ovaries** (sing., **ovary**) in the female, lie dorsal to the liver in the pleuroperitoneal cavity. Note the **caudal ligament** between the posterior ends of the kidneys. It arises from the vertebrae and passes to the tail. The posterior mesenteric artery pierces the ligament on its way from the dorsal aorta to the **digitiform gland**.

Male Urogenital System

The kidney of the male (Figure 3.33) has two distinct functional parts. The anterior part is related to sperm transport. It includes tubules that help transport sperm in the most anterior part of the kidney, often referred to as the **epididymis**, and tubules that form secretions that help transport sperm. These tubules lie in approximately the middle third of the kidney, a region known as **Leydig's gland**. The posterior thickened end of the kidney functions in excretion.

Each testis is supported by the **mesorchium**, a mesentery that suspends the organ from the middorsal wall of the body cavity. Within the anterior part of the mesorchium are small tubules (**ductuli efferentes**, which may be observed as strand-like structures) that extend from the testis to tubules in the epididymis of the kidney, which in turn are connected to the **archinephric duct**. Sperm take this route from the testes to reach the archinephric duct (originally a duct of the kidney) and then

be transported posteriorly for eventual release from the body (see later). In a mature male the archinephric duct is tightly convoluted and embedded in the ventral surface of the kidney. In an immature male it is nearly straight and resembles the oviduct of the female (see later).

As the archinephric duct approaches the cloaca, it straightens and expands into the **seminal vesicle**. Trace it posteriorly into the **sperm sac**, a small, anteriorly blindly ending pouch that lies on the ventral surface of the seminal vesicle. Free the anterior end of the sperm sac from the seminal vesicle and slit the sperm sac ventrally. Probe its roof and note the entrance of the seminal vesicle. Probe the posterior part of the sperm sac. Left and right sperm sacs unite to form the **urogenital sinus**, a median space that continues posteriorly through the **urogenital papilla** (noted earlier; see page 64 and also Figure 3.14), which is used for the passage of urine and sperm.

As noted earlier, urine is produced almost entirely (and probably exclusively in the adult male) by the posterior part of the kidney. However, it does not enter the archinephric duct or seminal vesicle. Instead, a thin, delicate **accessory urinary duct** carries urine to the urogenital sinus. The entrance of the duct is posterior to the entrance of the seminal vesicle into the sperm sac, but is difficult to locate. The accessory urinary duct extends along the medial margin of the kidney. Expose it by carefully lifting up the medial border of the kidney and delicately dissecting along its dorsomedial surface.

The remaining reproductive structures of the male are associated with the pelvic fin. The **clasper** is an intromittent organ inserted into the cloaca and oviduct of the female during copulation. A groove lies along the medial side of the clasper. Sperm is released from the urogenital papilla and travels along the groove to the cloaca of the female. Associated with each clasper is the elongated, sac-like **siphon**, which lies just deep to the skin on the ventral surface of the pelvic fin. Expose the siphon by removing the skin, and then cut open the siphon. Using a thin probe, search its posteromedial region for a passageway. Follow it as it passes along the groove of the clasper. The siphon expels fluid through this route that contributes to seminal fluid.

Key Terms: Male Urogenital System

accessory urinary duct
archinephric duct
caudal ligament
clasper
digitiform gland (rectal gland)
ductuli efferentes
epididymis
kidneys
Leydig's gland
mesorchium

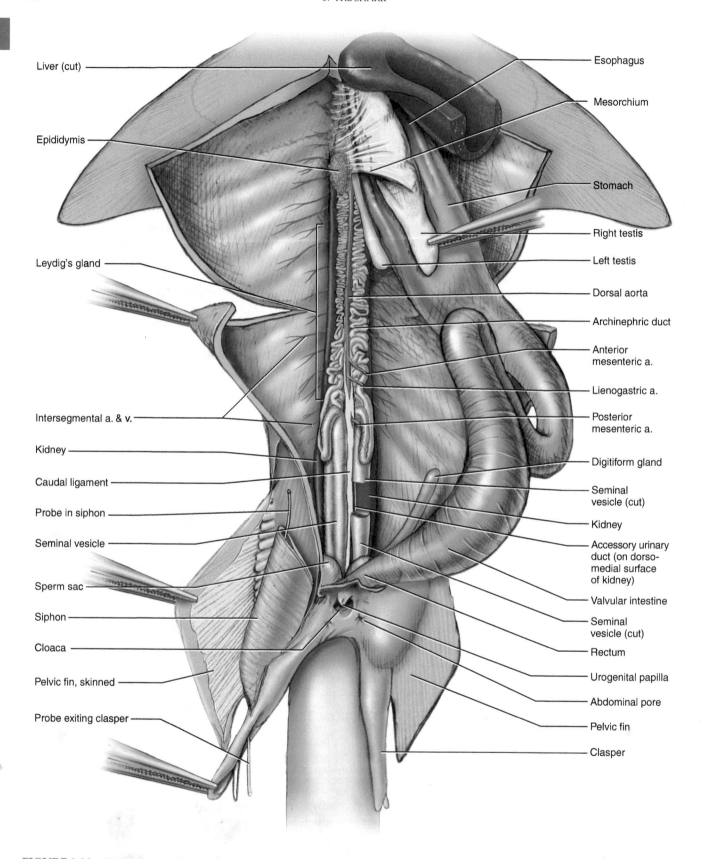

Liver (cut)

Epididymis

Leydig's gland

Intersegmental a. & v.

Kidney

Caudal ligament

Probe in siphon

Seminal vesicle

Sperm sac

Siphon

Cloaca

Pelvic fin, skinned

Probe exiting clasper

Esophagus

Mesorchium

Stomach

Right testis

Left testis

Dorsal aorta

Archinephric duct

Anterior
mesenteric a.

Lienogastric a.

Posterior
mesenteric a.

Digitiform gland

Seminal
vesicle (cut)

Kidney

Accessory urinary
duct (on dorso-
medial surface
of kidney)

Valvular intestine

Seminal
vesicle (cut)

Rectum

Urogenital papilla

Abdominal pore

Pelvic fin

Clasper

FIGURE 3.33 Pleuroperitoneal cavity in ventral view, showing the urogenital system of the male shark. Parts of the viscera have been removed. The right testis is reflected to the left. The anterior end of the right siphon has been cut away. The probe inserted into the siphon shows the path leading out of the siphon on the clasper.

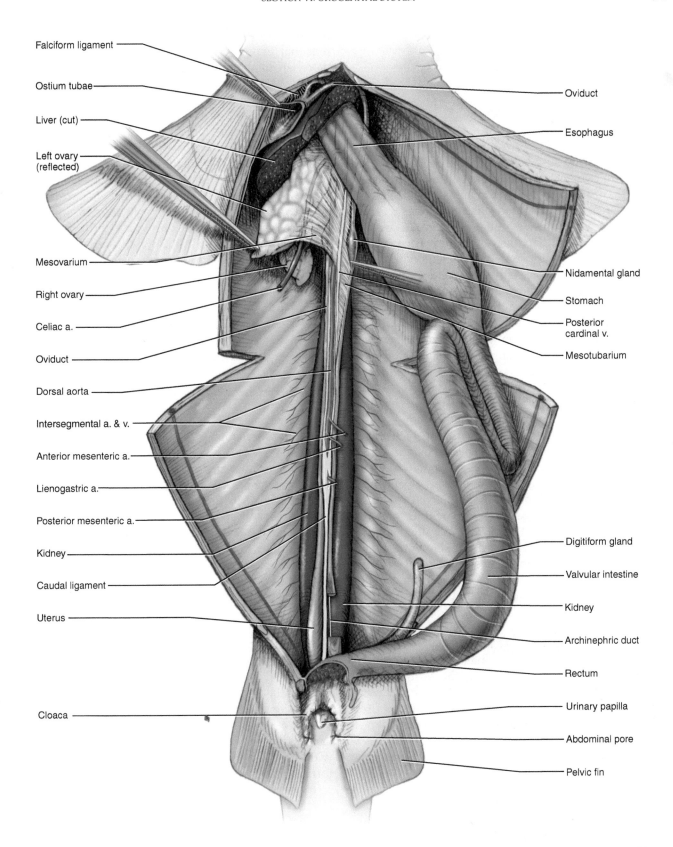

Falciform ligament

Ostium tubae

Liver (cut)

Left ovary (reflected)

Mesovarium

Right ovary

Celiac a.

Oviduct

Dorsal aorta

Intersegmental a. & v.

Anterior mesenteric a.

Lienogastric a.

Posterior mesenteric a.

Kidney

Caudal ligament

Uterus

Cloaca

Oviduct

Esophagus

Nidamental gland

Stomach

Posterior cardinal v.

Mesotubarium

Digitiform gland

Valvular intestine

Kidney

Archinephric duct

Rectum

Urinary papilla

Abdominal pore

Pelvic fin

FIGURE 3.34 Pleuroperitoneal cavity in ventral view, showing the urogenital system of the female shark. Parts of the viscera have been removed. The left ovary is reflected to the right.

seminal vesicle
siphon
sperm sac
testis (pl., **testes**)
urogenital papilla
urogenital sinus

Female Reproductive System

The **ovaries** are suspended from the roof of the anterior pleuroperitoneal cavity by the **mesovarium** (Figure 3.34). They are smooth and oval in immature females, but in mature females their surfaces have swellings that represent ova in various stages of development. An **oviduct** lies on the ventral surface of each **kidney**. Each oviduct is a narrow tube unsupported by a mesentery in immature specimens, but more conspicuous and supported by a **mesotubarium** in mature individuals. Trace an oviduct anteriorly. It passes dorsal to the ovary and then curves ventrally around the anterior surface of the liver. The left and right oviducts pass within the falcifom ligament and join to form a common opening, the **ostium tubae**, in the free, posterior edge of the ligament. It is conspicuous in mature females and can easily be probed, but may be difficult to find in immature females. Note that there is no direct connection between the ovaries and the ostium: Eggs are shed into the pleuroperitoneal cavity and must make their way to the ostium, where they can enter an oviduct.

Trace the oviducts posteriorly and note that each eventually enters the cloaca through an opening on either side of the **urinary papilla** (noted earlier; see page 64 and also Figure 3.14). An oviduct has two enlarged regions. The first, appearing as a slight anterior swelling dorsal to the ovary, is the **nidamental gland**, which secretes a thin membrane around groups of eggs as they pass through the oviduct. It is also the location where eggs are fertilized. The second enlargement, the **uterus**, occurs posteriorly. The uterus is greatly enlarged (Figure 3.35a,b) in pregnant specimens because it contains the developing pups (see later).

The anterior part of the female kidney is greatly reduced, and unlike in the male, the kidney is not involved in the reproductive system. However, an archinephric duct, which is formed from the kidney, is present. In the female it is uncoiled, much smaller than in the male, and transports urine. Accessory urinary ducts are not present in female *Squalus*, but may be present in females of other sharks.

Before considering further the anatomy of a gravid female and a pup, it is worth noting some general features of the reproductive mode in bony fishes as compared to chondrichthyans, and the different modes that occur among sharks and their relatives. The general reproductive mode in bony fishes is oviparity, in which numerous eggs are laid and fertilized outside the body.

The eggs in such oviparity are provided with very little yolk and hatch in an undeveloped or larval condition that requires considerable further time (weeks or months) for complete development (Carrier, Pratt, & Castro, 2004).

Many chondrichthyans (including all skates, holocephalimorphans, and some sharks; Mabragaña et al., 2011) are oviparous, but a second strategy, viviparity, occurs in others. In viviparity eggs develop within the female and the young are born as miniature adults (further details are provided later). Whether oviparous or viviparous, however, all elasmobranchs depart from the simple oviparous mode of bony fishes: Elasmobranchs produce relatively few eggs provided with considerable yolk, practice internal fertilization (made possible by the presence of claspers in males), and the young hatch or are born as miniature adults (Carrier et al., 2004). There is no maternal care in any chondrichthyan.

Oviparous elasmobranchs deposit a leathery egg case or capsule (the egg case of some species is often termed a mermaid's purse) around an egg. Such encapsulated eggs are then laid on the substrate, in many cases attached to structures such as seaweed. The embryos are entirely nourished by the yolk stored in a yolk sac. In some, slits may develop on either side of the egg case after a few weeks of development and allow for ventilation and oxygenation (see Carrier et al., 2004). An example of an egg case, of the smallnose fanskate *Sympterygia bonapartii*, is illustrated in Figure 3.36d. The long, threadlike horns help secure the egg case to the substrate or underwater structures. Among elasmobranchs, each egg case usually contains a single embryo, but exceptions occur—in some species up to four embryos may be enclosed within an egg case (Ebert & Winton, 2010).

This reproductive strategy is also termed *lecithotrophic* (from the Ancient Greek lekithos, egg yolk, and trophe, food), referring to the fact that nutrients are obtained entirely from yolk reserves. Variation occurs among lecithotrophic vertebrates—in some, the eggs are simply extruded onto the seafloor, whereas in others the eggs are retained for a variable period of time (thus at least part of embryonic development occurs within the female's body), and in still others development occurs and is completed within the mother's body, resulting in a live birth. The term viviparity is applied to this last mode—that is, in which development is completed within the mother's body. All of these developmental forms in elasmobranchs start with yolk reserves.

However, viviparity is a general term. While denoting the retention of the embryo within the female for the entire period of development, it encompasses a continuous range of conditions that occur in different species. Among these are elasmobranchs in which the embryo depends entirely or almost entirely on the stored yolk for organic nutrients, and is thus lecithotrophic. This condition in commonly referred to as yolk sac viviparity—the embryo remains in the mother's body and nutrition is predominantly or entirely lecithotrophic—and is the dominant reproductive mode in Chondrichthyes (Mabragaña et al., 2011). In some cases limited exchange of other substances may occur through uterine villi that line the uterus and contact the surface of the yolk sac; this is considered a type of placental analogue (see Homberger & Walker, 2003).

By contrast, some viviparous species are *matrotrophic* (from the Latin mater, mother, and Ancient Greek trophe, food) rather than lecithotrophic. Matrotrophy is a strategy in which the egg's energy reserves are supplemented during gestation by maternally derived nutrients (Conrath & Musick, 2012). There is a wide range of matrotrophic modes. In some species the embryo is largely dependent upon the mother's tissues via a yolk sac placenta: After several weeks of development, the distal end of the yolk sac vascularizes and comes into intimate contact with uterine tissues, thus shunting nutrients to the developing embryo (Carrier et al., 2004). Other forms of matrotrophy do not involve a well-defined placental connection; rather, maternal supplementation occurs through different means. One type of such supplementation occurs through secretions produced by the uterus (mucoid histotrophy and lipid histotrophy; these are generally termed placental analogues). In another type, supplementation occurs through ingestion by the embryo of other eggs that are in the uterus (oophagy).

Thus, there is a wide range of viviparous modes: at one extreme the embryo relies essentially on yolk reserves (lecithotrophy); at the other extreme the embryo receives considerable nutrition from the mother through a yolk sac placenta (one form of matrotrophy). In between are "intermediate" forms of matrotrophic nutrition, such as oophagy and placental analogues. It should be emphasized, however, that even in cases of matrotrophy, the development of the embryo is initially dependent on yolk (Conrath & Musick, 2012). Carrier et al. (2004) and Conrath and Musick (2012) provide more detailed discussions of the various conditions.

Squalus is characterized by yolk sac viviparity (thus, lecithotrophy): the pups are retained in the female's body and rely almost entirely on yolk for nourishment. As noted earlier, eggs pass from an ovary and through the ostium tubae (Figure 3.35c) to enter the oviduct. The ostium tubae is enlarged in a mature female and the falciform ligament is thus elongated and may be easily torn during dissection. As eggs pass through the oviduct, the nidamental gland secretes a thin shell around them, but it is resorbed after several months (Carrier et al., 2004), so the individual pups are easily observable in the uterus. The egg case is very thin and envelops several eggs; because the eggs stack one next to each other and form

a long tubular structure inside each uterus, it is called a "candle" in *Squalus*.

If you have a gravid female, cut open one of the uteri to observe the developing embryos (Figure 3.35d). If not, observe them in another student's specimen. Pups vary in length depending on the stage of gestation. The gestation period in *Squalus* is about 2 years (Conrath & Musick, 2012), at the end of which pups are about 23–29 cm in length (Homberger & Walker, 2003). The pup shown in Figure 3.36a–c is approximately 12 cm in length, which is typical of the size after about a year of development (between 12 and 20 cm in length, according to Homberger & Walker, 2003).

Remove a pup from the uterus to consider it in more detail (Figure 3.36). As noted earlier, it is nourished mainly with yolk, much of it stored in the **external yolk sac**, which extends through the ventral surface of the embryo via a **stalk**. Cut open the pleurperitoneal cavity and observe that the stalk is continuous with the **internal yolk sac**. During early development, yolk is digested by cells lining the yolk sac. The digested nutrients are absorbed into the vitelline veins, which adhere to the stalk, and thereby pass to the embryo. Later in development, ciliary action moves the yolk into the embryo's intestine for digestion and absorption (Homberger & Walker, 2003), with the yolk usually transferred from the external yolk sac to the internal yolk sac (Conrath & Musick, 2012). By about the time of birth, yolk in the external yolk has been exhausted and a small reserve remains in the internal yolk sac (Homberger & Walker, 2003).

If time permits, examine the embryo, ideally with a dissecting microscope, for many of the structures already noted in the adult. Among those that are easily observable are the lobes of the liver, ventral lobe of the pancreas, stomach, including the pylorus, spleen, intestine, including the duodenum and valvular intestine, digitiform gland, and kidneys (Figure 3.36b,c). You should also able to observe the connection of the internal yolk sac to the intestine (Figure 3.36c).

Key Terms: Female Reproductive System

external yolk sac
internal yolk sac
kidneys
mesotubarium
mesovarium
nidamental gland (shell gland)
ostium tubae
ovary (pl., **ovaries**)
oviduct
stalk
urinary papilla
uterus

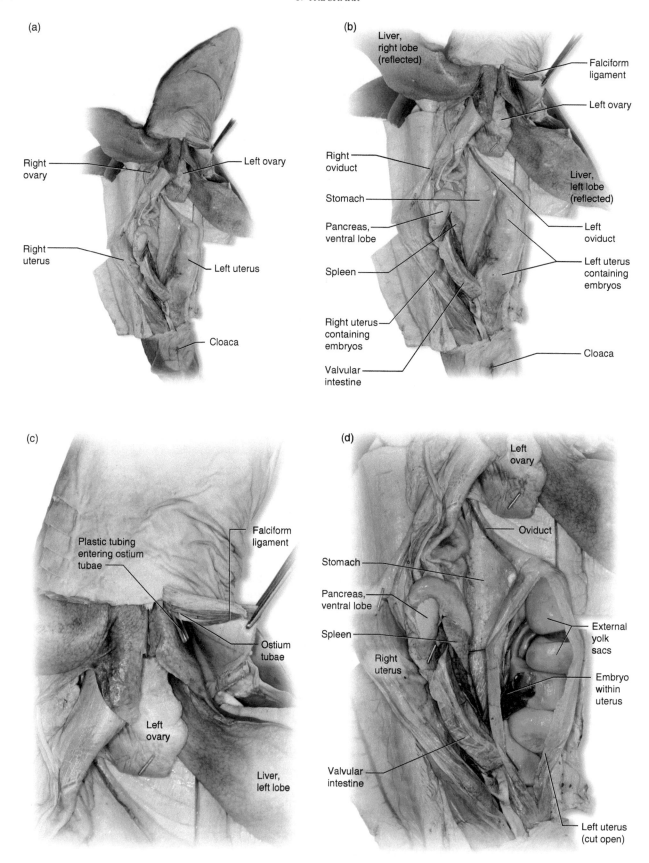

FIGURE 3.35 Pregnant dogfish in ventral view: (a) and (b) overview of structures of the pleuroperitoneal cavity; (c) detail of anterior part of pleuroperitoneal cavity to show ostium tubae; (d) detail of left uterus, cut open to reveal developing pups.

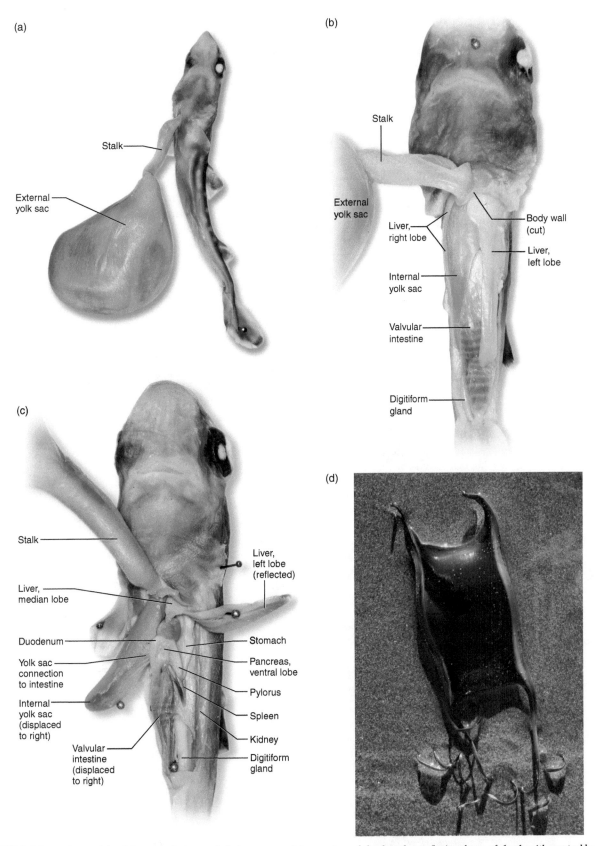

FIGURE 3.36 Embryo of shark in ventral view and skate egg case: (a) overview of shark embryo; (b,c) embryo of shark with ventral body wall removed to show details of internal structures; (d) an egg case (length about 8 cm, not including the thread-like horns) of the smallnose fanskate *Sympterygia bonapartii* recovered along the southern coast of Argentine Patagonia.

SECTION VII: SENSORY ORGANS

Ampullae of Lorenzini

The **ampullae of Lorenzini** (Figures 3.15 and 3.37) are modified parts of the lateral line system (see later) and primarily sensitive to electrical fields (they can help a shark sense prey by detecting the electrical fields generated by activities of the prey). They form a series of tube-like structures just beneath and parallel to the skin. The ampullae are concentrated on the head, particularly on the ventral and dorsal surfaces of the snout and posterior to the eye. Squeeze the snout. Thick fluid emerges from the ampullae through pores in the skin. Remove a portion of skin from the snout and top of the head between the eyes and observe the pattern formed by the ampullae.

Lateral Line System

The lateral line system is composed of a series of canals and sensory receptors that provide sensory information by detecting disturbances in water. The position of the **lateral line canal**, extending midlaterally along the length of the body and into the tail, was noted earlier (page 64). The canal lies within the skin. Make a cut in the skin, perpendicular to the canal, and examine the cut surface. The canal appears as a small hole. The canal leads to other canals in the head. If you skinned the head to find the ampullae of Lorenzini, you should be able to observe, about midway between the eye and the dorsal midline, a groove for one of these canals (the supraorbital) heading toward the snout (Figure 3.37). Another canal (the infraorbital) passes down behind the eye and then forward toward the snout. Trace the canals on the head, if time permits.

Nose

The **olfactory sacs** function in olfaction (Figures 3.38, 3.43, and 3.45). Reexamine the nares and note that each is subdivided into lateral incurrent and medial excurrent apertures that direct the flow of water into and out of

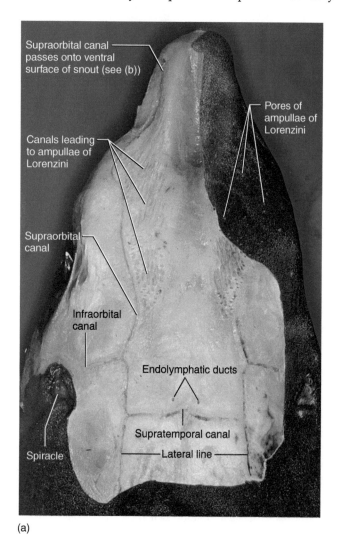

(a)

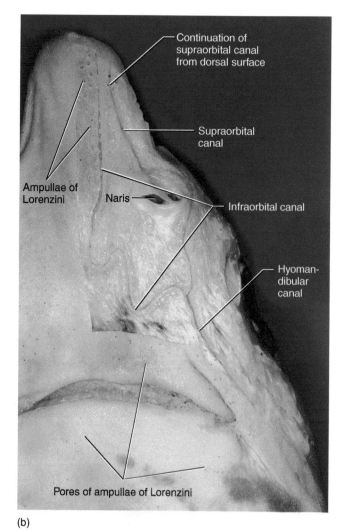

(b)

FIGURE 3.37   Head of the shark partially skinned in (a) dorsal and (b) ventral views to reveal the lateral line system.

the olfactory sac. Skin the region adjacent to one of the nares. Carefully pick away and remove the cartilaginous nasal capsule to expose the olfactory sac. It is a delicate, light-colored, spherical structure. Cut into the sac and note within it the parallel folds, or lamellae, on which the sensory receptors lie. The sac has no connection with the oral cavity.

Eye

Lift the edge of an eyelid and note its soft and thin inner surface, which consists of a portion of soft, modified skin. The skin folds over and fuses onto the surface of the **eyeball** as the transparent conjunctiva. Begin the exposure of the eyeball by removing the soft tissue of the upper eyelid. Then cut two or three very thin slices from the top of the supraorbital crest, holding the scalpel horizontal. This will clearly reveal the semicircular margin

of the **orbit**, the space housing the eyeball and several other structures. Continue to remove the crest. Make a few vertical slices and follow the semicircular margin as you make your way toward (but not to) the medial wall of the orbit. Stop after you have removed a thickness of about 3 mm of cartilage. The eyeball is covered by a gelatinous connective tissue; carefully remove this.

Observe the eyeball's medial surface as well as some of the extrinsic muscles of the eyeball that extend between it and the orbital wall. Identify the **superficial ophthalmic nerve**, a strand about one-third the width of the muscles, which passes anteroposteriorly along the dorsomedial wall of the orbit. Once the nerve is located, continue to remove cartilage to expose the orbit more completely, but do not cut into the otic capsule. Remove any other connective tissue, but avoid injuring the nerves in the orbit. You now have a dorsal view of the orbit and its contents (Figure 3.38).

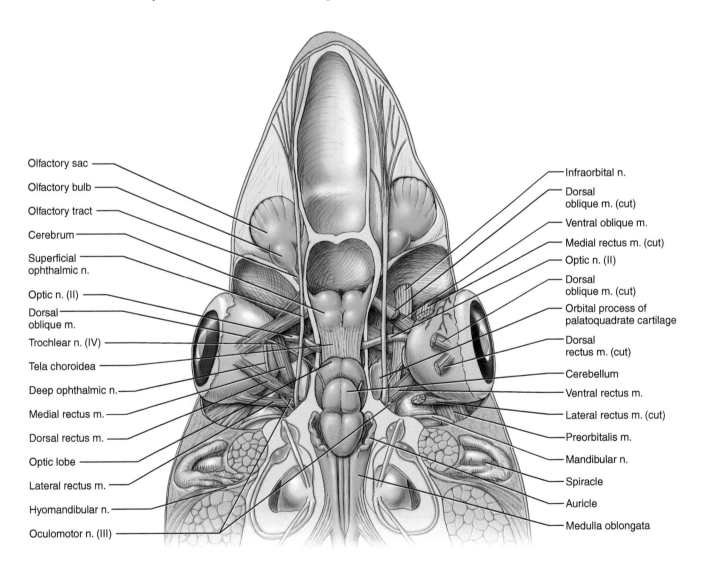

FIGURE 3.38 Head of the shark in dorsal view showing the chondrocranium shaved down to reveal the brain, sensory organs, and nerves.

The two groups of extrinsic eye muscles are the oblique and rectus muscles. There are two oblique muscles, which pass posterolaterally from the anteromedial wall of the orbit. In dorsal view the **dorsal oblique muscle** is clearly evident. To find the **ventral oblique muscle**, push the lateral part of the eyeball posteriorly and peer below the dorsal oblique.

The four rectus muscles radiate toward the eyeball from the posteromedial wall of the orbit. Three are clearly evident in dorsal view. The **lateral rectus muscle** extends almost directly laterally and attaches to the posterior surface of the eyeball. The **medial rectus muscle** passes anteriorly and attaches to the medial surface of the eyeball. The **dorsal rectus muscle** extends between the lateral and medial rectus muscles and attaches to the top of the eyeball. The fourth muscle, the **ventral rectus muscle**, will be seen more completely once the eyeball has been removed. For now, probe gently between the dorsal and lateral rectus muscles to find the ventral rectus. Avoid injuring the thin nerves that wind around the muscles.

By probing between the medial and dorsal rectus muscles, observe how the **optic pedicle** provides support for the eyeball (Figure 3.39). Tug the eyeball laterally and identify the **deep ophthalmic nerve**, a thin, whitish strand passing anteroposteriorly through the orbit and adhering to the medial surface of the eyeball. Using needle and forceps separate the nerve from the eyeball. Next, look deep within the orbit, where the ventral oblique and medial rectus muscles converge toward the eyeball. Note the thick **optic nerve** extending laterally from the medial wall of the orbit to the eyeball.

Remove the eyeball by cutting the oblique and rectus muscles near their insertions. Be careful not to injure the thin, whitish strands that pass to or near the muscles. These are nerves that must still be identified. Separate the eyeball from the optic pedicle. Also, cut through the optic nerve and the tissue between the eyeball and lower

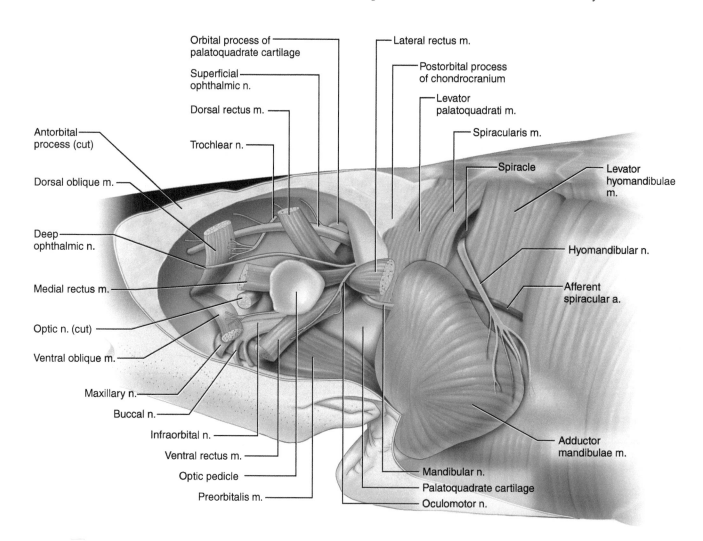

FIGURE 3.39 Head of the shark in left lateral view. The head has been skinned and the left eyeball removed to show muscles and nerves in the orbit.

eyelid. Finally, pick away connective tissue ventrally in the orbit.

Observe the orbit in lateral view and review the structures already identified: oblique and rectus muscles, superficial and deep ophthalmic nerves, optic nerve and optic pedicle (Figure 3.39). Lift the ventral rectus and observe the thick **infraorbital nerve** passing anteriorly and slightly ventrally (Figures 3.38 and 3.39). It is nearly as wide as any of the extrinsic muscles. Carefully picking away tissue where the nerve meets the anteroventral orbital wall reveals that the infraorbital nerve subdivides. As the nerve passes through the orbit, it crosses over the **preorbitalis muscle** (Figures 3.41 and 3.43), noted earlier as a modification of the adductor mandibulae. The preorbitalis muscle is about twice as wide as the extrinsic muscles and passes anteromedially across the floor of the orbit.

Trace the path of the superficial ophthalmic nerve and note the orbital process of the palatoquadrate (Figure 3.39; see also page 57 and Figure 3.3) that lies medial to the posterior half of the nerve. Next, follow the course of the deep ophthalmic nerve. It passes forward in the orbit, ventral to the dorsal rectus and dorsal oblique muscles. Lift the dorsal oblique muscle and observe the deep ophthalmic nerve and stalk of the optic pedicle. Just ventral to the deep ophthalmic is the **oculomotor nerve**. The latter passes almost immediately ventrally and around the margin of the ventral rectus. The lateral rectus muscle is innervated by the **abducens nerve**, which you will probably not see now (see page 105). Gently tug the dorsal oblique muscle anteriorly and note the thin strand-like **trochlear nerve** passing to it. Lastly, identify the **mandibular nerve**. It lies on the posterior wall of the orbit, almost directly posterior to the lateral rectus muscle, and extends laterally.

Examine the removed eyeball (Figure 3.40) and note the insertions of the extrinsic muscles. The outer surface of the eye is the fibrous tunic. Its lateral portion is modified into the transparent **cornea**. The remainder is the mainly cartilaginous **sclera**, which helps support the eyeball. The pigmented structure visible through the cornea is the **iris**, which has a circular opening, the **pupil**, at its center. Section the eyeball transversely and refer to Figure 3.41 to help identify the following structures. The **lens** is the hard spherical structure. Carefully cut through the pigmented tissue, the **ciliary body** (see later), that attaches to the lens, but only cut around half of the circumference of the lens so that it remains attached to the half of the eyeball that you will examine. The vascular tunic is the dark pigmented layer internal to the sclera, and most of it consists of the **choroid**. The medial wall of the eye, where the optic pedicle attaches, contains the **suprachoroidea**, a vascular tissue between the choroid and sclera.

The iris, noted just earlier, is a modified part of the vascular tunic. It extends between the cornea and the

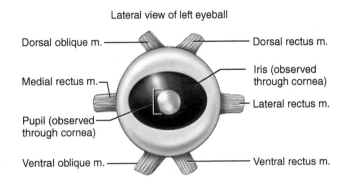

Lateral view of left eyeball

Dorsal oblique m. — — Dorsal rectus m.

— Iris (observed through cornea)

Medial rectus m. —

Pupil (observed through cornea) — — Lateral rectus m.

Ventral oblique m. — — Ventral rectus m.

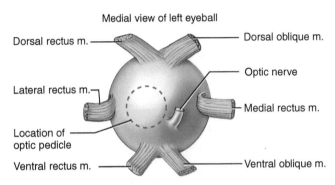

Medial view of left eyeball

Dorsal rectus m. — — Dorsal oblique m.

— Optic nerve

Lateral rectus m. —

— Medial rectus m.

Location of optic pedicle —

Ventral rectus m. — — Ventral oblique m.

FIGURE 3.40   Left eyeball of the shark in lateral and medial views.

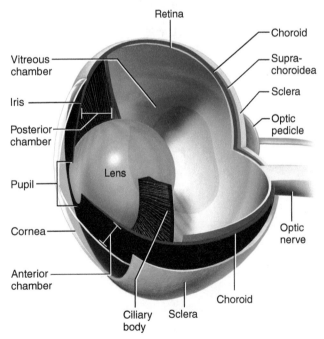

Retina

Choroid

Supra-choroidea

Vitreous chamber

Sclera

Iris

Optic pedicle

Posterior chamber

Lens

Pupil

Optic nerve

Cornea

Anterior chamber

Choroid

Ciliary body   Sclera

FIGURE 3.41   3-D cutaway illustration of the left eyeball of the shark in posterolateral view revealing internal structures.

lens; the pupil is the opening in the iris. The iris contains some of the eyeball's intrinsic musculature. These smooth muscles act to control the size of the pupil, thus regulating the amount of light that enters the eyeball. Another modified part of the vascular tunic forms the ciliary body, which holds the lens in place. It is the tissue you cut through to free the lens from one half of the eyeball. The ciliary body also contains some of the eyeball's intrinsic musculature that here helps control the shape of the lens. The light-colored tissue internal to the choroid is the **retina**, the light sensitive layer that contains the photoreceptors responsible for absorbing light. Note that it is an incomplete layer, extending to about the base of the ciliary body. It is not tightly interconnected with the choroid, and when the eyeball is cut open the retina tends to become detached. The large cavity of the eyeball medial to the lens is the **vitreous chamber**. It is filled by a gelatinous mass, the **vitreous humor**, which helps maintain the eyeball's shape and holds the retina in place. The smaller cavity lateral to the lens is subdivided by the iris into the **anterior chamber** and the **posterior chamber**, and filled with the watery **aqueous humor**.

Ear

The paired inner ears of the dogfish are organs of balance or equilibrium and are embedded in the otic capsules of the chondrocranium. Each ear consists of a series of ducts and sacs, collectively termed the membranous labyrinth. The ducts and sacs are suspended in a series of canals and chambers, the cartilaginous labyrinth, within the otic capsule. The membranous labyrinth is filled with a fluid termed *endolymph*. Movement of the endolymph within the canals leads to perception of the orientation and position of the body.

The membranous labyrinth is formed of three thin, semicircular ducts and the sacculus (Figures 3.42 and 3.43). The **anterior** and **posterior semicircular ducts** are vertically oriented. The third duct, the **lateral semicircular duct**, lies mainly horizontally. The **sacculus**,

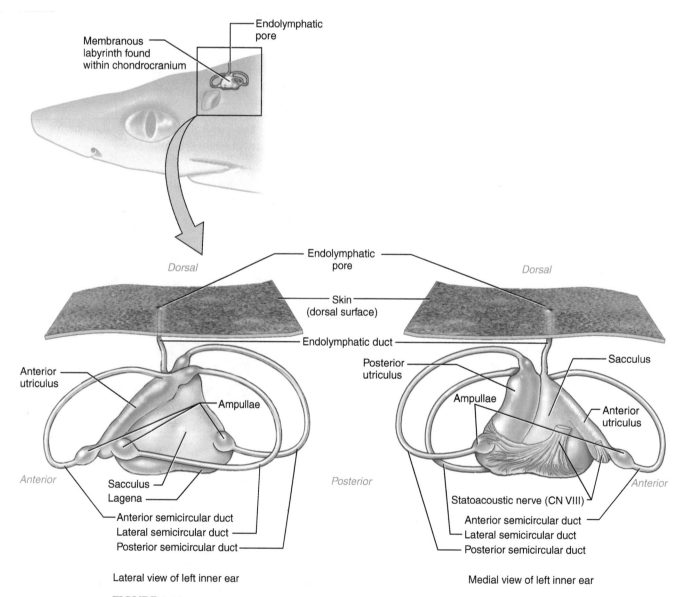

FIGURE 3.42 Membranous labyrinth of the left inner ear of the shark in lateral and medial views.

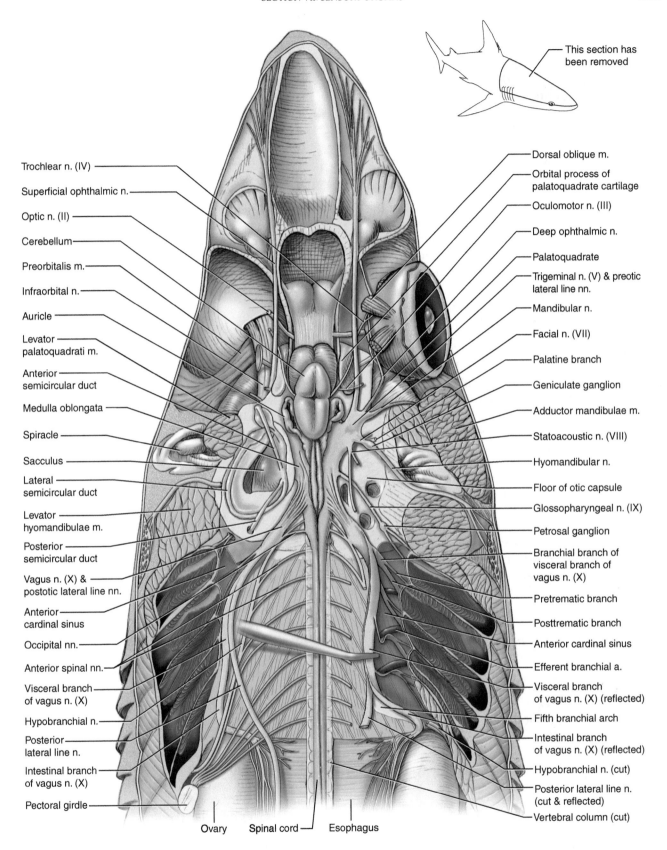

Trochlear n. (IV)
Superficial ophthalmic n.
Optic n. (II)
Cerebellum
Preorbitalis m.
Infraorbital n.
Auricle
Levator palatoquadrati m.
Anterior semicircular duct
Medulla oblongata
Spiracle
Sacculus
Lateral semicircular duct
Levator hyomandibulae m.
Posterior semicircular duct
Vagus n. (X) & postotic lateral line nn.
Anterior cardinal sinus
Occipital nn.
Anterior spinal nn.
Visceral branch of vagus n. (X)
Hypobranchial n.
Posterior lateral line n.
Intestinal branch of vagus n. (X)
Pectoral girdle

This section has been removed

Dorsal oblique m.
Orbital process of palatoquadrate cartilage
Oculomotor n. (III)
Deep ophthalmic n.
Palatoquadrate
Trigeminal n. (V) & preotic lateral line nn.
Mandibular n.
Facial n. (VII)
Palatine branch
Geniculate ganglion
Adductor mandibulae m.
Statoacoustic n. (VIII)
Hyomandibular n.
Floor of otic capsule
Glossopharyngeal n. (IX)
Petrosal ganglion
Branchial branch of visceral branch of vagus n. (X)
Pretrematic branch
Posttrematic branch
Anterior cardinal sinus
Efferent branchial a.
Visceral branch of vagus n. (X) (reflected)
Fifth branchial arch
Intestinal branch of vagus n. (X) (reflected)
Hypobranchial n. (cut)
Posterior lateral line n. (cut & reflected)
Vertebral column (cut)

Ovary   Spinal cord   Esophagus

FIGURE 3.43   Head and branchial region of the shark in dorsal view. Cartilage and soft tissue have been shaved down to reveal brain, nerves, and sensory organs. The right side auditory region has been shaved down farther ventrally than the left side. The left eyeball has been removed.

a large, triangular, sac-like structure, extends between the vertical ducts and medial to the lateral duct. The **endolymphatic duct** extends dorsally from the sacculus and reaches the exterior surface of the head through the endolymphatic pores (Figure 3.42). Two cylindrical chambers, the **anterior** and **posterior utriculi**, are closely associated with the sacculus. Each utriculus communicates with the sacculus by way of small openings. The ends of each duct attach to one of the utriculi. Those of the anterior and lateral semicircular ducts attach to the anterior utriculus, those of the posterior semicircular duct to the posterior utriculus.

The ear may be examined by shaving away the cartilage of the otic capsule. Before beginning, study a preparation of the ears. Such preparations usually have the chondrocranium embedded in an acrylic block. The form and position of the canals and chambers, injected with red latex, are clearly discernible.

Start your dissection by removing skin and musculature from around an otic capsule. Using a fresh scalpel blade, shave away thin slices of cartilage from one of the otic capsules, beginning dorsally. The slices should be thin enough so that you clearly see the blade through the cartilage. As you progress ventrally, you will probably need to remove more musculature laterally, as well as the spiracle. The dorsal portions of the anterior and posterior semicircular canals will become apparent through the cartilage. Carefully continue to remove cartilage to cut into the canals. The thin, tube-like semicircular ducts of the membranous labyrinth will then be exposed. As you proceed ventrally, you will also uncover the large central cavity housing the sacculus. The cavity lies medial to the lateral semicircular canal and increases in size ventrally. Careful dissection will reveal the sacculus, but it is often collapsed on the floor of the cavity. The position of the **lagena**, a posteroventral extension of the sacculus, may be discerned by probing the floor of the sacculus.

Key Terms: Sensory Organs

abducens nerve
ampullae of Lorenzini
anterior chamber
anterior semicircular duct
anterior utriculus
aqueous humor
choroid
ciliary body
cornea
deep ophthalmic nerve
dorsal oblique muscle
dorsal rectus muscle
endolymphatic duct
eyeball

infraorbital nerve
iris
lagena
lateral line canal
lateral rectus muscle
lateral semicircular duct
lens
mandibular nerve
medial rectus muscle
oculomotor nerve
olfactory sacs
optic nerve
optic pedicle
orbit
posterior chamber
posterior semicircular duct
posterior utriculus
preorbitalis muscle
pupil
retina
sacculus
sclera
superficial ophthalmic nerve
suprachoroidea
trochlear nerve
ventral oblique muscle
ventral rectus muscle
vitreous chamber
vitreous humor

SECTION VIII: BRAIN AND CRANIAL NERVES

The **brain** and **cranial nerves** (Figures 3.43–3.47; Table 3.2) should be exposed by removing thin slices of cartilage from the roof of the chondrocranium, as was done for the ear. Earlier dissections of the ear and eye probably destroyed some of the cranial nerves that must still be identified, so look for these structures mainly on the intact side of the head. Work carefully in removing the cartilage forming the dorsal roof of the orbit to avoid injuring the superficial ophthalmic nerve. The nerve was noted on page 97, but its connections to the rest of the nervous system may now be followed.

The brain sits in the large cranial cavity within the posterior part of the chondrocranium. Begin exposing the brain posteriorly, between the otic capsules. As you work your way forward, peer into the anterior part of the cranial cavity and you may be able to see the thin, strand-like **epiphysis** (Figures 3.44–3.47) extending dorsally from the brain to the epiphyseal foramen (see page 54 and Figure 3.2) in the roof of the chondrocranium. Continue to shave the roof and as much of the lateral walls of the cavity as you can without injuring the

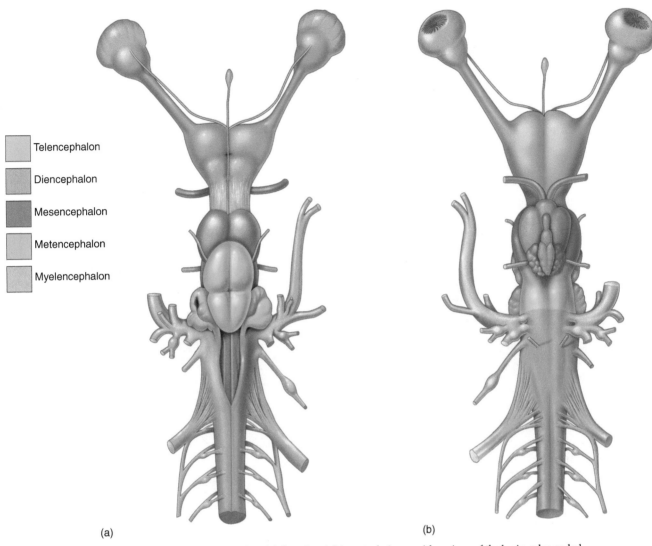

Telencephalon

Diencephalon

Mesencephalon

Metencephalon

Myelencephalon

(a)   (b)

FIGURE 3.44   Brain of the shark in (a) dorsal and (b) ventral views, with regions of the brain color-coded.

nerves. When you have finished opening the cranial cavity, the dorsal surface of the brain will be revealed. Work carefully, as the brain is easily scrambled by poking with instruments. Also remove thin frontal slices, mainly of muscle tissue, posterior to the chondrocranium until you expose the vertebral column. Shave down the neural arch dorsally and laterally. Stop when you can clearly see, through the cartilage, the **spinal nerves** extending laterally from the **spinal cord**.

Brain

The brain is subdivided into the following five regions, in anterior to posterior order: **telencephalon, diencephalon, mesencephalon, metencephalon,** and **myelencephalon** (Figure 3.44). The telencephalon includes the **olfactory bulbs** anteriorly (Figure 3.45). The olfactory sacs (see page 96) are anterior to the bulbs

(see Figure 3.45). The bulbs narrow into the **olfactory tracts**, which in turn extend into the **cerebral hemispheres**. Together these constitute the **cerebrum**. The diencephalon lies posterior to the cerebral hemispheres. Its roof is the **epithalamus**, its sides the **thalamus**, and its floor the **hypothalamus**. The epiphysis attaches posteriorly on the epithalamus. Most of the epithalamus is formed by a thin, vascular **tela choroidea**. Anteriorly, the tela choroidea attaches to the cerebrum. The **third ventricle**, the cavity of the diencephalon, lies beneath the tela choridea and will be seen presently.

The mesencephalon includes a pair of rounded **optic lobes**. The main structure of the metencephalon is the large **cerebellum**, which partially overhangs the optic lobes anteriorly and the myelencephalon posteriorly. The **auricles** of the cerebellum project anterolaterally from the posterior end of the metencephalon. The myelencephalon includes nearly all of the **medulla oblongata**,

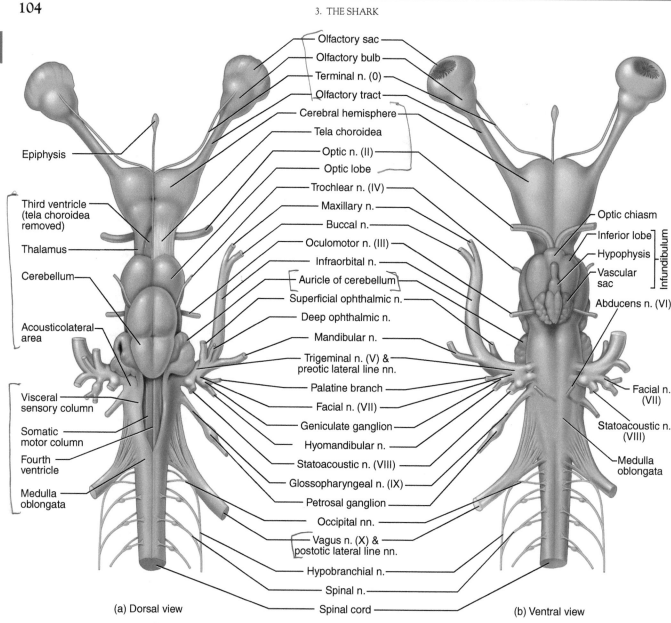

FIGURE 3.45 Brain and cranial nerves of the shark in (a) dorsal and (b) ventral views.

the posterior part of the brain that narrows into the spinal cord. The roof of the medulla oblongata is covered by a tela choroidea, which extends anteriorly to cover the roof of each auricle as well. Removing the tela choroidea reveals the **fourth ventricle**, the cavity of the medulla. The internal floor and sides of the medulla have several longitudinal ridges and grooves. The two large midventral ridges, one on either side of the midventral groove, are the **somatic motor columns**. Lateral to each column is a deep longitudinal depression. A second, large pair of ridges extends longitudinally on the wall of the medulla lateral to these depressions. These are the **visceral sensory columns**. Deep within the depression (thus between the somatic motor column and visceral sensory column) is a much smaller column, the **visceral motor column**. Finally, there is a large longitudinal ridge dorsal to each visceral sensory column. This is the **somatic sensory**

column, and its surface has the form of small, bead-like swellings. Anteriorly the somatic sensory column becomes enlarged and forms the **acousticolateral area**.

Cranial Nerves (CNN 0, I–X, and Lateral Line Nerves)

This section mainly examines the cranial nerves, but the occipital and the anterior spinal nerves must also be considered to properly comprehend the pattern and distribution of the nerves arising from the brain. Many of the nerves attach to the ventral surface of the brain, but their proximal ends will be examined shortly. Thus, unless otherwise instructed, do not jump ahead and study the other structures of the brain. Remove as much cartilage as possible from the lateral wall of the cranial cavity and of the neural arch without damaging the

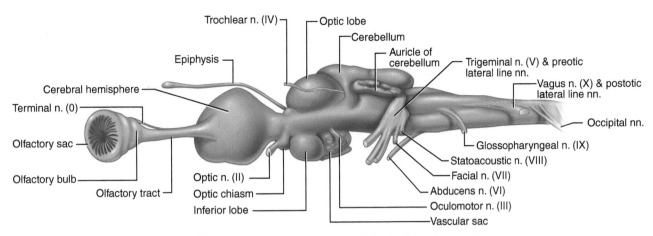

FIGURE 3.46   Brain and cranial nerves of the shark in left lateral view.

nerves. Also, clean the orbit of connective tissue, as was done for the other eye during examination of the orbit.

The cranial nerves of vertebrates were first studied in humans, and the sequence and names of the nerves were based on the pattern in humans. Twelve nerves were initially recognized and they were given both name and Roman numeral designations (see Table 7.5). A small anterior nerve, the terminal nerve, was discovered later to be widespread among vertebrates, including humans (see Peña-Melián et al., 2019). As it is anterior to CN I, it was designated as "0" (zero—which is not a Roman numeral!). These cranial nerves and their designations became commonly accepted, as their pattern is generally applicable to all amniotes.

This sequence, however, does not apply equally well to anamniotes, in which the final two nerves, the **accessory (XI)** and **hypoglossal (CN XII) nerves**, are not definitively recognized. Further, many anamniotes also have a set of six additional cranial nerves associated with the lateral line and electroreceptor organs. As these six nerves, termed **lateral line nerves**, are lacking in amniotes, they were never numbered and were considered to be components of some of the conventionally recognized cranial nerves. However, the lateral line nerves have their own roots and they are currently considered separate nerves, even though their separate emergence from the brain is difficult to identify grossly. They enter the brain very near cranial nerves VII and X and so may be referred to as either preotic or postotic lateral line nerves. More distally, the fibers of some do accompany those of other cranial nerves (see Table 3.2). For these reasons, they will not, with one exception, be considered in detail.

The most anterior cranial nerve is the **terminal nerve (CN 0)**, a thin nerve passing along the medial side of the olfactory tract. It is often difficult to find, but look for it where the olfactory tract meets the cerebrum. The **olfactory nerve (CN I)** is formed by many fine fibers that pass from the olfactory sac into the olfactory bulb, but they will not be seen grossly. The **optic nerve (CN II)** has already been seen (see page 98). Trace it from its attachment to the medial surface of the eyeball into the cranial cavity and then to its attachment on the diencephalon.

The **oculomotor nerve (CN III)** arises from the ventral surface of the mesencephalon and branches out to innervate four eye muscles (ventral oblique, and dorsal, ventral, and medial rectus muscles). Examine the ventral oblique and find the branch of the oculomotor that passes to it, and then trace the nerve posteriorly (see Figure 3.39). The nerve passes ventral to the ventral rectus muscle, then curves dorsally and gives off a branch to the ventral rectus muscle. It continues dorsally, crosses anterior to the base of the lateral rectus muscle and dorsal to the base of the dorsal rectus muscle, and passes medially. It gives off branches to the dorsal and medial rectus muscles and then enters the chondrocranium. The oculomotor nerve can be seen leaving the brain by pushing the brain gently to the side. The slender **trochlear nerve (CN IV)** extends anterolaterally from the dorsal surface of the mesencephalon. It crosses over the optic lobe, and passes to innervate the dorsal oblique muscle.

The **abducens nerve (CN VI**; this is discussed here, out of sequence, so the nerves discussed in the next paragraph can be treated together) arises from the ventral surface of the medulla. Its origin will be seen later. The abducens passes anterolaterally to innervate the lateral rectus muscle. It can be found on the ventral surface of this muscle.

The next three nerves, the **trigeminal (CN V)**, **facial (CN VII)**, and **statoacoustic (CN VIII) nerves** share a close origin from the surface of the brain. In addition, the roots of the preotic lateral line nerves arise from between the roots of the trigeminal and facial nerves. All of these nerves arise in this sequence from the anterior part of the medulla, beginning just behind the auricles of the cerebellum. However, they emerge so closely together that it is difficult to identify them separately. Their peripheral distributions, however, can be readily traced. The preotic lateral line nerves include the **anterodorsal lateral line nerve (ADLLN)**, the **anteroventral lateral line nerve (AVLLN)**, and the **otic lateral line nerve (OLLN)**.

Some branches of the trigeminal, facial, and lateral line nerves merge to form larger nerves or trunks. Although some branches of the trigeminal, facial, and lateral line nerves were observed during dissection of the orbit, use the following description to review them carefully and ascertain their pattern of distribution.

The trigeminal[1] is a large nerve that divides into four branches on emerging from the cranial cavity into the orbit. These are the **superficial ophthalmic**, **deep ophthalmic**, **mandibular**, and **maxillary branches**. The first branch is accompanied by the superficial ophthalmic branch of the ADLLN, the fourth branch by the buccal branch of the ADLLN (Table 3.2). The superficial ophthalmic branch is the most dorsal and passes just ventral to the dorsal margin of the orbit. The deep ophthalmic branch passes through the orbit but adheres to the dorsomedial surface of the eyeball. The mandibular branch extends laterally along the posterior wall of the orbit, almost directly posterior to the lateral rectus muscle. The maxillary branch contributes to the **infraorbital nerve**, the large nerve passing along the floor of the orbit and crossing the preorbitalis muscle. Anteriorly, the infraorbital divides into separate maxillary (medially) and buccal (laterally) nerves.

The facial nerve has two branches. One is the **hyomandibular branch**, which forms the hyomandibular nerve with the AVLLN. The hyomandibular nerve was observed as it passed over the external surface of the levator hyomandibulae muscle (page 70). Trace it now from the brain, carefully cutting away portions of the musculature, ear, and spiracle. Near its origin, the hyomandibular branch bears a swelling, the **geniculate ganglion**. The second branch of the facial nerve is the **palatine branch**, which participates in the innervation of the lining of the oral cavity.

The **statoacoustic nerve** (**CN VIII**) is a short nerve that innervates the ear. Remove cartilage of the otic capsule to see branches passing, in particular, to the ampullae of the semicircular ducts and the sacculus.

The **glossopharyngeal nerve** (**CN IX**) arises posterior to the statoacoustic nerve. It extends posterolaterally through the floor of the otic capsule to the first pharyngeal slit. Pick away the cartilage of the otic capsule to follow the nerve. Note the swelling, the **petrosal ganglion**, along the nerve just before it emerges from the capsule. Very near the dorsal margin of the pharyngeal slit, the glossopharyngeal divides into pretrematic and posttrematic branches (remember that *trema* refers to the slit). The glossopharyngeal also has a small pharyngeal branch, but do not look for it.

The three postotic lateral line nerves are the **middle lateral line nerve** (**MLLN**), the **supratemporal lateral line nerve** (**STLLN**), and the **posterior lateral line nerve** (**PLLN**). Their roots emerge from the brain between the glossopharyngeal (CN IX) and vagus (CN X; see later) nerves. The peripheral distribution of the PLLN will be traced later.

The **vagus nerve** (**CN X**) mainly innervates the remaining pharyngeal slits and the viscera. Note the series of fan-like rootlets emerging from the medulla just posterior to the glossopharyngeal nerve. These include the roots of the postotic lateral line nerves, more anteriorly, and of the vagus, more posteriorly. However, it is not practical to attempt to separate them. The roots of the vagus and the PLLN merge and pass posterolaterally through the otic capsule. Follow them posteriorly, removing cartilage and soft tissue as required. The vagus and the PLLN separate just medial to the first pharyngeal slit. The PLLN lies medial to the vagus and curves posteromedially as it extends between the epaxial and hypaxial musculature to innervate the lateral line canal in the trunk.

The vagus has two main parts, the visceral and intestinal branches. The visceral branch may be observed as it passes over the pharyngeal slits and gives rise to four **branchial branches**, one each to all but the most anterior pharyngeal slit (which is innervated by the glossopharyngeal nerve). Reflect the visceral branch of the vagus and follow the branchial branches laterally (Figure 3.43). They lie along the floor of the anterior cardinal sinus (see Figure 3.43 and pages 71 and 87). Like the glossopharyngeal, each branchial branch subdivides into pretrematic, posttrematic, and pharyngeal branches, but do not search for the pharyngeal branch. The intestinal branch continues posteriorly after the last of the branchial branches. Medial to the last branchial slit, it turns sharply medially. Dissect carefully here, as other nerves cross it dorsally (see later). The intestinal then continues ventrally into the pleuroperitoneal cavity, passing initially along the esophagus. This portion may also be seen by turning your specimen on its dorsal surface, slitting open the posterior cardinal sinus, and examining the dorsomedial wall of the sinus.

Return to the nerves crossing the intestinal branch of the vagus, noted in the preceding paragraph. The large nerve is the **hypobranchial nerve**, which passes ventrally to innervate the hypobranchial musculature. A smaller nerve, a spinal nerve, passes just posterior to the hypobranchial in this region. Trace the hypobranchial anteromedially as it passes deep to the PLLN. Note that the hypobranchial becomes gradually thinner as you trace it toward the brain. It initially arises from the brain as two or three occipital roots (see later).

The nerves immediately posterior to the vagus of the shark, and most anamniotes, are not directly comparable to those of more derived vertebrates. In the shark these nerves merge from roots that arise from the transition between the medulla and spinal cord. They are thus formed from roots that are occipital and spinal. As a

---

[1]The Trigeminal nerve (CN V) is termed the trigeminal because in amniotes it has three branches. These branches are named but also designated as V_1, V_2, and V_3. The trigeminal of anamniotes, however, has four branches.

result, they are not entirely within the chondrocranium and thus are not "cranial." There is ambiguity because the posterior end of the cranium is phylogenetically variable among anamniotes. However, they are in part homologous with cranial nerves of higher vertebrates and so are considered here.

Usually, the first two slender roots that emerge posterior to the vagus unite to form the **occipital nerve**, which then partially merges with the vagus nerve. The occipital nerve continues posteriorly and receives contributions from the first few spinal nerves, which arise posterior to the occipital nerve. The union of the occipital and spinal nerves is the hypobranchial nerve, as noted earlier. Each spinal nerve results from the union of a dorsal root and a ventral root that arise from the spinal cord. The roots unite a short distance from the spinal cord, and the dorsal root bears a swelling, or ganglion. In amniotes the transition between the head and the trunk becomes fixed and the occipital nerves clearly arise from the brain within the skull as a "cranial" nerve termed the hypoglossal nerve (CN XII) (see Table 7.5 and Figure 7.80). Thus, the hypobranchial nerve of the shark (and other anamniotes) is homologous with the hypoglossal of amniotes. As noted above, the accessory nerve (CN XI; see Table 7.5 and Figure 7.80) of higher vertebrates (mainly derived from the vagus) is not represented as a distinct nerve in the shark.

Ventral View of the Brain

When you are familiar with the cranial nerves, the brain may be removed from the chondrocranium. Cut the olfactory tracts and then across the spinal cord just posterior to the medulla. Lift the anterior part of the brain, locate and cut the optic nerves. Continue to cut each of the cranial nerves, leaving as long a stump as possible. Lift the brain laterally and note the ventral extension, the **hypophysis**, just behind the optic nerves. Cartilage posterior to the hypophysis will have to be removed in order to free the brain entirely without damage.

Examine the ventral surface of the brain, noting its regions as described earlier (page 103), and the cranial nerves, particularly those that arise from ventral surface— the optic, oculomotor, and abducens. The optic nerves converge toward the anterior part of the hypothalamus and form the **optic chiasm**, where the optic nerves cross over to the opposite side of the brain. The rest of the hypothalamus is formed mainly by the **infundibulum**, which bears several important structures. The paired inferior lobes of the infundibulum lie just posterior to the optic chiasm. Posterior to each inferior lobe, the infundibulum continues as a **vascular sac**. Between the left and right inferior lobes and vascular sacs lies the hypophysis, noted earlier. The hypophysis is usually torn during removal of the brain. If this occurs in your specimen, examine another student's specimen.

Sagittal Section of the Brain

The brain and spinal cord are hollow, with several connected cavities or ventricles in the brain itself, and a narrow **central canal** in the spinal cord. In life, the ventricles and central canal are filled with cerebrospinal fluid. Some of the ventricles have already been noted (see page 103), but to examine all the ventricles and their relationships to each other, cut the brain in half by making a midsagittal section using a new scalpel blade. Observe one of the halves in sagittal view and briefly review the regions of the brain before studying the ventricles (Figure 3.47).

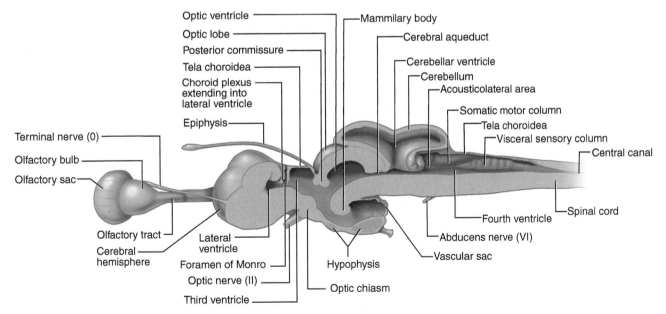

FIGURE 3.47 Right half of the brain of the shark, showing sagittal surface.

TABLE 3.2   The Cranial Nerves and Their Branches of the Shark.

Number	Nerve	Branches	Nerve Formed
CN 0	Terminal n.		
CN I	Olfactory n.		
CN II	Optic n.		
CN III	Oculomotor n.	Various branches, mainly to the ventral rectus, and dorsal, ventral, and medial oblique muscles	
CN IV	Trochlear n.		
CN V	Trigeminal n.	Superficial ophthalmic branch ———	
		Deep ophthalmic branch	
		Mandibular branch	
		Maxillary branch ——— → Infraorbital n.	Superficial ophthalmic n.

Preotic lateral line nerves:

Number	Nerve	Branches	Nerve Formed
ADLLN	Anterodorsal lateral line n.	Buccal branch ———	
		Superficial ophthalmic branch ———	
OLLN	Otic lateral line n.		
AVLLN	Anteroventral lateral line n. —————————————————		
CN VI	Abducens n.		Hyomandibular n.
CN VII	Facial n.	Hyomandibular branch ———	
		Palatine branch	
CN VIII	Statoacoustic n.		
CN IX	Glossopharyngeal n.	Pretrematic branch	
		Posttrematic branch	
		Pharyngeal branch	

Postotic lateral line nerves:

Number	Nerve	Branches	Nerve Formed
MLLN	Middle lateral line n.		
STLLN	Supratemporal lateral line n.		
PLLN	Posterior lateral line n.	Lateral branch	
		Dorsal branch (not discussed in text)	
		Ventral branch (not discussed in text)	
CN X	Vagus n.	Visceral branch: four branchial branches, each with pretrematic, posttrematic, and pharyngeal branches	
		Intestinal branch	
		Accessory branch (not discussed in text)	
	Occipital nn. —————————————————		Hypobranchial n.
	Anterior spinal nn. —————————————————		

The third ventricle (see page 103) is the cavity of the diencephalon. It communicates anteriorly with each of the **lateral ventricles** (one in each of the cerebral hemispheres) through a short passage, the **foramen of Monro**. As noted by Homberger and Walker (2003), the anterior end of the third ventricle is marked by a downward fold, the velum transversum, of the tela choroidea. A **choroid plexus** (a vascular tuft) extends from the anterior wall of the velum transversum into each of lateral ventricles. The choroid plexus is involved in the production of cerebrospinal fluid. Although the velum transversum is not indicated in Figure 3.47, its position may be noted as the choroid plexus extends forward from its anterior surface. A long, narrow passage, the **cerebral aqueduct**, extends from the posterior part of the third ventricle to reach the **fourth ventricle**, located within the medulla oblongata. Along the way, the cavities within the optic lobes, the **optic ventricles**, and the cerebellum, the **cerebellar ventricle**, communicate with the aqueduct.

Key Terms: Brain and Cranial Nerves

abducens nerve (CN VI)
accessory nerve (CN XI, not recognized in shark) (spinal accessory)
acousticolateral area
anterodorsal lateral line nerve (ADLLN)
anteroventral lateral line nerve (AVLLN)
auricles of **cerebellum**
brain
branchial branches of **visceral branch** of **vagus nerve**
buccal branch of **anterodorsal lateral line nerve (ADLLN)**
central canal
cerebellar ventricle
cerebellum
cerebral aqueduct (aqueduct of Sylvius)
cerebral hemispheres
cerebrum
choroid plexus
cranial nerves
deep ophthalmic branch of **trigeminal nerve** (profundus nerve)
diencephalon
epiphysis
epithalamus
facial nerve (CN VII)
foramen of Monro (interventricular foramen)
fourth ventricle
geniculate ganglion
glossopharyngeal nerve (CN IX)
hyomandibular branch of **facial nerve**
hyomandibular nerve

hypobranchial nerve
hypoglossal nerve (CN XII, not recognized in shark)
hypophysis
hypothalamus
infraorbital nerve
infundibulum
lateral ventricles
mandibular nerve (= **mandibular branch** of **trigeminal nerve)**
maxillary branch of **trigeminal nerve**
medulla oblongata
mesencephalon
metencephalon
middle lateral line nerve (MLLN)
myelencephalon
occipital nerve
oculomotor nerve (CN III)
olfactory bulbs
olfactory nerve (CN I)
olfactory tracts
optic chiasm
optic lobes
optic nerve (CN II)
optic ventricle
otic lateral line nerve (OLLN)
palatine branch of **facial nerve**
petrosal ganglion
posterior lateral line nerve (PLLN)
postotic lateral line nerves
preotic lateral line nerves
somatic motor column
somatic sensory column
spinal cord
spinal nerves
statoacoustic nerve (CN VIII) (vestibulocochlear, octaval)
superficial ophthalmic branch of **anterodorsal lateral line nerve (ADLLN)**
superficial ophthalmic branch of **trigeminal nerve**
superficial ophthalmic nerve
supratemporal lateral line nerve (STLLN)
tela choroidea
telencephalon
terminal nerve (CN 0)
thalamus
third ventricle
trigeminal nerve (CN V)
trochlear nerve (CN VI)
vagus nerve (CN X)
vascular sac
intestinal branch of **vagus nerve**
visceral branch of **vagus nerve**
visceral motor column
visceral sensory column

4

The Perch

INTRODUCTION

The yellow perch, *Perca flavescens*, is an actinopterygian. Besides the sharks and their relatives, there are two groups of living fish-like vertebrates: Actinopterygii, the ray-finned fishes, and Sarcopterygii, which includes the lobe-finned fishes and their tetrapod relatives. The latter is a relatively small group, at least in terms of their fish-like forms, and includes the coelacanths and lungfishes, barely a handful of species, which is a rather dismal record as far as fish go. Conversely, the actinopterygians are a huge success story, both in terms of diversity and numbers, with about 30,000 species known (although this number varies depending on author). The fins of actinopterygians are supported by slender, rod-like rays or lepidotrichia radiating from the base of the fin; the musculature controlling the fin is largely within the body wall.

As noted in Chapter 1, the vast majority of this diversity occurs among the derived actinopterygian clade Teleostei. This group has undergone extensive radiations to produce fishes that have invaded nearly every aquatic niche. There are far too many groups to discuss here, but we may mention Elopomorpha (eels and tarpons), the commercially important Clupeomorpha (anchovies and herrings), and Paracanthopterygii, another commercially important group that includes codfishes. The perch belongs to Perciformes, a member of Acanthopterygii, the spiny-finned fishes. Another interesting group is Ostariophysi, which includes catfishes, characins, and minnows. These fish share a unique sound detection system, the Weberian apparatus, which involves the swim bladder and modified elements of the first few trunk vertebrae. The vertebrae act as a conduction system between the swim bladder and the inner ear at the back of the skull. Vibrations of the swim bladder caused by sound waves are transmitted by the bony elements to a posterior extension of the ear's membranous labyrinth, stimulating the auditory center of the brain.

SECTION I: SKELETON

Skull

Examine a mounted skeleton of the perch and differentiate the **head**, **trunk**, and **caudal** regions (Figure 4.1). Unlike Chondrichthyes, Actinopterygii have well-developed bony skeletons. This is especially evident in the head, where numerous dermal (as well as endochondral) bony elements produce a complex skull. It is difficult and impractical to attempt identification of all the bones unless detailed preparations of skulls in various stages of disarticulation are available. Instead it is more productive to focus on various features that are particular to bony fishes, such as the bones of the jaw mechanism and opercular regions. Several other easily identifiable bones are included in this discussion for context.

Begin by locating the opercular bones on the pharyngeal region that cover the gills (Figures 4.1 and 4.2). The **operculum** is the large, triangular bony plate. Several bones surrounding the operculum can be easily identified. The J-shaped **preoperculum** lies anterior to the operculum, and the **suboperculum** lies posteroventral to the operculum. Ventral to the preoperculum is the **interoperculum**. The four **branchial arches** that support the gills lie deep to the opercular bones. Peer beneath the opercular to observe these arches.

Next, locate the large, circular **orbit** that houses the eyeball. Some of the deeper bones of the skull, such as the parasphenoid, are visible within the orbit. A series of bones surrounds and helps form the orbit. Several are large and easily identifiable, such as the **frontal**, which forms much of the skull roof, the **parietal**, posterior to the frontal, and the **lacrimal**, which forms the anteroventral margin of the orbit. The **maxilla** is the slender, edentulous bone articulating with the lacrimal. Its widened posterior end extends laterally and is embedded in soft tissue. Note that it does not form part of the margin of the mouth. The **premaxilla**, which bears teeth on its anteroventral surface, is the most anterior

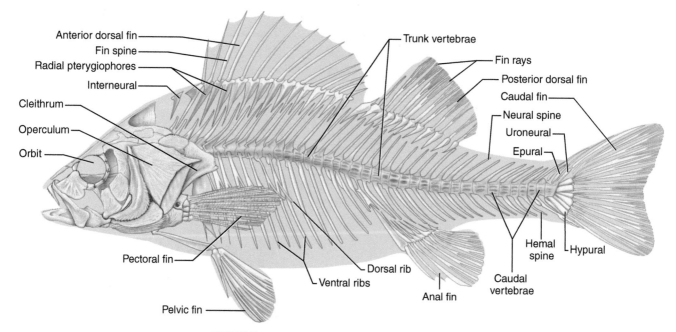

FIGURE 4.1   Skeleton of the perch in left lateral view.

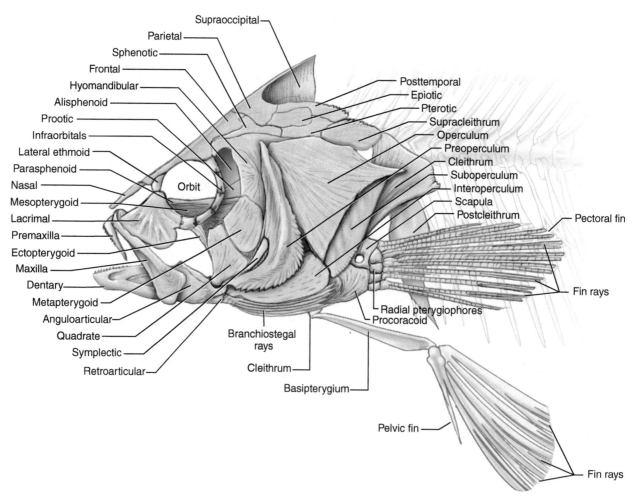

FIGURE 4.2   Skull and branchial skeleton of the perch in left lateral view.

bone of the upper jaw. It articulates with the premaxilla from the opposite side, and the **nasal** and maxilla posteriorly. A main feature of derived actinopterygians is their ability to protrude the premaxilla during opening of the mouth. The maxilla acts as a lever for the jaw muscles in helping to protrude the premaxilla forward. We should note that the names of several of the bones discussed here follow the convention currently used for actinopterygians, but the bones are not homologous with those of the same name in other vertebrates discussed in this manual. For example, the bone termed the frontal here is homologous with the parietal of tetrapods, whereas the parietal is homologous with the postparietal of other tetrapods.

A series of three main bones, lying between the orbit and preoperculum, extends anteroventrally. These bones are, in dorsal to ventral order, the **hyomandibular**, **metapterygoid**, and **quadrate**. They contribute to the suspensorium, the apparatus that supports the jaws on the rest of the skull. The hyomandibular and metapterygoid support the quadrate, which forms the articulation with the lower jaw. The lower jaw is formed by three bones. The **anguloarticular** forms most of the posterior end of the lower jaw and articulates with the quadrate of the upper jaw. The angular portion is a large, flat, dermal, and superficial component, whereas the endochondral articular is medial to the angular and so is not easily visible in lateral view. The **dentary**, which bears teeth, may be seen anterior to the anguloarticular. The **retroarticular** is a small endochondral ossification posteroventrally on the lower jaw associated with the ossification of the articular.

Key Terms: Skull

anguloarticular
branchial arches
dentary
frontal
head
hyomandibular
interoperculum
lacrimal
maxilla
metapterygoid
nasal
operculum
orbit
premaxilla
preoperculum
quadrate
retroarticular
tail
trunk
suboperculum

Postcranial Skeleton

The vertebral column includes **trunk vertebrae** anteriorly and **caudal vertebrae** posteriorly (Figure 4.1). The main part of a vertebra is the **centrum**, as also occurs in the shark (Figure 3.6). All the vertebrae bear elongated **neural spines** dorsally. The caudal vertebrae also bear elongated **hemal spines** ventrally. The trunk vertebrae bear ribs. There are two types of ribs in the perch, the **dorsal** and the **ventral ribs**. The ventral ribs are much more prominent and are usually present in prepared specimens. These curved structures, which form in the myosepta adjacent to the body cavity (coelom), extend ventrally. More delicate dorsal ribs extend laterally. They are attached ligamentously to the posterior surface of the more anterior ventral ribs, and are often missing in prepared specimens.

Examine the **anterior** and **posterior dorsal fins** on the middorsal line (Figure 4.1). The fins are supported by thin, elongated **fin rays**. These rays may be ossified, such as those supporting the anterior dorsal fin; in this case, they are usually termed *spines*. Other fin rays, termed *soft fin rays*, are flexible and unossified and may branch distally. Each fin ray is supported at its base by a **radial pterygiophore**, a ventrally tapered bony element. The tip of a radial pterygiophore extends ventrally into the connective tissue between two neural spines. Note that the radial pterygiophore series is continuous—they extend all the way along the dorsum beneath and between the anterior and posterior dorsal fins. The separation between these fins, in fact, is due to reduction of the spines between them. The posterior dorsal fin is structured similarly to the anterior dorsal fin, except that only the first two fin rays are spines; the remainder are soft fin rays.

The **anal fin** lies along the midventral line, opposite the position of the posterior dorsal fin. As in the latter, all but the first two fin rays are soft fin rays. Radial pterygiophores support the fin rays, essentially as in the dorsal fins, but taper dorsally. The anterior few (usually two) radial pterygiophores fuse into a large element that extends dorsally to attach to one or two ventral ribs, which here are quite reduced in size. It is posterior to this point of attachment that the series of hemal spines begins.

The **caudal fin** is supported by soft fin rays, which are arranged to form a homocercal tail. This type of fin is superficially symmetrical with about equal numbers of fin rays dorsal and ventral to the longitudinal axis extending posteriorly from the vertebral column. However, the body axis itself turns abruptly dorsally, so it is not symmetrical internally, although this is not easy to detect with the naked eye (see later). This upward turn is easily noted in less derived actinopterygians, such as sturgeons. It may still be appreciated in derived forms such as the perch by the orientation of the **uroneurals**, the last few neural spines, and the position of the

hypurals immediately posterior to them. The hypurals, also present ventral to the uroneurals, are the flattened hemal spines of the last few caudal vertebrae and provide most of the support for the fin rays of the caudal fin. Several neural spines, termed **epurals** and unattached to vertebrae, provide some support for the dorsal part of the fin.

Lastly, examine the paired fins (Figures 4.1 and 4.2). The **cleithrum** is the main supporting element of the pectoral girdle. It is a large, dorsoventrally elongated bone that lies mainly deep to the operculum and suboperculum and extends to the ventral midline to articulate with the cleithrum from the other side of the body. Dorsally it articulates with the **supracleithrum**, which in turn articulates with the **posttemporal**. The latter is attached to the posterior end of the skull. This chain of connections links the head and shoulder girdle. The **scapula** and **procoracoid** are the large skeletal elements of the shoulder girdle that directly support the fin. The scapula lies dorsal to the procoracoid. Their anterior ends lie deep to and are covered by the cleithrum. The **postcleithrum** is an elongated triangular bone that extends dorsoventrally. Its widened end lies deep to the cleithrum. It tapers ventrally, passes medial to the fin, and extends toward the pelvic fin. The radial pterygiophores, of which there are usually four, extend from the scapula or procoracoid and distally articulate with the soft fin rays supporting the fin. Movement is possible proximally (between the scapula or procoracoid and radial pterygiophores) and distally (between the radial pterygiophores and fin rays).

The pelvic girdle is formed by paired **basipterygia**. Each is a triangular plate of bone oriented anteroposteriorly, with the base located posteriorly. As it passes anteriorly, the basipterygium tapers and passes dorsal to the articulation between the right and left cleithra. The fin rays of each fin attach directly to the posterior end of a basipterygium and, except for the medial ray, are soft fin rays.

Key Terms: Postcranial Skeleton

anal fin
anterior dorsal fin
basipterygium (pl., **basipterygia**)
caudal fin
caudal vertebrae
centrum
cleithrum (pl., **cleithra**)
dorsal ribs
epurals
fin rays
hemal spines
hypurals
neural spines
postcleithrum
posterior dorsal fin

posttemporal
procoracoid
radial pterygiophores
scapula
supracleithrum
trunk vertebrae
uroneurals
ventral ribs

SECTION II: EXTERNAL ANATOMY

The external anatomy of the perch (Figure 4.3) is similar in several aspects to that of the dogfish shark. The body, which may be subdivided into **head**, **trunk**, and **tail** regions, is generally streamlined, not surprising in a swimming fish, and there are several fins. The constricted region connecting the trunk and tail regions is the **caudal peduncle**. Several differences are immediately apparent, however. The skin, for example, has numerous, readily apparent **scales**. Also, there is only a single opening on each side of the body for the exit of water from the pharynx, and the positions of the paired fins are quite different.

Examine the head (Figure 4.3). On each side it bears a large **eye**, lacking lids. Posterior to it, the preopercular region, containing the bones that help support the jaws, and the opercular region, containing the opercular bones covering the gills, are easily recognizable. The large **mouth** is terminal in the perch but may be slightly dorsal or ventral in other teleosteans. Note that the **maxilla**, a bone of the upper jaw, is free posteriorly, embedded in a fold of skin, and lacks **teeth** (see also Figure 4.2). The **premaxilla** is recognizable also, at the anterior end of the upper jaw, and can slide back and forth, thus allowing the perch (and most other teleosteans) to protrude its jaws. Teeth are present on the premaxilla, as well as on the lower jaw or **mandible**. **Nares** (sing., **naris**) can be found anterior to each eye. On each side of the head there are two nostrils, one anterior and the other posterior, opening into the nasal cavities. Water enters the nasal cavity through the anterior nostril and exits through the posterior nostril.

On the trunk and caudal peduncle, the prominent **lateral line** forms a distinct ridge along the scales. Other canals occur on the head but are much less conspicuous. Using forceps, pull a scale out from the trunk (Figure 4.4). Most of the scale is embedded in the skin, and only a small posterior portion is exposed. This posterior end, termed the bony portion (though it is distinct from true bone), has numerous small tooth-like projections called *ctenii*. This type of scale is termed **ctenoid** (comb-like), based on the structure of its posterior end (other teleosteans may have circular or cycloid scales because they lack ctenii; some lack scales). The embedded portion of the

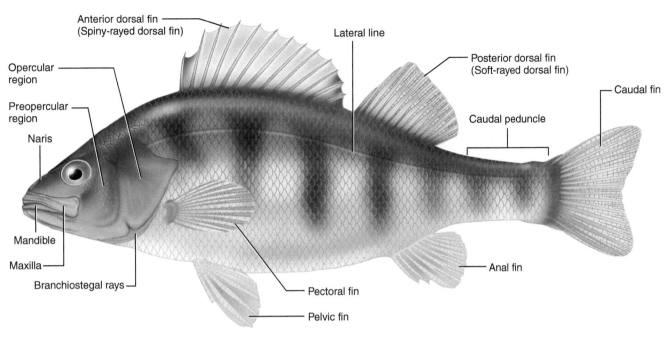

FIGURE 4.3   External features of the perch in left lateral view.

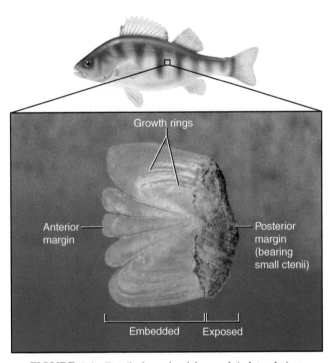

FIGURE 4.4   Detail of a scale of the perch in lateral view.

scale is made of fibrous connective tissue. Scales grow as the fish ages. The concentric growth rings on the embedded portion of the scale can be used to age an individual.

There are four median fins: the **anterior dorsal, posterior dorsal**, **anal**, and **caudal fins** (Figure 4.3). Identify the anterior and posterior dorsal fins along the dorsal midline. The anterior dorsal fin is larger and is supported by ossified fin rays, as noted earlier. Most of the supporting fin rays in the posterior dorsal fin are unossified and

flexible. The anal fin is on the ventral midline, just anterior to the tail, and is supported mainly by soft fin rays. The homocercal caudal fin is superficially symmetrical. Note the paired fins, the **pectoral** and **pelvic fins**. Although some teleosteans (for example, the catfish) have these fins in positions comparable to those in the shark, in the perch the pectoral fin is displaced dorsally and the pelvic fin is displaced anteriorly.

Lastly, examine the posterior openings of the urogenital and digestive tracts. Unlike the shark, the perch does not have a cloaca. Instead the digestive tract has a separate opening, an **anus**, the large, circular opening anterior (Figures 4.7–4.9) to the anal fin. The **urogenital aperture** is considerably smaller and less evident, and lies immediately posterior to the anus (Figures 4.7–4.9). In some females (see later), however, the urogenital opening may be as large as, and even larger than, the anus.

Key Terms: External Anatomy

anal fin
anterior dorsal fin
anus
caudal fin
caudal peduncle
ctenoid
eye
head
lateral line
mandible
maxilla
mouth
naris (nostril)

pectoral fin
pelvic fin
posterior dorsal fin
premaxilla
scales
tail
teeth
trunk
urogenital aperture

SECTION III: MOUTH, ORAL CAVITY, AND PHARYNX

Examine the **mouth**, which forms the anterior opening of the digestive tract. It is terminal, a position common in fish that swim to overtake their prey. Note the posterior end of the maxilla. Embedded in soft tissue, it is free to move laterally, an important feature during expansion of the oral cavity during feeding. Note the marginal series of **teeth** in the upper and lower jaws. In addition to the marginal series, there are **palatal teeth** in the roof of the **oral cavity**, and **pharyngeal teeth**, both upper and lower, in the posterior part of the **pharynx**. Postpone identifying them until the pharynx is opened (see later).

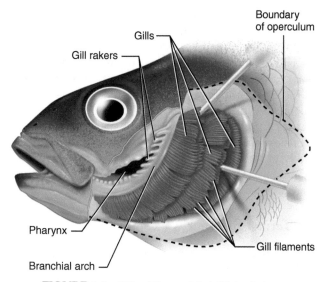

FIGURE 4.5   Gills of the perch in left lateral view.

Expose the oral cavity and pharynx by removing the opercular bones as follows. Lift the free, posterior end of the operculum. Insert one blade of a stout pair of scissors beneath the surface of the operculum, at approximately the midheight of the posterior margin. Keeping the blade close to the deep surface of the operculum, cut through the bones, heading toward and through the angle of the mouth. Spread the flaps

to observe the **gills**, each composed of numerous **gill filaments**, which are involved with gas exchange with the water flowing over them. Then cut away, bit by bit, the opercular flaps covering the gills until you have exposed the region, as shown in Figure 4.5. The most anterior of the four **branchial arches**, and the gill it supports, should be plainly visible. Manipulate the arches and gills to identify the remaining three arches. **Gill rakers** should be plainly visible on the first arch. The rakers are projections that extend inward across the pharyngeal slit. They help in feeding, preventing prey (and other debris) from passing through the pharyngeal openings and escaping.

In gas exchange, water passes through the mouth and oral cavity into the pharynx. It then passes over the gills and into the opercular chamber (which corresponds to the parabranchial chambers seen in the shark), the space between the gills and operculum, and then leaves posteriorly through the opening at the posterior end of the operculum. The floor of the opercular chamber is supported by the **branchiostegal rays**.

There are five passages or slits through the pharynx: three are between the four branchial arches, one is anterior to the first arch, and another is posterior to the last arch. Each arch bears a double set of filaments. This is similar to the condition in the **holobranch** of the shark, in which gill lamellae are present on the anterior and posterior surfaces of the interbranchial septum (see page 75). In the holobranch of the perch, however, the septum is absent, so the gill filaments are positioned almost side by side. Still, they correspond to the anterior and posterior sets of lamellae of the shark, with the anterior filaments being **posttrematic** ("after the slit") and the posterior being **pretrematic** ("before the slit"). The perch thus has four holobranchs. Unlike in the shark, there is no hemibranch (and, of course, no pseudobranch, as the spiracle is absent).

Key Terms: Mouth, Oral Cavity, and Pharynx

branchial arches
branchiostegal rays
gill filaments
gill rakers
gills
holobranch
mouth
oral cavity
palatal teeth
pharyngeal teeth
pharynx
posttrematic
pretrematic
teeth

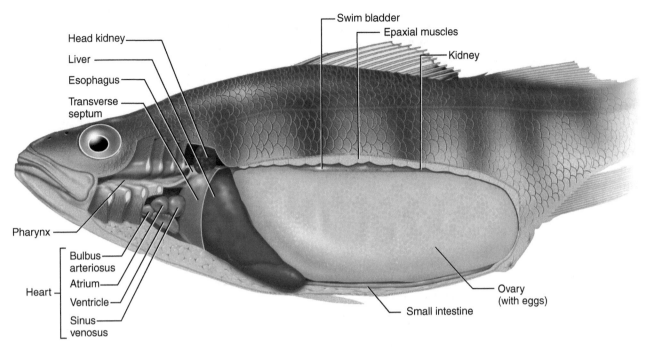

FIGURE 4.6 Left lateral view of the perch with body wall cut away to reveal pharynx and pleuroperitoneal cavity.

SECTION IV: PLEUROPERITONEAL CAVITY AND VISCERA

To expose the **pleuroperitoneal cavity** and viscera, make an incision along the midventral line, proceeding as follows. In a female with an enlarged urogenital aperture, insert a scissor blade into the anterolateral margin of the aperture. Make a small incision, approximately 0.5 cm, in this direction, and then turn back anteromedially toward the midline, anterior to the anus. This will avoid injury to the digestive tract. In a specimen with a small urogenital aperture, proceed as just described but begin by inserting the scissor blade into the anus. Continue to cut anteriorly along the midline, keeping close to the deep surface of the body wall, past the pelvic girdle. Turn your incision dorsally to pass posterior to the attachment of the pectoral fin on the body. Cut dorsally past the pectoral fin to a point just posterior to the eye.

Return to the posterior end of the midventral incision. Cut dorsally for about 1 cm, then reflect the flap of body wall and examine the dorsal part of the cavity. You should see a dark, membranous sac, the **swim bladder** (Figures 4.6–4.8). Probe it gently. Resume cutting through the body wall until you reach the level just ventral to the swim bladder (this point will be ventral to the position of the dorsalmost point of the anterior vertical incision). Then cut anteriorly, more or less parallel to the swim bladder, to join the anterior vertical incision. Your incision should veer slightly dorsally as you cut. In addition to the musculature of the body wall, you will also cut through the ventral ribs. Once finished, remove the

section of body wall and examine the underlying structures. The viscera are covered by **visceral peritoneum**, whereas the cavity itself is lined by **parietal peritoneum**.

Many female specimens will possess a very large **ovary** that seems to fill most of the pleuroperitoneal cavity (Figure 4.6). In such specimens you will probably be able to see only the darkly colored **liver** at the very anterior end of the cavity. Find the narrow **small intestine** ventral to the ovary. It is normally midventral, but may be displaced to one side by a massive ovary. Note the swim bladder dorsal to the ovary. In specimens with smaller gonads, several other visceral structures may be observed without further dissection. The small intestine is easily identifiable on the floor of the cavity. Posterior to the liver, you should observe the short, thick **stomach**, and perhaps the **spleen** and **gall bladder** if you have opened the right side of the cavity.

Once you have identified these structures, remove more of the body wall, preferably in small pieces, as follows. Remove the wall anterior to the liver, keeping close to deep surface of the wall and continuously checking that you are not destroying underlying structures. Immediately anterior to the liver is the **transverse septum**, a thin membrane that separates the pleuroperitoneal cavity and the **pericardial cavity**. Cut through the transverse septum along its attachment to the body wall to expose the **heart** within the pericardial cavity.

At this stage, continue the midventral incision anteriorly to the posterior margin of the mouth, as shown in Figures 4.8 and 4.9. The musculature is thicker here, so cut carefully to avoid damaging the heart. You may wish to use a scalpel. Also cut away the lateral portions of the branchial arches to expose the pharynx. Then remove more of the lateral body wall dorsal to the liver. This will expose a small, dark, lobulated mass, the **head kidney**, lying immediately dorsal to the liver. Once you have exposed the head kidney, use a scalpel to cut a parasagittal section through the musculature dorsal to the swim bladder. This will allow you to expose the **kidneys**, which lie against the dorsal wall of the cavity, dorsal to the swim bladder. Be careful in using the scalpel. It is worth removing a row or two of scales along the path you intend to cut. If this method proves too awkward, find the kidneys by removing the swim bladder, but do so after you have examined the remaining structures described next.

The preparation described above is time-consuming, but it reveals the pattern, context, and arrangement of the various systems and their structures in a single view. Although you can begin with any of the structures, it is best to examine the gonads first, because in many specimens they will be so large that they will have to be removed.

The ovary of the female will vary considerably in size with the reproductive cycle of the fish and may be massive, filled with eggs (Figure 4.6). The ovary of the perch is a secondarily fused single structure (although this is not true of most teleosteans, which retain paired ovaries) that is enveloped during embryonic development by bilateral peritoneal folds. This envelopment continues posteriorly and meets a funnel-like internal elongation of the urogenital aperture, lying just posterior to the anus. This combination (i.e., of the peritoneum and internal elongation of the aperture) forms an **ovarian duct** for passage of the eggs. The ovarian duct, however, is not homologous to that (usually considered a Müllerian duct) used for the same purpose in other vertebrates. Gently tugging the posterior end of the ovary will pull taut the funnel-shaped posterior end of the ovarian duct and make it easier to distinguish. In other teleosteans the ovarian duct is formed differently. For example, a common pattern is that the ovary contains an internal cavity, formed by envelopment of a small part of the coelomic cavity during embryonic development. The eggs are shed into this cavity, the lining of which extends posteriorly to form an ovarian duct. In most teleosteans, therefore, the eggs are released directly into a tube, the ovarian duct, rather than into the pleuroperitoneal cavity, as occurs in almost all other vertebrates.

Cut transversely through the ovary, approximately 3 cm from its posterior margin, and carefully remove the anterior portion. This will leave a cone-shaped posterior end in place. Gently reflect it ventrally and delicately dissect between the ovary and swim bladder, now clearly visible, to expose the small, light-colored, elongated, and oval **urinary bladder** (Figure 4.7). The bladder continues

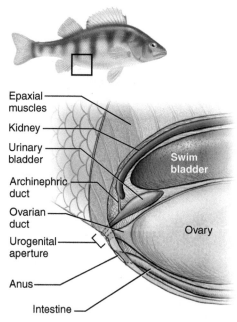

FIGURE 4.7 Schematic illustration showing the urogenital structures of the female perch in right lateral view.

posteroventrally into the urinary opening of the urogenital aperture, but postpone tracing it.

The paired **testes** of the male perch are lobulated, light colored, and posteriorly tapered structures (Figure 4.8). Each testis has its own duct, the **testicular duct**, which carries only sperm. The testicular duct is a specialization of teleosteans (although not all teleosteans, such as salmonids, possess a duct for the testis; instead, sperm are released by the testes into the body cavity and leave the body through pores) and is not comparable to the archinephric duct observed in sharks. It is fairly small and thus difficult to find without a dissecting microscope. Right and left testicular ducts unite near the posterior end of the testes into a single duct that leads out of the body through the urogenital aperture just posterior to the anus. The opening of the duct can be distinguished from that of the **archinephric duct** (see later) with a magnifying glass. The urinary bladder lies dorsal to the posterior end of the testes.

Turn your attention to the anterior end of the animal (Figures 4.8 and 4.9). Identify the oral cavity—look now for the teeth described earlier—and follow it posteriorly into the pharynx. The pharynx leads into the wide, short, and straight **esophagus** that passes posteriorly into the stomach. The stomach is **T**-shaped, with a broad horizontal portion and a short, vertical, pyloric portion forming the stem of the **T**. The coiled intestine follows the pyloric portion. Note the three finger-like projections, the **pyloric ceca**, at the anterior end of the intestine. These are typically present in teleosteans, though their number varies. The anterior part of the intestine, the **duodenum**, is somewhat wider than the remaining distal portion.

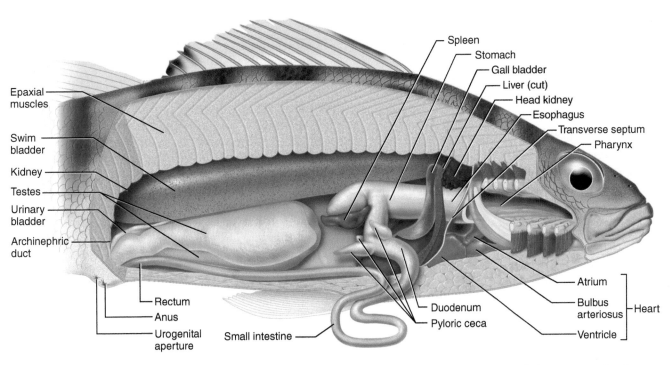

FIGURE 4.8 Cutaway view of the male perch in right lateral view to reveal structures of the pharynx and pleuroperitoneal cavity.

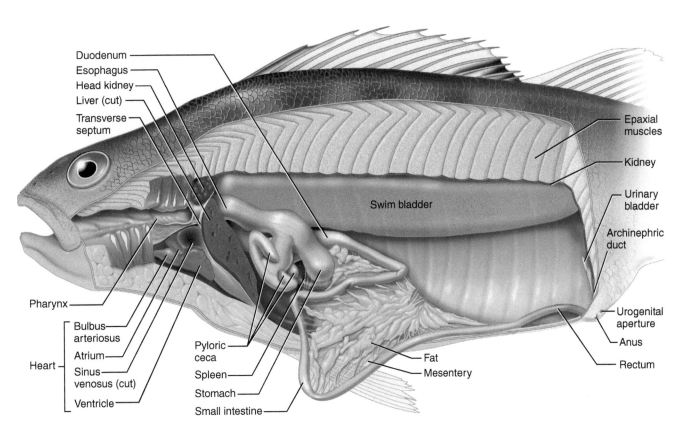

FIGURE 4.9 Cutaway view of the female perch, with ovary removed, in left lateral view to reveal structures of the pharynx and pleuroperitoneal cavity.

Although its terminal portion may be referred to as a **rectum**, it is not sharply demarcated from the rest of the intestine.

Note the structures associated with the digestive tract. The large, massive liver has already been noted. It may be necessary to remove part of it, as shown in Figures 4.8 and 4.9, in order for the stomach and other organs to be seen properly. The gall bladder is a small, elongated sac. The bile duct leads to the duodenum, but it is difficult to find. The **pancreas** cannot be seen grossly. Bits of pancreatic tissue are scattered throughout the mesentery, often embedded in the fatty tissue there (Figure 4.9). The spleen, not properly an organ of the digestive system but concerned with production of blood cells, is an elongated, dark structure near the posterior end of the stomach.

The swim bladder is the large, hollow sac lying, as noted earlier, dorsally in the body cavity. It is not enclosed by the peritoneum, and so is retroperitoneal in position. It is a hydrostatic organ used to control buoyancy. Its inflation decreases the fish's density, thus increasing buoyancy. Its deflation has the opposite effect. The bladder develops as an outgrowth of the anterior part of the digestive tract, and in many teleosteans it retains an open duct connection to the esophagus, a condition termed *physostomous*. In the perch, however, the connection between the bladder and gut is lost, termed the *physoclistous* condition.

The kidneys lie dorsal to the swim bladder and are thus also retroperitoneal. They are long, narrow, ribbon-like structures with somewhat scalloped lateral margins lying on either side of the dorsal midline of the body cavity. Posteriorly, the kidneys curve ventrally, following the contour of the body cavity (Figure 4.7). Each kidney is drained by an archinephric duct (in the males of some species it may also receive sperm, but the more common condition is that represented by the perch, in which a separate testicular duct serves for sperm passage). The right and left archinephric ducts enter the urinary bladder. Urine exits the body through a single duct leading to the urinary opening of the urogenital aperture. Dissection of this region to reveal the ducts is difficult without a microscope and considerable patience. Figure 4.7 indicates the structures and their relationships.

Finally, examine the heart, which has already been exposed and noted. As in the shark, the heart is an S-shaped, four-chambered structure that receives venous blood posteriorly and pumps it anteriorly into the gills (Figures 4.8 and 4.9). The most posterior chamber is the **sinus venosus**, which directs blood into the **atrium** lying immediately anterior to it. From the atrium, blood enters the **ventricle**, which lies ventrally. The ventricle pumps blood through the fourth chamber, the **bulbus arteriosus**, which leads into the ventral aorta. Afferent branchial arteries branch off the ventral aorta, leading blood through the gills. Efferent branchial arteries recollect the blood into the dorsal aorta, which distributes it to the various parts of the body. Unless you have an injected specimen, the vessels will be difficult to follow.

Key Terms: Pleuroperitoneal Cavity and Viscera

archinephric duct (Wolffian duct)
atrium
bulbus arteriosus
duodenum
esophagus
gall bladder
head kidney
heart
kidneys
liver
ovarian duct
ovary
pancreas
parietal peritoneum
pericardial cavity
pleuroperitoneal cavity
pyloric cecum (pl., **ceca**)
rectum
sinus venosus
small intestine
spleen
stomach
swim bladder
testes
testicular duct
transverse septum
urinary bladder
ventricle
visceral peritoneum

The Mudpuppy

INTRODUCTION

The common mudpuppy, *Necturus maculosus*, is a member of Caudata, which together with Anura (frogs and toads) and Gymnophiona (caecilians or apodans) form Lissamphibia (see Chapter 1). Caudata, including the salamanders and newts, are the least specialized lissamphibians in body form and locomotion. The body is elongated and stout, with well-developed axial musculature and tail. Salamanders use their limbs in combination with the side-to-side body undulations characteristic of fish and thus may resemble the earliest land vertebrates in locomotion. However, this subject is under intense investigation and this view of early tetrapod locomotion may be more appropriate for the later stages in the water-to-land transition rather than in the earlier stages (see Pierce et al., 2013). By contrast, frogs (see Chapter 6) are characterized by a shortened body and specialized saltatory locomotion, while the limbless gymnophionans inhabit terrestrial to aquatic habitats.

There are several clades of salamanders, of which Plethiodontidae includes by far the most species. Salamanders range from being fully aquatic to fully terrestrial. Members of several clades do not metamorphose. *Necturus* is commonly included with *Proteus* in Proteidae, a small group of fully aquatic, neotenic (i.e., retaining juvenile features) salamanders that retain their larval, filamentous external gills, two pairs of pharyngeal slits, and caudal fins as adults. *Necturus* includes six species. *N. maculosus* is apparently closely related to *N. lewisi*, the Red River mudpuppy. The remaining species of *Necturus* are commonly referred to as waterdogs.

SECTION I: SKELETON

Cranial Skeleton

Skull

Examine the skeleton of the head on isolated specimens (Figures 5.1–5.3), as well as on a mounted specimen

(Figure 5.4), to identify the **skull**, **mandible**, and **hyoid apparatus**. The skull (Figure 5.1) is formed from a dermal skull roof, chondrocranium (Figure 5.2), and palatal complex, itself formed from dermal bones and a remnant of the splanchnocranium (see later). Other parts of the splanchnocranium form the mandible (Figure 5.2) and the hyoid apparatus (see Figures 5.1 and 5.3).

The dermal skull roof covers the brain and major sense organs dorsally. The ventral surface, or underside, of the skull is formed mainly by the palatal complex. Between the dermal roof and palatal complex is the chondrocranium, which remains mainly cartilaginous and has only small exposures on the outside of the skull. In the shark (see page 54) the chondrocranium forms a nearly complete enclosing and supportive structure for the brain and major sense organs. The chondrocranium is highly developed in the shark and other chondrichthyans, but this is necessary because of the near absence of bone in these vertebrates. In most other vertebrates, bone covers the brain dorsally and forms the floor of the skull, so the chondrocranium is a relatively minor structure. This is particularly true in *Necturus*, in which the chondrocranium (Figure 5.2) retains larval characteristics, as do many other parts of the body, and is even less developed than in most vertebrates.

Examine an isolated skull. In dorsal view (Figure 5.1a) the paired **premaxillae** are seen anteriorly, followed by the paired **frontals** and **parietals**. Note the long slender anterior extension of the parietal that passes along the lateral margin of the frontal. The cartilaginous **antorbital processes**, which are part of the chondrocranium (Figure 5.2), project laterally near the anterior end of the parietal bones, but these delicate elements are often missing in prepared specimens. *Necturus* lacks maxillae, elements that are commonly present in the skull of tetrapods, including most other caudates. The maxilla is functionally replaced by a modified **vomer**, which forms the anterolateral margin of the skull. Posterior to the antorbital process, the **palatopterygoid** forms the central portion of the lateral margin.

The otic capsules lie at the posterolateral corners of the skull. The paired bones that contribute to each capsule are the **prootics** and the **opisthotics** (Figure 5.1a–c). The prootic lies lateral to the parietal near the suture of the parietals and frontals, and the opisthotic lies posterior to the **squamosal**. A cartilaginous region, containing the **fenestra ovalis**, lies between the prootic and opisthotic. The squamosal extends anterolaterally from the opisthotic, forming the posterolateral margin of the skull. The **quadrate** articulates with the anterior end of the squamosal and inclines anteroventrally (Figure 5.1c). It contacts the pterygoid anteriorly (Figure 5.1a,b) and

forms the jaw joint at its articulation with the mandible. The **quadrate cartilage** forms a slender bridge extending from the anterior end of the quadrate toward the lateral margin of the parietal (Figure 5.1a,c).

Just anterior to the opisthotic, and thus essentially covering the cartilaginous region between the prootic and opisthotic, is the disc-shaped **columella**. The columella covers the fenestra ovalis, mentioned earlier. A small stylus projects from its lateral surface. The columella, with its stylus, is a small delicate bone that transmits vibrations to the inner ear. It may be missing in your specimen, in which case the fenestra ovalis should

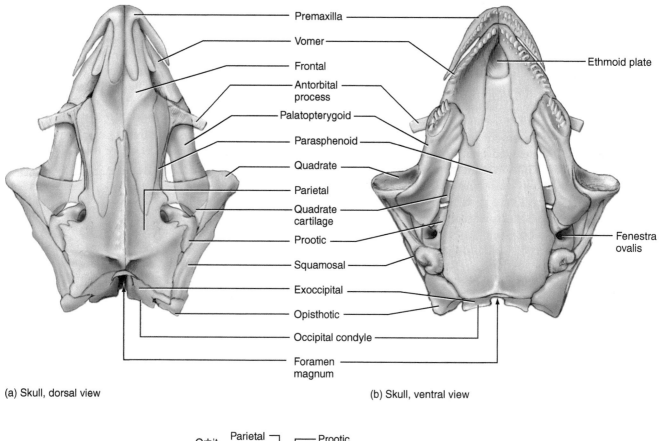

(a) Skull, dorsal view

(b) Skull, ventral view

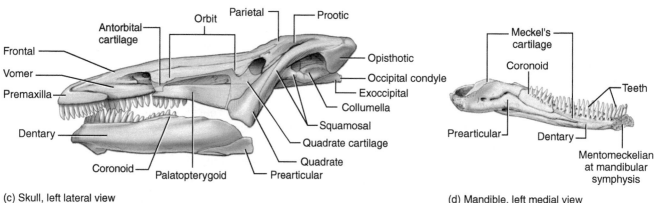

(c) Skull, left lateral view

(d) Mandible, left medial view

FIGURE 5.1 Skull and mandible of the mudpuppy: (a) skull in dorsal view; (b) skull in ventral view; (c) skull and mandible in left lateral view; and (d) left mandible in medial view.

be clearly visible. The **foramen magnum** (Figures 5.1a,b) is the large opening on the posterior surface of the skull through which the spinal cord passes. The **exoccipitals**, each bearing an **occipital condyle**, form the skull lateral and ventral to the foramen magnum.

Examine the skull in ventral view (Figure 5.1b) and identify the premaxillae, vomers, and palatopterygoids. Note that the premaxillae and palatopterygoids each bear a short row of relatively large anterior teeth, whereas the vomers bear a longer row of teeth. Close inspection reveals a row of smaller teeth lying along the lingual margins of the larger teeth. The large, triangular, and plate-like **parasphenoid** forms nearly all of the rest of the skull's ventral surface. A portion of the cartilaginous **ethmoid plate** is visible anterior to the parasphenoid. The palatopterygoid, identified earlier, apparently represents a compound element formed by the pterygoid and palatine, as noted by Trueb (1993). The palatine, which is usually present as an independent element in the palatal region of tetrapod skulls, is not individually recognizable in caudates (at least in adults).

The chondrocranium should be studied in a separate preparation (Figure 5.2), usually set in an acrylic block, even though many of its elements have already been observed. The posterior half of the chondrocranium consists of the otic capsules, with its cartilaginous portion set between the bony prootic and opisthotic, and the exoccipitals. Left and right otic capsules are connected dorsally by a delicate cartilaginous bridge, the **synotic tectum**. The **basal plate** forms a similar bridge between left and right exoccipitals. A **parachordal plate** extends anteriorly from each otic capsule. The **trabecular cartilages** are the slender rod-like elements extending anteriorly from the parachordals. The antorbital cartilages

(noted earlier) project laterally from each trabecular cartilage. Beyond the antorbital cartilages the trabecular cartilages converge to form the ethmoid plate (noted earlier) and then send a pair of **trabecular horns** anteriorly between the nasal capsules, which are extremely delicate and usually not preserved.

Just anterior and lateral to each otic capsule, many preparations include the bony and cartilaginous components of the quadrate. These have already been observed, but remember that they are part of the palatal complex rather than the chondrocranium. They are in part homologous to the palatoquadrate cartilage, and thus belong to the mandibular arch, which is part of the splanchnocranium.

Mandible

The mandible is formed by four elements (Figure 5.1c,d). **Meckel's cartilage** is a cartilaginous structure extending through the interior of the mandible. It is visible mainly in medial view (Figure 5.1d), particularly posteriorly, where it forms the articular surface for the quadrate of the upper jaw. Commonly (but not in *Necturus*), it is this posterior region where the articular ossifies to help form the quadrate–articular jaw joint common in nonmammalian tetrapods. Anteriorly, Meckel's cartilage ossifies as the **mentomeckelian** at the mandibular symphysis.

Meckel's cartilage is mostly covered by the **dentary**, **coronoid**,[1] and **prearticular**.[2] The dentary covers most of the lateral surface and bears a long row of marginal teeth (Figure 5.1c,d). The coronoid, which bears a shorter row of marginal teeth, is broadly exposed in medial view, but has a tiny exposure laterally along the dorsal margin of the central part of the mandible. The prearticular covers much of the posterior part of the mandible in medial view, tapering anteriorly as a wedge on the dentary.

Hyoid Apparatus

The hyoid apparatus (Figure 5.3) is large and composed of the hyoid arch and parts of the first three branchial arches. The hyoid arch is the largest and most anterior, and supports the tongue. It is composed on each side by the small **hypohyal** and, more laterally, the larger **ceratohyal**. The median **basibranchial 1** extends posteriorly from the hypohyals to the base of the first branchial arch, which is also composed on each side by two large

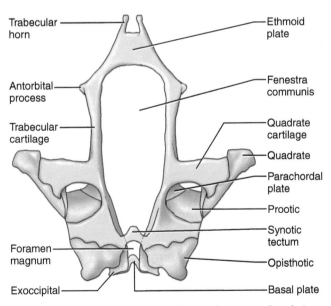

Trabecular horn

Antorbital process

Trabecular cartilage

Foramen magnum

Exoccipital

Ethmoid plate

Fenestra communis

Quadrate cartilage

Quadrate

Parachordal plate

Prootic

Synotic tectum

Opisthotic

Basal plate

FIGURE 5.2   Chondrocranium of the mudpuppy in dorsal view.

[1]In several dissection manuals (such as Homberger and Walker, 2003; Wischnitzer and Wischnitzer, 2006), the coronoid is termed the splenial. Our designation follows Deullman and Trueb (1994).

[2]This element is labeled as the articular in the dissection manuals noted immediately prior. However, Deullman and Trueb (1994) indicated that the medial dermal bone in this region is the prearticular, and that the angular (commonly present as an independent element in nonmammalian tetrapods) is fused posteriorly to the prearticular.

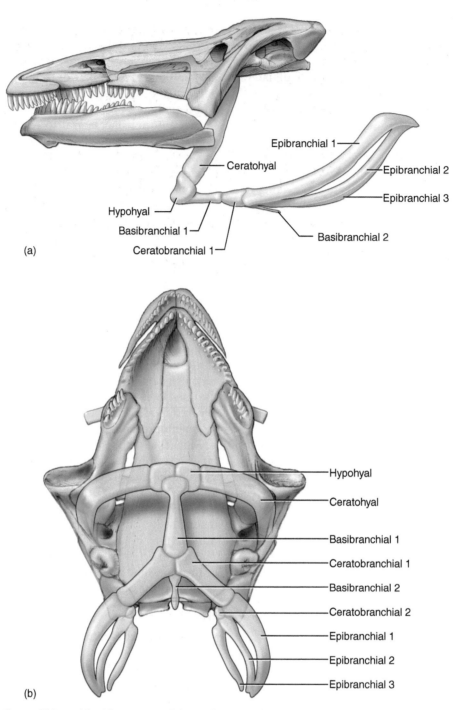

FIGURE 5.3   (a) Skull, mandible, and hyoid apparatus of the mudpuppy in left lateral view; and (b) skull and hyoid apparatus of the mudpuppy in ventral view.

elements, **ceratobranchial 1** and **epibranchial 1**. The more slender **basibranchial 2** extends posteriorly from the left and right ceratobranchial 1. **Ceratobranchial 2** is a tiny, nodular element at the base of **epibranchial 2**. **Epibranchial 3** lies posterior to epibranchial 2. The remaining branchial arches are apparently represented by tiny cartilaginous elements, but these are rarely preserved in prepared specimens.

Key Terms: Cranial Skeleton

antorbital plate
basal plate
basibranchial
ceratobranchial
ceratohyal
columella

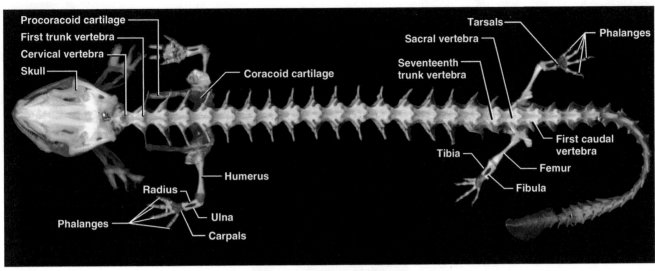

FIGURE 5.4 Skeleton of the mudpuppy in dorsal view.

coronoid (splenial)
dentary
epibranchial
ethmoid plate
exoccipital
fenestra ovalis (oval window)
foramen magnum
frontal
hyoid apparatus
hypohyal
mandible
Meckel's cartilage (mandibular cartilage)
mentomeckelian
occipital condyle
opisthotic
palatopterygoid
parachordal plate
parasphenoid
parietal
prearticular
premaxilla
prootic
quadrate
quadrate cartilage
skull
squamosal
synotic tectum
trabecular cartilage
trabecular horn
vomer

Postcranial Skeleton

Axial Skeleton

The axial skeleton consists of four types of vertebrae (Figure 5.4). Most anteriorly is a single **cervical vertebra**, the **atlas**, that articulates with the occipital condyles of the skull. In addition to this set of articulations is another: the anteroventral process of the atlas (the odontoid process) projects into the foramen magnum and articulates with the lateral margins of the foramen. Posteriorly, the atlas articulates with the first of the long series of **trunk vertebrae** (Figures 5.4 and 5.5). The latter bear **transverse processes**, which extend from the **centrum** and articulate with small, Y-shaped, posterolaterally directed **ribs**. The **neural canal** passes dorsal to the centrum. The **neural arch** forms the roof of the canal and bears a **neural process**. **Prezygapophyses** and **postzygapophyses** are present for articulation with the preceding vertebra and succeeding vertebra, respectively. A single **sacral vertebra** (Figures 5.4 and 5.8) articulates with the pelvic girdle by way of its ribs. The **caudal vertebrae** (Figure 5.6), each bearing a **hemal arch**, follow the sacral vertebra. Most lack ribs but bear transverse processes.

Appendicular Skeleton

The **pectoral girdle** (Figure 5.7) is composed of left and right halves that overlap ventrally but do not fuse. The ossified **scapula** is a short bone that dorsally bears the **suprascapular cartilage**. The **glenoid fossa**, the depression that articulates with the forelimb, is on the ventral surface of the scapula. The slender, elongated **procoracoid cartilage** projects anteriorly from the glenoid fossa. The **coracoid cartilage** forms the broad, ventral, plate-like part of the girdle; it is these cartilages that partly overlap.

As is typical of tetrapods, the forelimb consists of the brachium, antebrachium, and manus in proximal to distal order. The brachium is supported by the **humerus**. In the antebrachium the **radius** is the anteromedial bone, and the **ulna** is the posterolateral bone. The manus includes six cartilaginous **carpals**, followed distally by

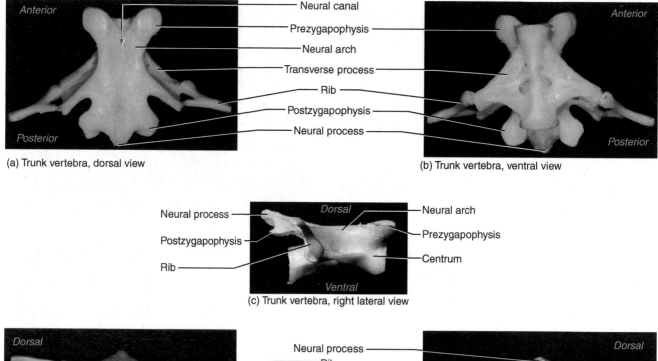

FIGURE 5.5   Trunk vertebra of the mudpuppy: (a) dorsal, (b) ventral, (c) right lateral, (d) anterior, and (e) posterior views.

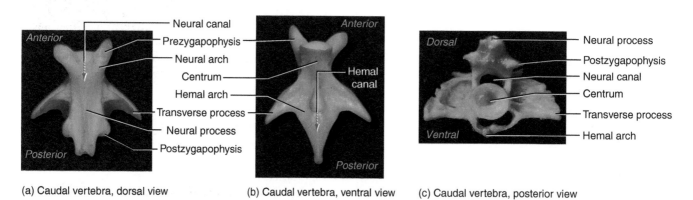

FIGURE 5.6   Caudal vertebra of the mudpuppy: (a) dorsal, (b) ventral, and (c) posterior views.

four digits. The most proximal segment of each digit is a **metacarpal**, and the more distal elements are **phalanges**. There is some ambiguity as to the use of the term digit. It is considered by some to include only the phalanges, the free elements of the manus (or pes), and by others to include the metacarpals (or metatarsals) plus the phalanges. Here, we follow the latter scheme, as outlined by Kardong (2018).

The **pelvic girdle** (Figure 5.8), like the pectoral, is mainly cartilaginous. The **ilium** is ossified and extends ventrally

from its articulation with the rib of the sacral vertebra toward the **acetabulum**, the depression that articulates with the hind limb. The girdle is formed ventrally by the expansive **puboischiadic plate**. Examine the plate in ventral view. The elongated, triangular **pubic cartilage** is the anterior portion. The posterior portion is the **ischiadic cartilage**, which contains a pair of ossifications termed **ischia**. Note the **obturator foramen** just anterior to each acetabulum.

The hind limb is also formed of three segments, the thigh, crus, and pes. The **femur** is the single bone of

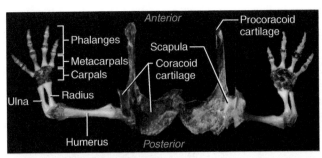

(a) Pectoral girdle, dorsal view

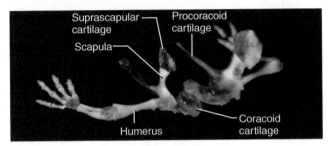

(b) Pectoral girdle, left posterodorsolateral view

FIGURE 5.7 Pectoral girdle and forelimbs of the mudpuppy: (a) dorsal and (b) posterodorsolateral views.

the thigh. The **tibia** and **fibula** lie anteriorly and posteriorly, respectively, in the crus. The pes is formed proximally by six cartilaginous **tarsals** and distally by four ossified digits. The proximal bone of each digit is a **metatarsal**, while the distal elements are **phalanges**.

Key Terms: Postcranial Skeleton

acetabulum
atlas
carpals
caudal vertebrae
centrum (vertebral body)
cervical vertebra
coracoid cartilage
femur
fibula
glenoid fossa
hemal arch
humerus
ilium
ischiadic cartilage
ischium (pl., **ischia**)
metacarpal
metatarsal
neural arch (vertebral arch)
neural canal (vertebral canal)
neural process (spinous process)
obturator foramen
pectoral girdle

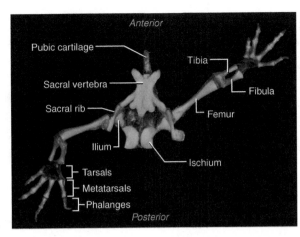

(a) Pelvic girdle, dorsal view

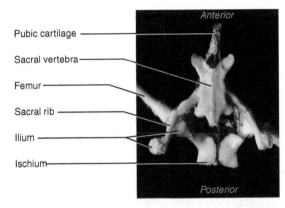

(b) Pelvic girdle, dorsal view (closeup)

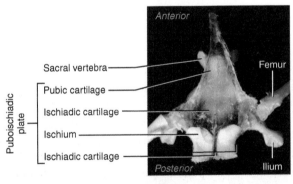

(c) Pelvic girdle, ventral view (closeup)

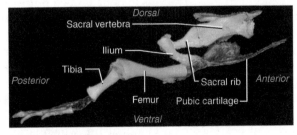

(d) Pelvic girdle, right lateral view

FIGURE 5.8 Pelvic girdle and hind limbs of the mudpuppy: (a) dorsal view, (b) close-up of pelvic girdle in dorsal view, (c) close-up of pelvic girdle in ventral view, and (d) right lateral view.

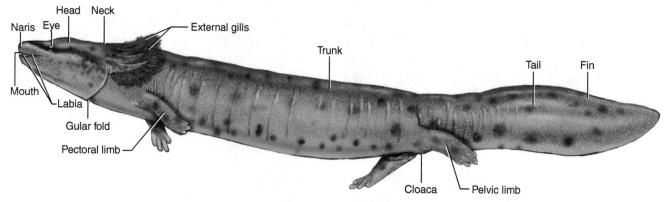

FIGURE 5.9   External features of the mudpuppy in left lateral view.

pelvic girdle
phalanges
postzygapophyses
prezygapophyses
procoracoid cartilage
pubic cartilage
puboischiadic plate
radius
ribs
sacral vertebra
scapula
suprascapular cartilage
tarsals
tibia
transverse process
trunk vertebrae
ulna

SECTION II: EXTERNAL ANATOMY

Perhaps the mudpuppy's (Figure 5.9) most striking larval feature is the presence of three pairs of **external gills**, located just posterior to the flattened **head**. Also retained are two pairs of **pharyngeal slits**, one between the first and second gills, the other between the second and third gills. The slits may be probed but will be observed later (Figure 5.25). Note the prominent transverse **gular fold**.

The **mouth** is large and bounded by **labia** (sing., labium). The lidless **eyes** are small, as are the widely separated **nares** (sing., naris), which communicate with the oral cavity (see later). A short **neck** is present between the head and the long **trunk**. Posteriorly, the large, flattened **tail** bears a small, marginal **fin**, which lacks the supporting rays present in fishes.

Paired **pectoral** and **pelvic limbs** are small but bear the three segments typical of terrestrial tetrapods. The **cloaca**, marking the posterior end of the trunk, lies ventrally between the pelvic limbs. In males the cloacal

aperture is surrounded by small projections or papillae. The region around the cloaca is swollen due to the presence of the **cloacal gland**. Lastly, note the smooth, scaleless skin, which has an important gas exchange function. A lateral line system is present as well, but is not obvious.

Key Terms: External Anatomy

cloaca
cloacal gland
external gills
eyes
fin
gular fold
head
labia (lips)
mouth
nares (nostrils)
neck
pectoral limbs
pelvic limbs
pharyngeal slits
tail
trunk

SECTION III: MUSCULAR SYSTEM

The musculature of the mudpuppy is more complex than that of fish. Among the many changes involved in the transition from an entirely aquatic to a mainly terrestrial mode of life are those involved in locomotion and support. Although the mudpuppy is aquatic, it has a tetrapod body plan and the musculature, particularly of the appendages and trunk, has become differentiated to serve these specialized functions. In addition, the musculature that controls the jaws and tongue has also become specialized.

Study of the musculature first involves removing the skin. Although the specimen in the figures of this

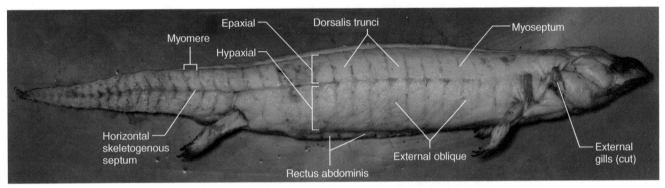

FIGURE 5.10   Musculature of the mudpuppy in right lateral view.

chapter has been skinned completely, you need only skin one half of the body. Make a short, shallow incision in the abdominal region, just to one side of the midventral line, preferably on that side of the body that has already been opened for injection (if your specimen has been so prepared). Use a needle or thin probe to lift the skin from the underlying musculature. Once you are sure that you have cut through only skin, continue the incision, lifting the skin with a probe as you do so, anteriorly to the level of the gular fold and posteriorly past the cloaca and onto the tail.

Begin to separate the skin from the musculature using a probe or needle. Although it is easier to separate the skin from the musculature than in the dogfish, proceed cautiously, as the muscles of the mudpuppy are delicate. Work your way toward the dorsal surface of the opposite side of the body. Anteriorly, make a transverse cut, back across the midventral line and onto the other side of the body, following the margin of the gular fold. Do not remove the skin from the fold at this time. Continue the cut around the posterior edge of the last external gill. Then, curve anteriorly and down along the anterior edge of the first external gill toward the angle of the jaw. Separate the skin from the musculature as you cut. Posteriorly, make a transverse cut, back across the midventral line, and onto the other side of the body.

Make an anteroposterior incision just to one side of the midventral line of the head. Extend the incision, using scissors as described earlier, posteriorly to the margin of the gular fold, and anteriorly to the leading edge of the mandible. Then cut the skin along the margin of the gular fold and along the ventral edge of the mandible, around the angle of the jaw and along the margin of the upper jaw. When you reach the level of the eye, cut dorsally to the posterior edge of the eye and then proceed slightly anterior to the eye toward the dorsal midline. Continue to separate skin from muscles until you have removed the skin from one half of the body.

A classification of the muscles of the mudpuppy is presented in Table 5.1. The classification is based ultimately on embryonic derivation, which makes comparison of different muscles among vertebrates (that is, the establishment of homologies) much easier. It is fairly straightforward to trace the evolutionary changes among different muscles when comparing among similar vertebrates (for example, different sorts of fishes) but is less easy when trying to homologize muscle of vertebrates that have considerably different body types, such as a fish and a tetrapod. On the other hand, strict adherence to such a classification is not the most convenient method to work through a dissection to identify the muscles. Thus, the muscles are described here according to position.

Muscles of the Trunk and Tail

Examine the musculature of the trunk and tail in lateral view (Figure 5.10). It is subdivided, on each side of the body, by the **horizontal skeletogenous septum** into dorsal **epaxial** and ventral **hypaxial** musculature. Compared with that of the dogfish, the epaxial musculature is relatively reduced, in part reflecting its diminished importance in locomotion, and the septum is thus more dorsal in position. The musculature consists of **myomeres** separated by **myosepta** as in fishes, but note that the myomeres are relatively simple rather than assuming the complexly folded pattern observed in fishes. The musculature is subdivided into left and right portions by a middorsal septum (Figure 5.12) and ventrally by the linea alba (Figure 5.14).

The epaxial musculature is subdivided by myosepta, but the more superficial bundles mainly fuse together to form the **dorsalis trunci**, a longitudinal muscle extending anteroposteriorly from the trunk into the tail, where it becomes laterally compressed (Figure 5.10). Deeper portions of the epaxial musculature include bundles, the interspinal muscles, which extend between vertebrae, but these are not illustrated here (though they are listed in Table 5.1).

The hypaxial musculature of fishes comprises a single sheet, but in tetrapods is subdivided into several portions. Among these are three sheet-like flank layers that

TABLE 5.1   Classification of Mudpuppy Musculature.

AXIAL MUSCLES

Epaxial Muscles
  Dorsalis trunci (posterior portion)
  Interspinalis muscles

Epibranchial Muscles
  Dorsalis trunci (anterior portion)

Hypaxial Muscles
  Caudocruralis
  External oblique
  Internal oblique
  Ischiocaudalis
  Levator scapulae (acts on pectoral girdle)
  Rectus abdominis
  Subvertebralis
  Thoraciscapularis (acts on pectoral girdle)
  Transversus abdominis

Hypobranchial Muscles
  Prehyoid Muscles
    Genioglossus
    Geniohyoideus
  Posthyoid Muscles
    Omoarcual (acts on pectoral girdle)
    Pectoriscapularis (acts on pectoral girdle)
    Rectus cervicus

BRANCHIOMERIC MUSCLES

Mandibular Muscles
  Adductor mandibulae anterior
  Adductor mandibulae externus
  Intermandibularis

Hyoid Muscles
  Branchiohyoideus
  Depressor mandibulae
  Interhyoideus
  Sphincter colli

Branchial Muscles
  Cucullaris (acts on pectoral girdle)
  depressores arcuum
  Dilatator laryngis
  Levatores arcuum
  Subarcuales
  Transversi ventrales

APPENDICULAR MUSCLES

Pectoral Muscles
  Coracobrachialis
  Coracoradialis
  Dorsalis scapulae
  Extensor antebrachii et carpi radialis
  Extensor antebrachii et carpi ulnaris
  Extensor digitorum communis (of forelimb)
  Flexor antebrachii et carpi radialis
  Flexor antebrachii et carpi ulnaris
  Flexor superficialis longus
  Humeroantebrachialis
  Latissimus dorsi
  Pectoralis
  Procoracohumeralis
  Subcoracoscapularis
  Supracoracoideus
  Triceps

Pelvic Muscles
  Caudofemoralis
  Extensor digitorum communis (of hind limb)
  Extensor fibularis
  Flexor primordialis communis
  Ilioextensorius
  Iliofemoralis
  Iliofibularis
  Iliotibialis
  Ischiofemoralis
  Ischioflexorius
  Puboischiofemoralis externus
  Puboischiofemoralis internus
  Puboischiotibialis
  Pubotibialis

cover the pleuroperitoneal cavity laterally. The most superficial of these layers is the **external oblique**, the fibers of which extend posteroventrally (Figures 5.10 and 5.11a). Deep to it is the **internal oblique**, the middle layer. Expose it by making a shallow dorsoventral cut through the external oblique of one of the myomeres about midway between the forelimb and the hind limb (Figure 5.11a). Work slowly, making additional shallow cuts and gently probing the fibers until you can see a change in the fiber direction. The fibers of the internal oblique extend posterodorsally and thus are oriented at nearly right angles to those of the external oblique. When you reach this layer, you may more readily cut away portions of the external oblique to expose the internal oblique more broadly. The deepest layer is the **transversus abdominis**. To expose it, repeat the procedure just described to cut through the internal oblique. You may wish to expose the internal oblique by dissecting a different myomere. This has been done for two adjacent myomeres in Figure 5.11a. The fiber direction of the transversus abdominis is different again, in this case nearly dorsoventral.

Other subdivisions of the hypaxial musculature are the **subvertebralis** and the **rectus abdominis**. The former represents a dorsal subdivision, extending from vertebra to vertebra and attaching to the lateral surface of each centrum and the transverse processes of the vertebrae. The best way to observe it is in ventral view with the body cavity exposed, as is shown in Figure 5.11b. However, do not look for the subvertebralis until you have studied the structures within the cavity. When you are ready to do so, examine the musculature on the dorsal

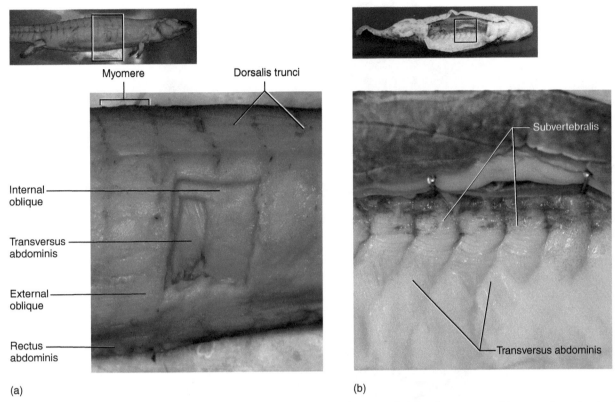

FIGURE 5.11 Hypaxial muscles of the mudpuppy: (a) flank muscles in lateral view, with deeper layers exposed in successive cut-outs of body wall; and (b) subvertebralis muscle, with body cavity opened.

wall of the body cavity. Observing the fibers clearly will require you to carefully remove the peritoneum, the thin layer of serous epithelium that lines the body cavity. The fibers of the subvertebralis extend longitudinally. Laterally adjacent to it is the transversus abdominis, the deepest layer of the flank muscles, already noted in the previous paragraph. The rectus abdominis, clearly visible in external view, is the ventral subdivision of the hypaxial musculature. It extends anteroposteriorly just lateral to the linea alba (Figure 5.22a).

Muscles of the Head and Pectoral Region

Examine the musculature of the head and pectoral region (Figures 5.12–5.14). The muscles that close the jaw occupy the lateral and dorsal surfaces of the head largely posterior to the eyes (Figures 5.12 and 5.13). The bulky lateral muscle is the **adductor mandibulae externus**, which is partially subdivided. It arises from the quadrate and squamosal of the skull and inserts on the mandible. Medial to it is the **adductor mandibulae anterior**, which lies adjacent to the middorsal line and extends anteriorly, medial to the eye. It arises from the frontal and parietal and inserts on the mandible. Posterolateral to the adductor mandibulae externus is the **branchiohyoideus**, considered further later.

Posterior to the head muscles, observe the dorsalis trunci extending forward to the head (Figure 5.12).

The delicate dorsal portions of several muscles may be seen on either side of the dorsalis trunci in the gill and pectoral regions. Although these are labeled in Figure 5.12, they are better observed in lateral view (Figure 5.13a), as described later. If you haven't already done so, cut off the external gills of one side close to their base.

Examine the pectoral region in lateral view (Figure 5.13a). Dorsal to the forelimb are two fan-shaped muscles. The larger, more posterior **latissimus dorsi** arises from the fascia covering the dorsalis trunci. Its converging fibers extend mainly anteroventrally to insert on the humerus, and it functions to retract the forelimb. Anterior to it, the smaller **dorsalis scapulae** arises from the suprascapular cartilage and passes ventrally, its fibers converging to insert on the humerus. It serves to pull the humerus anteriorly. Two smaller, strap-like muscles lie anterior to the dorsalis scapulae. The **cucullaris** is immediately anterior to it, followed by the **pectoriscapularis**. The former arises from the fascia of the dorsalis trunci, and its nearly parallel fibers extend slightly posteroventrally to insert on the scapula, serving to pull it anterodorsally. The fibers of the pectoriscapularis extend slightly more obliquely from the epibranchial cartilage to insert on the scapula, pulling it anteriorly. The fibers of the **rectus cervicus** may be observed passing anteriorly from beneath the ventral margin

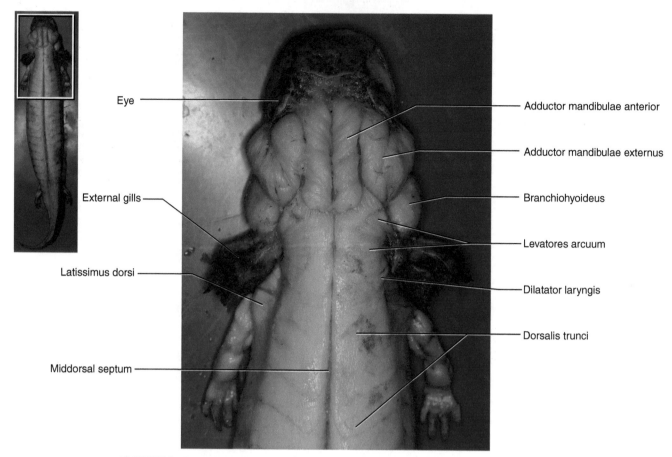

Eye

External gills

Latissimus dorsi

Middorsal septum

Adductor mandibulae anterior

Adductor mandibulae externus

Branchiohyoideus

Levatores arcuum

Dilatator laryngis

Dorsalis trunci

FIGURE 5.12   Musculature of the head and pectoral region of the mudpuppy in dorsal view.

of the pectoriscapularis. The **procoracohumeralis**, with fibers also extending anteroposteriorly, lies ventral to the rectus cervicus and is considered later.

A fan-shaped series of delicate muscles lies dorsal to the base of the external gills (Figure 5.13a). The more posterior fibers constitute the **dilatator laryngis**. It arises from the epaxial musculature, crosses the origin of the cucullaris, and passes medially to insert on the laryngeal cartilages. It functions to open the glottis. Anterior to the dilatator laryngis are several slips that constitute the **levatores arcuum** muscles. These arise, like the dilatator laryngis, from the fascia covering the dorsalis trunci but insert on the branchial arches to raise the gills.

The broader **branchiohyoideus** lies anterior to the base of the external gills. This muscle arises from the first branchial arch. Its fibers extend anteroventrally, and near the angle of the jaw pass deep to the fibers of the **sphincter colli** and then extend anteriorly, on the ventral surface, to insert on the ceratohyal cartilage. It functions to retract the hyoid apparatus. Its ventral portion will be examined later. The **depressor mandibulae**, which opens the jaw, lies between the adductor mandibulae externus and branchiohyoideus. To see it, retract and look between these two muscles, as shown in Figure 5.13b. It arises from the squamosal

and parietal of the skull and inserts on the angular of the mandible.

Examine the head and pectoral region in ventral view (Figure 5.14). Three muscles lie superficially on the ventral surface of the head. These are the **intermandibularis**, most anteriorly, followed by the **interhyoideus** and then the sphincter colli. They are difficult to distinguish, but careful observation reveals subtle differences in their fiber direction. The intermandibularis is a thin, approximately triangular sheet with nearly transverse fibers. It arises from the mandibular ramus and inserts along a midventral raphe, functioning to raise the floor of the oral cavity. The fibers of the interhyoideus, also a thin sheet, extend slightly oblique to those of the intermandibularis. Its anterior portion is partly covered by the intermandibularis. The interhyoideus also inserts on a midventral raphe. It serves to raise the throat region. The sphincter colli is the posterior part of the interhyoideus (some authors apply interhyoideus to both). It was noted earlier in connection with the branchiohyoideus. In ventral view it forms a wide band of fibers. It arises from the fascia of the branchiohyoideus and inserts, like the interhyoideus, on a midventral raphe.

During skinning, the skin within the gular fold was left in place. Now, carefully separate and remove this portion. Be particularly careful where it adheres to

5

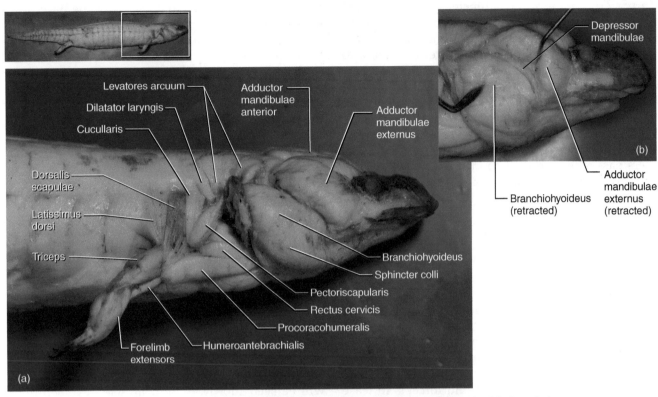

FIGURE 5.13   Musculature of the head and pectoral region of the mudpuppy in right lateral view.

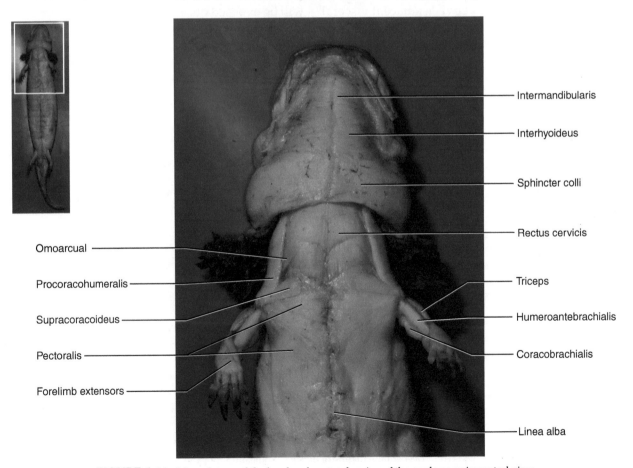

FIGURE 5.14   Musculature of the head and pectoral region of the mudpuppy in ventral view.

the posterior edge of the sphincter colli. Then, slide a probe beneath the posterior border of this muscle, just to one side of the midventral line. Slowly work the probe to the anterior part of the intermandibularis, being careful not to injure the deeper musculature. When you have separated the superficial muscles from the underlying muscles, cut along the probe with small scissors. Carefully reflect the superficial muscles (Figure 5.15). The large muscle revealed is the ventral part of the branchiohyoideus. Follow it now as its fibers pass anteromedially to insert on the ceratohyal cartilage. The long, strap-like **geniohyoideus** will be revealed along the midventral line extending antero-posteriorly between the mandibular symphysis and the basibranchial 2 of the hyoid apparatus. It functions to pull the hyoid apparatus anteriorly and lowers the mandible. Probe near the anterior end of the geniohyoideus, deep to its lateral margin, to uncover the **genioglossus**, a small muscle with fibers extending posteriorly and slightly laterally. The **rectus cervicis** is also clearly revealed.

Three small muscles lie ventrally on the branchial region. Spread apart and look deep between the branchiohyoideus and rectus cervicis muscles. Dissect carefully, using needle and forceps, to tease away tissue (Figure 5.16). The **subarcuales** comprise three muscles. They extend between adjacent visceral arches and function to adduct the gills. The **transversi ventrales** cross the posterior end of the subarcuales and pull the gills ventromedially. Smaller muscles, the depressores arcuum (not illustrated), pass from the third to fifth branchial arches to the base of the external gills, serving to depress them. They are difficult to find, and probably were removed with the skin.

Extending posteriorly and slightly laterally alongside the rectus cervicis is the procoracohumeralis, which was noted earlier (Figure 5.14). Arising from the procoracoid process of the pectoral girdle, this muscle inserts on the humerus and pulls it anteriorly. The **omoarcual** is a delicate muscle arising from the rectus cervicis medial to about the midlength of the procoracohumeralis. Lift the medial border of the latter to better expose the omoarcual. Its fibers extend posterolaterally to insert on the procoracoid process to pull the shoulder anteriorly. Two larger muscles cover the pectoral region. The more anterior and smaller **supracoracoideus** arises from the ventral surface of the coracoid plate. Its fibers converge posterolaterally to insert on the humerus. The larger, fan-shaped muscle posterior to the supracoracoideus is the **pectoralis**, arising mainly from the linea alba and inserting on the humerus. The supracoracoideus and pectoralis are humeral adductors. The posterior portion of the supracoracoideus is covered by the pectoralis, and will be seen shortly. A third muscle lies deep to the supracoracoideus. This is the **coracoradialis**, which will be exposed once the remaining shoulder and forelimb muscles have been studied.

The brachium is mainly covered by the **triceps, coracobrachialis**, and **humeroantebrachialis** (Figures 5.14 and 5.17–5.19). The triceps is the largest, occupying the

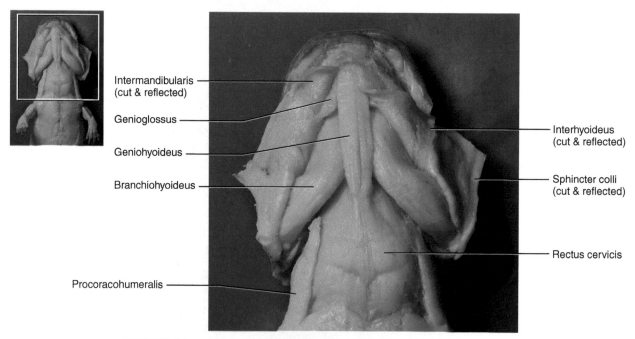

Intermandibularis
(cut & reflected)

Genioglossus

Geniohyoideus

Branchiohyoideus

Procoracohumeralis

Interhyoideus
(cut & reflected)

Sphincter colli
(cut & reflected)

Rectus cervicis

FIGURE 5.15   Deeper musculature of the head of the mudpuppy in ventral view.

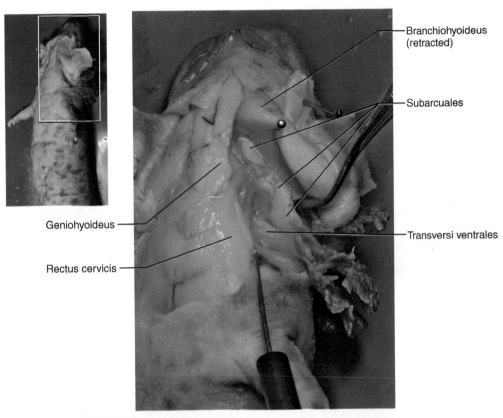

FIGURE 5.16   Subarcual musculature of the mudpuppy in ventral view.

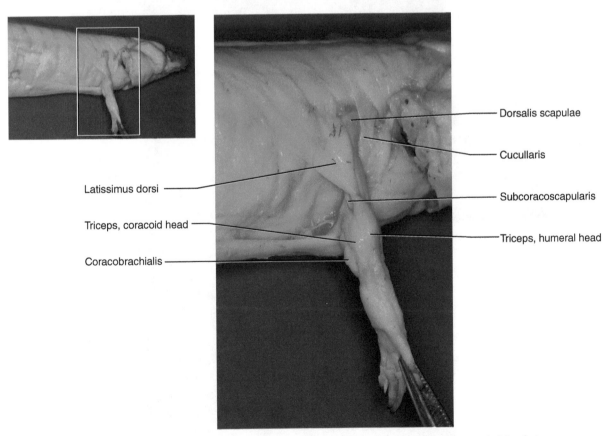

FIGURE 5.17   Musculature of the dorsal surface of the forelimb of the mudpuppy as seen in lateral view.

THE DISSECTION OF VERTEBRATES

dorsal surface of the brachium, and consists of two easily recognizable parts. Its humeral head, arising from the humerus, is dorsolateral (Figures 5.17 and 5.19), and the coracoid head, arising from the coracoid, is more medial (Figures 5.17 and 5.18). These unite, inserting by a common tendon proximally on the ulna, and serve to extend the antebrachium. The coracobrachialis is ventromedial (Figures 5.14 and 5.18). It extends between the coracoid plate and distal end of the humerus, and adducts the forelimb. The humeroantebrachialis, anterior

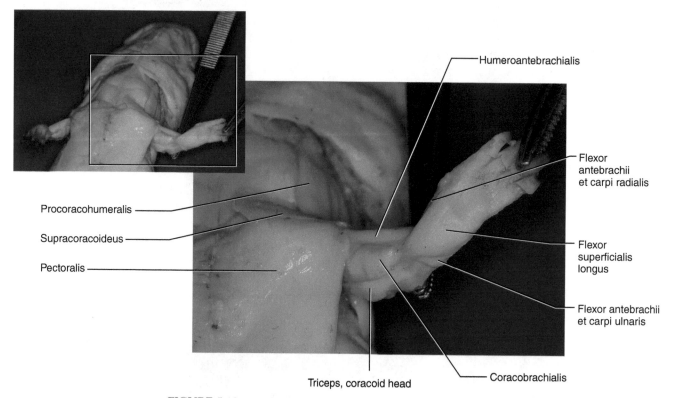

FIGURE 5.18   Forelimb musculature of the mudpuppy in ventral view.

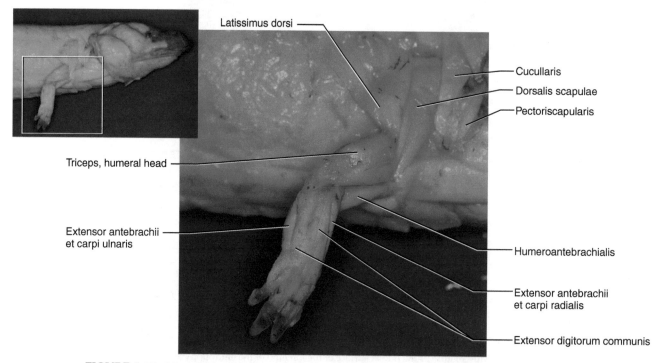

FIGURE 5.19   Musculature of the dorsal surface of the forelimb of the mudpuppy as seen in lateral view.

to the coracobrachialis, extends between the humerus and proximal end of the radius and flexes the antebrachium. The small **subcoracoscapularis** extends between the scapula and coracoid plate and the humerus. This muscle separates the proximal portions of the humeral and coracoid heads of the triceps, and may be seen between them (Figure 5.17).

The antebrachium bears the flexors (Figure 5.18) and extensors (Figure 5.19) of the wrist and digits. The flexors, on the ventral surface, are the **flexor superficialis longus, flexor antebrachii et carpi radialis**, and **flexor antebrachii et carpi ulnaris**. The extensors are on the dorsal surface of the antebrachium, and include the **extensor digitorum communis**, **extensor antebrachii et carpi radialis**, and **extensor antebrachii et carpi ulnaris**.

Next, examine the deeper musculature associated with the forelimb. Approach the shoulder region dorsally. Cut across and reflect the latissimus dorsi and gently pull the suprascapular cartilage laterally. Clear tissue from near its posterior margin to reveal the narrow **thoraciscapularis** (Figure 5.20). It arises from the surface of the hypaxial musculature and inserts on the scapula, serving to retract and depress the scapula. While holding the suprascapular cartilage laterally, examine and carefully clear tissue from just anterior to its anterodorsal margin. You will reveal a delicate muscle, the **levator scapulae**. It extends anteriorly, passing deep to the posterior margin of the cucullaris, toward its origin on the posterior surface of the skull. Its function is to pull the scapula anteriorly. Pull the cucullaris forward to obtain a better view.

In terrestrial or metamorphosed salamanders, the levator scapulae muscle becomes involved with the ear in sound transmission and is termed the opercular muscle.

Now that the other muscles of the pectoral region and forelimb have been examined, the coracoradialis, mentioned earlier, may be considered. To find it, first identify its tendon, which lies between the humeroantebrachialis and coracobrachialis (Figure 5.21). Trace the tendon proximally as it passes deep the margin of the pectoralis. Once you have an idea of where this tendon heads, carefully separate the pectoralis from the underlying musculature, and cut and reflect it, as shown in Figure 5.21. You may now appreciate the full extent of the supracoracoideus. Lift this muscle as well, by carefully passing a probe deep to it. Then cut and reflect it, as also shown in Figure 5.21, to reveal the coracoradialis.

Muscles of the Pelvic Region

Examine the pelvic region in ventral view (Figure 5.22). The large, triangular muscle, one on either side of the midventral line, is the **puboischiofemoralis externus**. Its fibers extend from its origin on the puboischiadic plate to insert on the femur. Posterior to it is the **puboischiotibialis**, which in the male lies just anterior to the cloacal gland. It also arises from the puboischiadic plate, but extends to the tibia. Both muscles adduct the hind limb.

If your specimen is a male, cut through the midline of the cloacal gland, lift one of the halves, and carefully

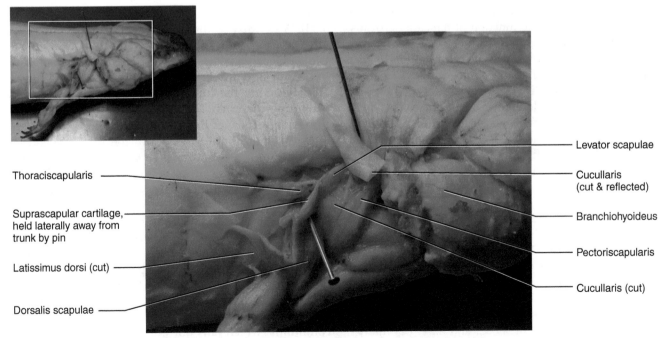

Thoraciscapularis

Suprascapular cartilage, held laterally away from trunk by pin

Latissimus dorsi (cut)

Dorsalis scapulae

Levator scapulae

Cucullaris (cut & reflected)

Branchiohyoideus

Pectoriscapularis

Cucullaris (cut)

FIGURE 5.20   Deep shoulder musculature of the mudpuppy in lateral view.

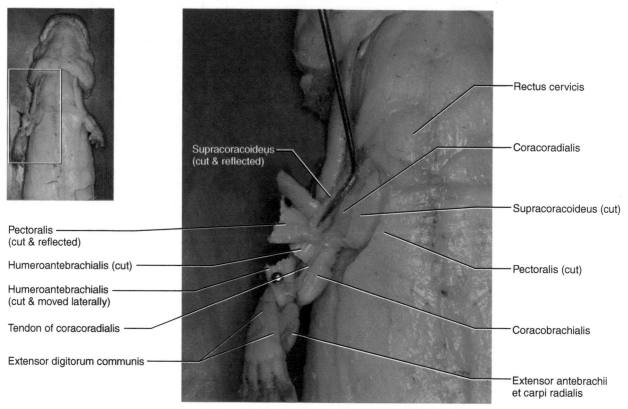

Supracoracoideus
(cut & reflected)

Rectus cervicis

Coracoradialis

Supracoracoideus (cut)

Pectoralis
(cut & reflected)

Humeroantebrachialis (cut)

Pectoralis (cut)

Humeroantebrachialis
(cut & moved laterally)

Tendon of coracoradialis

Coracobrachialis

Extensor digitorum communis

Extensor antebrachii
et carpi radialis

FIGURE 5.21   Deeper musculature of the pectoral region of the mudpuppy in ventral view.

remove it to expose the underlying musculature. Two elongated muscles are easily observed next to the cloaca, the smaller and more medial **ischiocaudalis** and the **caudocruralis** lateral to it (Figure 5.22a). The former, arising from the ischium, inserts on the caudal vertebrae to flex the tail, which is also the function of the caudocruralis, but this muscle arises from the surface of the puboischiotibialis. A third narrow muscle, the **caudofemoralis**, lies deep and slightly lateral to the caudocruralis. To view it, retract the caudocruralis (Figure 5.22b; see also Figures 5.23 and 5.24). The caudofemoralis arises from the caudal vertebrae and inserts on the femur to retract the thigh. A small, triangular muscle, the **ischiofemoralis**, lies deep to the puboischiotibialis. Carefully cut across the latter near the midline, as shown in Figure 5.22b, to expose the ischiofemoralis. Arising from the ischium, it inserts on and retracts the femur.

Lying laterally adjacent to the puboischiofemoralis externus and puboischiotibialis is the long, narrow **pubotibialis** (Figure 5.22). It arises from the pubic cartilage and inserts on the tibia, serving to adduct the hind limb. The **puboischiofemoralis internus** lies just lateral to the pubotibialis. Arising from the puboischiadic plate, it inserts on the femur and pulls the hind limb anteriorly. The **ischioflexorius** (Figures 5.22 and 5.24) lies mainly posterior to the puboischiotibialis on the medioventral

surface of the hind limb. The ischioflexorius arises from the ischium and inserts on the fascia of the crus, serving to flex it and the pes.

Three muscles may be clearly observed superficially on the dorsal surface of the thigh (Figure 5.23). Anteriorly is the puboischiofemoralis internus, already noted, followed by **iliotibialis**, and then the **ilioextensorius** posterior to it. The last two arise from the ilium and insert on the tibia, with the iliotibialis abducting the hind limb and the ilioextensorius extending the crus. The **iliofibularis** follows, on the posterior surface of the thigh (Figure 5.24). Also arising from the ilium, it inserts on the fibula to abduct the hind limb. A small, deeper muscle, the **iliofemoralis**, lies just posterior to the proximal end of the iliofibularis. Dissect in this region, removing nerves and blood vessels, to expose it.

The flexor (Figure 5.22b) and extensor (Figure 5.23) muscles lies on the crus. Its ventral surface bears the flexors, including the **flexor primordialis communis**. The ischioflexorius helps flex the pes, as noted earlier. The dorsal surface of the crus bears the extensors, with the **extensor tibialis** anteriorly and the **extensor fibularis** posteriorly. Between them is the **extensor digitorum communis**, which fans out into slips that insert on the metatarsal, serving to extend the pes and digits.

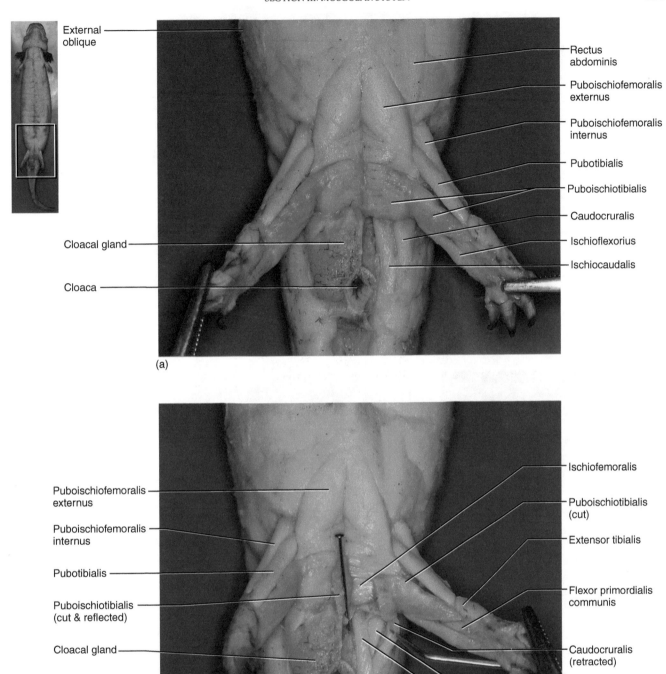

External oblique

Rectus abdominis

Puboischiofemoralis externus

Puboischiofemoralis internus

Pubotibialis

Puboischiotibialis

Caudocruralis

Ischioflexorius

Ischiocaudalis

Cloacal gland

Cloaca

(a)

Puboischiofemoralis externus

Puboischiofemoralis internus

Pubotibialis

Puboischiotibialis (cut & reflected)

Cloacal gland

Cloaca

Ischiofemoralis

Puboischiotibialis (cut)

Extensor tibialis

Flexor primordialis communis

Caudocruralis (retracted)

Caudofemoralis

Ischiocaudalis

(b)

FIGURE 5.22 Superficial (a) and deeper (b) musculature of the pelvic region of the mudpuppy in ventral view.

Key Terms: Muscular System

adductor mandibulae anterior
adductor mandibulae externus
branchiohyoideus
caudocruralis (caudopuboischiotibialis)

caudofemoralis
coracobrachialis
coracoradialis
cucullaris
depressor mandibulae
dilatator laryngis

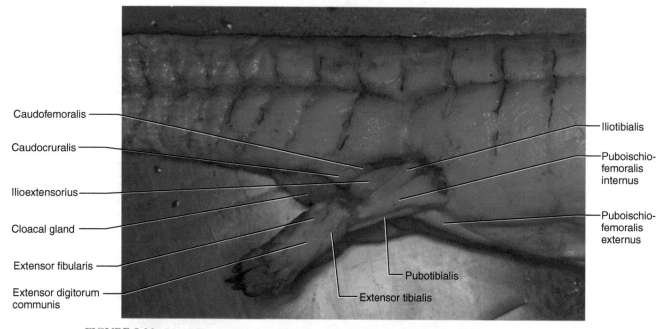

Caudofemoralis

Caudocruralis

Ilioextensorius

Cloacal gland

Extensor fibularis

Extensor digitorum
communis

Iliotibialis

Puboischio-
femoralis
internus

Puboischio-
femoralis
externus

Pubotibialis

Extensor tibialis

FIGURE 5.23 Musculature of the dorsal surface of the hind limb of the mudpuppy as seen in lateral view.

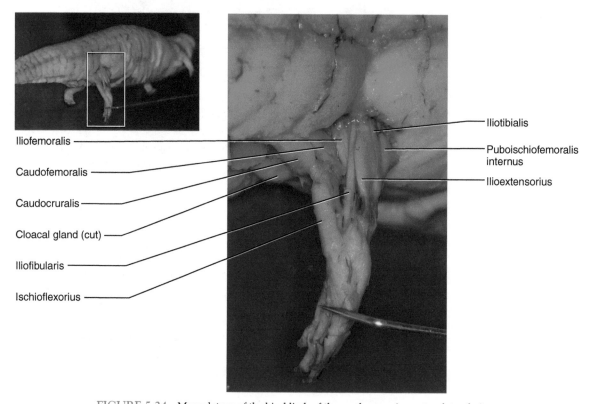

Iliofemoralis

Caudofemoralis

Caudocruralis

Cloacal gland (cut)

Iliofibularis

Ischioflexorius

Iliotibialis

Puboischiofemoralis
internus

Ilioextensorius

FIGURE 5.24 Musculature of the hind limb of the mudpuppy in posterolateral view.

dorsalis scapulae
dorsalis trunci
epaxial musculature
extensor antebrachii et carpi radialis
extensor antebrachii et carpi ulnaris

extensor digitorum communis (of forelimb)
extensor digitorum communis (of hind limb)
extensor fibularis
extensor tibialis
external oblique

flexor antebrachii et carpi radialis
flexor antebrachii et carpi ulnaris
flexor primordialis communis
flexor superficialis longus
genioglossus
geniohyoideus
horizontal skeletogenous septum
humeroantebrachialis
hypaxial musculature
ilioextensorius
iliofemoralis
iliofibularis
iliotibialis
interhyoideus
intermandibularis
internal oblique
ischiocaudalis
ischiofemoralis
ischioflexorius
latissimus dorsi
levator scapulae
levatores arcuum
myomeres
myosepta
omoarcual
pectoralis
pectoriscapularis
procoracohumeralis
puboischiofemoralis externus
puboischiofemoralis internus
puboischiotibialis
pubotibialis
rectus abdominis
rectus cervicis
sphincter colli
subarcuales
subcoracoscapularis
subvertebralis
supracoracoideus
thoraciscapularis
transversi ventrales
transversus abdominis
triceps

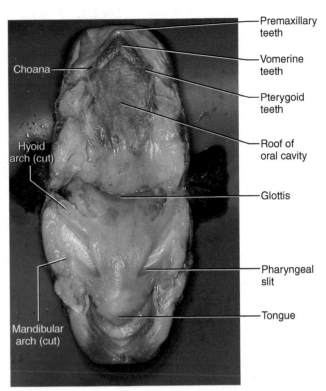

FIGURE 5.25   Oral cavity and pharynx of the mudpuppy.

SECTION IV: MOUTH, ORAL CAVITY, AND PHARYNX

Open the **oral cavity** and **pharynx** by cutting from each corner of the mouth posteriorly through the gills. Swing open the lower jaw, as shown in Figure 5.25, to expose the oral cavity and pharynx. These regions are not distinctly demarcated. Note the teeth, arranged in V-shaped rows in the anterior part of the oral cavity. In the upper jaw, the most anterior row contains **premaxillary teeth**. Just behind these teeth are the **vomerine teeth** medially and the **pterygoid teeth** laterally. A **choana**, the internal opening of the nasal passage, may be seen on each side of the roof of the oral cavity, just lateral to the pterygoid teeth. In the lower jaw there is a single row of teeth. Most of them are **dentary teeth**; a few **coronoid teeth** lie near the posterior end of the row. The **tongue**, supported by the **hyoid arch**, lies behind these teeth. Identify the sectioned surfaces of the hyoid arch and **mandibular arch**. The entrances of the pharyngeal slits into the pharynx may be noted on each side. Probe to verify that there are two slits. Midventrally on the floor of the pharynx, between the entrances to the slits, is the **glottis**, a small, slit-like opening into the respiratory system.

Key Terms: Mouth, Oral Cavity, and Pharynx

choana
coronoid teeth (splenial teeth)
dentary teeth
glottis
hyoid arch
mandibular arch
oral cavity
pharynx
premaxillary teeth

pterygoid teeth
tongue
vomerine teeth

SECTION V: PLEUROPERITONEAL CAVITY AND VISCERA

Dissection of the pleuroperitoneal cavity in *Necturus* allows examination of the viscera and respiratory and urogenital structures. Many vessels will also be noted at this time. Enter the pleuroperitoneal cavity by making a longitudinal incision about 0.5 cm to the left of the mid-ventral line. If the vessels of your specimen have been injected, its abdominal wall will have been cut. If so, continue along this cut to expose the cavity. Cut anteriorly until you reach the level of the posterior margin of the forelimbs, at which point you will have reached the posterior margin of the coracoid cartilage. Cut posteriorly to the posterior margin of the hind limbs, cutting through the puboischiadic plate.

Gently spread the abdominal wall (Figure 5.26). The large, elongated, dark structure midventrally is the **liver**. Lift the cut edge of the body wall to observe the **falciform ligament** extending between the wall and the ventral surface of the liver. The ligament is much larger than in the other vertebrates considered here. Note the **ventral abdominal vein** passing along the midventral abdominal wall and toward the liver through the free posterior margin of the ligament. Examine the posteroventral part of the cavity to observe the thin-walled **urinary bladder**. It is supported from the midventral abdominal wall by the **median ligament**. A **vesicular vein** from the bladder passes in the anterior free margin of this ligament to join the ventral abdominal vein.

Carefully cut through the falciform ligament and make two transverse cuts through the body wall, one on either side, to create fours flaps that can be spread apart. (Much of the ventral body wall has been removed in Figures 5.27–5.29, but you need not do so to observe the following structures.) The long, light-colored, tubular **stomach** (Figure 5.26) lies dorsal and slightly to the left of the liver. Lift the stomach to see that it is supported by the **mesogaster** (or greater omentum). Spread apart the liver and stomach. The **gastrohepatic ligament** is the small, triangular sheet stretching between their anterior portions (Figure 5.27).

The elongated **spleen** hangs from the posterior left side of the stomach (Figures 5.27 and 5.28). The **gastrosplenic ligament** extends between these organs (Figure 5.27). The stomach ends abruptly at the **pyloric sphincter**, a marked constriction beyond which the digestive tract continues as the long, coiled **small intestine**, followed by the short, straight **large intestine**. The **mesentery** supports the small intestine. Spread the coils of the small intestine to observe it. The **mesorectum** supports the large intestine (Figure 5.27).

The first loop of the small intestine is the **duodenum** (Figures 5.26 and 5.27). The **hepatoduodenal ligament** extends between the duodenum and the dorsal surface of the liver (Figure 5.27). Several organs and vessels lie in this region, but may be difficult to discern. Examine this region in several views as follows, referring to Figures 5.27–5.29. The **pancreas** lies along the duodenum (and partly within the hepatoduodenal ligament). To help identify it, pull the stomach to the left and reflect the liver to the right to expose its dorsal surface (Figure 5.27). The pancreas is irregular, but note that part of it extends anteriorly toward the spleen. The **lienogastric vein** lies partly embedded within this lobe and extends toward the spleen.

Observe the long, thin translucent left **lung** lying dorsal to the stomach. The right lung is similar in form; look for it between the stomach and the liver. You should discern the right and left lungs, apparently lying side by side. For now, distinguish between them but do not tear through any mesenteries. The liver is supported anterodorsally by the **hepatocavopulmonary ligament**. The **posterior vena cava** is the large vessel passing through the posterior end of this ligament (Figure 5.27). Note the large **hepatic portal vein** embedded in the dorsal part of the liver.

Next, reflect the stomach to the right so that it lies on the liver's dorsal surface, as shown in Figure 5.28. Note the relationships among the liver, stomach, spleen, and pancreas, as well as their associated mesenteries. Lift the left lung. It is supported by a narrow **pulmonary ligament**, which is connected to the mesogaster.

Next, let the viscera fall back in place, and then reflect the liver to the left as shown in Figure 5.29. Examine the posterodorsal part of the liver for the **gallbladder**, a thin, translucent, greenish sac. Gently lift it and examine the region where it attaches to the liver. The **cystic duct** leaves the gallbladder but is joined by several (and usually fairly narrow) **hepatic ducts** from the liver to form the **common bile duct**, which passes partly through the substance of the pancreas to reach the duodenum. The bile duct is similar to that of the shark and cat, but is much shorter. Note again the positions of the posterior vena cava and lienogastric and mesenteric veins. Examine the right lung. Its pulmonary ligament, wider than that of the left lung, is part of the hepatocavopulmonary ligament.

Key Terms: Pleuroperitoneal Cavity and Viscera

common bile duct
cystic duct
dorsal gastric artery
duodenum

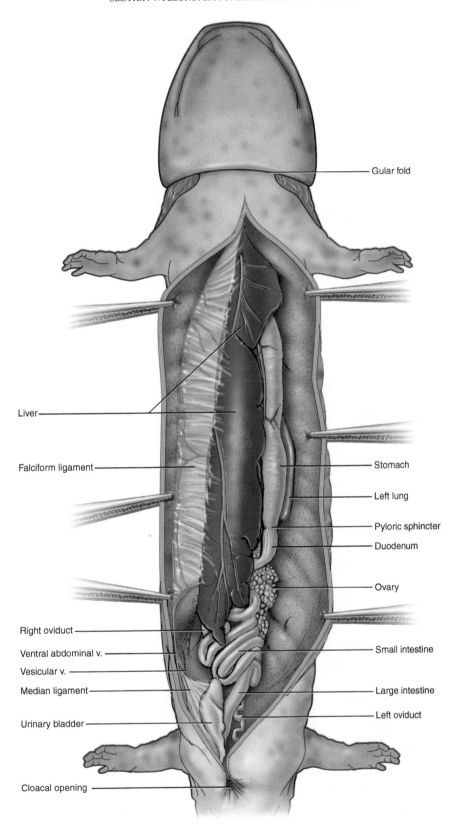

FIGURE 5.26 Pleuroperitoneal cavity of the mudpuppy in ventral view.

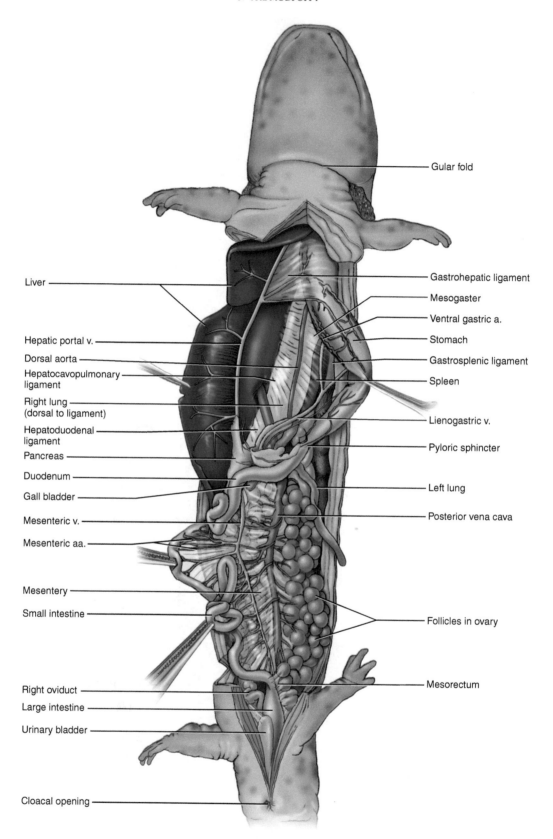

Gular fold

Liver

Gastrohepatic ligament

Mesogaster

Ventral gastric a.

Hepatic portal v.

Stomach

Dorsal aorta

Gastrosplenic ligament

Hepatocavopulmonary
ligament

Spleen

Right lung
(dorsal to ligament)

Hepatoduodenal
ligament

Lienogastric v.

Pancreas

Pyloric sphincter

Duodenum

Gall bladder

Left lung

Mesenteric v.

Posterior vena cava

Mesenteric aa.

Mesentery

Small intestine

Follicles in ovary

Right oviduct

Mesorectum

Large intestine

Urinary bladder

Cloacal opening

FIGURE 5.27 Pleuroperitoneal cavity of the mudpuppy in ventral view, with the liver reflected to the right and stomach pulled to the left.

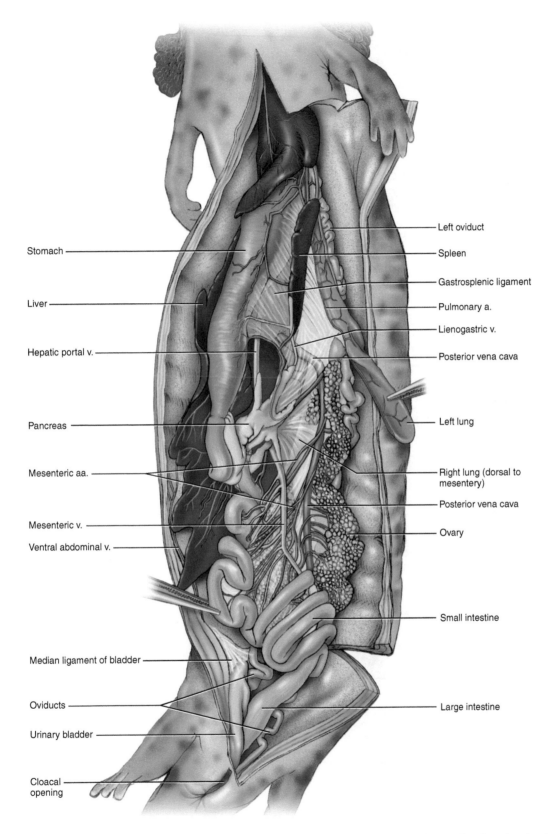

Stomach

Liver

Hepatic portal v.

Pancreas

Mesenteric aa.

Mesenteric v.

Ventral abdominal v.

Median ligament of bladder

Oviducts

Urinary bladder

Cloacal opening

Left oviduct

Spleen

Gastrosplenic ligament

Pulmonary a.

Lienogastric v.

Posterior vena cava

Left lung

Right lung (dorsal to mesentery)

Posterior vena cava

Ovary

Small intestine

Large intestine

FIGURE 5.28   Pleuroperitoneal cavity of the mudpuppy in ventral view, with the liver and stomach reflected to the right.

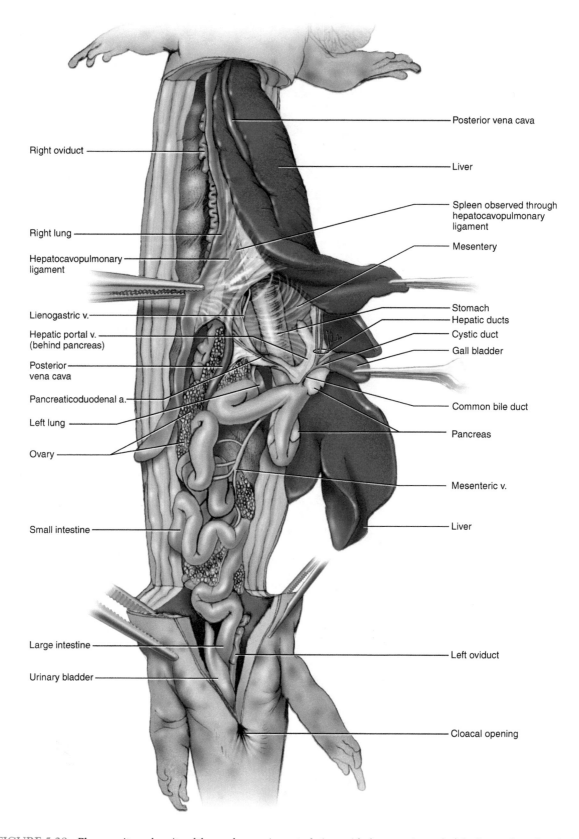

FIGURE 5.29   Pleuroperitoneal cavity of the mudpuppy in ventral view, with the posterior end of the liver reflected to the left.

falciform ligament
gallbladder
gastrohepatic ligament
gastrosplenic ligament
hepatic ducts
hepatic portal vein
hepatocavopulmonary ligament
hepatoduodenal ligament
large intestine
lienic artery (splenic artery)
lienogastric artery (gastrosplenic artery; or gastric artery, as used by Homberger and Walker, 2003)
lienogastric vein (gastrosplenic vein)
liver
lung
median ligament
mesentery
mesogaster (greater omentum)
mesorectum
pancreas
posterior vena cava
pulmonary ligament
pyloric sphincter
small intestine
spleen
stomach
urinary bladder
ventral abdominal vein
ventral gastric artery
vesicular vein

SECTION VI: UROGENITAL SYSTEM

Urinary and reproductive structures should be examined next, but do not injure mesenteries and vessels in doing so. Postpone detailed tracing of the structures until the vessels have been studied. Lift the coils of the small intestine in the posterior part of the pleuroperitoneal cavity to locate one of the paired gonads.

Male Urogenital System

In males the **testis** is an elongated organ posteriorly in the pleuroperitoneal cavity lying ventral to the **kidney** (Figure 5.30). It is supported by the **mesorchium**. The kidney, longer than the testis, is considerably wider posteriorly than anteriorly. Its narrow anterior part is genital in function, while the wider posterior portion is urinary.

The **archinephric duct** is a longitudinal tube that extends along the lateral margin of the kidney. The portion along the genital part of the kidney is coiled, but its more posterior part, lying along the urinary portion of the kidney, is straight. Numerous small collecting tubules may be observed leading to the duct from the kidney. Trace the duct posteriorly to its entrance into the cloaca. In males the duct carries sperm from the testis, as well as urine from the urinary portion of the kidney. Sperm reaches the duct by way of **ductuli efferentes**, which lead into the genital portion of the kidney and thence to the archinephric duct. The vestigial **oviduct** is a dark thread-like structure along the lateral edge of the archinephric duct. It has no connection with the latter and continues anteriorly on its own. The margins of the cloaca bear numerous **cloacal papillae**. Skin the region on one side of the cloaca to expose the **cloacal gland**, which consists of numerous tiny tubules and is involved in clumping sperm to form spermatophores.

Female Urogenital System

In females the elongated **ovary**, supported by the **mesovarium**, may be quite large (Figure 5.31). The presence of numerous eggs within follicles gives the ovary a lobulated or granular appearance, in contrast to the more regular surface of the testes. The follicles and eggs vary in size depending on their stage of maturity, being quite large in some specimens (Figure 5.27) and smaller in others (Figures 5.29 and 5.31). The archinephric duct lies along the lateral margin of the kidney but is straight and narrower than in the male. In the female it carries only urine from the kidney. Follow it posteriorly to its entrance into the cloaca.

The **oviduct**, supported by the **mesotubarium**, is the long, prominent, and convoluted tube lying between the ovary and kidney and extending nearly the length of pleuroperitoneal cavity. At its anterior end is the open, funnel-shaped **ostium tubae**, into which the eggs pass after they have been released into the coelom by the ovary. Follow the oviduct posteriorly to its entrance into the cloaca. Cloacal glands and papillae are absent in the female.

Key Terms: Urogenital System

archinephric duct
cloacal gland
cloacal papillae
ductuli efferentes
kidney
mesorchium
mesotubarium
mesovarium
ostium tubae
ovary
oviduct
testis

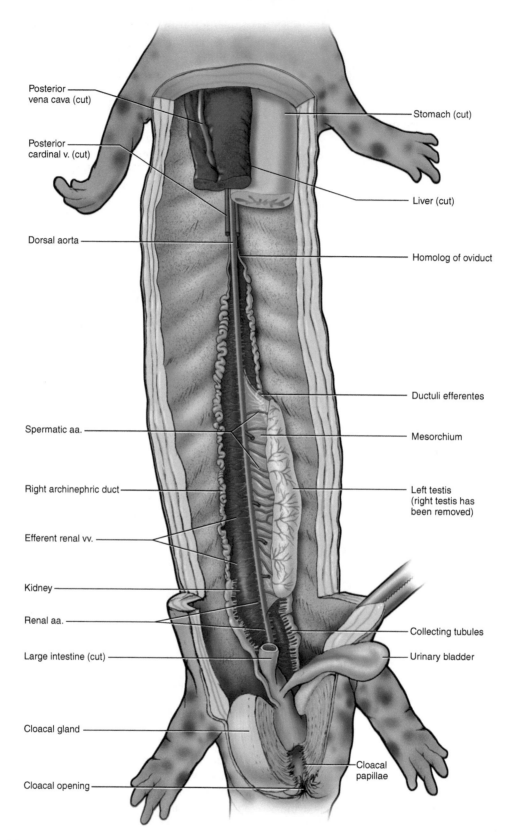

FIGURE 5.30 Pleuroperitoneal cavity of the male mudpuppy in ventral view, showing the urogenital system. Portions of the viscera and the venous system have been removed.

5

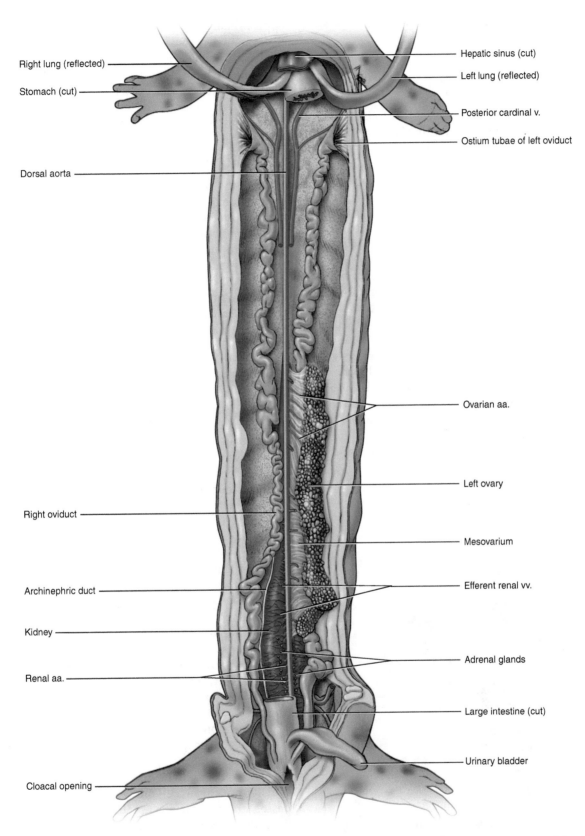

Right lung (reflected)

Stomach (cut)

Dorsal aorta

Right oviduct

Archinephric duct

Kidney

Renal aa.

Cloacal opening

Hepatic sinus (cut)

Left lung (reflected)

Posterior cardinal v.

Ostium tubae of left oviduct

Ovarian aa.

Left ovary

Mesovarium

Efferent renal vv.

Adrenal glands

Large intestine (cut)

Urinary bladder

FIGURE 5.31 Pleuroperitoneal cavity of the female mudpuppy in ventral view, showing the urogenital system. Portions of the viscera and venous system have been removed.

SECTION VII: CARDIOVASCULAR SYSTEM

Heart

The **pericardial cavity** lies just anterior to the liver. It is enclosed by the **pericardial sac** and contains the **heart** (Figure 5.32). Continue the midventral incision of the abdominal wall anteriorly, cutting through the coracoid cartilage, to expose the pericardial cavity. In doing so you will also cut through the **transverse septum**, the partition separating the pericardial and pleuroperitoneal cavities. Do not injure the **posterior vena cava**, which passes through the septum to reach the pericardial cavity. Carefully remove the musculature ventral to the cavity. The cavity is lined by **parietal pericardium**, while the heart itself is covered by **visceral pericardium**.

The largest and most conspicuous part of the heart is the **ventricle**, which occupies the posteroventral part of the pericardial cavity (Figure 5.33). The ventricle is partially separated into a smaller left side and a larger right side by an incomplete interventricular septum. Lift the posterior end of the ventricle to observe the **sinus venosus** (Figure 5.34). The atrium lies anterodorsal to the ventricle and is partially divided into **left** and **right atria**. (These are recognized and labeled as separate left and right atria in the figures of this section, largely because blood returning from the lungs, through the pulmonary veins [see later] enters the left side. Some authors,

such as Homberger and Walker, 2003, recognize, also appropriately, the atrium as single, given that the partition between the left and right sides is partial.) The **conus arteriosus** is the narrow tube extending anteriorly between the atria from the right side of the ventricle. It continues anteriorly as the wider **bulbus arteriosus**. The latter is the enlarged and modified base of the **ventral aorta** (compare with that of the shark, as in Figure 3.27).

Venous System

Trace the posterior vena cava, identified earlier, as it passes from the liver through the transverse septum. On entering the pericardial cavity, it subdivides into left and right **hepatic sinuses**, large vessels that extend anterolaterally into the sinus venosus (Figures 5.33–5.35). Probe just anterolateral to the entrance of the hepatic sinus on one side to observe the **common cardinal vein**, which enters the posterolateral end of the sinus venosus. The common cardinal vein mainly receives vessels that drain the head and forelimbs, such as the **anterior cardinal vein** (referred to as the jugular vein by some authors) and **lingual veins** as well as the **subclavian vein**, which is formed by the union of the **brachial** and **cutaneous veins**. The common cardinal also receives the **posterior cardinal vein** and **lateral vein**, which extends along the lateral side of the trunk. These vessels all join the common cardinal in close proximity (Figures 5.33 and 5.35). The anterior cardinal vein is formed by the confluence of the **external jugular vein**, which passes dorsal to the gills and onto the side of the head, and the smaller **internal jugular vein**, which drains the roof of the oral cavity and pharynx (Figure 5.35).

Various veins within the pleuroperitoneal cavity have already been identified. The posterior vena cava was identified first as it passed to the liver through the hepatocavopulmonary ligament and then through the transverse septum; reidentify this vein. Follow it as is passes through the liver, from which it receives several **hepatic veins**. Posteriorly, the posterior vena cava passes middorsally in the pleuroperitoneal cavity, extending between the kidneys (Figure 5.35). It receives numerous **efferent renal veins** from the kidneys, as well as **testicular** or **ovarian veins** from the gonads (Figure 5.35).

The **pulmonary vein** extends anteroposteriorly along the surface of the lung, collecting oxygenated blood from the lung and returning it to the heart during air breathing. The vein should be injected with blue latex but occasionally appears red. Follow the pulmonary vein as it passes anteriorly. It curves medially and passes dorsal to the posterior vena cava and hepatic sinus, where it joins the pulmonary vein from the lung of the other side of the body (Figures 5.33 and 5.35). Their union forms a short, single vein (common pulmonary vein) that extends

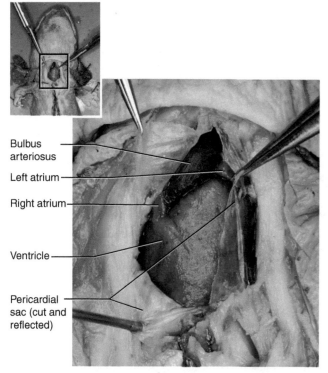

Bulbus arteriosus

Left atrium

Right atrium

Ventricle

Pericardial sac (cut and reflected)

FIGURE 5.32 Pericardial cavity of the mudpuppy in ventral view, opened to reveal the heart.

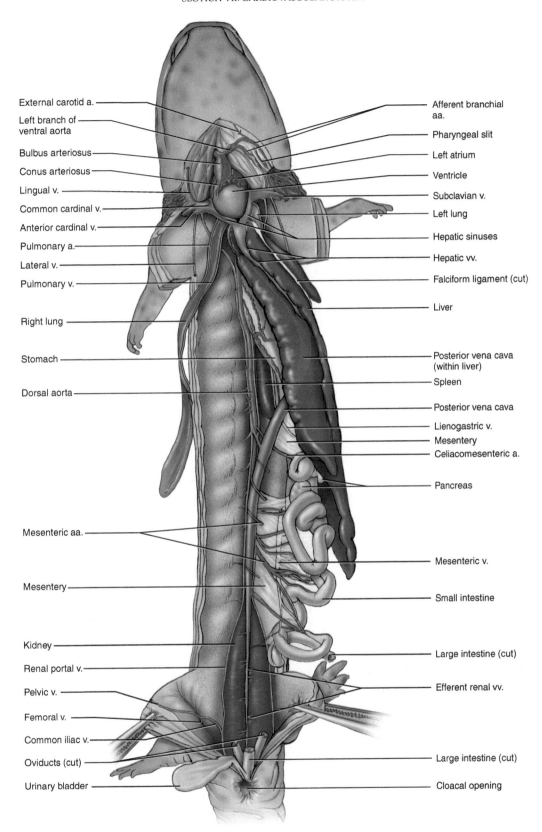

External carotid a.

Left branch of
ventral aorta

Bulbus arteriosus

Conus arteriosus

Lingual v.

Common cardinal v.

Anterior cardinal v.

Pulmonary a.

Lateral v.

Pulmonary v.

Right lung

Stomach

Dorsal aorta

Mesenteric aa.

Mesentery

Kidney

Renal portal v.

Pelvic v.

Femoral v.

Common iliac v.

Oviducts (cut)

Urinary bladder

Afferent branchial
aa.

Pharyngeal slit

Left atrium

Ventricle

Subclavian v.

Left lung

Hepatic sinuses

Hepatic vv.

Falciform ligament (cut)

Liver

Posterior vena cava
(within liver)

Spleen

Posterior vena cava

Lienogastric v.

Mesentery

Celiacomesenteric a.

Pancreas

Mesenteric v.

Small intestine

Large intestine (cut)

Efferent renal vv.

Large intestine (cut)

Cloacal opening

FIGURE 5.33 Pleuroperitoneal cavity of the mudpuppy in ventral view, revealing heart and vessels. Liver is pulled to the left.

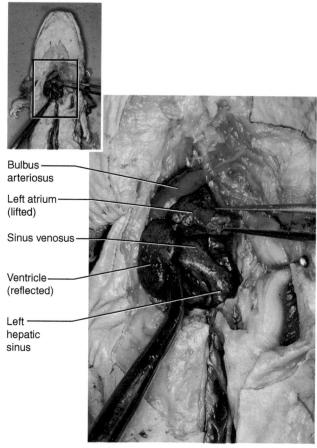

Bulbus
arteriosus

Left atrium
(lifted)

Sinus venosus

Ventricle
(reflected)

Left
hepatic
sinus

FIGURE 5.34 Pericardial cavity of the mudpuppy in ventral view, exposing the heart. The ventricle and atrium are lifted and moved aside to reveal the sinus venosus.

anteriorly before entering the left side of the atrium of the heart.

A **renal portal vein** extends longitudinally along the dorsolateral margin of each kidney and delivers blood to the capillaries of the kidney via the tiny **afferent renal veins** (Figure 5.35). Follow one of the renal portal veins posteriorly. At about the level of the hind limbs, each renal portal vein receives a **common iliac vein**. Posterior to the kidneys, the left and right renal portal veins form by the bifurcation of the median **caudal vein** from the tail. Anterior to the kidneys, the renal portal veins become the **posterior cardinal veins**. In this region the posterior cardinals anastomose with the posterior vena cava via the converging **subcardinal veins** (Figure 5.35). Anteriorly, each posterior cardinal vein drains into a common cardinal vein.

The **ventral abdominal vein** was observed passing through the free posterior margin of the falciform ligament (Figure 5.26). Return to this vessel. As you will see presently, its anterior end joins the **hepatic portal vein**. For now, follow the ventral abdominal vein posteriorly along the midventral abdominal wall. As noted earlier, it receives the **vesicular vein** from the urinary bladder. The

ventral abdominal vein is formed, farther posteriorly, by the union of left and right **pelvic veins**. Follow one of the pelvic veins posterolaterally. At the level of the hind limbs, it receives the **femoral vein**. At this point, note that the common iliac vein, observed earlier as it entered the renal portal vein, joins the pelvic and femoral veins. The iliac is thus a short vein extending between the renal portal vein and the union of the pelvic and femoral veins.

Observe the mesentery by spreading the coils of the small intestine. The **mesenteric vein** extends within the mesentery (Figure 5.33). Follow it as it passes anteriorly toward the pancreas. Next, locate the lienogastric vein, identified earlier as extending between the pancreas and spleen. It is formed by the confluence of **gastric veins** from the stomach and the **lienic vein** from the spleen. The union of the mesenteric and lienogastric veins, within the pancreas, forms the **hepatic portal vein**, which passes along the dorsal surface of the liver before entering this organ (Figure 5.35). The ventral abdominal vein joins the hepatic portal vein a short distance anterior to the union of the mesenteric and lienogastric veins.

Arterial System

Return to the bulbus arteriosus. As it leaves the front of the pericardial cavity, it bifurcates into left and right arches, or branches, of the ventral aorta (Figure 5.36). Each of these diverges laterally and subdivides into two branches. They are probably not injected, but usually have remnants of blood so that they can, with diligence, be followed. The most anterior branch is the **first afferent branchial artery**, which follows the first branchial arch and enters the first external gill. The second branch bifurcates into the **second** and **third afferent branchial arteries**. Follow them, respectively, into the second and third external gills. The afferent branchial arteries carry deoxygenated blood to the gills for oxygenation. The **external carotid artery**, which should be injected, is a narrow vessel extending anteromedially along the anterior margin of the first afferent branchial artery. Although it is in close proximity, and indeed connected by tiny anastomoses, to the first afferent branchial artery, the external carotid artery is a branch of the first efferent branchial artery (see later), and therefore carries blood that has already passed through the gills for oxygenation.

Remove the lining on the roof of the oral cavity and pharynx. Medial to the external gills are the left and right **radices** (sing., **radix**) of the **dorsal aorta** (Figure 5.37). Each radix arches toward the dorsal midline and joins its fellow to form the dorsal aorta, which continues middorsally along the trunk and tail. Follow the radix laterally. Its first branch is the **vertebral artery**, which passes anteriorly and almost immediately enters the musculature. Farther laterally, a short trunk, the **carotid duct**, enters the anterior surface of the radix. The duct connects to the following two arteries. The **first efferent branchial artery** extends anteromedially from the first external gill. The **internal carotid artery** curves anteromedially from the duct, and helps supply the facial region and brain. The external carotid artery, observed in association with the afferent branchial arteries, helps supply the floor of the oral cavity and pharynx. It arises from the first efferent branchial artery just after the latter emerges from the gill. Return to the radix and continue to follow it laterally. The **pulmonary artery** arises from its posterior surface and passes to the lung. The pulmonary artery is injected with red latex, and the pulmonary vein should be injected with blue latex, although occasionally it also appears red. To distinguish between the pulmonary artery and pulmonary vein, trace each of them anteriorly. The pulmonary artery curves slightly laterally and connects to the radix, whereas the pulmonary vein passes medially and then dorsal to the posterior vena cava (as already noted earlier). Beyond the pulmonary artery, the **second** and **third efferent branchial arteries**, which emerge from the second and third gills, respectively, unite to form the radix.

The efferent branchial arteries recollect blood that has passed through the gills and is therefore oxygenated (as also occurs in the shark). From the description of the arteries given in the previous paragraph, it should be evident that the pulmonary artery, being a branch of the radix of the aorta, itself formed by the union of the efferent branchial arteries, carries blood that has already passed through the gills and is therefore oxygenated. The mudpuppy is distinct in this regard from other typical air-breathing tetrapods, in which the pulmonary artery leads from the heart and carries deoxygenated blood to the lungs for oxygenation. Embryologically, the pulmonary artery develops from the sixth aortic arch, which passes from the dorsal to ventral aorta, and is interrupted in its course by the capillaries of the gill, which thus define the afferent (ventral) and efferent (dorsal) portions of the arch, or branchial arteries. In most other lung-bearing tetrapods, the ventral portion of the sixth aortic arch is retained as the connection from the heart to the lungs (that is, to the pulmonary artery), and the efferent portion of the arch is either retained as a reduced vessel, termed the duct arteriosus, or, more commonly, lost in the adult. In the mudpuppy, however, the reverse occurs: the afferent portion of the arch is lost and the efferent portion, the duct arteriosus, is retained. This arrangement, of course, is tied to the presence of both gills and lungs in the mudpuppy, and it allows the mudpuppy "options" for oxygenation. If, for example, the external gills are unable to take up sufficient oxygen because of low water oxygen saturation, blood can be routed to the lungs for further oxygenation when the mudpuppy surfaces to breathe; and then returned, as usual in tetrapods, to the heart by way of the pulmonary vein. When oxygen uptake through the external gills is sufficient, then the mudpuppy can bypass the lung circulation, shunting blood to the rest of the body. See Homberger and Walker (2003) for further details of and differences between the circulatory patterns in the mudpuppy and metamorphosed terrestrial urodeles.

As in the other vertebrates so far studied, the dorsal aorta supplies the structures of the trunk by paired and unpaired branches. Its most anterior paired vessels are the **subclavian arteries**. Each subclavian extends almost directly laterally, giving off several small arteries before continuing into the forelimb as the **brachial artery**. The next few posterior arteries are single median arteries that mainly supply the viscera (Figure 5.38). The most anterior of these, the **gastric artery**, arises near the anterior end of the stomach. It gives rise to several arteries, some passing to the stomach, principally the **dorsal** and **ventral gastric arteries**, and one, the **lienic artery**, passing to the anterior part of the spleen. The larger **celiacomesenteric artery** arises near the posterior end of the stomach and soon subdivides into various branches that primarily supply the spleen (through another lienic artery, this one passing to the posterior end of the spleen), stomach, duodenum, and liver. The dorsal aorta then gives off several unpaired **mesenteric arteries** that pass through the mesentery and supply the intestines. In addition, the aorta gives off paired **testicular** or **ovarian arteries** to the testes or ovaries that pass through the mesorchium or mesovarium, respectively, and paired **renal arteries** that pass into the kidneys. Dorsal to the posterior part of the kidneys, large, paired **iliac arteries** extend laterally. Each branches into three vessels, an anterior **epigastric artery** (onto the body wall), a lateral **femoral artery** (into the hind limb), and a posterior **hypogastric artery** (to the cloaca and urinary bladder). The small, paired **cloacal arteries** are the last branches of the aorta, which then continues into the tail as the **caudal artery**.

At this point, now that you have examined the vessels and viscera, recall the suggestion on page 130 that you observe the subvertebralis muscle (Figure 5.11b).

Key Terms: Cardiovascular System

afferent branchial arteries
afferent renal arteries
anterior cardinal vein
brachial artery
brachial vein
bulbus arteriosus
carotid duct
caudal artery
caudal vein
celiacomesenteric artery
cloacal arteries
common cardinal vein

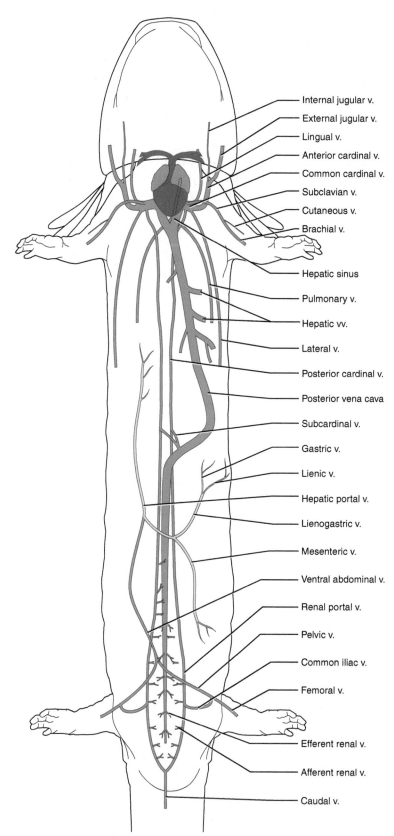

Internal jugular v.
External jugular v.
Lingual v.
Anterior cardinal v.
Common cardinal v.
Subclavian v.
Cutaneous v.
Brachial v.
Hepatic sinus
Pulmonary v.
Hepatic vv.
Lateral v.
Posterior cardinal v.
Posterior vena cava
Subcardinal v.
Gastric v.
Lienic v.
Hepatic portal v.
Lienogastric v.
Mesenteric v.
Ventral abdominal v.
Renal portal v.
Pelvic v.
Common iliac v.
Femoral v.
Efferent renal v.
Afferent renal v.
Caudal v.

FIGURE 5.35   Schematic illustration showing the pattern of the venous system of the mudpuppy superimposed on a ventral view of the body outline.

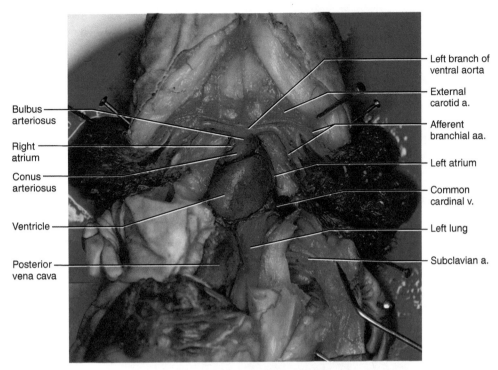

Bulbus arteriosus

Right atrium

Conus arteriosus

Ventricle

Posterior vena cava

Left branch of ventral aorta

External carotid a.

Afferent branchial aa.

Left atrium

Common cardinal v.

Left lung

Subclavian a.

FIGURE 5.36   Ventral view of the heart and ventral aorta of the mudpuppy.

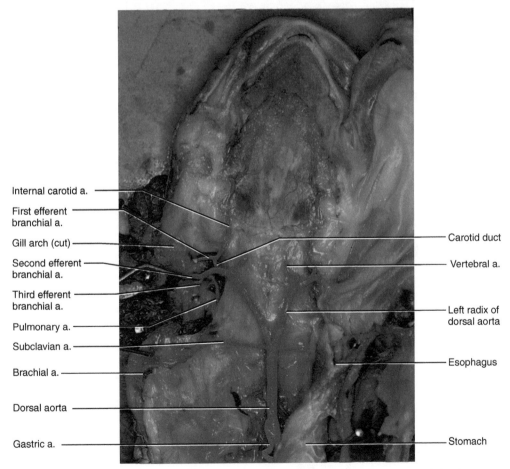

Internal carotid a.

First efferent branchial a.

Gill arch (cut)

Second efferent branchial a.

Third efferent branchial a.

Pulmonary a.

Subclavian a.

Brachial a.

Dorsal aorta

Gastric a.

Carotid duct

Vertebral a.

Left radix of dorsal aorta

Esophagus

Stomach

FIGURE 5.37   Dorsal aorta and branches of the mudpuppy in ventral view. The right side of the mouth has been cut and the lower jaw swung open.

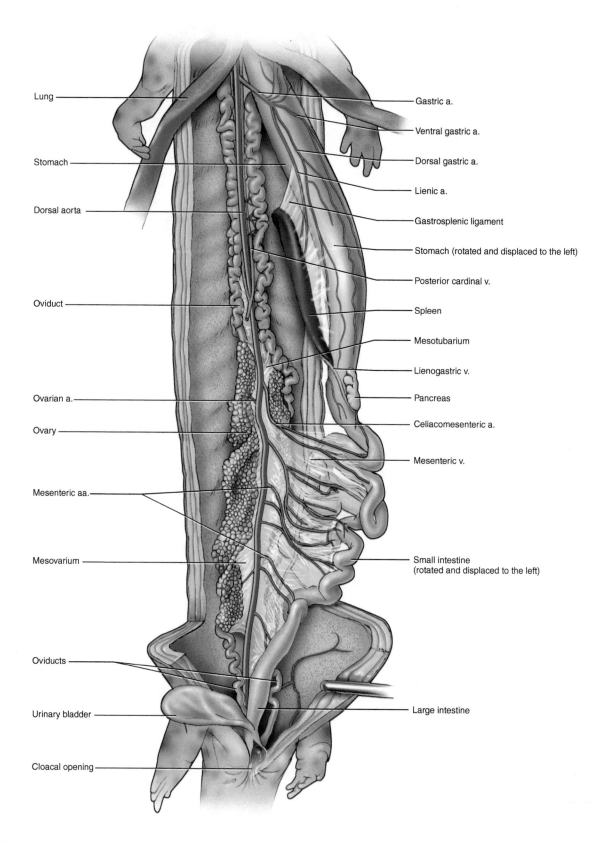

FIGURE 5.38 Pleuroperitoneal cavity of the mudpuppy in ventral view, showing the pattern of the arterial system posterior to the heart. Portions of the venous system have been removed.

5

common iliac vein
conus arteriosus
cutaneous vein
dorsal aorta
efferent branchial artery
efferent renal veins
epigastric artery
external carotid artery
external jugular vein
femoral artery
femoral vein
gastric artery
gastric veins
heart
hepatic portal vein
hepatic sinuses
hepatic veins
hypogastric artery
iliac artery
internal carotid artery
internal jugular vein
lateral vein
left atrium
lienic vein (splenic vein)
lienogastric vein (gastrosplenic vein)
lingual vein
mesenteric arteries
mesenteric vein

ovarian artery
ovarian vein
parietal pericardium
pelvic veins
pericardial cavity
pericardial sac
posterior cardinal vein
posterior vena cava
pulmonary artery
pulmonary vein
radices (sing., radix)
renal arteries
renal portal vein
right atrium
sinus venosus
subcardinal vein
subclavian artery
subclavian vein
testicular artery
testicular vein
transverse septum
ventral abdominal vein
ventral aorta
ventricle
vertebral artery
vesicular vein
visceral pericardium

6

The Frog

INTRODUCTION

The North American bullfrog, *Rana catesbeiana*, is a member of Anura, which together with Caudata (salamanders and newts) and Gymnophiona (caecilians or apodans) are included in Lissamphibia (see Chapter 1). Frogs have a long fossil history, with the Early Triassic *Triadobatrachus massinoti* recognized as among the earliest frogs. As with lissamphibians generally, anurans tend to have permeable, scaleless skin kept moist by numerous mucous glands, which allows for considerable cutaneous respiration. Frogs have highly specialized locomotory features that make them instantly recognizable. Most obvious perhaps are that the body is rigid, short and wide, the hind limbs are long and familiarly though not exclusively used for jumping, and the tail is absent (Figure 6.8). It is to the last feature that the group owes its name: Anuran is derived from the Ancient Greek words meaning "without" (*an*) and "tail" (*oura*).

These specializations, among others, provide ample proof of the risks involved in viewing living vertebrates as primitive or somehow intermediate between other vertebrate grades. They (and the other lissamphibians) are in fact highly derived tetrapods. Anurans are the most successful lissamphibians, including more than 5200 species and comprising nearly 90% of all lissamphibians (Frost et al., 2006), living on all continents except Antarctica. They have diversified into numerous and markedly different ecological types.

Anurans generally have an aquatic tadpole or larval stage and undergo metamorphosis to produce the radically different adult form, but different reproductive strategies have evolved. Some, such as some members of Pipidae, a group of specialized aquatic frogs, produce eggs that develop directly into juvenile frogs, whereas other pipids have aquatic larvae. Some species of *Nectophrynoides* (Bufonidae, the true toads) are viviparous, and in *Gastrotheca* (Hylidae) the juvenile frogs develop directly in pouches in the female's skin. In the species *Rhinoderma darwini* (Rhinodermatidae), the tadpoles complete their development in the vocal sacs of the male.

Jumping, using long hind limbs, is the stereotyped frog locomotory behavior. Again, considerable specialization exists among anurans in this regard. For example, pipids are specialized aquatic frogs with webbed feet used for propulsion through water. Several clades (e.g., Centrolenidae, Hylidae, and Rhacophoridae) include arboreal frogs, which can move by quadrupedal walking or climbing as well as by leaping. Hemisotidae include burrowing frogs that dig headforemost, a behavior reflected by their heavily ossified skulls. Several terrestrial frogs tend to hop or walk rather than jump, such as Bufonidae, which tend to have heavy or robust bodies with relatively short legs.

Although frogs are commonly used in vertebrate dissection courses, it is worth remembering that, as vertebrates, they are neither primitive nor typical—they are just readily available. However, their frequent use, particularly for some species, has been a factor in their decline in many areas. Perhaps instructors might consider the use of the abundant Marine Toad instead. On the other hand, in some countries, such as Argentina, the bullfrog is bred for commercial purposes and has become a pest that is eliminating native species.

R. catesbeiana belongs to Ranidae, although the systematics of this group are not resolved and it may be paraphyletic (see Frost et al., 2006). *R. catesbeiana* is a native North American frog with a fairly wide natural distribution and has been introduced in parts of Asia, South America, and Europe. Bullfrogs vary considerably in size, but length tends to be between 10 and 17 cm, although many will be 20 cm in length (measured from the tip of the snout to the cloaca). They live in water and so are found near lakes, ponds, and rivers.

SECTION I: SKELETON

Skull, Mandible, and Hyoid Apparatus

The skeleton of anurans demonstrates quite dramatically the misleading assumption that many beginning

The Dissection of Vertebrates, Third Edition
https://doi.org/10.1016/B978-0-12-410460-0.00006-1

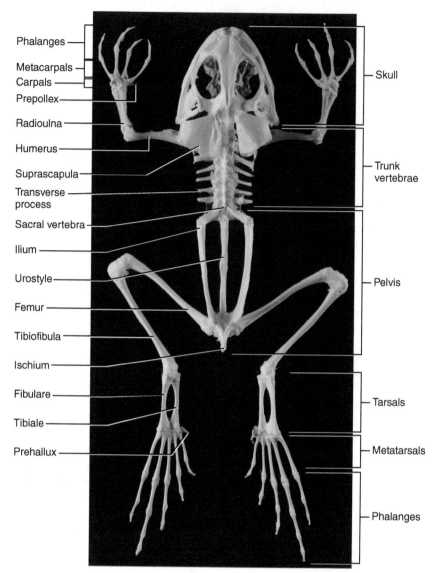

Phalanges
Metacarpals
Carpals
Prepollex
Radioulna
Humerus
Suprascapula
Transverse process
Sacral vertebra
Ilium
Urostyle
Femur
Tibiofibula
Ischium
Fibulare
Tibiale
Prehallux

Skull
Trunk vertebrae
Pelvis
Tarsals
Metatarsals
Phalanges

FIGURE 6.1   Dorsal view of the skeleton of the frog.

students have about lissamphibians—that they are intermediate between fishes and "higher" tetrapods, and so are simpler versions of reptiles and mammals. One glance at the highly specialized skeleton of a frog should suffice to dispel such views. The skeleton described here is of *R. catesbeiana*, but its features apply to anurans generally. Further images are available in Gosselin-Ildari (2004).

Lissamphibians generally tend to have specialized and reduced (and in some cases largely cartilaginous) skeletons, and that of frogs is no exception (Figure 6.1). There has indeed been considerable loss of bone and decrease in ossification over basal tetrapods (as well as "higher" tetrapods), which is clearly evident in the broad, flattened, and fenestrated **skull** (Figures 6.1 and 6.2). Among the bones commonly present in tetrapods but absent in anurans are the lacrimal, prefrontal, postfrontal, and opisthotic. Two particularly large openings

are the **orbits** dorsally and the **interpterygoid vacuities** ventrally on the palate.

Examine the skull in dorsal view (Figure 6.2a). Its margin is approximately parabolic. On each side, this margin is composed of the small median **premaxilla**, the long **maxilla**, and the shorter **quadratojugal**, in anterior to posterior order. Ventrally, the premaxilla and maxilla bear a single row of small teeth, the **premaxillary** and **maxillary teeth**, respectively (Figure 6.2b,c). The **vomers** lie anteriorly just behind the premaxillae. They bear **vomerine teeth**.

The paired **nasals** are broad and flattened medially and contact each other on the dorsal midline (Figure 6.2a,c,d). Each has a narrow process that extends lateroventrally, forming the anterior margin of the large orbits, and contacts an ascending process of the maxilla. A **naris**, the opening into the nasal cavities, lies anterior to each orbit. The paired **frontoparietals** are elongated, flattened

6

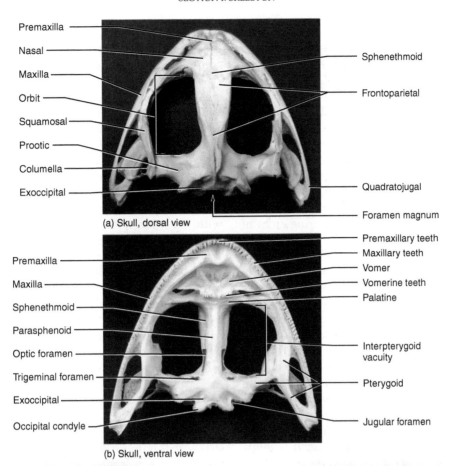

Premaxilla

Nasal

Maxilla

Orbit

Squamosal

Prootic

Columella

Exoccipital

Sphenethmoid

Frontoparietal

Quadratojugal

Foramen magnum

(a) Skull, dorsal view

Premaxilla

Maxilla

Sphenethmoid

Parasphenoid

Optic foramen

Trigeminal foramen

Exoccipital

Occipital condyle

Premaxillary teeth

Maxillary teeth

Vomer

Vomerine teeth

Palatine

Interpterygoid vacuity

Pterygoid

Jugular foramen

(b) Skull, ventral view

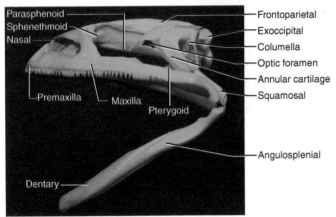

Parasphenoid

Sphenethmoid

Nasal

Premaxilla

Maxilla

Pterygoid

Dentary

Frontoparietal

Exoccipital

Columella

Optic foramen

Annular cartilage

Squamosal

Angulosplenial

(c) Skull and mandible, left lateral view

Frontoparietal

Nasal

Naris

Maxilla

Annular cartilage removed

Squamosal

Pterygoid

(d) Skull, left dorsolateral view

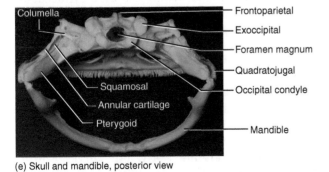

Columella

Squamosal

Annular cartilage

Pterygoid

Frontoparietal

Exoccipital

Foramen magnum

Quadratojugal

Occipital condyle

Mandible

(e) Skull and mandible, posterior view

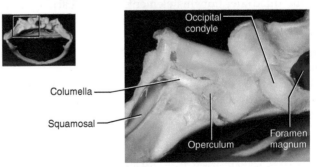

Occipital condyle

Columella

Squamosal

Operculum

Foramen magnum

(f) Detail of otic region, posterior view

FIGURE 6.2   Skull and mandible of the frog: (a) dorsal and (b) ventral views of the skull; (c) left lateral view of the skull and mandible; (d) left dorsolateral view of the skull; (e) posterior view of the skull and mandible; and (f) detail of left otic region in posterior view.

bones that meet along the dorsal midline to form most of the cranial roof. Much of the side and ventral parts of the braincase are formed by the **sphenethmoid**, which is exposed mainly in ventral and lateral views (Figure 6.2b,c). It is essentially tubular, but its anterior part expands laterally. Examine the skull in ventral view to see this anterior part. Here, the **palatines**[1] extend as transverse bars on either side of the sphenethmoid to reach the maxillae. A small portion of the sphenethmoid that helps form the roof of the braincase is exposed dorsally, wedged between the nasal and frontoparietal bones.

The **parasphenoid** is approximately cruciate or **t**-shaped (Figure 6.2b). The anteriorly tapering stem of the **t** covers the sphenethmoid ventrally. The top part of the stem projects posteriorly toward the **exoccipital** bones. Each transverse arm of the **t** extends laterally toward a **pterygoid** bone. The paired **exoccipital** bones form the posterior part of the skull (Figure 6.2e,f). They enclose the **foramen magnum**, the large opening for passage of the spinal cord. Ventrally, each exoccipital bears an **occipital condyle** for articulation with the atlas. The pterygoid is triradiate, or **y**-shaped, with its three arms extending out to contact other skeletal elements (Figure 6.2b). Its anterior arm extends anteriorly and contacts the maxilla and nasal, while its posterior arm extends posteriorly, curving gently laterally, to the angle of the jaw (see later). The medial arm is shortest. It extends toward the **prootic** (see later) and transverse stem of the parasphenoid.

Examine the posterior end of the skull in lateral view and identify the **T**-shaped **squamosal** (Figure 6.2c,d). The top of the **T** is curved, with one arm extending anteroventrally, the other posterodorsally. Some prepared specimens provide an unobstructed view of the squamosal (Figure 6.2d), due to the absence of the **annular cartilage** that supports the **tympanic membrane**. Other specimens retain the cartilage, and the distal end of the **columella** (Figure 6.2c; see also Section III) can be seen within the area it circumscribes. The long stem of the **T** of the squamosal is oriented posteroventrally toward the angle of the jaw, where the jaw joint is formed. The main element that forms the upper part of the jaw joint is the **quadrate**, but this element, usually cartilaginous in anurans, is usually not visible in adults because it is covered laterally by the ventral part of the squamosal, and medially and posteromedially by the posterior arm of the pterygoid.

The medial end of the columella (or stapes) expands into a (partly cartilaginous) footplate that covers the anterior part of the fenestra ovalis, the entrance into the

inner ear, whereas its lateral, slender portion is termed the stylus (Figure 6.2e,f). Another term used for the columella is plectrum (see Duellmann and Trueb, 1994). Anurans (and some salamanders) have a second skeletal element that covers the posterior part of the fenestra ovalis. This is the cartilaginous **operculum** (Figure 6.2f), which is coupled with the columella (for further details see Mason et al., 2003; Mason and Narins, 2002).

Examine the skull in dorsal view. The irregularly shaped prootics contain the inner ear. The prootic extends between the squamosal laterally, and the exoccipital and frontoparietal medially. Usually the prootic fuses with the exoccipital (a combined element sometimes referred to as the otoccipital, as noted by Duellmann and Trueb, 1994). Anteriorly the prootic helps form the posterior wall of the orbit and margin of the **optic foramen** (Figure 6.2b,c) and contains a large opening, the **trigeminal foramen** (Figure 6.2b; termed the prootic foramen by Duellmann and Trueb, 1994), through which the trigeminal (Cranial Nerve V) and facial (Cranial Nerve VII) nerves pass. Posteriorly the prootic and exoccipital form the **foramen ovale** (or jugular foramen according to Trueb, 1973), the opening just beside the occipital condyle, for passage of the glossopharyngeal and vagus nerves (Cranial Nerves IX and X, respectively).

The **mandible** is a slender, parabolic, edentulous structure (Figures 6.2c,e and 6.3). As in the skull, several bones (e.g., the coronoid, articular, and prearticular) commonly present in nonmammalian tetrapods are missing. Teeth are also absent from the lower jaw, as in nearly all anurans. The mandible includes **Meckel's cartilage**, nearly all of which is covered by two large elements, the **dentary** anterolaterally and the **angulosplenial** medially and posteriorly. The dentary is a thin, flange-like bone. The angulosplenial is larger. Its posteromedial surface bears the medially directed flange-like **coronoid process**. Posteriorly the angulosplenial articulates with the quadrate of the skull. Portions of Meckel's cartilage, if preserved, may be observed in the dorsal, trough-like groove of the angulosplenial. Anteriorly, the **mentomeckelian** ossifies at the mandibular symphysis. The right and left mentomeckelians are attached anteriorly at the mandibular symphysis by a syndesmotic (i.e., fibrous) connection. Trueb (1993) indicated that in frogs, the mentomeckelian is derived from the infrarostral (a structure of the larval chondrocranium), and the mentomeckelian of frogs and salamanders may not be homologous.

The **hyoid apparatus** is mainly a thin, broad cartilaginous plate, or body, on the floor of the oral cavity that supports the tongue and larynx (Figure 6.4). Several processes project from it. The **anterior cornu** (pl., **cornua**) initially extends anteriorly, but curves sharply posterodorsally to attach to the skull. In Figure 6.4 only part of the left anterior cornu is preserved. The **posterior**

---

[1]Trueb (1993) indicated that the element termed the palatine may not be homologous with the palatine of other tetrapods, as it seems to arise as a lingual process of the maxilla. This author referred to the process as the neopalatine.

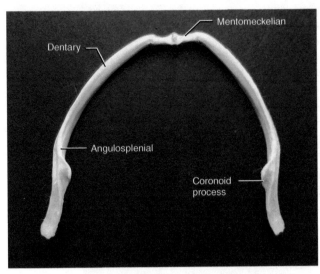

FIGURE 6.3 Mandible of the frog in dorsal view.

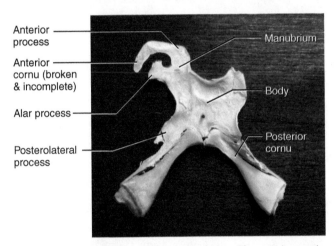

FIGURE 6.4 Hyoid of the frog in dorsal view. The anterior portion of the right side is missing.

cornua are bony rods extending from the posterior margin. They begin medially but diverge to contact the larynx.

Key Terms: Skull, Mandible, and Hyoid Apparatus

annular cartilage
anterior cornu
columella
coronoid process
dentary
exoccipital
foramen magnum
foramen ovale
frontoparietal
hyoid apparatus
interpterygoid vacuity
mandible

maxilla
maxillary teeth
Meckel's cartilage
mentomeckelian
naris
nasal
occipital condyle
operculum
orbit
palatine
parasphenoid
posterior cornu
premaxilla
premaxillary teeth
prootic
pterygoid
quadratojugal
skull
sphenethmoid
squamosal
trigeminal foramen
tympanic membrane
vomerine teeth
vomer

Postcranial Skeleton

The postcranial skeleton also shows evidence of extreme modification, associated mainly with the highly specialized locomotor mechanism characteristic of most anurans. The **vertebral column** is reduced, with only nine free **vertebrae** (Figure 6.5). Most anteriorly in this series is the **atlas**, which articulates with the occipital condyles of the skull. The last free vertebra is a **sacral vertebra**, attaching to the **pelvic girdle** (see later). Extending posteriorly from the sacral vertebra is the elongated, rodlike **urostyle**, which is formed by the fusion of several postsacral vertebrae. The vertebrae have prominent **transverse processes**, but ribs are lacking.

The appendicular skeleton is well developed. The **pectoral girdle**, largely ossified, contains several elements (Figures 6.1 and 6.6). The **scapula** is a nearly vertical plate-like structure. Extending dorsomedially from it is the **suprascapula**, which has a prominent and usually calcified cartilaginous portion medially. Ventrally, there are two large paired elements. The more anterior and slender **clavicles**, dermal elements, extend almost directly medially from the scapula. The larger, more posterior **procoracoids** form a plate-like base to the girdle. The **glenoid fossa**, for articulation with the **humerus** (see later), is formed mainly by the scapula and procoracoid. An anterior median element, the **omosternum**, lies anteriorly. The cartilaginous **episternum** extends anteriorly from it. A posterior median element, the **sternum**,

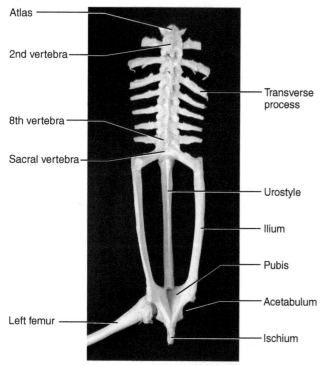

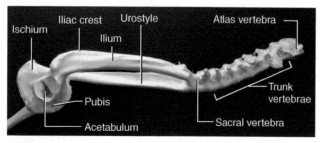

FIGURE 6.5   Vertebral column and pelvis of the frog in (a) dorsal and (b) right lateral views.

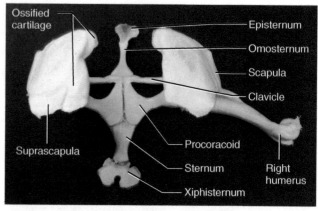

(a) Pectoral girdle, dorsal view

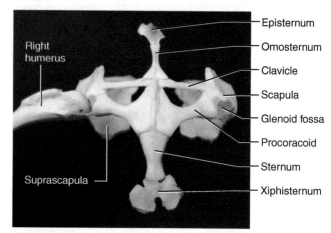

(b) Pectoral girdle, ventral view

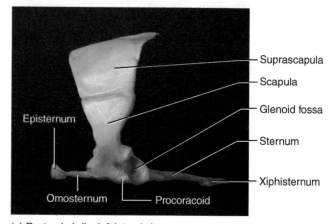

(c) Pectoral girdle, left lateral view

FIGURE 6.6   Pectoral girdle of the frog in (a) dorsal, (b) ventral, and (c) left lateral views.

articulates with the procoracoids. A cartilaginous **xiphisternum** extends posteriorly from it.

The forelimb includes three segments, the most proximal of which is the brachium (see later), which is supported by the humerus; this bone extends laterally from the glenoid fossa (Figures 6.1 and 6.6). The next segment, the antebrachium, is supported by the **radius** and **ulna**, which fuse to form the **radioulna** (Figures 6.1 and 6.7). The **manus** includes a proximal series of small, nodular **carpals**, followed by four complete **digits**, including **metacarpals II–V** and **phalanges** (as noted in Chapter 5, the inclusion of a metacarpal as part of a digit follows Kardong, 2018). The two medial digits each bear two phalanges, and the lateral two bear three phalanges each. A small **prepollex** extends medially from the carpals and may represent a reduced metacarpal.

In the **pelvic girdle**, the **pelvis** is formed on each side by the **ilium**, **ischium**, and **pubis** (Figures 6.1

and 6.5). The slender ilium is an elongated, anteriorly directed element, with a well developed **iliac crest**. The ischium and pubis together outline a semicircle in lateral view, the ischium forming the more posterior portion. The **acetabulum** is a conspicuous depression for articulation with the **femur**. The hind limb consists of the proximal femur, followed by the

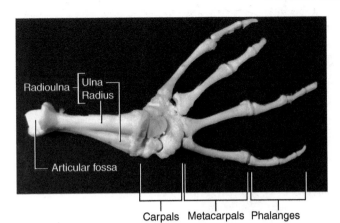

FIGURE 6.7   Right antebrachium and manus of the frog in dorsal view.

tibia and fibula, fused to form the tibiofibula. The pes has rather typical metatarsals and phalanges, but a modification in the tarsal region produces another functional segment to the hind limb, a feature that is related to the saltatory locomotion of frogs. Here, the two proximal tarsals are elongated to form a medial tibiale and lateral fibulare (=calcaneum), which are partly fused at their ends. The tibiale and fibulare are commonly viewed as equivalent to the astragalus and calcaneum, respectively, the larger tarsal elements of amniotes. Although this does indeed seem to be the case for the fibulare, the tibiale is not homologous with the astragalus. As noted by O'Keefe et al. (2006), the amniote astragalus is apparently a fusion of as many as four anamniote tarsal elements. The distal tarsals have the more typical nodular form. There are five digits, with the first being the shortest and the fourth being the longest. Digits I and II each bear two phalanges; III and V, three; and IV, four. A small prehallux, similar to the prepollex, extends medially from the tarsal region.

Key Terms: Postcranial Skeleton

acetabulum
atlas
carpals
clavicle
digits
episternum
femur
fibula
fibulare (= calcaneum)
glenoid fossa
humerus
iliac crest
ilium
ischium
manus

metacarpal
metatarsal
omosternum
pectoral girdle
pelvic girdle
pelvis
pes
phalanges
prehallux
prepollex
procoracoid
pubis
radioulna
radius
sacral vertebra
scapula
sternum
suprascapula
tibia
tibiale
tibiofibula
transverse processes
ulna
urostyle
vertebrae
vertebral column
xiphisternum

SECTION II: EXTERNAL ANATOMY

As with the underlying skeleton, many features of the external anatomy are highly modified in association with the highly specialized saltatory mode of locomotion. The most obvious features are that frogs have very short, wide bodies and very large hind limbs and lack a tail (Figure 6.8). With respect to the shortness of the trunk and relative size of the hind limbs, frogs exhibit among the most extreme specializations of any vertebrate. The forelimb, including the brachium, antebrachium, and manus, is typical in form and proportions to those of other tetrapods. In the hind limb, the thigh and crus are also typical, but the pes is extremely elongated, a characteristic due to the marked modification of two proximal tarsals (Figure 6.1; see Section I), which produces an additional functional segment. There are four digits in the manus. The pes has five digits, which are webbed, as in many swimming forms. The thumb hypertrophies in males during the breeding season to help hold the female during amplexus. Claws are absent, as in amphibians generally.

The skin, as in many anurans, is thin and highly glandular. These features are associated with the considerable degree of respiration through the skin in most anurans. The mouth is very large, but otherwise normal.

FIGURE 6.8   External features of the frog.

The **nares** are rather small, located anteriorly and close together on the dorsal surface of the snout. The **eyes** are fairly large and project out from the top of the head, but in preserved specimens they are often retracted and covered by the small **eyelids**, which are not independently moveable. A **nictitating membrane** is present (Figure 6.9a).

Posterior to each eye, the conspicuous and circular **tympanic membrane** represents the ear externally (Figure 6.9a). Actually, the membrane itself lies deep to the skin and may be separated from it. As noted earlier, the membrane is supported by the annular cartilage (Figure 6.9b). Males are easily distinguished from females by the size of the tympanic membrane. In females it is about the same size as the eye, whereas in males it is much larger than the eye.

The **cloaca** is present posteriorly (Figure 6.13). However, due to the absence of a tail, it appears to be located somewhat dorsally.

Key Terms: External Anatomy

antebrachium
brachium
cloaca
crus
eyelid
eyes
manus
nares (sing., **naris**)
nictitating membrane
pes
skin
thigh
tympanic membrane

SECTION III: MOUTH, ORAL CAVITY, AND PHARYNX

Open the mouth by cutting through the angle of the jaw on each side, so that you can open the mouth to reveal the **oral cavity** and **pharynx**, as shown in Figure 6.10. Note the large **tongue** on the floor of the oral cavity. The tongue is attached anteriorly and folded back into the oral cavity, so that its distal, bifid end lies posteriorly. The tongue is extended to catch insects by rotating it dorsally around its anterior attachment. A single row of small teeth, often easier to feel than to see, lies around the margin of the upper jaw. As described above, these teeth are mostly **maxillary teeth**. The few **premaxillary teeth** are near the midline. The teeth lie external to the **maxillary groove**, which extends around the margin of the upper jaw. The **pterygoid ridge** is lingual to the groove. A row of **vomerine teeth** is present on each of the vomers. These teeth lie farther posteriorly near the midline of the roof of the oral cavity.

A series of openings enter the oral cavity. The **choanae** are prominent and lie lateral to the vomerine teeth. The large posterolateral openings are the **auditory tubes**, which lead to the middle ear cavities. The floor of the orbits lies between the choanae and auditory tubes. Pressure on this area will force the eyeballs up into their open position. Ventrally in the oral cavity, posterior to the tongue, is the slit-like **glottis**, which leads to the **lungs** (see later). The glottis is on a small projection, the **laryngeal prominence**, which is formed by cartilages. Laterally on the floor of the oral cavity, on each side of the anterior end of the glottis, is a small opening in males that leads to a vocal sac. The sacs are used in calling during the mating season, but are difficult to find.

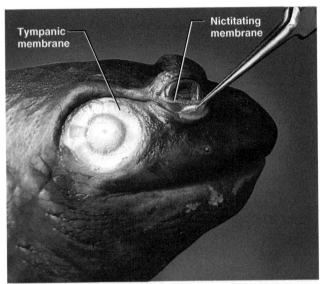

(a) Tympanic membrane and nictitating membrane, right lateral view

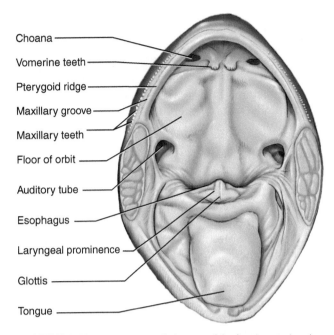

FIGURE 6.10   Oral cavity and pharynx of the frog in anterior view.

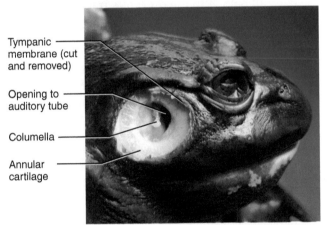

(b) Columella with tympanic membrane removed, right lateral view

FIGURE 6.9   Auditory region of the frog in right lateral view.

lungs
maxillary groove
maxillary teeth
oral cavity
pharynx
premaxillary teeth
pterygoid ridge
tongue
vomerine teeth

The entrance into the **esophagus** is posterior to the glottis. Probe it gently to verify that it does indeed extend posteriorly.

Break the tympanic membrane to expose the middle ear cavity (Figure 6.9b). The **columella** (see page 162), which has a cartilaginous distal portion, passes through this cavity to the membrane. Probe the cavity to verify that it connects with the oral cavity through the auditory tube.

Key Terms: Mouth, Oral Cavity, and Pharynx

auditory tube (Eustachian tube)
choanae
columella
esophagus
glottis
laryngeal prominence

SECTION IV: PLEUROPERITONEAL CAVITY, VISCERA, AND UROGENITAL SYSTEM

Using a scalpel and just to one side of the midventral line, make a small, shallow incision—just large enough to insert a scissor blade—through the skin (and only through the skin), which is very thin. Avoid damaging the underlying musculature. Using scissors, continue the incision anteriorly approximately to the level of the axilla, and posteriorly to approximately midway across the width of the hind limb. From its anterior limit, extend the incision laterally to pass just posterior to the forelimb and reach nearly to the frog's dorsum. Similarly, from the posterior limit, cut the skin around the margin of the hind limb. Repeat these steps for the skin on the other side of the body, finally producing two flaps that can be reflected laterally.

Begin reflecting the skin on one side. It tends to pull away easily but does adhere more strongly in several places. These represent the attachments of lymphatic sacs. Scrape away the connecting tissue, but stay close to the skin in doing so. You will reveal the pectoral and abdominal musculature. Proceed cautiously on the ventrolateral surface, where the **musculocutaneous vein** extends anteroposteriorly—leave the vein on the musculature; it lies just lateral to the lateral edge of the pectoralis major (Figure 6.11). Anteriorly, it veers medially and passes deep to the pectoralis major. Follow the vein posteriorly as it reflects onto the deep surface of the skin, and note that it is formed by the coalescence of the numerous veins draining the skin.

Examine the abdominal musculature. On the midline, you will note the path of the **ventral abdominal vein**, which actually lies within the abdominal cavity and will be exposed presently. Make two anteroposterior incisions through the musculature. The first will be approximately 0.5 cm to one side of the midventral line, to avoid damaging the ventral abdominal vein. For convenience, make this incision on the same side of the body on which the musculocutaneous vein was exposed. Make the second incision parallel and just medial to that portion of the musculocutaneous vein on the abdominal wall. Then make a transverse cut anteriorly and another posteriorly so that you may remove the rectangular block of musculature. This will expose approximately half of the pleuroperitoneal cavity. For the musculature of the other side, simply cut transversely through the musculature from the middle of the median anteroposterior incision, and reflect the resulting two flaps.

Follow the ventral abdominal vein as it passes into the cleft between the **right** and **left lateral lobes** of the liver, the large dark mass lying anteriorly in the pleuroperitoneal cavity. The liver is relatively wide and short, conforming to the shape of the body. Its lobes are usually subdivided to varying degrees. The left lobe usually extends farther posteriorly due to the development of its posterior lobe, and the right lobe of the liver usually covers a smaller median lobe of the liver. Spread the lateral lobes of the liver to reveal the spherical, sac-like **gall bladder**, which lies just posterior to the passage of the ventral abdominal vein (Figure 6.12). Extending from this vein and the liver to the midventral body wall is the **falciform ligament**. Lift the body wall, and break through the falciform ligament to reveal the **pericardium**, a sac-like structure that contains the **heart** (see later) nestled between the anterior ends of the lateral lobes of the liver.

Examine the structures posterior to the liver. In the female, the irregularly shaped **ovaries** are generally conspicuous, "speckled" structures containing developing follicles that are usually visible. The ovaries vary in size, depending on stage of the reproductive cycle, and may be massive, occupying a large part of the pleuroperitoneal cavity (compare Figures 6.11–6.13). The small, ovoid **testes** of the male are much less apparent, being confined to their relatively dorsal position and thus covered by other viscera. They will be described shortly. In both sexes, each gonad is associated with a conspicuous **fat body** (Figures 6.12–6.14), which is subdivided into numerous digitiform lobes that are often pressed up against the sides of the pleuroperitoneal cavity. Stored nutrients in the fat bodies are used primarily to nourish the developing gametes. The size of the fat bodies thus varies greatly with the stage of reproductive cycle. In females, the paired **oviducts** are large, highly convoluted tubes occupying much of the rest of the ventral part of the pleuroperitoneal cavity. Stretches of the digestive tract (see later) may be exposed among the coils of the oviducts, but these are generally narrower and slightly darker. In males, large oviducts will not be present; the coils all belong to the digestive tract.

If your specimen is a female with very large or massive ovaries, remove them to provide a better view of the remaining structures, as shown in Figure 6.13. Grasp an ovary, reflect it laterally, and remove it by cutting through its mesentery, the **mesovarium** (Fig. 6.13).

Examine the viscera. Find the **stomach**, tucked deep to the lateral side of the left lobe of the liver (Figure 6.12). It extends posteriorly on the left side of the pleuroperitoneal cavity. Proximally, it leads to the short, thick **esophagus**. Lift the proximal end of the stomach to reveal the left **lung**, far anterior in the pleuroperitoneal cavity (Figures 6.13 and 6.14). The right lung may be found dorsal to the right lobe of the liver. The lungs appear as small, contracted sacs in preserved specimens, but they are generally larger in live frogs. Return to the stomach. Distally it narrows and turns abruptly to the right and leads to the **intestine**.

The intestine may be subdivided into the narrow, coiled **small intestine** followed by a short, wide **large intestine** that leads to the **cloaca** (Figure 6.13). The **pancreas** lies in the mesentery between the **duodenum**, the first part of the small intestine, and the stomach (Figure 6.12). The **spleen** is a dark, ovoid body lying in the mesentery farther distally and dorsal to the small intestine. Examine the posterior part of the pleuroperitoneal cavity to find the large, thin-walled **urinary bladder**. It empties into the ventral surface of the tube-like cloaca (Figures 6.12 and 6.14).

Examine an oviduct in a female specimen (Figure 6.13). Follow it anteriorly. As it passes anteriorly, it becomes narrower but remains highly coiled, and then, as it passes dorsal to the lung, straightens to reach its opening, the **ostium tubae**, which lies just lateral to the pericardium and faces ventromedially. Ova enter the oviduct through the ostium tubae and then pass posteriorly through the oviduct.

At its other end, the oviduct widens and straightens to form the **ovisac**, which may contain masses of eggs. The ovisacs empty into the dorsal surface of the cloaca, just proximal to the level of the entrance of the urinary bladder.

The prominent **kidneys** lie on the dorsal wall of the pleuroperitoneal cavity (Figures 6.13 and 6.15). They are dark, flattened, ovoid structures. The large vessel between them is the **posterior vena cava** (see later). **Adrenal glands** lie along the ventral surface of the kidneys and usually appear as lighter-colored bands. Along the lateral margin of the posterior end of each kidney lies an **archinephric duct**, which leads posteriorly into dorsal surface of the cloaca, and very near the entrance of the ovisac in the female. The archinephric duct transports only urine in the female, but carries both urine and sperm in the male. The whitish strands that appear dorsal to the kidneys and extend posteriorly are part of the sciatic plexus and give rise to the nerves of the hind limb (Figure 6.13). To follow the nerves more anteriorly, break through the peritoneum so that a kidney may be lifted from the dorsal body wall.

The ovaries in the female have already been identified. Identify the testes in a male. They lie on the ventral surface of the kidneys, and each is supported by its mesentery, the **mesorchium** (Figure 6.14). The testes are small, smooth, and ovoid structures, their light color in sharp contrast with that of the kidneys. Sperm pass from the testes through inconspicuous **ductuli efferentes** in the mesorchium, and then enter the kidney to reach the archinephric duct.

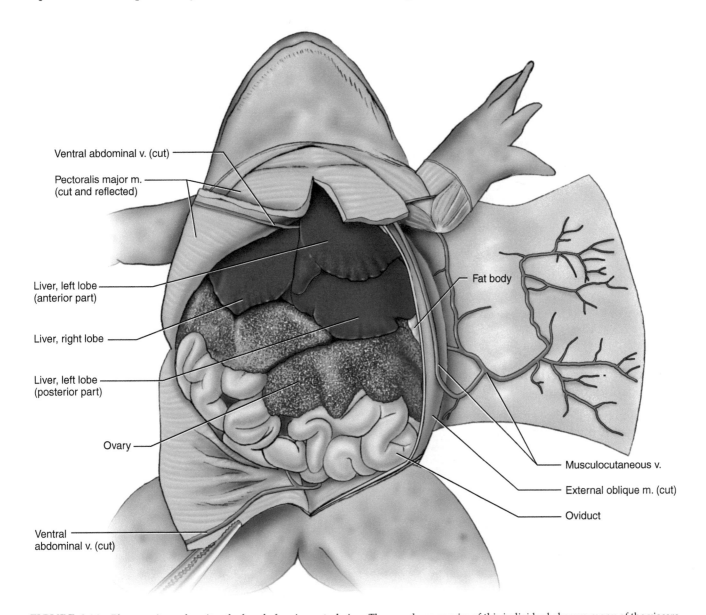

Ventral abdominal v. (cut)
Pectoralis major m. (cut and reflected)
Liver, left lobe (anterior part)
Liver, right lobe
Liver, left lobe (posterior part)
Ovary
Ventral abdominal v. (cut)
Fat body
Musculocutaneous v.
External oblique m. (cut)
Oviduct

FIGURE 6.11 Pleuroperitoneal cavity of a female frog in ventral view. The very large ovaries of this individual obscure many of the viscera.

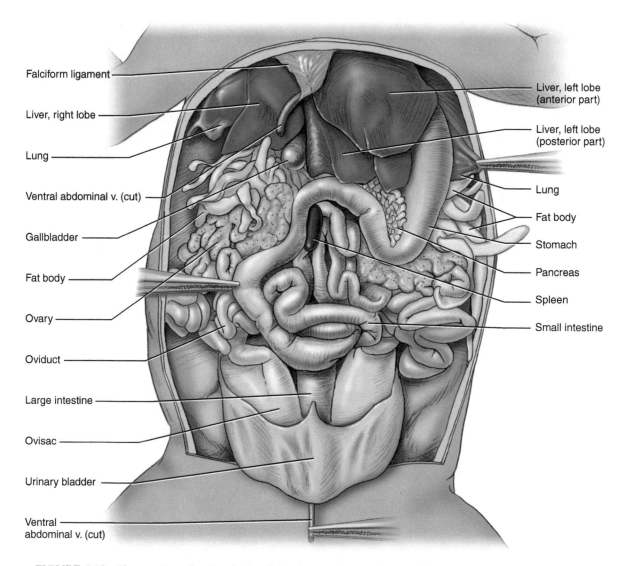

Falciform ligament

Liver, right lobe

Lung

Ventral abdominal v. (cut)

Gallbladder

Fat body

Ovary

Oviduct

Large intestine

Ovisac

Urinary bladder

Ventral abdominal v. (cut)

Liver, left lobe (anterior part)

Liver, left lobe (posterior part)

Lung

Fat body

Stomach

Pancreas

Spleen

Small intestine

FIGURE 6.12 Pleuroperitoneal cavity of a female frog in ventral view. The smaller ovaries expose many of the viscera.

Key Terms: Pleuroperitoneal Cavity, Viscera, and Urogenital System

adrenal gland
archinephric duct (Wolffian duct)
cloaca
ductuli efferentes
duodenum
esophagus
falciform ligament
fat body
gall bladder
heart
intestine
kidney
large intestine
liver, right and **left lateral lobes**

lung
mesorchium
mesovarium
musculocutaneous vein
ostium tubae
ovary
oviduct
ovisac
pancreas
pericardium
posterior vena cava
small intestine
spleen
stomach
testis
urinary bladder
ventral abdominal vein

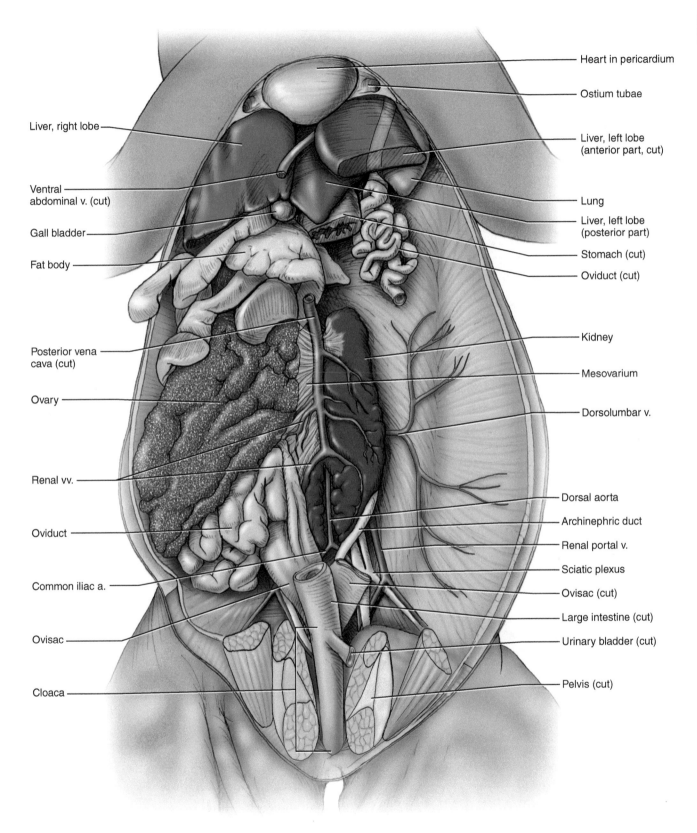

Liver, right lobe

Ventral abdominal v. (cut)

Gall bladder

Fat body

Posterior vena cava (cut)

Ovary

Renal vv.

Oviduct

Common iliac a.

Ovisac

Cloaca

Heart in pericardium

Ostium tubae

Liver, left lobe (anterior part, cut)

Lung

Liver, left lobe (posterior part)

Stomach (cut)

Oviduct (cut)

Kidney

Mesovarium

Dorsolumbar v.

Dorsal aorta

Archinephric duct

Renal portal v.

Sciatic plexus

Ovisac (cut)

Large intestine (cut)

Urinary bladder (cut)

Pelvis (cut)

FIGURE 6.13   Pleuroperitoneal cavity of a female frog in ventral view. Several structures have been removed from the left side to expose the urogenital system.

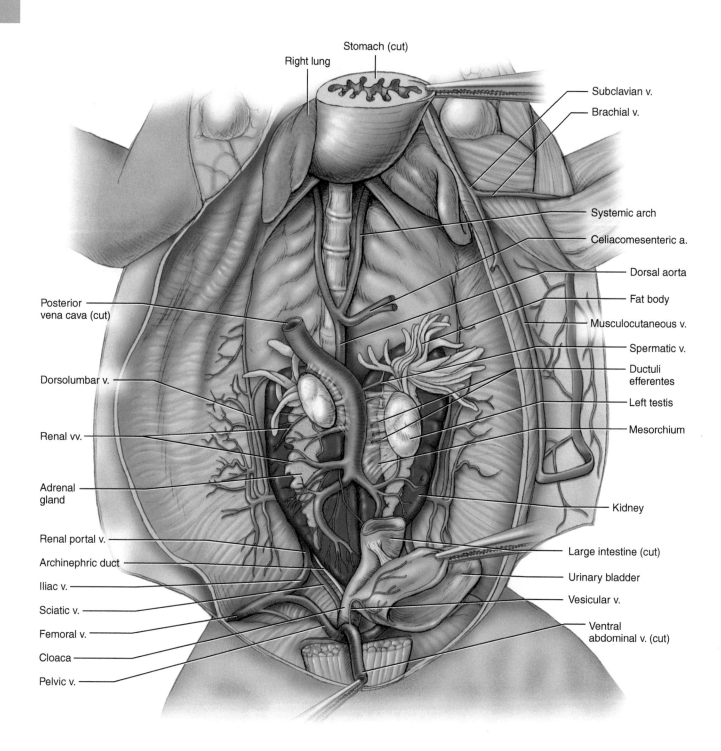

FIGURE 6.14　Pleuroperitoneal cavity of a male frog in ventral view, with many of the viscera removed to expose the urogenital system.

SECTION V: CARDIOVASCULAR SYSTEM

Remove the musculature ventral to the **pericardium** to expose the vessels anterior to the **heart**, as shown in Figure 6.15. Make a longitudinal, midventral slit through the pericardium to open the **pericardial cavity** and expose the heart. Several of the heart's components are plainly visible in ventral view. Its most prominent structure is the single **ventricle**, which lies in the posterior half of the pericardial cavity. Lift the

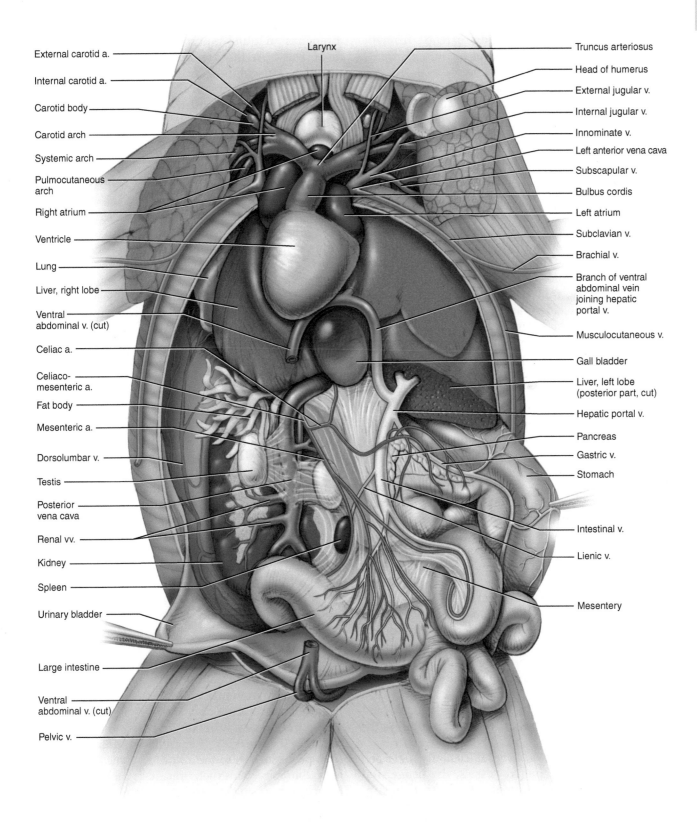

External carotid a.

Internal carotid a.

Carotid body

Carotid arch

Systemic arch

Pulmocutaneous arch

Right atrium

Ventricle

Lung

Liver, right lobe

Ventral abdominal v. (cut)

Celiac a.

Celiaco-mesenteric a.

Fat body

Mesenteric a.

Dorsolumbar v.

Testis

Posterior vena cava

Renal vv.

Kidney

Spleen

Urinary bladder

Large intestine

Ventral abdominal v. (cut)

Pelvic v.

Larynx

Truncus arteriosus

Head of humerus

External jugular v.

Internal jugular v.

Innominate v.

Left anterior vena cava

Subscapular v.

Bulbus cordis

Left atrium

Subclavian v.

Brachial v.

Branch of ventral abdominal vein joining hepatic portal v.

Musculocutaneous v.

Gall bladder

Liver, left lobe (posterior part, cut)

Hepatic portal v.

Pancreas

Gastric v.

Stomach

Intestinal v.

Lienic v.

Mesentery

FIGURE 6.15   Pleuroperitoneal and pericardial cavities of a male frog in ventral view.

THE DISSECTION OF VERTEBRATES

ventricle to see the **sinus venosus** (Figure 6.16). The **right** and **left atria** (sing., **atrium**) are the conspicuous structures anterior to the ventricle (Figure 6.15). Between them, the **bulbus cordis** extends from the ventricle anteriorly and slightly to the left. This anterior part of the heart may be covered by fat and connective tissue. Carefully pick away and remove them. In an injected specimen the structures are clearly identifiable and easy to expose. In uninjected specimens the vessels are harder to identify and the thin-walled atria readily torn, so proceed cautiously.

The bulbus cordis leads into the **truncus arteriosus**, which continues a short distance anteriorly before bifurcating into right and left branches, each of which extends anterolaterally and gives rise to three large arteries (Figures 6.15 and 6.17). Follow one of the branches, and identify these arteries and their branching patterns. The most anterior branch is the **carotid arch**. It divides into a small branch, the **external carotid artery**, that extends anteriorly to supply the tongue and lower jaw, and a bulbous **carotid body** that continues as the **internal carotid artery** to supply the eye, brain, and upper jaw. The most posterior branch is the **pulmocutaneous arch**. It curves sharply posteriorly, giving off a **cutaneous artery** extending laterally to supply the skin, and then continues to the lung as the **pulmonary artery**. The middle branch is the **systemic arch**. It supplies most of the rest of the body with blood, and so is the largest of the three branches. Initially the systemic arch passes nearly laterally, but it soon arches strongly dorsally. As it passes dorsal to the lung it curves medially and posteriorly, giving off two branches, the **occipitovertebral artery** and the **subclavian artery**, which continues into the brachium as the **brachial artery** (Figure 6.17). After giving off these two branches, the systemic arch enters the pleuroperitoneal cavity.

Within the cavity each systemic arch passes posteromedially. Just anterior to the level of the kidneys, the right and left systemic arches unite to form the **dorsal aorta**, which continues posteriorly along the middorsal wall of the cavity (Figures 6.14 and 6.17). Immediately after its origin, the dorsal aorta sends off a large branch, the **celiacomesenteric artery**, to the abdominal viscera. This vessel soon bifurcates into the **celiac artery**, which mainly supplies the liver, gall bladder, stomach, and pancreas, and the **mesenteric artery**, which supplies the intestines and spleen.

Between the **kidneys**, several (usually between four and six) smaller, paired vessels, the **urogenital arteries**, extend laterally from the dorsal aorta to supply the kidneys and gonads, as well as fat bodies and urogenital ducts (Figure 6.17). Near the posterior end of the kidneys the dorsal aorta bifurcates into right and left **common iliac arteries** (Figures 6.13 and 6.17).

In this region the arteries lie dorsal to the veins, which are also more prominent. Follow one of the common iliac arteries posteriorly. It gives off two smaller branches in quick succession from its lateral surface. These are the **hypogastric artery**, which mainly supplies musculature in this region and the urinary bladder; and the **femoral artery**, which helps supply several muscles and the skin in this region. The common iliac artery then continues into the hind limb as the **sciatic artery** (Figure 6.17). Its many branches supply the leg.

Return to the heart, lift the ventricle, and examine the sinus venosus, which leads into the right atrium (Figure 6.16). Note the large vessels, the **right anterior vena cava** and the **left anterior vena cava**, extending along the lateral edge of the atria and passing into the sinus venosus. These venae cavae collect blood from the head and forelimbs, as well as the skin. Many of the vessels that enter the venae cavae lie dorsal to the arterial vessels and should be injected with blue latex. In a few specimens, however, these will have been infiltrated by the latex of the arteries and will be partly or completely red.

Trace an anterior vena cava. The pattern described here is the general pattern, but there is variation. Indeed, the branching patterns of the right and left venae cavae may vary. The anterior vena cava collects blood from several vessels and empties into the sinus venosus. The main vessels forming the anterior vena cava are the **external jugular, innominate**, and **subclavian veins** (Figures 6.15 and 6.18). These three may join together. The external jugular vein extends almost directly anteriorly and drains the tongue and lower jaw. For most of its length it passes nearly parallel and just medial to the external carotid artery. The innominate vein may be quite short or may extend laterally for a longer distance before receiving its tributaries, the **internal jugular** and **subscapular veins**. The internal jugular, draining the eye, brain, and upper jaw, extends anterolaterally, whereas the subscapular vein, draining mainly the muscles associated with the pectoral girdle, passes nearly laterally. Finally, the subclavian vein passes posterolaterally, formed by the confluence of the **brachial vein** from the forelimb and the **musculocutaneous vein**, noted earlier, from the pectoral musculature and the deep surface of the skin.

The **pulmonary veins** return blood to the heart from the lungs (Figure 6.18). The pulmonary vein, one on each side, passes dorsal to the anterior vena cava. Right and left pulmonary veins then unite just anterior to the heart to form a short single vessel that extends posteriorly into the left atrium.

Much of the blood posterior to the heart (that from the skin being the main exception) is returned via the **posterior vena cava**, the large vessel extending between

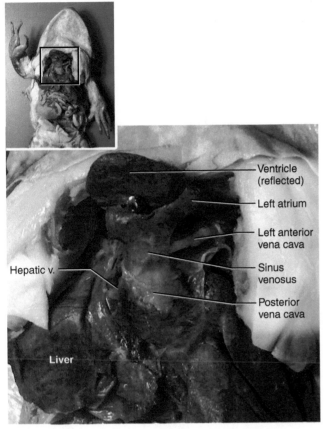

Ventricle (reflected)

Left atrium

Left anterior vena cava

Sinus venosus

Posterior vena cava

Hepatic v.

Liver

FIGURE 6.16   Heart of the frog in ventral view, with the ventricle reflected.

the kidneys (Figures 6.13–6.15). Here it receives several **renal veins** from the kidneys (the fat bodies generally drain into the more anterior renal veins) and **spermatic veins** from the testes or **ovarian veins** from the ovaries. The posterior vena cava extends anteriorly through the liver, receiving from it several **hepatic veins** (Figure 6.18), to reach the posterior end of the sinus venosus. The posterior vena cava also receives blood from much of the hind limbs and dorsal body musculature by way of the paired **renal portal veins**, which enter the kidneys (Figures 6.13, 6.14, and 6.18). This blood then makes its way through the kidneys to reach the posterior vena cava via the **renal veins**.

Examine a renal portal vein. It extends mainly along the dorsolateral surface of the kidney. It is formed posteriorly by the union of the **iliac** and **sciatic veins** (Figures 6.14 and 6.18), the latter from the medial side of the thigh. The iliac vein is formed by the **femoral vein**, the large vessel from the lateral side of the thigh, and the **pelvic vein**. A connection between the femoral and sciatic veins, the **communicating iliac vein**, extends from the femoral, curving dorsally and then medioventrally to meet the sciatic vein (Figure 6.18). The pelvic vein extends ventromedially to join the pelvic vein from the other side of the body. Their union forms the **ventral abdominal vein**,

identified earlier, but which will be described shortly. A small vesicular vein, draining the urinary bladder, enters the ventral abdominal vein just after its origin. The renal portal vein passes anteriorly, sending numerous branches into the kidney. The **dorsolumbar vein**, which drains the dorsal and lateral abdominal walls, consists of numerous branches that collect usually into a single vessel that enters the renal portal vein at about the mid-length level of the kidney (Figures 6.13–6.15 and 6.18).

As noted earlier, the ventral abdominal vein passes anteriorly along the midventral wall of the pleuroperitoneal cavity and extends between the lobes of the liver. It then arches dorsally and then posteriorly. It gives off three branches, two of which enter the right and left lateral lobes of the liver, and the third continuing to join the **hepatic portal vein**, which, as described later, enters the liver (Figures 6.15 and 6.18). Blood from the hind limb may thus return to the heart through the posterior vena cava by several routes. It may pass through the renal portal system or the hepatic veins. In the latter instance it may pass through the branches of the ventral abdominal vein that enter the liver directly or through the hepatic portal vein via that branch of the ventral abdominal vein joining the hepatic portal vein.

The hepatic portal vein drains the abdominal viscera. It is formed mainly by the following vessels: the **gastric vein**, which collects several vessels and drains the stomach and part of the esophagus, and the **intestinal vein**, formed by vessels that drain most of the small intestine and large intestine (Figures 6.15 and 6.18). A **lienic vein** from the spleen also empties into it. The hepatic portal then receives a branch from the ventral abdominal vein before entering the liver.

Key Terms: Cardiovascular System

brachial artery
brachial vein
bulbus cordis
carotid arch
carotid body
celiac artery
celiacomesenteric artery
common iliac artery
communicating iliac vein
cutaneous artery
dorsal aorta
dorsolumbar vein
external carotid artery
external jugular vein
femoral artery
femoral vein
gastric vein
heart
hepatic portal vein

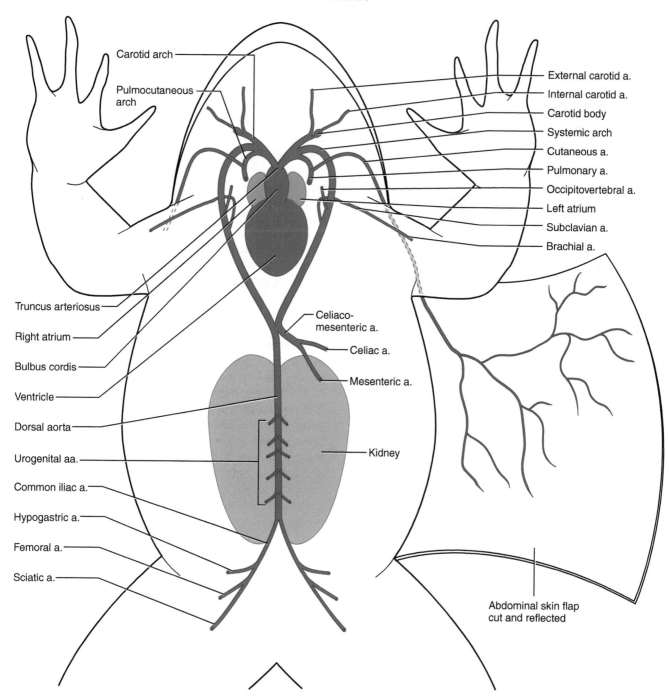

Carotid arch
Pulmocutaneous arch
External carotid a.
Internal carotid a.
Carotid body
Systemic arch
Cutaneous a.
Pulmonary a.
Occipitovertebral a.
Left atrium
Subclavian a.
Brachial a.
Truncus arteriosus
Right atrium
Bulbus cordis
Ventricle
Dorsal aorta
Urogenital aa.
Common iliac a.
Hypogastric a.
Femoral a.
Sciatic a.
Celiaco-mesenteric a.
Celiac a.
Mesenteric a.
Kidney
Abdominal skin flap cut and reflected

FIGURE 6.17　Schematic illustration of the arterial system of the frog in ventral view superimposed on the body outline.

hepatic veins
hypogastric artery
iliac vein
innominate vein
internal carotid artery
internal jugular vein
intestinal vein
kidney
left anterior vena cava
left atrium

lienic vein
mesenteric artery
musculocutaneous vein
occipitovertebral artery
ovarian vein
pelvic vein
pericardial cavity
pericardium
posterior vena cava
pulmocutaneous arch

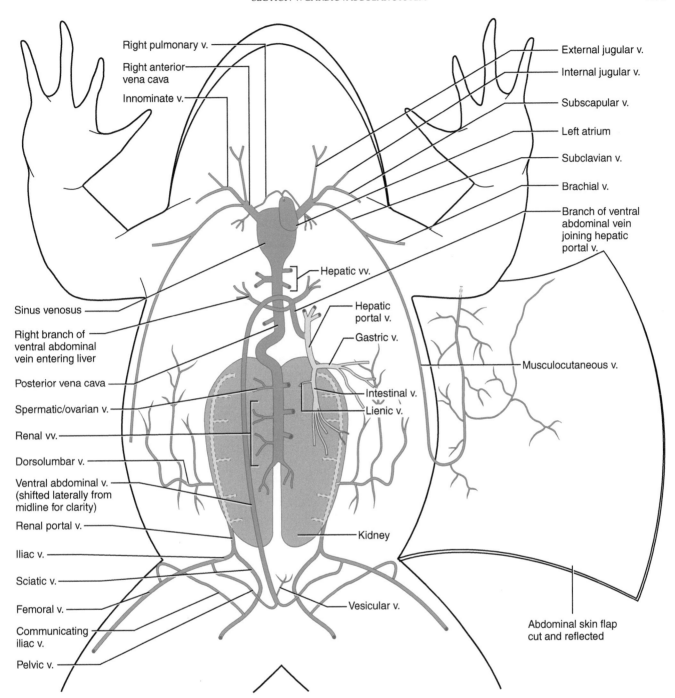

FIGURE 6.18 Schematic illustration of the venous system of the frog in ventral view superimposed on the body outline. Right atrium, ventricle, and bulbus cordis of heart removed.

pulmonary artery
pulmonary vein
renal portal vein
renal vein
right anterior vena cava
right atrium
sciatic artery
sciatic vein
sinus venosus

spermatic veins
subclavian artery
subclavian vein
subscapular vein
systemic arch (aortic arch)
truncus arteriosus
urogenital arteries
ventral abdominal vein
ventricle

7

The Cat

INTRODUCTION

The domestic cat, *Felis domestica*, belongs to Mammalia, a synapsid lineage that has its roots in the Triassic Period. Mammals can be readily diagnosed based on living representatives. Among their more distinguishing features are the presence of an insulating covering of hair, a diaphragm that separates the thoracic and abdominopelvic cavities, and mammary glands (to which they owe their name). Such soft tissue features, however, fossilize only under exceptional circumstances, and recognition of the most ancient mammals has been based on other features. Paleontologists have traditionally relied mainly on three skeletal features in determining whether a fossil species has achieved the mammalian condition: (1) the presence of three middle ear ossicles (nonmammalian synapsids have one, or rarely, two); (2) a jaw joint between the squamosal bone of the skull and the dentary bone of the lower jaw (in nonmammalian synapsids, the jaw joint is between the quadrate and articular bones, or in a few cases, all four of these bones participate); and (3) each half of the lower jaw consisting of a single bone, the dentary (in basal mammals and nonmammalian synapsids, each half includes more than one bone).

Modifications leading to the typical mammalian condition occurred in several advanced cynodont lineages, and the point at which the "mammalian threshold" was crossed has traditionally been recognized by possession of a jaw joint between the squamosal and dentary. It is not clear whether all other features usually considered "mammalian" were acquired simultaneously. Fossil evidence suggests they were not. Thus, use of the jaw joint criterion to diagnose Mammalia is arbitrary—any of several other features could be employed equally as well. For this reason, paleontologists have moved away from character-based definitions of Mammalia (or any higher-level taxon) and have come to rely on ancestry as the criterion for defining a group. As noted in Chapter 1, cladistics relies on shared, derived characters to reveal ancestry. The name Mammalia is now usually restricted to the group that includes the living groups of mammals, their common ancestor, and all extinct descendants of this ancestor. It is true that application of the name Mammalia to this particular group is also arbitrary, but this system does not rely on characters and promotes a more stable system of grouping organisms.

The two major subdivisions of living mammals are Monotremata and the much more diverse Theria, or "higher mammals." The monotremes are restricted to Australia and New Guinea and include the duckbill platypus and two species of echidnas. Although they have hair and mammary glands, they retain the general amniote reproductive strategy of laying and incubating eggs. They possess several derived characters, however. For example, adult platypuses lack teeth, and the rostrum is covered by a leathery bill or beak.

Theria is subdivided into Metatheria (marsupials) and Eutheria (placental mammals). The metatherians were much more widespread and diverse earlier in their history. Living metatherians, while still fairly diverse, are restricted primarily to Australia and South America, with a fairly recent incursion, the opossum, into North America. Metatherians are also known as marsupials or pouched mammals due to the presence of an abdominal marsupium (pouch) or fold in the female. Typically, the young are born in an immature condition and make their way from the vulva to the marsupium, where they latch on to a nipple and complete their development. This is the reproductive strategy of the more familiar marsupials such as the kangaroo, wallaby, and koala. However, not all marsupials follow this strategy. Some, such as the caenolestids, or shrew-opossums, lack a pouch, and others (the bandicoot) have a placenta similar to that of the placental mammals.

The eutherians are the most diverse and widespread of mammals, present on all continents save Antarctica (with modern humans the exception, of course). They have undergone several major radiations during their history and are currently among the more dominant or conspicuous elements of most terrestrial and aquatic habitats. The degree of habit and structural modifications is

The Dissection of Vertebrates, Third Edition
https://doi.org/10.1016/B978-0-12-410460-0.00007-3

truly remarkable. Eutherians include the relatively generalized shrews, fossorial moles, and armored armadillos. Several groups of herbivorous types have evolved, including rodents, the most numerous of mammals; bovids, such as the common cattle, bison and sheep, and the swift and graceful gazelles; the sleek and powerful equids, such as horses and zebras; and the huge rhinoceroses, hippopotami, and elephants, the largest of living land animals. The chiropterans, or bats, have mastered the skies and are the second most numerous group of mammals. Several eutherian lineages have invaded the seas. Among these are the cetaceans (whales, dolphins, and porpoises; now grouped together with artiodactyls as Cetartiodactyla, although some authors still prefer Artiodactlya for this group) and sirenians (manatees and dugongs), which are completely aquatic, and the pinnipeds (walrus and seals), which live both on land and in the water. The primates, a group that includes humans, are generally arboreal specialists.

Another great group of placental mammals includes the carnivorans, to which the cat belongs (carnivoran is the informal term for the mammalian clade Carnivora, whereas carnivore refers to a mainly or strictly meat-eating animal). Living carnivorans include ursids (bears), canids (dogs and their kin), mustelids (weasels and kin), procyonids (raccoons), hyaenids, felids (cats), and viverrids (genets), as well as the pinnipeds just mentioned. The felids include some of the fiercest and largest mammalian predators, such as tigers and lions. The domestic cat is among the smaller members of the group.

SECTION I: SKELETON

Cranial Skeleton

Skull

As with other vertebrates, the skull (Figure 7.1) of the cat protects and supports the brain and sense organs and is used in food gathering and processing. It is a single structure formed from various centers of ossification, but may be conveniently divided into a facial or rostral region, including the nose, orbits, and upper jaws, and a cranial region, including the braincase and ear. The lower jaw, while not usually considered part of the skull itself, is included in the following description.

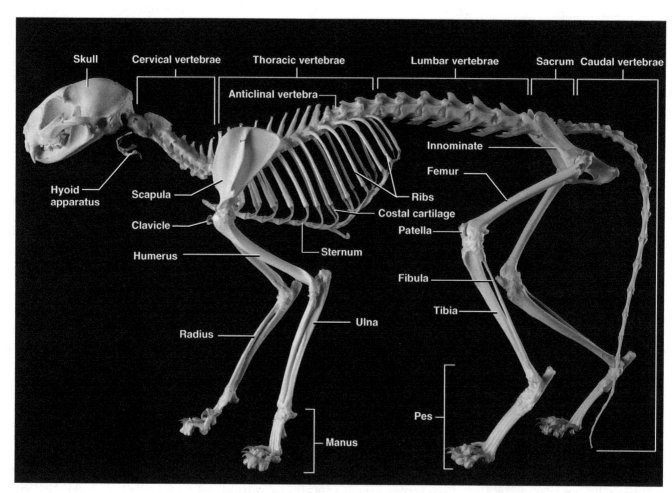

FIGURE 7.1   Skeleton of the cat in left lateral view.

(a)

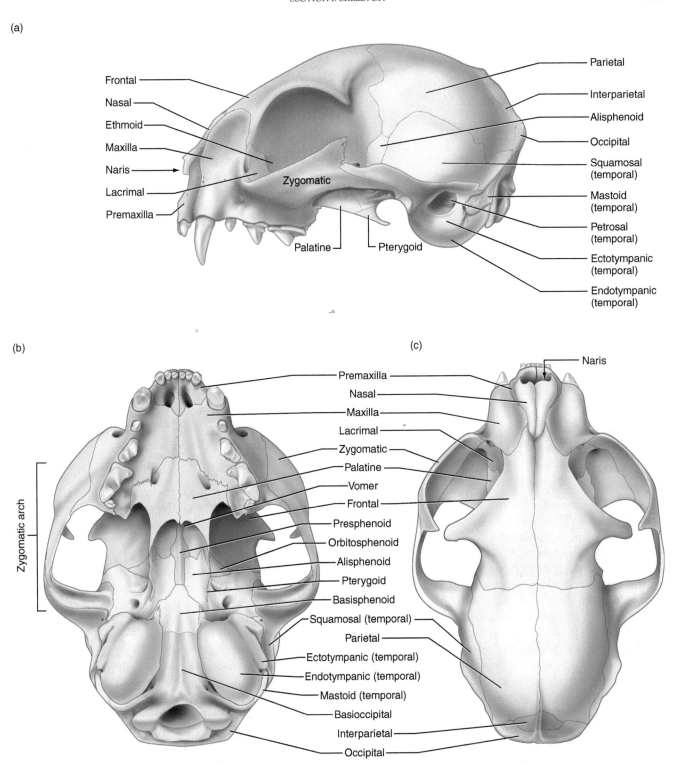

Frontal

Nasal

Ethmoid

Maxilla

Naris

Lacrimal

Premaxilla

Zygomatic

Palatine

Pterygoid

Parietal

Interparietal

Alisphenoid

Occipital

Squamosal
(temporal)

Mastoid
(temporal)

Petrosal
(temporal)

Ectotympanic
(temporal)

Endotympanic
(temporal)

(b)

(c)

Premaxilla

Nasal

Maxilla

Lacrimal

Zygomatic

Palatine

Vomer

Frontal

Presphenoid

Orbitosphenoid

Alisphenoid

Pterygoid

Basisphenoid

Squamosal (temporal)

Parietal

Ectotympanic (temporal)

Endotympanic (temporal)

Mastoid (temporal)

Basioccipital

Interparietal

Occipital

Naris

Zygomatic arch

FIGURE 7.2   Skull of the cat in (a) left lateral, (b) ventral, and (c) dorsal views with bones color-coded.

Examine a skull and identify its main features (Figure 7.2a–c). Anteriorly and ventrally, note the marginal series of teeth. Directly above the middle anterior teeth is a large opening, the naris, that leads into the nasal cavity. In life, the opening is separated by a cartilaginous septum so that two external nares or nostrils are present. On either side of the skull, dorsal to the posterior teeth, are the large orbits, the cavities that house the eyeballs. Posterior to each orbit is the expansive cranial wall. This surface is the **temporal fossa**. It protects the brain and

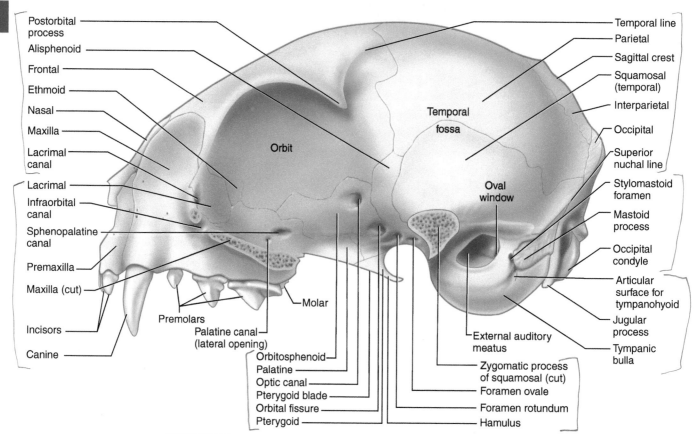

FIGURE 7.3   Skull of the cat in left lateral view with zygomatic arch removed.

serves as the origin for the temporal musculature, which helps to close the jaw. Phylogenetically, the fossa was derived mainly from a depressed region of the skull; thus comes its designation as a *fossa*, even though in the cat, as in other mammals, it is convex.

The ventral margins of the orbit and fossa are marked by the laterally bowed zygomatic arch, which serves mainly for the origin of the masseteric musculature, another group of jaw-closing muscles (Figure 7.4). Just posterior to the posterior root of the zygomatic arch is the external auditory meatus, the opening and passageway leading to the middle and inner ear (Figures 7.3 and 7.4). The posterior surface of the skull, the **occiput**, is pierced ventrally by the large **foramen magnum**, through which the spinal cord passes to the brain.

Examine the skull in ventral view (Figure 7.4). The flat region between the teeth is the **hard palate**. At its posterior end is the median internal opening of the nasal passage. In life there are two openings, the **choanae**. Farther posteriorly lie the paired, oval, and strongly convex **tympanic bullae**. The bulla forms the floor of the middle ear.

Nearly all the bones of the skull are paired. First identify the paired bones along the dorsal midline. Anteriorly, the small, triangular **nasals** form the dorsal margin of the naris. Posterior to the nasals are the large

frontals. These extensive bones contribute to the dorsal and lateral surfaces of the skull. Laterally they form most of the orbital walls. Note the **postorbital process** that projects ventrally from each frontal to demarcate the posterior margin of the orbit. Farther posteriorly are the **parietals**. These large bones form most of the dorsal and lateral parts of the cranium. The small **interparietal** bone is wedged between the posterior ends of the parietals. Usually in older individuals the interparietal fuses to the surrounding bones. The **temporal lines** are faint ridges that mark the boundary of muscular attachment on the temporal fossa. These lines curve posteromedially from the postorbital process of the frontals. They continue posteriorly on the parietals, converging toward the **sagittal crest**, which lies mainly on the interparietal. The crest may be high and sharp or represented only by a rugose ridge.

The **occipital bone** forms the occiput, the skull's posterior surface, and contributes to the **basicranium**, (i.e., the base of the cranium). Its most notable features are the foramen magnum, noted earlier (Table 7.1 describes the functions of skull foramina), and the **occipital condyles**, which articulate with the first vertebra and support the head on the neck. The occipital is formed from four separate bones. They fuse early in life (in humans,

Incisors

Canine

Premolars

Molar

Choana

Styloform
process

Auditory
tube

Mandibular fossa

External
auditory meatus

Tympanic bulla

Stylomastoid
foramen

Mastoid
process

Jugular
foramen

Jugular process

Hypoglossal canal
(anterior opening)

Premaxilla

Palatine fissure

Maxilla

Infraorbital canal

Palatine canal
(ventral opening)

Zygomatic

Palatine

Vomer

Frontal

Postorbital process

Presphenoid

Orbitosphenoid

Alisphenoid

Foramen rotundum

Pterygoid

Foramen ovale

Basisphenoid

Squamosal (temporal)

Ectotympanic (temporal)

Endotympanic (temporal)

Basioccipital

Mastoid (temporal)

Occipital

Foramen magnum

Occipital condyle

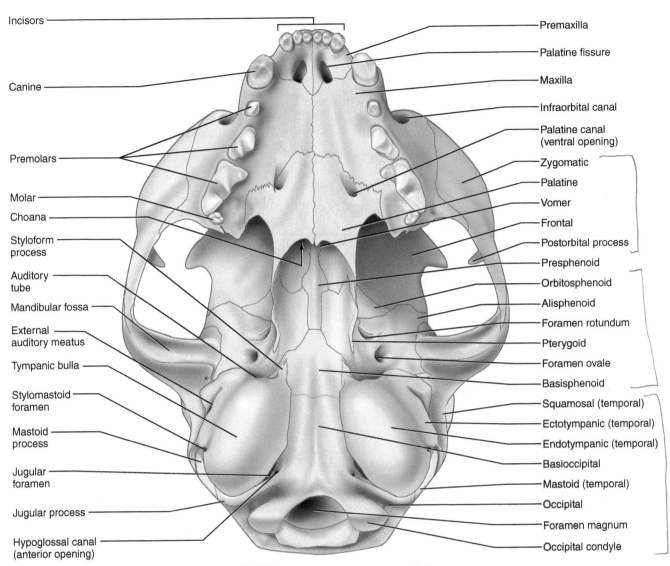

FIGURE 7.4   Skull of the cat in ventral view.

for example, they are fused together by about 6 years of age), and thus are not illustrated as separate bones, but it is convenient to recognize them separately for the following discussion. The dorsal part of the occipital is the **supraoccipital** (also known as the squamous potion of the occipital), which forms the dorsal margin of the foramen magnum. The supraoccipital bears the external occipital crest, a median ridge that extends ventrally from the dorsal margin of the occipital. The paired **exoccipitals** (or the lateral portions of the occipital) form the lateral parts of the occiput and the lateral margins of the foramen magnum. Each exoccipital carries an occipital condyle and extends laterally to contact the mastoid part of the temporal bone (see later) as well as ventrally over the posterior part of the tympanic bulla. The ventral end of the projection forms a blunt **jugular process**. The ventral margin of the foramen magnum is completed by the

basioccipital (or basilar portion of the occipital), which also contributes to the basicranium as it extends anteriorly between the tympanic bullae.

Return to the front of the skull (see Figures 7.3 and 7.4). The lateral and ventral margins of the nasal aperture are formed by the **premaxillae**. Each is a fairly small bone with a long, narrow vertical portion and a short ventral portion into which are implanted the front teeth or the upper **incisors**. There are three incisors in each premaxilla. Ventrally the premaxillae form the anterior part of the hard palate. Here, each premaxilla has medial and lateral branches separated by the **palatine fissure**.

The **maxillae** are larger bones that hold the rest of the teeth, and in ventral view form most of the hard palate. The large, curved **canine** is the first tooth in the maxilla. The following three teeth are **premolars**. Phylogenetically, they represent premolars 2 through 4

TABLE 7.1 Cranial and Mandibular Foramina of the Cat: Location and Function.

Name	Location	Main Structures Transmitted
Auditory tube, opening	Between tympanic bulla (endotympanic) and basisphenoid	Auditory tube emerges from middle ear, from which it passes to nasopharynx
Condyloid canal	Exoccipital	Vein
Cribriform foramina	Cribriform plate of ethmoid	Subdivisions of olfactory nerve (CN I)
Ethmoid foramen	Between frontal and orbitosphenoid	Branch of ophthalmic division of trigeminal nerve (CN V_1)
External auditory meatus	Ectotympanic	Passageway for outer ear
Foramen magnum	Between supraoccipital, basioccipital, and exoccipitals	Spinal cord, hypoglossal nerve (CN XII), basilar artery
Foramen ovale	Alisphenoid	Mandibular division of trigeminal nerve (CN V_3)
Foramen rotundum	Alisphenoid	Maxillary division of trigeminal nerve (CN V_2)
Hypoglossal canal	Exoccipital	Hypoglossal nerve (CN XII)
Infraorbital canal	Maxilla	Infraorbital branch, maxillary division of trigeminal nerve (CN V_2), infraorbital artery
Internal acoustic meatus	Petrosal	Facial (CN VII) and vestibulocochlear (CN VIII) nerves
Jugular foramen	Dorsally between petrosal and basioccipital; ventrally between endotympanic (tympanic bulla) and basioccipital)	Glossopharyngeal nerve (CN IX), vagus nerve (CN X), accessory nerve (CN XI), internal jugular vein; also, hypoglossal nerve (CN XII), after it emerges from the hypoglossal canal
Lacrimal canal	Lacrimal	Lacrimal duct
Mandibular foramen (posterior opening of mandibular canal)	Posteriorly on medial surface of dentary	Dentary branch of mandibular division of trigeminal nerve (CN V_3); blood vessels
Mental foramina (anterior openings of mandibular canal)	Anteriorly on lateral surface of dentary	Dentary branch of mandibular division of trigeminal nerve (CN V_3); blood vessels
Optic canal	Orbitosphenoid	Optic nerve (CN II) and ophthalmic artery
Orbital fissure	Between orbitosphenoid and alisphenoid	Oculomotor nerve (CN III), trochlear nerve (CN IV), abducens nerve (CN VI), ophthalmic division of trigeminal nerve (CN V_1), branch of maxillary division of trigeminal nerve (CN V_2)
Oval window	Petrosal	Receives footplate of stapes

Palatine canal
  Lateral opening
  Ventral opening
	
	Palatine (orbital portion)
	Palatine (palatine portion), near suture with maxilla
		
		Greater palatine branch, maxillary division of trigeminal nerve (CN V_2)

Palatine fissure	Between premaxilla and maxilla	Nasopalatine branch, maxillary division of trigeminal nerve (CN V_2), nasal artery
Round window	Petrosal	Covered in life by secondary tympanic membrane, which accommodates vibrations of fluid in inner ear
Sphenopalatine foramen	Palatine (orbital portion)	Sphenopalatine branch, maxillary division of trigeminal nerve (CN V_2), sphenopalatine artery
Stylomastoid foramen	Temporal, between mastoid and ectotympanic	Facial nerve (CN VII)

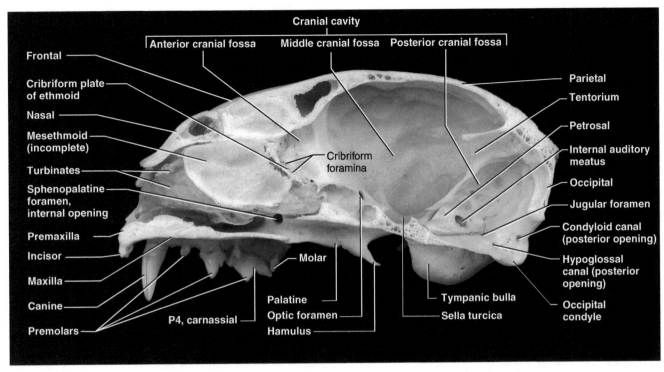

FIGURE 7.5   Right side of the skull of the cat shown in sagittal section.

(P2–P4). Cats lost P1 during their evolution, and a gap, the **diastema**, is present between the canine and the small, peg-like P2. P3 is a larger, triangular tooth that is followed by the very long, blade-like P4. Note the small, final tooth, **molar** 1 (or M1), set transversely and covered in lateral view by the posterior end of P4. The blade-like P4 is the main meat-shearing tooth, and is referred to as the **carnassial** (Figure 7.5). It meets the blade-like m1, the carnassial of the lower jaw.

In lateral view each maxilla makes a large contribution to the snout or **rostrum**, where it contacts the premaxilla, nasal, and frontal bones. It also contributes a small portion to the orbital margin, where it contacts the small, delicate **lacrimal**, which makes a small contribution to the anterior orbital wall. The conspicuous opening of the **lacrimal canal** lies at the anterior margin of the orbit. Finally, the maxilla passes ventral to the **zygomatic** and participates in the anterior root of the zygomatic arch. Note that the maxilla contributes a shelf that forms the anterior floor of the orbit. The **infraorbital canal** is the large passage, entirely within the maxilla, at the anterior end of the zygomatic arch.

The **palatines** complete the hard palate posteriorly. Ventrally each palatine is a horizontal plate of bone, and the internal opening of the nasal cavity lies posterior to these plates at the midline. The ventral opening of the small **palatine canal** lies very near the suture between the palatine and maxilla, approximately medial to the anterior part of P4. Posteriorly each palatine forms a thin, vertical plate of bone. This portion helps, together with

the **pterygoid**, to form the **pterygoid blades** or flanges, as well as the ventral part of the orbital wall (Figure 7.3). The region between the pterygoid blades forms the bony part of the tube-like nasopharynx (see page 239 and Figure 7.40). The orbital portion of the palatine (Figure 7.3) has two openings. The larger is the **sphenopalatine foramen**; the smaller is the lateral opening of the palatine canal.

In ventral view, locate the posterior end of the palatine bones (Figure 7.4). They contact a complex, composite element, usually termed the **sphenoid**, consisting of several ossifications. The main part, or body, is the **basisphenoid**. The pterygoid processes of the basisphenoid are the portions that contact the palatine and form the posterior end of the pterygoid blades. The narrow, hook-like posterior projection at the end of each pterygoid blade is the **hamulus**, to which a palatal muscle is attached. The main part of the basisphenoid contributes to the basicranium. It lies ventrally between the anterior parts of the tympanic bullae. The lateral, wing-like portion of the basisphenoid is the **alisphenoid**. It is exposed on the lateral surface of the skull, and will be discussed shortly. Return to the ventral surface and locate the **presphenoid**, the elongated, narrow, median bone lying between the pterygoid blades. The presphenoid also has a lateral extension, the **orbitosphenoid**, that is exposed in the ventral orbital wall. The connection between the presphenoid and orbitosphenoid is concealed by the palatine. The orbitosphenoid contacts the palatine, frontal, and alisphenoid bones. A tiny **ethmoid foramen** (see

Table 7.1) lies along the suture between the orbitosphenoid and frontal, but it may be difficult to see.

A series of four openings into the cranial cavity lies just dorsal to the pterygoid blades, on the lateral surface of the skull (Figure 7.3). Locate these foramina. The two anterior foramina are approximately twice as large as the two posterior foramina. The most anterior foramen is the **optic canal**, and it lies entirely within the orbitosphenoid. The second is the **orbital fissure**. Its margin is formed by the orbitosphenoid and the alisphenoid. The third and fourth foramina, respectively the **foramen rotundum** and **foramen ovale**, lie entirely within the alisphenoid.

Peer into the internal opening of the nasal cavity. Extending anteriorly from the presphenoid is the **vomer**. Its ventral surface is keeled. The vomer passes anteriorly and contacts the premaxillae and the maxillae. Examine the nasal aperture to view the vomer in anterior view, and note its narrow trough-like form. The vomer contributes to the bony nasal septum, which helps partition the nasal cavity into left and right parts.

The nasal cavity is filled with delicate, scroll-like **turbinate bones** (Figure 7.5). Most of these scrolls are formed from the maxilloturbinate, a paired bone that is connected to the medial wall of each maxilla. The turbinates dorsally in the nasal cavity are formed from the ethmoturbinates, which form the bulk of the **ethmoid**. In some cats a tiny portion of the ethmoid is exposed on the orbital wall, just posterior to the lacrimal (Figure 7.3). Other parts of the ethmoid are the **mesethmoid** and the **cribriform plate**. The mesethmoid is a median bony plate that, together with the vomer, helps form the bony nasal septum. The cribriform plate is a transverse plate, pierced by many small **cribriform foramina**, that forms the anteroventral wall of the cranial cavity. The cribriform plate can be viewed in a sagittally sectioned skull (Figure 7.5) or in one where the skull roof has been removed.

The **temporal bone** consists of three components that are fused together: the squamous, petrous, and tympanic portions. The **squamosals** are the large, flat bones ventral to the parietals that help complete the lateral wall of the braincase (Figure 7.3). Ventrally the squamosal sends out a projection, the zygomatic process, that extends laterally and then anteriorly to form the posterior part of the zygomatic arch. On the ventral surface of the lateral part of the zygomatic process is the **mandibular fossa** (Figure 7.4), a smooth, transverse groove that articulates with the lower jaw in forming the temporomandibular joint.

The petrous portion of the temporal includes the **petrosal** and **mastoid bones**. The petrosal contains the inner ear and may be seen through the external auditory meatus; its medial part should be viewed in a sagittally sectioned skull (Figure 7.5). The tiny bones or ossicles of the middle ear are the **malleus, incus**, and **stapes**. The middle ear is essentially between the external auditory meatus and the petrosal and is covered ventrally by the tympanic bulla. The middle ear bones sometimes remain in place. The malleus is the slender, elongated bone lying across the inner end of the external auditory meatus. The other two are more difficult to distinguish, but the stapes lies medially and fits into the **oval window** of the petrosal (Figure 7.3).

When the middle ear bones are absent, the two lateral foramina of the petrosal can be identified by peering into the external auditory meatus. Move the skull so that you have an anterolateral view into the middle ear. The foramina lie posteriorly on the petrosal. The more ventral one is the **round window** and opens posteriorly. The oval window, mentioned earlier, lies dorsal to the round window and opens laterally.

The mastoid bone is the only part of the petrous portion that is exposed externally. It overlaps the posterolateral surface of the tympanic bulla. Its ventral portion forms the **mastoid process**. (It is tiny in the cat, but you can feel it protruding just behind your ear.) The small **stylomastoid foramen** opens just anterior to the mastoid process (Figures 7.3 and 7.4). The mastoid bone continues dorsally on the occiput as a short wedge between the squamosal and occipital bones. The hyoid apparatus articulates with the tympanic bulla just ventral to the mastoid process (Figure 7.3).

The tympanic portion of the temporal includes the rounded, oval tympanic bulla, which is formed from two ossifications. The **ectotympanic** mainly forms the ring of bone surrounding the external auditory meatus but contributes to the bulla ventrolaterally. The rest of the bulla is formed by the **endotympanic**. Anteromedially the bulla sends a pointed wedge, the **styloform process**, onto the base of the skull. Lateral to the process is the large opening for the **auditory tube** (Figure 7.4).

Note the large **jugular foramen** on the base of the skull at the posteromedial margin of the tympanic bulla. It passes between the bulla, the basioccipital (medially), and the exoccipital (posteriorly). A second, much smaller opening lies on the posterior wall (i.e., the exoccipital) of the larger passage. This is the anterior opening of the **hypoglossal canal**. Identify the posterior opening by examining the medial wall of an occipital condyle (i.e., look into the foramen magnum) or a sagittal section (Figure 7.5). Pass a bristle through the hypoglossal canal to determine its course. Posterior and dorsal to the posterior opening of the hypoglossal canal is the posterior opening of the **condyloid canal**.

The zygomatic, contacting the maxilla anteriorly and the zygomatic process of the squamosal posteriorly, forms most of the zygomatic arch. It forms the ventral margin of the orbit and sends up a postorbital process that closely approaches the postorbital process of the

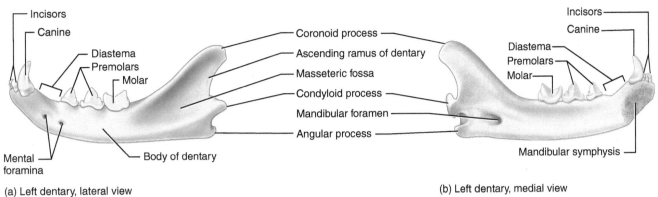

FIGURE 7.6   Left dentary of the cat in (a) lateral and (b) medial views.

frontal. Ventrally and laterally the zygomatic bears a crescentic, slightly concave region for origin of masseteric musculature.

Examine a sagittal section of the skull (Figure 7.5). The posterior half of the skull contains the large **cranial cavity** for the brain. The cavity is subdivided into a small **anterior cranial fossa** that receives the olfactory bulbs of the brain (see Figure 7.40), a **middle cranial fossa**, the largest subdivision, which mainly houses the cerebrum, and a **posterior cranial fossa**. The last two fossae are separated by a partial bony septum, the **tentorium**. A depression, the **sella turcica** (shaped like a traditional Turkish saddle in humans, but not in the cat), of the basisphenoid on the floor of the middle cranial fossa houses the hypophysis (see Figure 7.40). Note the petrosal, which houses the inner ear. The large opening is the **internal acoustic meatus** (Figure 7.5) for passage of the facial (CN VII) and vestibulocochlear (CN VIII) nerves (see Section VII).

Mandible

The **mandible**, or lower jaw, is formed on each side by a single bone, the **dentary** (Figure 7.6). Left and right dentaries articulate anteriorly at the **mandibular symphysis**. The horizontal part of the dentary, in which the teeth are implanted, is the **body**. Note the presence, normally, of three incisors, followed by a canine, which is separated from the three cheek teeth by a diastema. The cheek teeth include p3, p4, and the elongated, blade-like m1. The m1 is the carnassial of the lower jaw; hence the carnassial pair is P4/m1. Typically, two **mental foramina** (see later and Table 7.1 for function) are present anteriorly on the lateral surface of the body (Figure 7.6a).

The part posterior to the body is the **ascending ramus**, which has three processes. The **coronoid process** is the largest, extending dorsally. The temporal muscle inserts mainly on its dorsolateral and medial surfaces. The **masseteric fossa**, the large, triangular depression on the lateral surface of the coronoid process, serves as

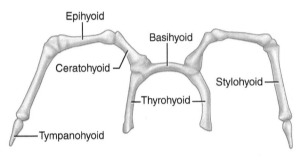

FIGURE 7.7   Hyoid apparatus of the cat.

part of the insertion area for the masseteric musculature. The transversely expanded **condyloid process,** forming the lower half of the temporomandibular joint, bears a semicylindrical facet for articulation with the mandibular fossa on the temporal. Manipulate the mandible on the skull, and note how the form of the joint permits mainly up and down (or orthal) movements of the jaw and restricts lateral motion. Identify the roughened area on the anteromedial surface of the condyloid process. It is the insertion site for the medial pterygoid muscle. The **angular process** is the projection of the posteroventral part of the dentary. Its medial surface serves as the insertion site for the lateral pterygoid muscle and its lateral surface for part of the masseteric musculature.

A large **mandibular foramen** lies anterior and slightly dorsal to the angular process and the medial surface of the ramus (Figure 7.6b). This foramen is the posterior opening of the mandibular canal, which extends anteriorly through the body for the passage of the blood vessels and the mandibular division of the trigeminal nerve. The mental foramina are the main anterior openings of the canal.

Hyoid Apparatus

The **hyoid apparatus** (Figures 7.1 and 7.7) is composed of several small bones, the phylogenetic remnants of some of the gill arches in fishes. It sits in the throat at the base of the tongue and supports the tongue and laryngeal muscles. It is composed of a median bar, the **basihyoid** or body, which is oriented transversely at the anterior end of the larynx. Two pairs of horns, or cornua, extend from the basihyal. The **lesser cornua** are the longer, anterior pair, whereas the **greater cornua** are the smaller, posterior pair. Each lesser cornu consists of a chain of four ossicles that curves anteriorly and dorsally to attach to the skull. These elements are the **ceratohyoid**, **epihyoid**, **stylohyoid**, and **tympanohyoid**, which has a ligamentous attachment to the temporal bone of the skull. The greater cornu is formed on each side by the **thyrohyoid**, which articulates with the thyroid cartilage of the larynx. The discrepancy between the size and names of the cornua derives from their condition in humans, in which the greater cornua are indeed larger, as the lesser cornu is represented only by the ceratohyoid.

Key Terms: Cranial Skeleton

alisphenoid
angular process
anterior cranial fossa
ascending ramus of dentary
auditory tube (Eustachian tube, pharyngotympanic tube)
basicranium
basihyoid
basioccipital
basisphenoid
body of dentary
canine
carnassial
ceratohyoid
choanae
condyloid canal
condyloid process
coronoid process
cranial cavity
cribriform foramina
cribriform plate
dentary
diastema
ectotympanic
endotympanic
epihyoid
ethmoid bone
ethmoid foramen
external auditory meatus
foramen magnum
foramen ovale
foramen rotundum
frontal
greater cornua
hamulus

hard palate
hyoid apparatus
hypoglossal canal
incisors
incus
infraorbital canal (antorbital canal)
internal acoustic meatus
interparietal
jugular foramen
jugular process
lacrimal bone
lacrimal canal
lesser cornua
malleus
mandible
mandibular foramen
mandibular fossa
mandibular symphysis
masseteric fossa
mastoid
mastoid process
maxilla (pl., maxillae)
mental foramina
mesethmoid
middle cranial fossa
molar
naris (pl., nares)
nasal
nasal aperture
occipital
occipital condyles
occiput
optic canal
orbital fissure
orbitosphenoid
orbits
oval window (fenestra ovalis; vestibular window; fenestra vestibuli)
palatine (incisive)
palatine canal
palatine fissure
parietals
petrosal (periotic, petromastoid)
posterior cranial fossa
postorbital process
premaxillae
premolar
presphenoid
pterygoid
pterygoid blades
rostrum
round window (fenestra rotundum; fenestra cochlea)
sagittal crest
sella turcica
sphenoid

7

sphenopalatine foramen
squamosal
stapes
styloform process
stylohyoid
stylomastoid foramen
temporal
temporal fossa
temporal lines
tentorium
thyrohyoid
turbinates
tympanic bullae
tympanohyoid
vomer
zygomatic arch
zygomatic (malar, jugal)

Cranial Skeleton: Supplement—Comparative Mammalian Skulls

Mammals are a highly diverse clade and the form of their skulls reflects this diversity (as is also true of the reptiles discussed in Chapter 8). Among the more extreme specializations, we might mention the elongated, tubular, and edentulous skulls of South American anteaters, the massive and highly pneumatized skulls of elephants, and the huge skulls of baleen whales such as the right whale. In the latter, the skull seems little more than upper and lower jaws, forming gigantic struts, linked to the back of the skull to allow feeding with and support of the baleen plates in the upper jaw. Such differences in form largely reflect evolutionary adaptations to different feeding habits, although in many mammals other factors (such as digging) may also play a role in shaping the form of the skull. Indeed, the jaws, particularly the mandible, and the rest of the skull may need to respond to different pressures. Whereas the mandible is clearly affected by feeding, the upper jaw and the skull may face conflicting demands, such as the size of the brain, bearing horns or antlers, and fighting behavior (as in rams).

We may gain an appreciation of mammalian diversity by considering the cranial skeletons of more commonly available mammals, such as of the sheep, *Ovis aries*, and the beaver, *Castor canadensis*. Here, we provide general comparative descriptions of the skulls and mandibles of these mammals, and assess both the similarities and differences between them and that of the cat, to evaluate the functional adaptations reflected by their form, or in other words, the relationship between form and function. The goal of this section is not necessarily to try to identify all the bones and structures as was done for the cat (although, by and large, most of these structures can readily be identified), and thus Key Terms are not

identified. Rather, the point is to consider some of the major differences in light of the particular uses required by these mammals. In the following discussion, examine skulls of these mammals and refer to Figures 7.S1 through 7.S4.

In comparing the skulls of the sheep, beaver, and cat, several obvious differences are apparent, but so are many similarities. Among the overall differences are that the skull of the sheep has a much longer and deeper rostrum (though not all carnivorans have as short a rostrum as the cat). The rostrum of the beaver is also elongated, though not to the degree as in the sheep, and the entire dorsal profile of the skull is flattened, rather than curved. In the sheep the orbit is partly enclosed posteriorly by a bony bar, rather than being openly continuous posteriorly with the temporal fossa. In both the sheep and beaver the zygomatic arch, particularly anteriorly, is stouter than in the cat. In the mandible, the coronoid process is relatively slender in the sheep and beaver, and larger in the cat, but the reverse is true of the angular process. The condyloid process is approximately at the same level as the tooth row in the cat, but raised dorsally above the tooth row in the sheep and beaver.

Consider the teeth. The cheek teeth in the cat (there are four, but other carnivorans may have more) seem relatively feeble, whereas those in the sheep (six teeth) and beaver (four teeth) seem more substantial. Those of the sheep and beaver are broad rather than sharp-edged as in the cat. The cheek teeth are separated from the front teeth by a much more pronounced diastema in the sheep and beaver (the diastema is relatively short even in carnivorans that have a longer snout than the cat). The beaver possesses a massive upper and lower incisor in each jaw quadrant, whereas the cat has three small incisors and a large canine, which is absent in the beaver. The sheep, on the other hand, lacks the upper incisors and canine, but has three small incisors and an incisiform canine on each side of the mandible. The six cheek teeth of the sheep comprise three premolars and three molars. The beaver's four cheek teeth include a single premolar followed by three molars.

On the other hand, there are many similarities among the three mammals. For example, the bones are generally similar in shape and position, although the proportions may differ. Identify the main bones of the skulls in the sheep and beaver, using the descriptions that follow, the indicated figures, and the knowledge gained through your detailed study of the cat skull. The nasals, frontal, and parietals are easily recognizable along the roof and lateral cranial wall of the skull. The frontal extends notably farther posteriorly in the sheep. As well, the frontal is pierced dorsally by the supraorbital foramen, which communicates with the orbit.

Another notable difference is that in the sheep particularly, but also the beaver, the parietal extends

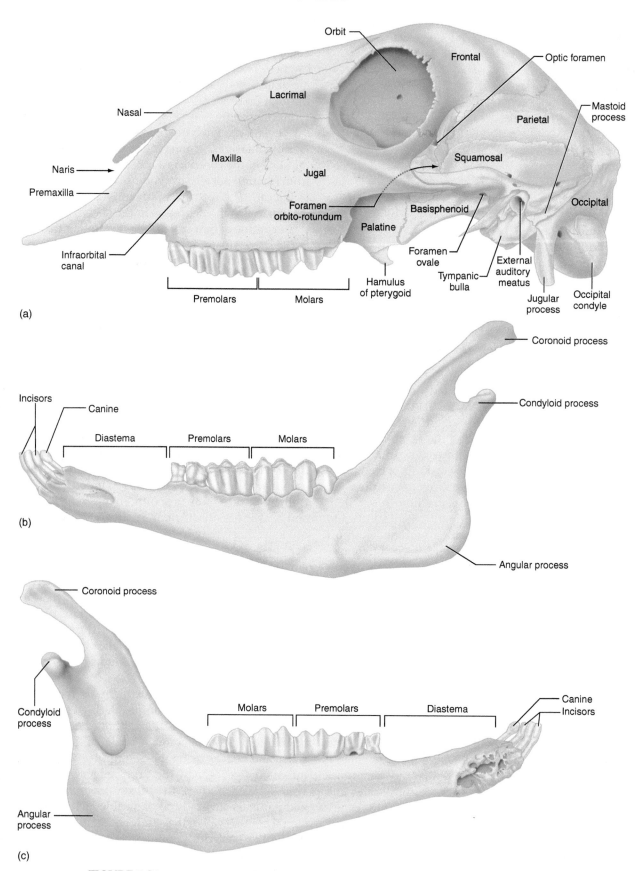

FIGURE 7.S1   Skull of the sheep in (a) lateral view, and dentary in (b) lateral and (c) medial views.

7

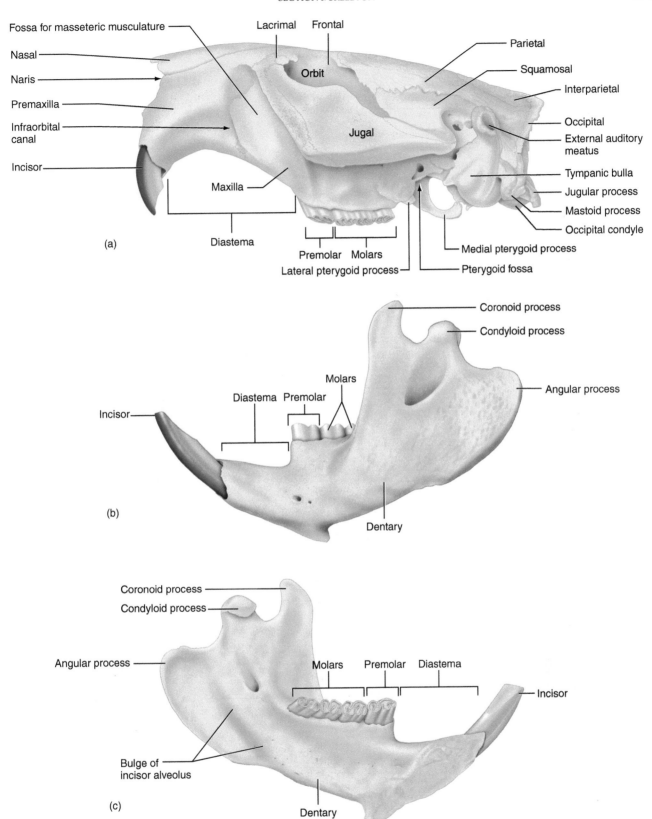

FIGURE 7.S2 Skull of the beaver in (a) lateral view, and dentary in (b) lateral and (c) medial views.

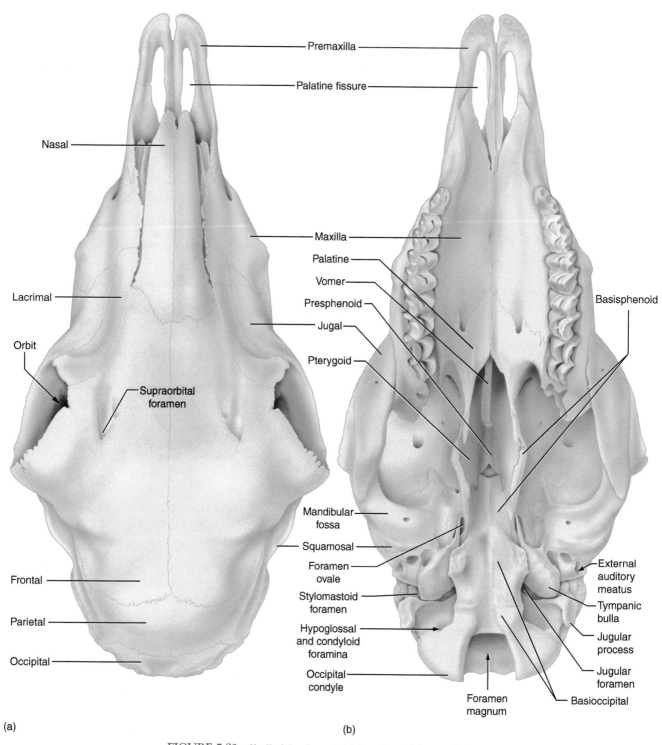

FIGURE 7.S3   Skull of the sheep in (a) dorsal and (b) ventral views.

considerably anterior relative to the posterior margin of the frontal. The frontal of the sheep bears a prominent descending postorbital process that meets an equally prominent ascending postorbital process of the zygomatic to form the postorbital bar. The nasals in both the sheep and beaver are much more prominent and elongated. The same is true of the premaxillae, which contribute more prominently to the palate and are particularly stout in the beaver. The maxilla occupies the usual position and bears the cheek teeth. In the sheep and beaver the maxilla is more extensive and tends to be deeper. A notable feature in the beaver is that the maxilla bears

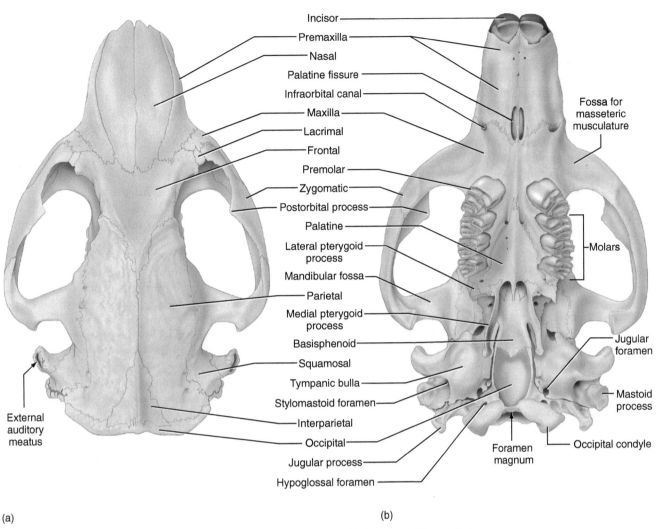

Incisor
Premaxilla
Nasal
Palatine fissure
Infraorbital canal
Maxilla
Lacrimal
Frontal
Premolar
Zygomatic
Postorbital process
Palatine
Lateral pterygoid process
Mandibular fossa
Parietal
Medial pterygoid process
Basisphenoid
Squamosal
Tympanic bulla
Stylomastoid foramen
Interparietal
Occipital
Jugular process
Hypoglossal foramen

External auditory meatus

Fossa for masseteric musculature

Molars

Jugular foramen

Mastoid process

Occipital condyle

Foramen magnum

(a)

(b)

FIGURE 7.S4   Skull of the beaver in (a) dorsal and (b) ventral views.

an extensive depression just anterior to the anterior root of the zygomatic arch (the premaxilla bears a small dorsal portion of this depression). The fossa is neatly delimited anteriorly by a distinct shelf-like ridge on the lateral surface of the maxilla and premaxilla. As noted later, the fossa in the beaver serves for insertion of part of the masseter muscle, one of the jaw-closing muscles. The infraorbital foramen is hidden in lateral view medial to the shelf, whereas in the sheep, as in the cat, the opening is visible. On the palate, the premaxillae of the sheep help enclose more prominent palatine fissures. Posteriorly the palatines follow the maxillae and help complete the palate, and then turn to form the usual blade-like walls of the pterygoid region. The lacrimal contributes to the anterior margin of the orbit in all three mammals, but it is particularly large in the sheep, extending far anteriorly on the lateral rostral wall. Also, it extends posteriorly as an inflated, pyramid-shaped portion, often with a lacy appearance in some individuals, that projects into

the anterior part of the orbit. The squamosal (part of the temporal) occupies its usual place on the posterolateral side of the skull and bears the posterior root of the zygomatic arch.

The occipital, as usual, forms the posterior surface of the skull. It encloses the foramen magnum and bears the paired occipital condyles that articulate with the vertebral column. The occipital is formed by the usual four elements as described in the cat, but they are not explicitly indicated in the figures of the sheep and beaver (except for the basioccipital in Figures 7.S3b and 7.S4b). The only major difference is that in the sheep the supraoccipital is excluded from the margin of the foramen magnum. Indeed, it is possible to see the suture of the supraoccipital with the exoccipitals even in older individuals. This suture takes the form of a V-shaped contact just dorsal to the foramen magnum. The interparietal forms a large triangular wedge between the occipital and parietals in the beaver, but is less conspicuous in the sheep.

Ventrally, the basioccipital helps form the floor of the braincase. The tympanic bullae lie to either side of this bone. In the beaver the bulla is relatively large and mediolaterally wide, similar to but not nearly as inflated as in the cat. The bulla is smaller in the sheep and transversely compressed, forming a more nearly dorsoventral structure. The external acoustic meatus is the opening into the ear. In contrast to the condition in the cat, it is borne by an elongated bony tube extending laterally from the rest of the temporal. It is particularly long in the beaver and less so in the sheep. The mastoid process, the external exposure of the petrosal, is prominent in the beaver, and less so in the cat and sheep. The jugular process, just lateral to the occipital condyle, is larger in the beaver and exceptionally so in the sheep. Several of the other entrances in this region into the skull, such as the stylomastoid, hypoglossal, and jugular foramina, are reasonably similar to those already identified in the cat. In the sheep the hypoglossal opening lies at the base of a deep depression (the condyloid fossa) between the occipital condyle and jugular process. As well, the condyloid foramen, which is smaller and lies lateral to the hypoglossal foramen, also opens directly from the floor of the fossa.

The basisphenoid follows anterior to the basioccipital, and as in the cat, has a lateral wing, the alisphenoid, that extends dorsally into the orbital wall. Anterior to it follow, in order, the presphenoid and the vomer, as already observed in the cat. Ventrolaterally the basisphenoid encloses the foramen ovale in the sheep, as in the cat, and continues to contact the palatine. The pterygoid is more broadly exposed in lateral view in the sheep, but the hamulus is about the same size. Anterodorsal to the foramen ovale (and hidden in lateral view by the zygomatic arch) is a large opening, the foramen orbito-rotundum (a confluence of the separate foramen rotundum and orbital fissure noted in the cat). Anterodorsal to this opening is the optic canal, about as large as the foramen ovale. In the beaver the relationships of the foramina are more complex and are not considered here. However, note the large lateral and medial pterygoid processes that extend posterior to the palatines. The medial is particularly prominent, extending posteriorly as a stout hook-like structure. Between the pterygoid processes, note the deep pterygoid fossa.

Several of the differences noted among the skulls of the cat, sheep, and beaver are adaptations reflecting differences in dietary habits. That the differences are so pronounced should be of no surprise, as the skull is instrumental in procuring and processing food. The cat represents a carnivore and a strict one at that, and many of its features are indicative of a carnivorous diet. Of the main jaw-closing muscles, for example, the temporalis muscle is larger than the masseter muscle, and the relatively large temporal fossa reflects this difference. So

too does the relatively large coronoid process (on which the temporalis mainly inserts) compared to the smaller angular process (on which part of the masseter muscle inserts) of the mandible.

Although the sheep and beaver represent different herbivore types, several of the features of their skulls reflect the processing of vegetation (more accurately, of food with certain physical properties, rather than all kinds of vegetation; other plant feeders, such as primates for example, have other mechanical requirements and the feeding apparatus is different from the pattern common in rodents and the ruminant artiodactyls discussed here). In herbivores the temporalis muscle tends to be smaller than the masseter muscle, and thus the temporal fossa is relatively smaller, and the angular process is larger than the coronoid processes in these mammals (indeed, the coronoid process is barely developed in some rodents). In addition, note the position of the condyloid process of the mandible in the herbivores as compared to the carnivore. This position reflects a more dorsal position of the joint between the skull and mandible. The effect of the different positions is that in the cat, with the joint approximately level with the tooth row, the teeth can be engaged sequentially, in a scissor-like manner, a suitable action for cutting and slicing. In the herbivores the position of the joint engages the cheek teeth simultaneously, which can thus form the long grinding and crushing batteries necessary for processing vegetation.

The form of the cheek teeth themselves is also revealing. In the cat, as already noted, these teeth are mainly laterally compressed with a single sharp edge, an efficient mechanism for cutting and slicing through flesh and tendons when the teeth pass adjacent to each other in a scissor-like fashion. Not all carnivorous mammals have such strictly blade-like teeth. A more flexible diet than in the cat is revealed by the dentition of canids (dogs and their relatives), which retain a sharp-edged carnassial pair but the teeth following the carnassials have crushing surfaces. In those that mainly crush bones, some of the cheek teeth are robust and tend to have broad occlusal surfaces, such as in hyenas.

In a herbivore, however, the cheek teeth tend to be morphologically similar to each other, with the premolars becoming molarized (i.e., molar-like), and they have relatively wide and complex occlusal surfaces bearing several sharp-edged ridges or lophs (again, this does not apply to all herbivores, as many, such as primates, do not possess such morphology). The lophs are composed of enamel, the hardest part of the tooth, and between the lophs is softer dentine, which wears more rapidly and thus forms depressions between the lophs. When the lophs move past each other, they act like two rasps that grate food between them. The main difference between the beaver and sheep is that the ridges are transversely oriented in the former and anteroposteriorly oriented in the latter, which reflects differences in the way the jaw moves in these two herbivores, and is considered later. The rasping effect of the lophs results in similar effects.

The need to grind tough vegetation, however, results in considerable wear of the teeth. Herbivores have adapted to this by evolving tall (or hypsodont) teeth. Such teeth are set in deep alveoli in the jaw bones, and thus the maxilla and mandible have become deeper to house the taller teeth. This is particularly evident in grazing mammals, such as the sheep, in which the orbits have even moved posteriorly to accommodate the roots of the upper teeth. In some instances (some rodents and rabbits) the roots of the cheek teeth do not close and the teeth are ever-growing (or hypselodont).

A characteristic feature of herbivorous mammals is that their mandibles have considerable freedom of movement, either side to side or backward and forward, reflecting the need to have the lophs of the cheek teeth move past each other. In carnivores, however, the mandible tends to be restricted mainly to up and down (or orthal) movements. The form of the jaw joint reflects these motions. In the cat the condyloid process is rounded and transversely widened, articulating with a concave and transversely wide mandibular fossa. This joint restricts lateral movement. In the sheep both the condyle and fossa are nearly flat, permitting considerable side-to-side movement. The beaver, and rodents in general, have specialized mandibular movements. In these mammals the lower jaw can function in two distinct positions. With the mandible in posterior position, the cheek teeth can be engaged in mastication to crush and grind, but in this position the tips of the incisors do not meet. Rodents, however, can pull the mandible forward to engage the incisors to gnaw food or clip vegetation (and in some cases, to dig), in which case the cheek teeth are not fully engaged. The form of the jaw joint reflects these possibilities. Here, the condyloid process is rounded and the mandibular fossa is elongated, which allow for anteroposterior (or propalinal) and transverse jaw movement. Another feature of rodent dentition that increases masticatory efficiency is that the upper and lower tooth rows are about equidistant from each other, whereas in most other mammals, the cat and sheep included, the upper and lower tooth rows are not the same distance apart, meaning the lower teeth are positioned medial to the upper teeth when the jaw is closed. The result is that the upper and lower teeth of only one side can occlude. In the beaver, however, both side upper and lower teeth meet at the same time, so mastication may be carried out by both side tooth rows through anteroposterior movements of the jaw—and this, of course, explains the transversely oriented lophs of this mammal.

Consider next the front teeth. In general, the incisors are used to obtain food, whether grasping vegetation or nipping prey. In the cat the incisors are all present and the canine is well developed as a pointed, conical tooth. This makes functional sense in several ways. Typically, the incisors are used to capture and manipulate prey (although they may also serve other functions, such as grooming), and the canines especially are used to stab and pierce in subduing and killing prey. This requires a quick biting movement from a relatively wide gape. The temporalis muscle is most mechanically efficient, given its position, in initiating such movements to provide a forceful bite at the front of the jaws. This explains the general tendency in carnivores for the temporalis being the larger of the two jaw-closing muscles discussed here.

In herbivores there is less need for a powerful temporalis, given that subduing and killing prey is not a concern. The emphasis in such mammals as the sheep is on constant mastication, mainly a function of the cheek teeth, for which the masseter muscle is ideally positioned. It is likely that the postorbital bar, noted earlier, is an adaption for absorbing stress during mastication. In the sheep the upper incisors and canines are lost, and this is generally true of ruminant artiodactyls. In their place, at the front of the oral cavity, on the ventral surface of the premaxilla, is a hard pad of fibrous tissue with a thick, horny epithelial covering, the dental pad or plate, against which the lower teeth produce their nipping and cutting actions. In the males of some antlerless ruminants, such as the mouse and musk deer, the canine is retained as a large tooth for fighting and display.

In the beaver, as in rodents generally, there is only a single pair of enlarged, curved, gnawing, and ever-growing incisors in the upper and lower jaws, the other incisors and canines having been lost during evolution. The teeth bear enamel only on their anterior surfaces, and as this material is harder than the dentine posterior to it, the enamel wears more slowly, producing the self-sharpening, chiseled edge characteristic of rodents. These teeth, as noted earlier, can be engaged by moving the mandible forward. To this end, rodents (the mountain beaver, *Aplodontia rufa*, which is not a true beaver, is an exception) exhibit a trend toward specialization of the masseter musculature so that at least one of its divisions takes origin on the rostrum, in contrast to the typical condition in mammals, where the masseter originates from the zygomatic arch.

An anterior attachment of part of the masseter results in a more nearly anteroposterior line of action, which increases the muscle's mechanical advantage in pulling the mandible forward. There are several patterns for such reorganization of the masseter. The beaver is characteristic of the sciuromorphous condition, in which the anterior part of the superficial masseter arises from the anterior surface of the zygomatic arch and the lateral surface of the rostrum—the depression that serves this origin was noted earlier. The other patterns are termed hystricomorphous and myomorphous. In the former, as in porcupines and the guinea pig, the anterior part of the medial masseter arises from the anterior surface of the rostrum and extends through a usually greatly enlarged infraorbital foramen. In the myomorphous condition, as in rats and mice, the anterior part of the superficial masseter arises from a markedly enlarged anterior portion of the zygomatic arch; the anterior part of the medial masseter arises from the rostrum and extends through a moderately enlarged infraorbital foramen.

A final consideration is the size of the diastema, which, as noted earlier, tends to be more extensive in many (but not all) herbivorous mammals. The longer diastema may be a function of the longer snout and jaws, but the reason for this is not clear. Suggestions include that the diastema is functionally important in mastication, allowing additional space for the tongue to manipulate food, but the longer snout may also reflect a need for a herbivore to select food with the incisors without injuring its eyes. Also, separating the incisors from the check teeth improves the specific function of both kinds of teeth. On the one hand, the sharper, cutting front teeth are relatively far removed from the pivot (the jaw joint), where speed is enhanced over force. On the other hand, the crushing cheek teeth are closer to the pivot, where force in enhanced over speed. In rodents the longer snout also helps accommodate the extremely large incisors. These teeth extend far back into the skull and mandible, curving posteriorly through the rostrum, and the bugle produced by the alveolus of the upper incisor on the lateral surface of the premaxilla is usually clearly evident. In the mandible, the incisor curves along the ventral margin of the mandible, and extends posterior to the cheek teeth. The bulge of its alveolus is also clearly evident in medial view.

Postcranial Skeleton

Vertebral Column

The **vertebral column** is composed of a series of movable bones, or **vertebrae**. The column is an important structure in support and locomotion and has evolved in mammals into five distinct regions. Examine a mounted skeleton to identify these regions (Figure 7.1). Most anteriorly, in the neck, are the **cervical vertebrae**, which are followed by the rib-bearing **thoracic vertebrae** in the trunk. Next are the **lumbar vertebrae**, lying dorsal to the abdomen and between the thoracic vertebrae and **sacral vertebrae**. The latter are fused into a solid unit and articulate with the pelvis. Lastly, the **caudal vertebrae** support the tail. An opening, the **intervertebral foramen**, is present on each side between adjacent vertebrae for the passage of a spinal nerve.

Each vertebra generally possesses several parts. For the following, examine a typical cervical vertebra (Figure 7.8). The **centra** (sing., **centrum**) or vertebral bodies, oval or kidney-shaped in section, form the main support of the vertebral column. The **neural canal**, through which the spinal cord passes, lies dorsal to the centrum. The neural canal is enclosed laterally and dorsally by the **neural arch**. The arch is composed on each side by a **pedicle**, an oblique plate forming the side of the arch, and a **lamina**, a horizontal plate forming the top of the arch. A **neural process** projects dorsally from the neural arch. A **transverse process** projects laterally on either side from near the junction of the centrum with a pedicle. Near the junction between the pedicle and lamina, each vertebra has anterior and posterior projections for articulation with other vertebrae. The anterior projection, the **prezygapophysis**, bears a facet that faces dorsally or dorsomedially. The facet of the posterior process or the **postzygapophysis** faces ventrally or ventromedially. Thus, a prezygapophysis slips under the postzygapophysis of the preceding vertebrae.

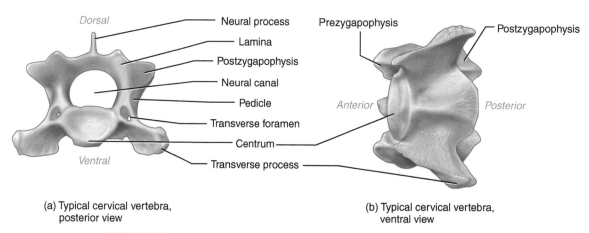

FIGURE 7.8   A cervical vertebra of the cat in (a) posterior and (b) ventral views.

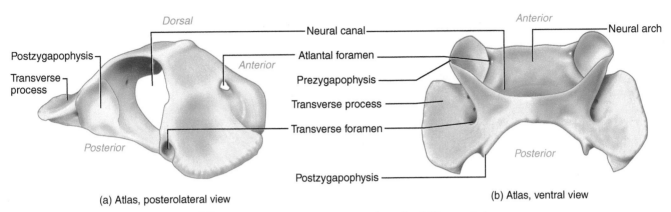

(a) Atlas, posterolateral view

(b) Atlas, ventral view

FIGURE 7.9   Atlas of the cat in (a) posterolateral and (b) ventral views.

These features are common to most vertebrae, though they may vary in size and proportion among vertebral types. Some vertebrae may also have additional processes. Examine a set of disarticulated vertebrae and distinguish the main types, as described next.

Cervical Vertebrae

The cat has seven cervical vertebrae, as do most mammals. The first two are specialized and markedly different from the remaining five, due to their function in support and movement of the head. The first vertebra is the **atlas** (Figure 7.9), after the mythological giant Atlas, who was charged with supporting the world on his shoulders. The atlas is easily recognizable. It is ring-like and lacks a neural process. Ventrally, the centrum is replaced by a narrow strut, and the neural arch is the more prominent component. The transverse processes are broad, wing-like structures. The prezygapophyses bear strongly curved articular surfaces for the occipital condyles. The joints between the occipital condyles and atlas permit mainly up and down or nodding movements of the head on the neck. The postzygapophyses bear nearly flattened surfaces for articulation with the second cervical vertebra, the **axis**. These joints allow mainly rotational movements of the head. The paired **atlantal foramina** pierce the neural arch transversely just dorsal to the prezygapophyses. Vessels (such as the craniodorsal artery, a branch of the vertebral artery) and the first cervical nerve pass through the foramen. The **transverse foramina**, despite their name, do not extend transversely but anteroposteriorly through the base of the transverse process. The transverse foramen, through which the vertebral artery passes, is characteristic of the cervical vertebrae, except for the last, in which it is usually absent.

The axis is also distinctive (Figure 7.10). It has a large and anteroposteriorly elongated neural process that hangs over the atlas. Also, it has a **dens**, a process that projects anteriorly between the prezygapophyses

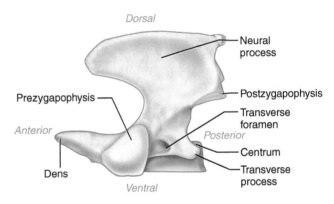

(a) Axis, left lateral view

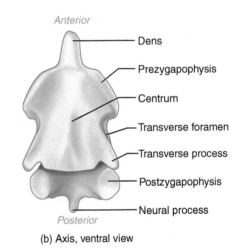

(b) Axis, ventral view

FIGURE 7.10   Axis of the cat in (a) left lateral and (b) ventral views.

and articulates with the atlas. This articulation participates in shaking movements of the head. Note the slender, posteriorly projecting transverse processes. The remaining cervical vertebrae are more typical vertebrae (Figure 7.8). Each possesses a relatively slender neural process that increases in height from the third cervical,

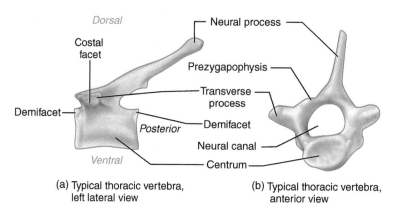

(a) Typical thoracic vertebra,
left lateral view

(b) Typical thoracic vertebra,
anterior view

FIGURE 7.11   A thoracic vertebra of the cat in (a) left lateral and (b) anterior views.

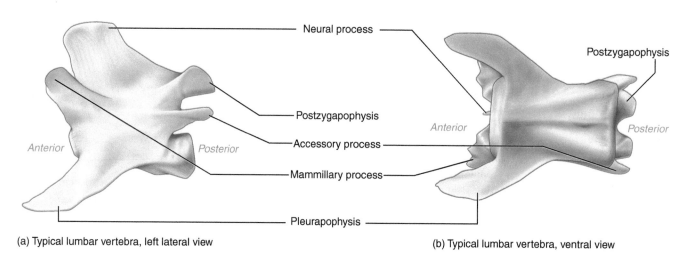

(a) Typical lumbar vertebra, left lateral view

(b) Typical lumbar vertebra, ventral view

FIGURE 7.12   A lumbar vertebra of the cat in (a) left lateral and (b) ventral views.

in which the process is very strongly reduced, to the seventh cervical.

Thoracic Vertebrae

The cat usually has 13 thoracic vertebrae. These are distinguishable by the presence of additional facets for articulation with ribs and very short transverse processes. The **capitulum** of a **rib** (see later) sits between and articulates with two successive centra, for which most thoracic vertebrae possess **costal demifacets** (Figure 7.11). These are anteriorly and posteriorly paired facets, so that a vertebra typically has four demifacets. The demifacets of successive vertebrae form (on each side) a complete or "full" articular surface for the capitulum of a rib. Thoracic vertebrae also have a **costal facet** on the ventral surface of each transverse process for articulation with the **tuberculum** of the corresponding rib.

The thoracic series undergoes various changes as it proceeds posteriorly. For example, the first thoracic has paired costal facets anteriorly and paired demifacets posteriorly. The last few thoracic vertebrae, usually the 11th through 13th, only have paired costal facets on the centrum (i.e., no demifacets). Also, the 11th thoracic has markedly reduced transverse processes, while the 12th and 13th lack transverse processes (reflecting the absence of a tuberculum on the last few ribs). The last few thoracic vertebrae have a slender **accessory process** (see Figure 7.12) that extends posteriorly from the pedicle, just lateral to each postzygapophysis. The accessory processes reinforce the articulation between the pre- and postzygapophyses.

Finally, note the change in the neural process throughout the thoracic series. Anteriorly, the process is high and inclines posteriorly. The height of the process tends to decrease posteriorly whereas inclination tends to increase until, depending on the individual, the 10th or 11th thoracic vertebra, at which point the process is markedly reduced and points anteriorly. This vertebra is known as the **anticlinal vertebra** (Figure 7.1) to denote the change in inclination. The neural process of the remaining thoracic vertebrae (as well as of the lumbar vertebrae) also inclines anteriorly and increases in

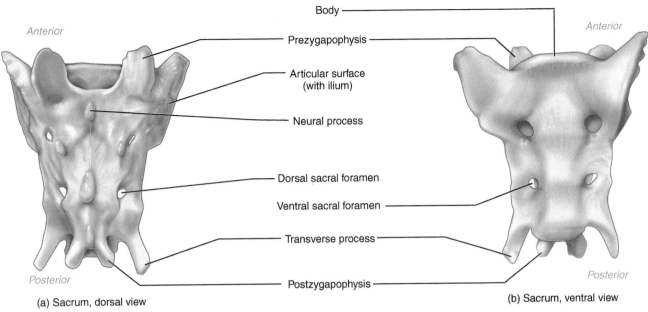

FIGURE 7.13   Sacrum of the cat in (a) dorsal and (b) ventral views.

height. The change in inclination of the neural processes allows the cat to extend and flex its back, especially during running.

Lumbar Vertebrae

The cat typically has seven lumbar vertebrae. They are characterized by their large size, which increases posteriorly through the lumbar series, and **pleurapophyses**, elongated, blade-like processes that sweep anteroventrally (Figure 7.12). A pleurapophysis represents a fusion of a transverse process and an embryonic rib. The prezygapophyses are extended dorsally beyond the articular surfaces into **mammillary processes** for attachment of epaxial muscles. Finally, note that accessory processes are present on all but the last two lumbar vertebrae.

Sacral Vertebrae

The cat has three sacral vertebrae that are fused together to form a single structure, the **sacrum**, which links the spine to the pelvic girdle (Figure 7.13). Anteriorly the sacrum has on either side a broad, rugose, and nearly circular surface for articulation with the innominate bones of the pelvis. The sacral vertebrae decrease in size posteriorly. Even though they are fused together, many of the structures can be recognized. Identify the three neural processes, and note the areas of fusion at the pre- and postzygapophyses. The pleurapophyses of the sacral vertebrae are expanded anteriorly and posteriorly and fuse with each other to enclose two pairs of **sacral foramina** on the dorsal and ventral surfaces of

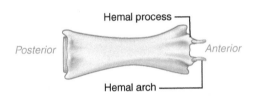

FIGURE 7.14   A caudal vertebra of the cat in ventral view.

the sacrum for the passage of dorsal branches of spinal nerves. The pleurapophysis of the first sacral forms most of the articular surface for the pelvis.

Caudal Vertebrae

Cats typically possess from 21 to 23 caudal vertebrae, which are the smallest of the vertebrae (Figure 7.14). They tend to become progressively smaller and less complex posteriorly. The anterior caudal vertebrae have zygapophyses, and neural and transverse processes, but the more posterior caudal vertebrae are elongated, cylindrical structures consisting almost entirely of the centrum. Beginning with the third or fourth caudal vertebra, you might note the presence of **hemal arches**, small V-shaped bones, lying at the anterior end of the ventral surface of the vertebra. Successive hemal arches enclose the hemal canal, through which caudal blood vessels pass. Each hemal arch articulates with a pair of small tubercles, the **hemal processes**, on the centrum. The small, delicate arches are usually lost during preparation, but their position can be determined by identifying the hemal processes.

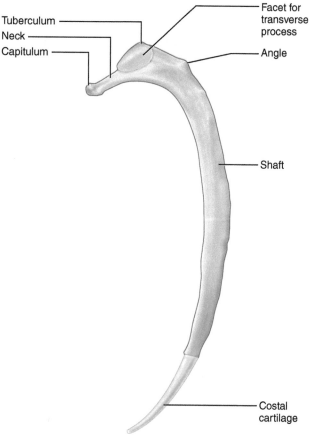

Tuberculum

Neck

Capitulum

Facet for transverse process

Angle

Shaft

Costal cartilage

FIGURE 7.15   A rib of the cat.

Ribs

The **ribs** form a strong but partly flexible cage that protects vital organs (e.g., heart and lungs) and participates in inspiration and expiration. Each rib articulates dorsally with the vertebral column and ventrally with a **costal cartilage**.

The 13 pairs of ribs (Figure 7.15) may be subdivided into three types: **vertebrosternal**, **vertebrochondral**, and **vertebral**, based on the attachment of their costal cartilages onto the **sternum** or breast bone (Figure 7.16). The first 9 ribs are vertebrosternal (or "true" ribs), meaning that the costal cartilages attach directly to the sternum. The next 3 are vertebrochondral (or "false" ribs) because their costal cartilages attach to the costal cartilage of another rib; that is, they do not attach directly to the sternum. The last rib is a vertebral rib because its short cartilage does not gain access to the sternum. It is thus also known as a floating rib (which is also considered to be a false rib).

Ribs differ in length but are generally similar in being curved, slender, rod-like bones. The **capitulum** or head bears surfaces for articulation with the demifacets of the thoracic vertebrae. This is followed by a short, constricted **neck** and then by the **tuberculum**, which has a

facet for articulation with the costal facet on the transverse process of a thoracic rib. A tuberculum is absent in the final two or three ribs. The **shaft** of the rib curves distally and bears a well-defined **angle** beyond the tuberculum. A costal cartilage is attached to the distal surface of each rib.

Sternum

The sternum (Figure 7.16) consists of eight **sternebrae** arranged anteroposteriorly midventrally on the thorax. The most anterior of the series is the spear tip-shaped **manubrium**. In some individuals the manubrium appears to be formed by the fusion of two elements. The next six sternebrae, constituting the **body** of the sternum, tend to be elongated, spindle-shaped elements. Finally, there is the elongated and tapering **xiphisternum**. The **xiphoid cartilage** attaches to its posterior end. The costal cartilages of the vertebrosternal ribs attach directly to the sternum. Typically, the pattern is that a costal cartilage attaches between adjacent sternebrae.

Forelimb
Scapula

The **scapula**, or shoulder blade (Figure 7.17a–b), is a flat, triangular bone. Its medial surface is nearly flat, whereas its lateral surface has a prominent **scapular spine**. Examine a mounted skeleton, and note that the apex of the scapula is directed ventrally. Identify the scapula's **anterior**, **dorsal**, and **posterior borders**. The **glenoid fossa** is the smooth, concave surface at the apex for articulation with the humerus. The delicate **coracoid process** projects medially from the anterior margin of the glenoid fossa and is the site of origin for the coracobrachialis muscle.

The medial surface bears the **subscapular fossa**. It is relatively flat, with a few prominent scar ridges indicating tendinous muscular insertions. Note the prominent ridge near the posterior border that demarcates a narrow, slightly concave surface for muscular attachment. On the lateral surface, the scapular spine rises prominently and separates the **supraspinous fossa** anteriorly from the **infraspinous fossa** posteriorly, both of which are fairly smooth surfaces. Ventrally the spine ends in the **acromion process**. Just dorsal to the acromion process is the posteriorly projecting **metacromion process**.

Clavicle

The **clavicle** is a small, slender, slightly curved bone suspended in the musculature associated with the forelimb (Figures 7.1 and 7.17c). It lies just anterior to the proximal end of the **humerus** (see next). As such, the clavicle of the cat does not articulate with other skeletal elements.

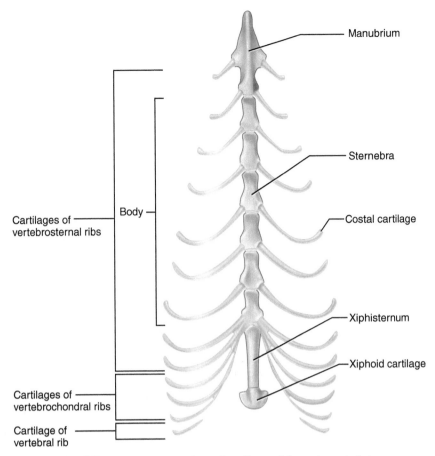

Manubrium

Sternebra

Costal cartilage

Xiphisternum

Xiphoid cartilage

Body

Cartilages of
vertebrosternal ribs

Cartilages of
vertebrochondral ribs

Cartilage of
vertebral rib

FIGURE 7.16   Sternum and costal cartilages of the cat in ventral view.

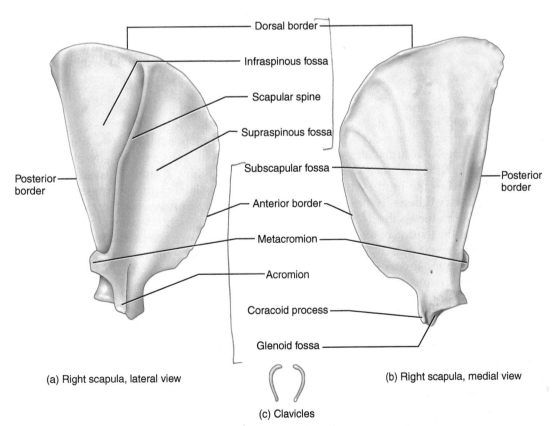

Dorsal border

Infraspinous fossa

Scapular spine

Supraspinous fossa

Subscapular fossa

Posterior
border

Posterior
border

Anterior border

Metacromion

Acromion

Coracoid process

Glenoid fossa

(a) Right scapula, lateral view

(b) Right scapula, medial view

(c) Clavicles

FIGURE 7.17   Right scapula of the cat in (a) right lateral and (b) medial views; and (c) clavicles of the cat.

THE DISSECTION OF VERTEBRATES

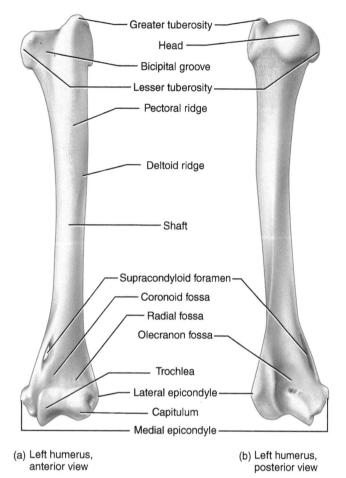

Greater tuberosity
Head
Bicipital groove
Lesser tuberosity
Pectoral ridge

Deltoid ridge

Shaft

Supracondyloid foramen
Coronoid fossa
Radial fossa
Olecranon fossa

Trochlea
Lateral epicondyle
Capitulum
Medial epicondyle

(a) Left humerus,   (b) Left humerus,
    anterior view   posterior view

FIGURE 7.18   Left humerus of the cat in (a) anterior and (b) posterior views.

Humerus

The **humerus** (Figure 7.18) is the bone of the brachium, or proximal part of the forelimb. It articulates proximally with the glenoid fossa of the scapula and distally with the radius and ulna. The **head** is the large, smooth, and rounded surface that articulates with the glenoid fossa of the scapula. It is best seen in proximal and posterior views. Lateral to the head is the **greater tuberosity**, whereas the **lesser tuberosity** lies medially. Anteriorly between the tuberosities is the deep **bicipital groove**, along which passes the tendon of the biceps brachii muscle. The **pectoral ridge** extends distally from the greater tuberosity on the anterior surface of the humeral **shaft**. The less prominent **deltoid ridge** extends distally and obliquely from the posterior part of the greater tuberosity. This ridge converges toward and meets the pectoral ridge about midway along the shaft.

Distally, there is a smooth, spool-shaped surface, the **condyle**, which is actually composed of two articular surfaces. The smaller, lateral surface, the **capitulum**, articulates with the **radius** (see later), whereas the larger, medial surface, the **trochlea**, articulates with the **ulna**

(see later). The **coronoid** and **radial fossae** are shallow depressions that lie just proximal to the trochlea and capitulum, respectively, on the anterior surface of the shaft. On the posterior surface, just proximal to the condyle, is the deep **olecranon fossa** that receives the olecranon of the ulna. The **medial** and **lateral epicondyles** (or entepicondyles and ectepicondyles, respectively) are the rugose prominences that lie on either side of the trochlea. The **supracondyloid foramen** is the oval passage lying proximal to the trochlea.

Ulna

The **ulna** (Figure 7.19) is the longer of the two bones in the forearm, or **antebrachium**. It has a prominent proximal portion, but its shaft tapers distally. Proximally the ulna articulates with the humerus and the radius. Distally it contacts the carpus or wrist. The **trochlear notch** is the deep semicircular surface for articulation with the trochlea of the humerus. The **olecranon** is the squared process extending proximal to the trochlear notch. It is the site of insertion of the tendon of the triceps brachii muscle. The **coronoid process** extends anteriorly from the distal base of the trochlear notch. The **radial notch**, for articulation with the head of the radius, is a curved surface that lies laterally along the base of the trochlear notch. The anterior part of the radial notch forms the lateral surface of the coronoid process. The roughened **interosseus crest** lies on the central portion of the anterolateral surface of the ulnar shaft. A sheet of connective tissue extends between the ulnar interosseus crest and the radial interosseus crest (see below) and helps stabilize the antebrachium. The ulna tapers distally and ends in the **styloid process**, which articulates with the lateral part of the carpus.

Radius

The **radius** (Figure 7.19) is the second bone of the antebrachium. In contrast to the ulna, the radius is more slender proximally and widens distally. As the radius articulates with the lateral part of the humerus and the medial part of the carpus, it crosses over the ulna when a cat assumes a normal standing position (i.e., the manus is *pronated*, meaning the palm faces the ground; the opposite position, with the palm directed up, is termed *supinated*). Proximally the radius consists of the head, which bears an oval, concave **fovea** that articulates with the capitulum of the humerus. The shape of the head allows the radius to rotate on the capitulum. Immediately distal to the fovea, the head bears a smooth, narrow strip, the **articular circumference**, that articulates with the radial notch of the ulna. The **neck** of the radius is a short segment between the head and the **bicipital tuberosity**, onto which the tendon for the biceps brachii muscle inserts. The roughened **interosseus crest** lies distally from the tuberosity on the medial edge of the shaft.

7

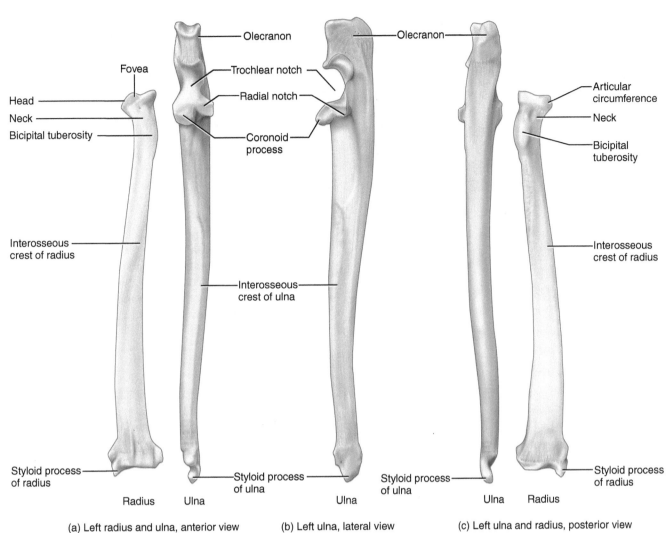

FIGURE 7.19 Bones of the left antebrachium of the cat; (a) radius and ulna in anterior view, (b) ulna in lateral view, and (c) radius and ulna in posterior view.

The distal end of the radius has a small articular surface for the ulna medially that allows the radius to turn on the ulna. The distal surface has a large, concave surface that articulates with the scapholunar bone of the wrist. Note the two grooves on the anterior surface for passage of extensor tendons, and the short **styloid process** projecting distally from the medial surface of the radius.

Manus

The **manus** (Figure 7.20) includes three portions—the **carpus** (wrist), **metacarpus**, and **phalanges**—in proximal to distal order. The cat has five **digits** (fingers), each of which is formed by a **metacarpal** and phalanges. The carpus consists of seven small, irregularly shaped bones set in two rows. In medial to lateral order, the proximal row consists of the **scapholunar** and **cuneiform**, with

the **pisiform** projecting from the ventral surface of the cuneiform (and thus not visible in Figure 7.20); the distal row consists of the **trapezium, trapezoid, magnum**, and **unciform**. These bones articulate distally with the metacarpals. Note how the bones of the carpus and the metacarpals are arranged in an interlocking pattern that restricts motion.

The metacarpals are included in the palm of the manus. Of the five metacarpals, metacarpal 1 is the most medial and shortest, being less than half the length of metacarpals 2 through 5. Each metacarpal consists of a base proximally, a shaft, and a head distally. Digit 1 has only the proximal phalanx, followed by the ungual phalanx, which bears a large claw. Digits 2 through 5 each have three segments, the proximal, intermediate (or middle), and distal (or ungual) phalanges; the latter also bear large claws.

Hind Limb

Pelvis

The pelvis or hip consists of paired **innominate** bones (Figure 7.21) that articulate with each other ventrally at the **pelvic symphysis** and with the sacrum dorsally. Examine a mounted specimen, and note the position and orientation of the innominate. Each innominate is composed largely from three bones, the **ilium**, **ischium**,

and **pubis**, although a fourth center of ossification, the **acetabular** bone, makes a small contribution. These bones are firmly fused together in the adult. Examine an innominate bone in lateral view and note the large ventral opening, the **obturator foramen**. Locate the **acetabulum**, the deep socket that receives the head of the femur (see below).

The ilium consists of a **body**, near the acetabulum, and an anterodorsally projecting **wing**. The **iliac crest** is the roughened, anterodorsal edge of the ilium. The rugose articular surface for the sacrum lies on the medial surface of the wing. The ischium extends posteriorly from the acetabulum and has an expanded termination, the **ischial tuberosity**. The pubis and the rest of the ischium are oriented ventromedially. Both bones contribute to the medial margin of the obturator foramen. Also, the ischium and pubis of each side of the body meet to form, respectively, the **ischial** and **pubic symphyses**, which together form the pelvic symphysis. The acetabular bone forms the thin, medial part of the acetabulum.

Femur

The **femur** (Figure 7.22) is the bone of the proximal part of the hind limb, or thigh. The **head** of the femur is a hemispherical surface that fits into the acetabulum of the innominate. The head is supported by the **neck**, which projects obliquely from the proximal end of the femur. Lateral to the head is the roughened, proximally projecting **greater trochanter**, which serves for attachments

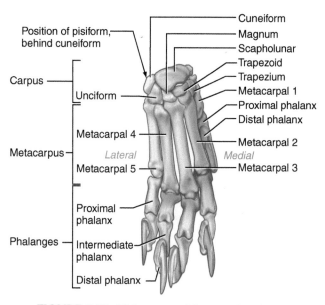

FIGURE 7.20   Right manus of the cat in dorsal view.

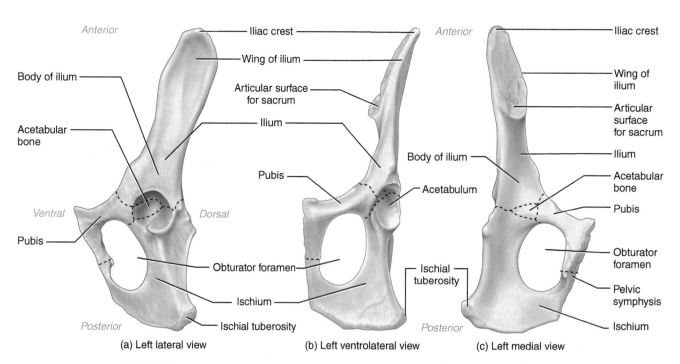

FIGURE 7.21   Left innominate of the cat in (a) lateral, (b) ventrolateral, and (c) medial views.

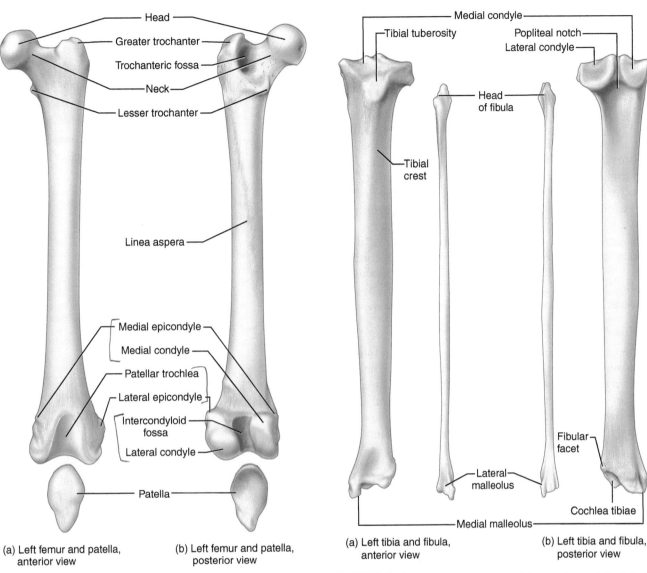

(a) Left femur and patella,
anterior view

(b) Left femur and patella,
posterior view

FIGURE 7.22   Left femur and patella of the cat in (a) anterior and
(b) posterior views.

(a) Left tibia and fibula,
anterior view

(b) Left tibia and fibula,
posterior view

FIGURE 7.23   Bones of the left crus of the cat; tibia and fibula in (a)
anterior and (b) posterior views.

of hip musculature. The deep depression posteriorly
between the trochanter and head is the **trochanteric
fossa**. On the posterior surface of the shaft, just distal
to the head, is the **lesser trochanter**. The intertrochan-
teric line connects the two trochanters, but is especially
defined along the posterior edge of the greater trochanter.
The most prominent muscular insertion site on the diaphysis is the **linea aspera**, the roughened ridge that extends
diagonally along the posterior surface of the femur.

The femur expands distally into the prominent and
posteriorly projecting **lateral** and **medial condyles**. Each
condyle bears a smooth, semicircular surface for articulation with the tibia. The **intercondyloid fossa** is the
depression posteriorly between the condyles. The rugose
areas for muscular attachment proximal to the condyles
are the **lateral** and **medial epicondyles**. The **patellar**

trochlea, for articulation with the patella, lies anteriorly
between the condyles. It is a smooth, shallow trough oriented proximodistally.

Patella

The **patella** (Figure 7.22), or kneecap, is a small, tear-
shaped sesamoid bone, with its apex directed distally. Its
anterior surface is roughened. Posteriorly it bears a smooth,
shallowly concave surface for articulation with the femur.

Tibia

The **tibia** (Figure 7.23) is the larger and medial bone of
the crus, or middle segment of the hind limb. Its proximal surface bears **lateral** and **medial condyles** that articulate with the femur. Just distal to the lateral condyle, on
the lateral surface and facing distally, is the small, nearly

oval facet for the head of the fibula (see later). On the posterior surface of the tibia, between the condyles, is the **popliteal notch**. A small muscle, the popliteus, lies in the notch and is a flexor of the knee joint. The **tibial tuberosity**, for insertion of the patellar ligament, lies anteriorly. The **tibial crest** continues distally from the tuberosity along the shaft.

The distal end of the tibia has two articular surfaces. The large surface on the distal surface, the **cochlea tibiae**, is for the astragalus, the tarsal bone with which the pes articulates with the hind limb. Note how this facet consists of two sulci separated by a median ridge. This structure restricts motion at the ankle almost entirely to a fore and aft direction, producing flexion and extension. If available, manipulate the tibia and pes to observe this. The small, nearly triangular **fibular facet**, for articulation with the fibula, faces posterolaterally and is contiguous with the lateral part of the cochlea tibiae. The **medial malleolus** is the distal extension of the tibia's medial surface. It forms the medial protrusion of the ankle.

Fibula

The **fibula** (Figure 7.23) is the slender, lateral, and shorter bone of the crus. The **head** is irregular and expanded. It bears a proximal facet for articulation with the tibia. The slender shaft widens distally. There are two distal facets, both toward the anterior half of the medial surface. The more proximal facet is for the distal articulation with the tibia. The distal facet articulates with the lateral part of the trochlea of the astragalus. The **lateral malleolus** projects distally from the posterolateral end of the fibula.

Pes

The **pes** (Figure 7.24) consists of **tarsals**, **metatarsals**, and **phalanges**. There are seven tarsals, but the two most proximal bones, the **astragalus** and **calcaneum**, are much larger than the others. The astragalus is the medial bone. It articulates proximally with the tibia and fibula. Note the form of the surface, or **trochlea tali**, for articulation with the tibia. The trochlea tali consists of medial and lateral keeled surfaces separated by a sulcus. The astragalus articulates ventrally with the calcaneum, which lies laterally. It is about twice as long as the astragalus and projects posteriorly as the heel. Distally the astragalus articulates with the **navicular**, while the calcaneum articulates with the **cuboid**. The navicular articulates distally with the **lateral**, **intermediate**, and **medial cuneiform** bones, and laterally with the cuboid. Note how the articulations among the tarsals and metatarsals are arranged to produce interlocking joints that tend to restrict movement. For example, the lateral cuneiform articulates with metatarsal III distally, but its medial surface articulates with both the intermediate cuneiform and metatarsal II. There are five metatarsals. The first is strongly reduced to a small nub that articulates with the medial cuneiform. The phalanges

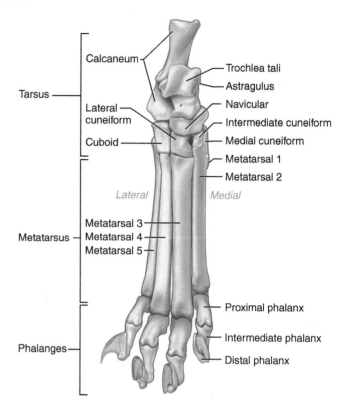

FIGURE 7.24   Right pes of the cat in dorsal view.

for digit 1 have been lost in the cat. The remaining metatarsals are stout, elongated elements and each articulates with a series of three phalanges, the proximal, intermediate (middle), and distal (ungual) phalanges.

Key Terms: Postcranial Skeleton

accessory process
acetabular
acetabulum
acromion process
angle
antebrachium (forearm)
anterior, dorsal, posterior borders of **scapula**
anticlinal vertebra
articular circumference
astragalus (talus)
atlantal foramen (pl., **foramina**; intervertebral foramen)
atlas
axis
bicipital groove
bicipital tuberosity
body of **ilium**
body of **sternum**
calcaneum (calcaneus)

capitulum of humerus
capitulum (head) of rib
carpus
caudal vertebrae
centra (sing., centrum)
cervical vertebrae
clavicle
cochlea tibiae
condyle
coracoid process
coronoid fossa
coronoid process
costal cartilage
costal demifacet
costal facet
cuboid
cuneiform (lateral, medial, intermediate)
deltoid ridge
dens
digits
femur
fibula
fibular facet
fovea
glenoid fossa
greater trochanter
greater tuberosity
head of femur
head of humerus
head of radius
hemal arches
hemal processes
humerus
iliac crest
ilium
infraspinous fossa
innominate (coxal bone; os coxae)
intercondyloid fossa
interosseous crest of radius
interosseous crest of ulna
intervertebral foramen
ischial symphysis
ischial tuberosity
ischium
lamina
lateral condyle
lateral epicondyle of femur
lateral epicondyle of humerus
lateral malleolus
lesser trochanter
lesser tuberosity
linea aspera
lumbar vertebrae
magnum
mammillary processes

manubrium
manus
medial condyles
medial epicondyle of femur
medial epicondyle of humerus
medial malleolus
metacarpals
metacarpus
metacromion process
metatarsals
navicular
neck of femur
neck of radius
neck of rib
neural arch
neural canal (vertebral canal)
neural process (spinous process)
obturator foramen
olecranon
olecranon fossa
patella
patellar trochlea
pectoral ridge
pedicle
pelvic symphysis
pes
phalanges (sing., phalanx)
pisiform
pleurapophyses
popliteal notch
postzygapophysis
prezygapophysis
pubic symphysis
pubis (pl., pubes)
radial fossa
radial notch
radius
rib
sacral foramina
sacral vertebrae
sacrum
scapholunar
scapula
scapular spine
shaft of humerus (diaphysis)
shaft of femur (diaphysis)
shaft of fibula (diaphysis)
shaft of rib
sternebrae
sternum
styloid process of radius
styloid process of ulna
subscapular fossa
supracondyloid foramen
supraspinous fossa

SECTION II: EXTERNAL ANATOMY

The body of the cat is covered almost entirely by **hair** and fur (together forming the pelage), a characteristic feature of mammals. Only the nose, **lips**, and thick pads of the feet are exceptions. The **head** and **trunk** are separated by a distinct **neck**, which allows considerable mobility of the head (Figure 7.25). Dorsally the trunk is termed the **dorsum**. Ventrally, it is subdivided into the **thorax** anteriorly and the **abdomen** more posteriorly. The **pelvis** is the region of the trunk associated with the hind limb, just anterior to the elongated **tail**.

On the head note the **pinnae** (sing., **pinna**), the mainly cartilaginous extension of the external ears. Upper and lower **palpebrae** (sing., **palpebra**) guard the large **eyes**. A third eyelid, the **nictitating membrane**, lies on the medial part of the eye. In life, it can be drawn laterally over the eye. The **nares** (sing., **naris**) are the external openings of the nose and are surrounded by a region of skin termed the **rhinarium**. The **mouth** is surrounded by lips. **Vibrissae** are stiff, tactile hairs or whiskers present mainly around the mouth and snout, but also over the eyes.

Each limb comprises three segments. In proximal to distal order, they are the **brachium**, **antebrachium**, and **manus** in the forelimb, and **thigh**,[1] **crus**, and **pes** in the

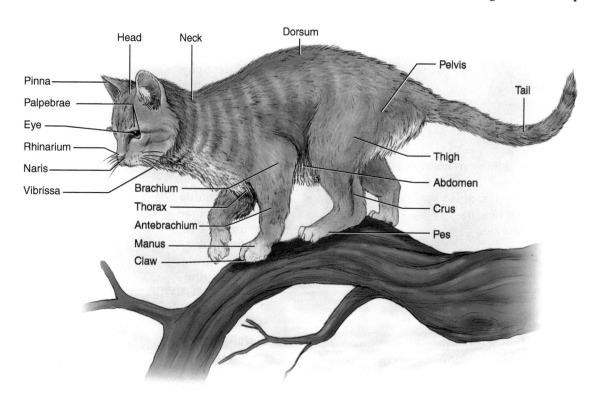

FIGURE 7.25　External features of the cat.

[1] The thigh may also be referred to as the femur, the latter also being the name of the bone within the region, as noted by Homberger and Walker (2003).

hind limb. Each digit, except digit 1 of the pes, ends in a retractable **claw**. The walking surface of the forepaws and hind paws each have epidermal thickenings of the skin termed **tori** (sing., **torus**) that act as pads during locomotion.

The **anus**, the posterior opening of the intestine, lies ventral to the base of the tail. The urinary and reproductive structures lie just anterior to the anus. In the male the **scrotum** is a sac-like projection that contains the testes. Anterior to it is the **penis**, which is probably retracted and thus not readily apparent. Its tip contains the common opening of the urogenital ducts. In the female the urinary and reproductive tracts open into the **urogenital aperture**.

Key Terms: External Anatomy

abdomen
antebrachium (forearm)
anus
brachium (arm)
claw
crus (shank)
dorsum (back)
eyes
hair
head
lips
manus
mouth
nares (sing., **naris**; nostrils)
neck (cervix)
nictitating membrane
palpebrae (sing., **palpebra**; eyelid)
pelvis
penis
pes
pinnae (sing., **pinna**)
rhinarium
scrotum
tail
thigh (femur)
thorax
tori (sing., **torus**)
trunk
urogenital aperture
vibrissae (sing., **vibrissa**; whisker)

SECTION III: MUSCULAR SYSTEM

Muscle Terminology

Tables 7.2–7.4 list the muscles covered here, including their areas of origin, insertion, and main actions. Muscles may be named according to several criteria.

Often a name refers to the muscle's area of origin and area of insertion, usually but not always in this order. For example, the spinodeltoid extends from the spine of the scapula to the deltoid ridge of the humerus; the sternomastoid from the sternum to the mastoid process of the temporal bone of the skull; xiphihumeralis from the xiphoid process of the sternum to the humerus. A name may also refer to the main position of a muscle. For example, the subscapularis essentially occupies all the subscapular fossa of the scapula; the temporalis the temporal fossa of the skull. In other cases, names are descriptors of action as well as shape and form. For example, the adductor femoris, which adducts the femur or thigh, is named for its action. Examples of names based on shape or form include the deltoid, which is generally triangular, shaped like the Greek letter delta (Δ); often, "-ceps" (from the Ancient Greek word for "head") appears in a muscle's name in reference to the number of heads that form the muscle—the triceps, for example, is formed by three heads, at least in humans, to which much of the terminology was first applied.

Connective Tissue and Fiber Direction

Various features and structures are associated with the muscles, but dissection and identification of the muscles requires knowing only the following features. When the skin has been removed, you will note that the muscles are covered by connective tissue. There are various kinds of connective tissue: **epimysium** covers or envelopes the surface of individual muscles; **loose connective tissue** lies between structures (such as muscles, nerves, or blood vessels) and helps fill the spaces between them; **fascia** is a very dense, sheet-like connective tissue associated in various ways with muscles; and **tendons** are dense connective tissue that connect muscles (mainly) to skeletal elements. When a tendon is very short, the muscle has a *fleshy attachment*. When it is longer, the muscle has a *tendinous attachment*. In many cases, the tendon is large and sheet-like; such a tendon is an **aponeurosis**, and the attachment is *aponeurotic*.

The visible fibers of a muscle will all tend, in the majority of cases, to have a very similar direction. It is very helpful to note fiber direction in distinguishing where one muscle ends and another begins, particularly when muscles overlap one another. This is an important clue in identifying muscles, as well as in knowing where to separate them.

Fiber direction, however, is often obscured by connective tissue, and thus it is essential that you remove the connective tissue until the direction is clearly revealed. You do not need to clean the whole surface of every muscle. It is sufficient to clear surfaces near where text directions indicate the edges of muscles should be. Most connective tissue can usually be picked away using forceps and a dissecting needle. It is sometimes, but rarely, necessary to use a scalpel in nicking away the connective tissue on the surface of a muscle. Separating muscles can usually be accomplished by using a blunt probe to tear through loose connective tissue. A scalpel should not (unless indicated) be used to cut between muscles. In clearing connective tissue, then, you will be mainly concerned with epimysium or loose connective tissue. You will not cut through aponeuroses or fascia, unless specifically directed to do so.

Key Terms: Muscular System

aponeurosis
epimysium
fascia
loose connective tissue
tendon

Subdivision of the Musculature

As for the dogfish and mudpuppy, the musculature of the cat may be subdivided into formal groups based on phylogenetic derivation. The major muscles can be identified to group based on embryology and innervation. However, the muscles of the cat have undergone considerable modification and are much more complex than in the dogfish. Thus, the musculature from any formal group usually cannot be studied at the same time, because some muscles lie superficially and others more deeply. Also, in many cases muscles have evolved different functions than in their remote predecessors. For example, the trapezius is originally a branchiomeric muscle, but in the cat it functions as an appendicular muscle. For the cat, then, it is more practical, much as it was for the mudpuppy as well, to dissect and identify the muscles as they are observed in superficial and deep views. The musculature is organized under informal headings for convenience in dissection. This should help organize your dissection and study of the muscles.

Skinning the Cat

To examine the muscles and other internal structures, the cat will have to be skinned, but the skin should not be completely removed. It is best to remove the skin from only part of the cat and leave it as a flap that can be wrapped around the cat at the end of each dissection. This helps protect the muscles and other structures and prolongs preservation of your specimen.

Before beginning, read the instructions below with your specimen in front of you, and assess where you will cut through the skin. Determine the course of the incisions according to these instructions and Figure 7.26.

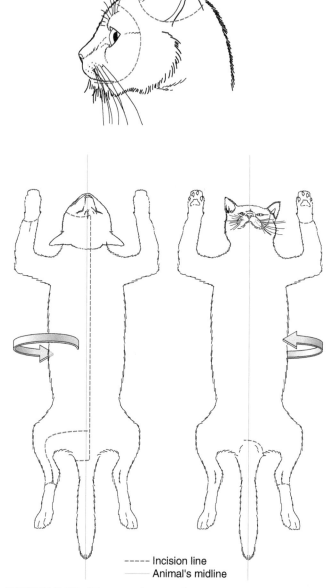

----- Incision line
—— Animal's midline

FIGURE 7.26   Diagrams showing the incision lines to be followed in skinning the cat.

Once skinning is completed, you will have a flap of skin still attached ventrally on one side of the chest, abdomen, and groin. For example, if your first longitudinal incision is on the right side of the middorsal line, then you will skin mainly the left side of the body, skinning the left arm and leg, and leaving the skin attached on the right side of the ventral surface of the cat. It does not matter which side of the body you choose to skin. However, avoid skinning a side if it has been damaged or had skin removed to facilitate injection of vessels. For example, in most specimens the skin on one side of the throat will probably have been removed and the musculature damaged when the arteries and veins were injected. Sometimes, an area in the abdomen will have been skinned and cut to inject the hepatic portal system.

You need not make all the incisions before you begin skinning. Using a new scalpel blade, make a shallow longitudinal incision along the back, about 1 cm to one side of the neural processes of the vertebrae. As you cut, ensure that you have cut only through the skin by reflecting the edge of the flap. Extend the cut from near the base of the tail forward onto the back, neck, and head past the pinnae. At about the level of the eyes, make a sideways incision toward the other side eye (that is, cut back across and past the middorsal line), angling the incision toward the back of the eye. Continue to cut toward the back of the mouth and onto the underside of the lower jaw. Follow around the mandible to the other side. Return to the base of the tail and extend the incision back across the midline, around and under the tail. Continue forward, anterior to and then around the external genitalia. Continue the incision to just past the midventral line. Make an encircling cut around the forelimb, about midway along the antebrachium. Do the same for the hind limb, just past the ankle. Then make another incision, on the lateral surface of the hind limb, between this encircling incision and the longitudinal incision on the back.

Skinning can be accomplished by various techniques. For much of the back, for example, the skin is often readily removed, and a blunt probe, forceps, or your fingers will do the job of tearing through the fibrous connective tissue. In other areas, the skin adheres more strongly, and requires a scalpel. When using a scalpel, hold the blade parallel with the surface of the body and use short strokes; often, simply nicking the connective tissue will suffice, especially if you pull the skin flap away from the body to tense the connective tissue. You will encounter small nerves and blood vessels passing to the skin. Cut through them. As you skin the trunk, you will notice thin and narrow bands of muscles that adhere to the underside of the skin. These bundles represent the cutaneous maximus, one of the cutaneous muscles, a large sheet that covers much of the trunk and is especially prominent ventrally and near the axilla (armpit). This muscle should be removed with the skin, except near the axilla. In this region, it adheres strongly to the shoulder muscles and should be cut.

In a pregnant or lactating female, the **mammary glands** (see page 299) are large, flattened, glandular masses in the thoracic and abdominal regions, and should be removed with the skin. If your specimen is a male, be careful in skinning the groin region. A large wad of fat, which contains the **spermatic cords** (see pages 220 and 296, and Figures 7.32, 7.36, and 7.69) that you will need to see later, lies beneath the skin in this region. Leave the wad of fat intact.

The skin on the neck and throat may be more difficult to remove. Dissect carefully so that you do not injure the vessels in the neck and throat. The skin of the head is very thick and difficult to remove. Use care so that you do not injure the various vessels, nerves, and ducts, and the salivary glands that lie just below and beside the pinna. It is not worth trying to skin the pinna; simply cut through it, leaving a stump about 1 cm long. Another cutaneous muscle will be found over much of the head and neck. Remove it with the skin.

In skinning the arm and leg, tear the connective tissue with a probe or forceps to separate the skin and muscles, but try not to tear the skin itself. Proceed from both the trunk down the limb as well as from the encircling cut up the limb. The skin may adhere closely to the muscles. Dissect especially carefully along the anterolateral surface of the brachium. The **cephalic vein** (see Figure 7.27) lies here and should not be removed with the skin.

Appendicular Musculature

Much of the musculature that you will dissect includes the appendicular muscles, i.e., those associated with the forelimb and hind limb. Thus, if either or both ends of a muscle attach to the scapula (or pelvis) or more distal limb element, it can be considered an appendicular muscle. These muscles will be organized into superficial and deep portions and will be observed with the cat positioned so as to give you lateral and ventral views of its body.

Muscles of the Forelimb

Table 7.2 lists the forelimb muscles and indicates their origin, insertion, and main functions.

Superficial Forelimb Muscles: Lateral View (Figure 7.27)

Examine the cat in lateral view. Identify the large, white **lumbodorsal fascia** covering the back in the lumbar region. The fascia is actually composed of two main sheets of tough connective tissue, one above the other. The **latissimus dorsi** is the wide muscle that originates mainly from the lateral margin of the fascia. Its fibers pass anteroventrally and converge toward the axillary region. It pulls the humerus posterodorsally. The muscle emerging from under the latissimus dorsi and extending onto the abdomen is the **external oblique** (see later).

Anterior to the latissimus dorsi are three trapezius muscles. The most posterior is the triangular **spinotrapezius**, which draws the scapula posterodorsally. From their middorsal origin, the fibers of the spinotrapezius pass anteroventrally and converge toward the scapular spine. Next anteriorly is the **acromiotrapezius**. It is wider, but very thin. Its fibers fan out from their insertion along the scapular spine toward the middorsal line, but

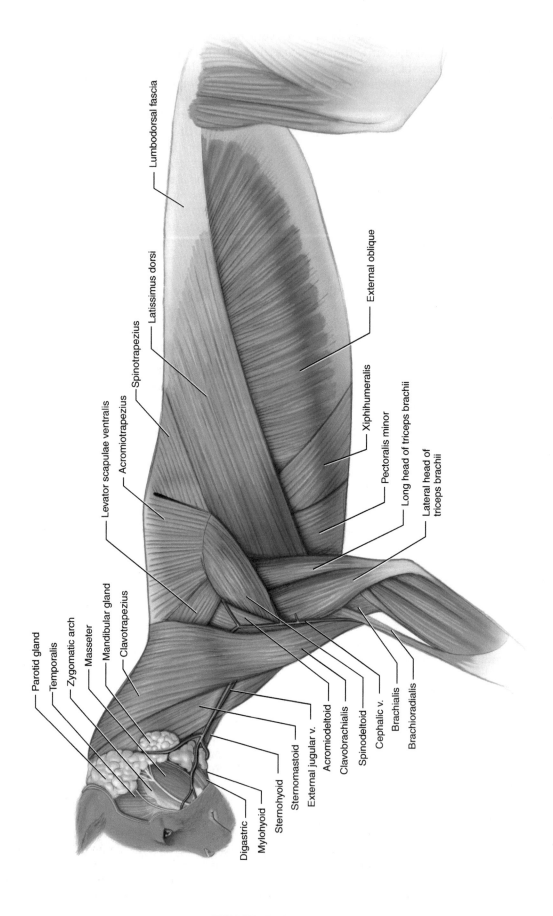

FIGURE 7.27 Muscles of the cat in left lateral view.

do not extend all the way to the middorsal line. Instead, they attach to a thin aponeurosis. The acromiotrapezius can stabilize the scapula or pull it toward the dorsal midline. The most anterior trapezius muscle is the **clavotrapezius**, which covers most of the neck laterally and pulls the scapula anterodorsally. It is a wide muscle that originates from the neck middorsally, extending from just behind the skull to the acromiotrapezius. Its fibers pass almost directly ventrally toward the front of the shoulder. In mammals with a well-developed clavicle, the clavotrapezius inserts on the clavicle. As the clavicle is greatly reduced in the cat, however, the clavotrapezius inserts partly on the clavicle and partly onto another muscle, the **clavobrachialis**, which continues ventrally to insert on the ulna. Actually, a raphe (a narrow band of collagen fibers to which muscle fibers may attach), as well as the clavicle, separates these muscles; as their fibers appear to be continuous across the raphe, they seem to form a single muscle. The clavicle is embedded in the musculature but may be discerned by palpation (see later).

The **sternomastoid** lies anterior to the clavotrapezius. It is not an appendicular muscle, but should be identified at this point. It extends from the mastoid process of the skull (see Section I), just posterior to the ear, along the anterior margin of the clavotrapezius, but then veers medially to insert on the sternum. The sternomastoid will be considered again later. For now, note the **parotid** and **mandibular glands**, also considered in detail later, that lie on its anterior part, and that the muscle is crossed by the large **external jugular vein**.

Examine the musculature of the shoulder and forelimb in lateral view. The **cephalic vein** (Figure 7.27) lies on the anterolateral surface of the brachium. The **spinodeltoid** muscle extends anteroventrally from the scapular spine, arising at the insertion of the acromiotrapezius. Its anterior end passes deep to the triangular **acromiodeltoid**. These muscles are flexors and lateral rotators of the humerus. The narrow **levator scapulae ventralis**, lying dorsal to the acromiodeltoid, extends anterodorsally and then passes deep to the posterior margin of the clavotrapezius. The **long head** of the **triceps brachii** covers the brachium posteriorly, while the **lateral head** of the triceps covers the brachium laterally. Separate the edge of the spinodeltoid from the triceps and note that these heads extend proximally, deep to the spinodeltoid. The **medial head** of the triceps is discussed later. All three heads insert on the olecranon process of the humerus and extend the antebrachium. Note the **brachialis**, a flexor of the antebrachium, on the anterolateral surface of the brachium. It is readily observable near the elbow just anterior to the lateral head of the triceps.

Note again the position of the cephalic vein. Distally, it crosses the clavobrachialis, whereas proximally it lies on the acromiodeltoid, and then curves anteriorly, deep to the clavotrapezius. Lift the posterior margin of the clavotrapezius, just anterior to the acromiodeltoid, and cut through the connective tissue, fat, and nerves that you encounter. Continue to separate this muscle and the clavobrachialis from the underlying musculature. As you do so, you will see the posteriorly oriented fibers of the pectoral muscles (see later) inserting on the humerus.

Working dorsally, continue to separate the clavotrapezius, where the levator scapulae ventralis passes deep to its margin. It is fairly easy to free the muscle here, but work carefully anterior to the acromiodeltoid. The clavicle is present in this region, and you should avoid injuring the **cleidomastoid** (Figure 7.28), which attaches on the clavicle. Palpate for the clavicle. Once you identify its position, stop dissecting the clavotrapezius.

Superficial Forelimb Muscles: Ventral View (Figure 7.28)

The pectoralis muscles, which draw the humerus toward the midventral line, lie superficially on the ventral surface of the thorax. Examine them in ventral view with the brachium held laterally. As a whole, this musculature is a triangular mass extending, on each side, from the ventral midline to the humerus. Carefully pick away at and clear connective tissue covering the muscles on one side to reveal four muscles. Once cleared, they can readily be distinguished by fiber direction. For the most part, the individual muscles are arranged in layers, one atop the other, and tend to adhere strongly to each other. Thus, do not attempt to separate the muscles by cutting them at their margins. Instead, work with a probe and needle to separate the muscles parallel to their surfaces. You do not need to separate each muscle completely; only go far enough to get an idea of the depth and extent of each.

The most superficial muscle is the **pectoantebrachialis**, a thin, narrow, anterior muscle that extends laterally. The **epitrochlearis** (which is not one of the pectorals) is a thin muscle covering the surface of the brachium posterior to the pectoantebrachialis. The epitrochlearis attaches to the olecranon of the ulna and thus extends the antebrachium. The **pectoralis major**, which may be subdivided in two portions, lies dorsal to the pectoantebrachialis. Its anterior part extends laterally and is covered almost entirely by the pectoantebrachialis (except for its anteriormost part, which is exposed in ventral view). Its posterior part extends obliquely, with its fibers passing anterolaterally from the midline.

The **pectoralis minor** is the muscle lying dorsal and posterior to the pectoralis major. The pectoralis minor of the cat is larger than the pectoralis major, but the names are derived from human terminology, in which the pectoralis major is indeed larger. The fibers of the pectoralis minor extend even more obliquely than those of the pectoralis major as they converge toward the proximal half of the humerus. The fourth muscle is the

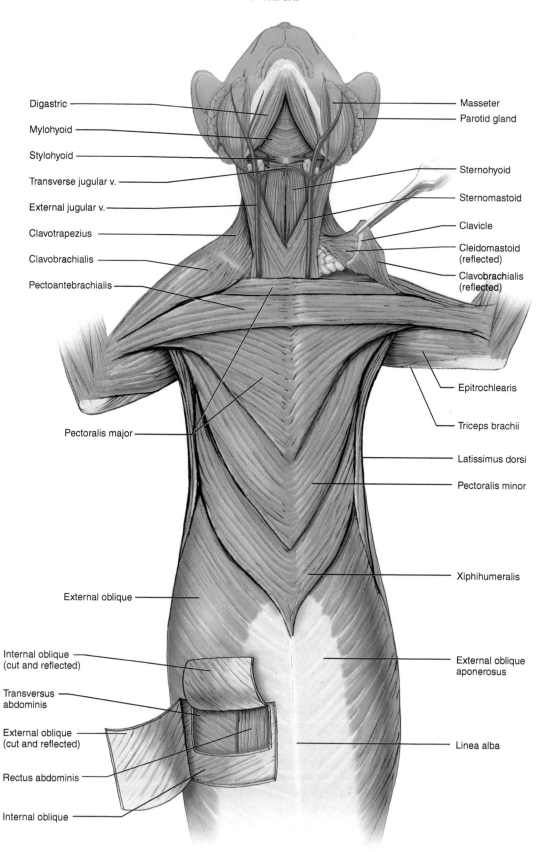

FIGURE 7.28   Muscles of the cat in ventral view.

xiphihumeralis, a long, thin, and narrow muscle that lies posterior and dorsal to the pectoralis minor. For the most part, the xiphihumeralis lies along the pectoralis minor posteriorly and attaches to the xiphisternum and xiphoid cartilage, which give it part of its name.

You have already noted how the fibers of the latissimus dorsi converge toward the axilla. Also note that the latissimus dorsi and the pectoralis minor adhere to one another. Carefully separate these muscles by picking away at the connective tissue that binds them, using a needle and forceps. If necessary, use a scalpel to cut them apart.

Identify the medial edge of the clavotrapezius along the neck and follow it posteriorly until it abuts against the anterior margin of the pectoralis musculature (Figure 7.28). From about this point, the muscle becomes the clavobrachialis, and its margin is bound tightly to the pectoralis major. Beginning medially, separate these muscles. You will soon be able to lift the clavobrachialis and discern the deeper part of the pectoralis, which passes diagonally beneath the clavobrachialis. You should now be able to separate the latter from the pectoralis much more easily.

Return to the medial margin of the clavotrapezius. Lift and separate it from the underlying musculature. You will uncover the cleidomastoid, mentioned earlier, which passes from the clavicle toward the skull. Probe deep to the sternomastoid to discern its course. Separate as much of the cleidomastoid from the clavotrapezius, but work delicately near the clavicle: These muscles attach to the clavicle in close proximity, and dissection often causes damages. Continue to pick away the fat and tough connective tissue along the medial margin of the clavotrapezius, being careful not to injure the external jugular vein. Also work deep and medial to the cleidomastoid. You should be able to separate and lift the clavotrapezius, clavobrachialis, and cleidomastoid as a unit from the underlying muscles. Clean the connective tissue and fat from the bend of the elbow, being careful not to injure the cephalic vein or the slender **brachioradialis** (Figure 7.27), which supinates the manus.

Deep Forelimb Muscles: Lateral View (Figure 7.29)

Return to the levator scapulae ventralis. Remove the fat between it and the clavotrapezius, clean away the connective tissue that covers it (i.e., the part deep to the clavotrapezius), and follow it anteriorly as it passes from the scapula toward the skull. Clean away connective tissue deep to the levator and separate it from the underlying muscles. Once you have isolated the levator, you should easily be able to remove fat and connective tissue from the anterior margin of the acromiotrapezius and free the muscle from the underlying musculature. Continue posteriorly and separate the spinotrapezius from underlying muscles. Then cut through the middle of the trapezius muscles, at right angles to the fibers, and reflect their ends.

Examine the anterior part of the latissimus dorsi beneath the spinotrapezius, and then cut the latissimus about 3 cm posterior to the brachium and reflect it. Clean off the connective tissue from the underlying musculature. Examine the muscles on the lateral surface of the scapula. The **supraspinatus** fills the supraspinous fossa, whereas the **infraspinatus** occupies the infraspinous fossa. These muscles insert on the greater tuberosity of the humerus. The supraspinatus extends the humerus, and the infraspinatus rotates it laterally. The **teres major** arises from the scapula, posteroventral to the infraspinatus. It inserts medially on the humerus, flexing and medially rotating it. Follow the long head of the triceps as it passes deep to the spinodeltoid and infraspinatus and find the **teres minor**, which flexes and laterally rotates the humerus. It lies along the anteroventral margin of the infraspinatus, from which it is initially difficult to distinguish. In Figure 7.29 the spinodeltoid is shown as cut and reflected to expose the teres minor, but it is not necessary to cut the spinodeltoid.

Pull the dorsal border of the scapula laterally and examine the muscles in dorsal view. The fibers of the **rhomboideus** extend anteromedially between the dorsal border of the scapula and the middorsal line. This muscle pulls the scapula toward the vertebral column. The **rhomboideus capitis**, extending separately as a narrow band toward the back of the skull, rotates and pulls the scapula anteriorly. The rhomboideus capitis lies on the surface of the **splenius** (see later), a large, flat muscle covering the neck dorsolaterally. Look ventral to the rhomboideus and note the **serratus ventralis**. It is the large muscle composed from various slips that extend between the scapula and thorax. The slips arise from the ribs and converge to insert on the vertebral border of the scapula. The serratus ventralis will be considered again below.

Deep Forelimb Muscles: Lateral View with Forelimb Abducted (Figure 7.30)

Earlier, the latissimus dorsi was separated from the pectoralis minor. Now separate the latissimus dorsi from the lateral thoracic wall. You will find considerable fat along with connective tissue adhering to the muscle. Remove the fat and connective tissue, but do not injure the blood vessels and nerves in the axilla. At first the shoulder region will not be apparent, but it can be palpated. As you pick away tissue, the scapula and its associated musculature will come into view. When you have cleaned the musculature associated with the scapula, pull the scapula laterally to see a series of strap-like muscular slips fanning out to the thorax from the scapula. These slips are part of the serratus ventralis.

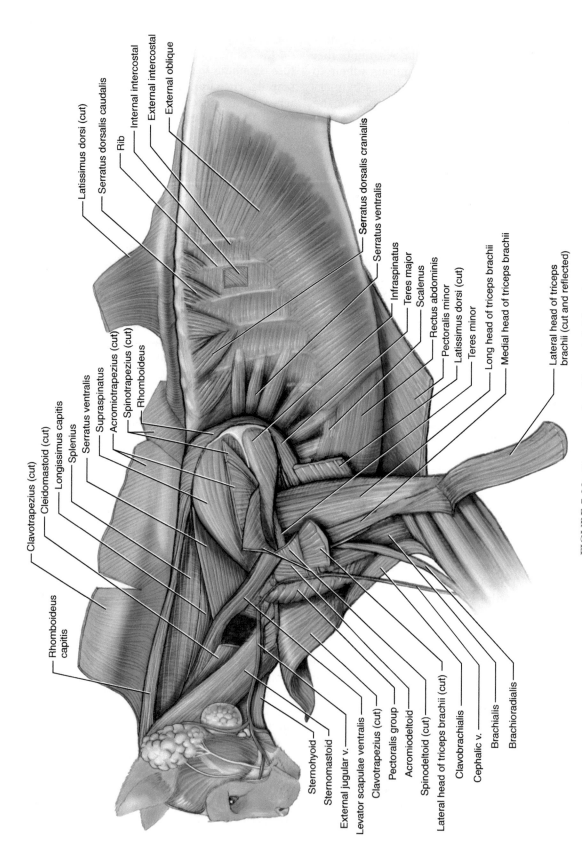

FIGURE 7.29 Deeper muscles of the cat in left lateral view.

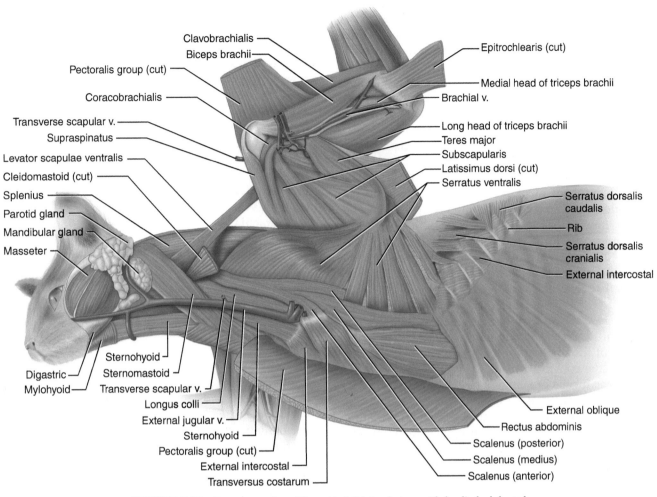

Clavobrachialis
Biceps brachii
Pectoralis group (cut)
Coracobrachialis
Transverse scapular v.
Supraspinatus
Levator scapulae ventralis
Cleidomastoid (cut)
Splenius
Parotid gland
Mandibular gland
Masseter

Epitrochlearis (cut)
Medial head of triceps brachii
Brachial v.
Long head of triceps brachii
Teres major
Subscapularis
Latissimus dorsi (cut)
Serratus ventralis
Serratus dorsalis caudalis
Rib
Serratus dorsalis cranialis
External intercostal

Digastric
Mylohyoid
Sternohyoid
Sternomastoid
Transverse scapular v.
Longus colli
External jugular v.
Sternohyoid
Pectoralis group (cut)
External intercostal
Transversus costarum

External oblique
Rectus abdominis
Scalenus (posterior)
Scalenus (medius)
Scalenus (anterior)

FIGURE 7.30 Deeper muscles of the cat in left lateral view, with forelimb abducted.

Using scissors and beginning posteriorly, make an anteroposterior cut through the pectoralis muscles about 1 cm to one side of the midline. Work your way first through the xiphihumeralis and pectoralis minor. As you do so, lift the musculature and clean away connective tissue beneath it to ascertain that are cutting through pectoralis musculature only. As you approach the pectoantebrachialis, lift the musculature and look for vessels and nerves passing out of the thorax and toward the shoulder. They may be embedded in some fat. Do not injure the vessels or nerves. The most apparent vessel is the **axillary vein**, a large vein injected with blue latex. It is accompanied by the **axillary artery**. These vessels quickly branch into several smaller vessels. The various large nerves are part of the brachial plexus.

Continue cutting through the pectoralis major and then reflect the flaps, removing the connective tissue that binds the xiphihumeralis and pectoralis minor to the thorax, so that you can reflect the pectoral musculature completely away from the thoracic wall. Then delicately clean away connective tissue and fat from around the vessels and nerves. The nerves appear as shiny, whitish strands (some rather large) passing toward the shoulder and brachium. Most lie dorsal to the vessels, but some are anterior to them.

If the musculature has been properly cleaned, it should be easy to follow the external jugular vein posteriorly. At the level of the anterior part of the shoulder, it receives a large tributary, the **transverse scapular vein** (possibly already cut). The external jugular continues posteriorly and joins the smaller internal jugular vein to form the bijugular trunk. The latter, in turn, joins the subclavian vein, which is the continuation of the axillary vein as it passes from the axilla into the thorax (see also Figure 7.54). When you have identified these vessels, which continue onto the forelimb, cut through the vessels, as well as nerves, leaving stumps about 1 cm long emerging from the thorax.

Position the brachium as shown in Figure 7.30, and continue cleaning the medial surface of the shoulder as well as the thoracic wall. Also, pick away connective tissue surrounding the vessels as they enter the shoulder musculature. Reidentify the latissimus dorsi, pectoralis muscles, and the serratus ventralis. Note how the fan-like arrangement of the serratus ventralis converges toward the dorsal border of the scapula. Note too that the posterior part of this muscle is clearly subdivided into separate slips, but its anterior fibers are arranged nearly as a continuous sheet.

The large muscle covering the medial surface of the scapula is the **subscapularis**. The fibers of its several bundles converge toward the apex of the scapula and insert on the lesser tuberosity of the humerus. It is an adductor of the humerus. The teres major initially appears as one of these bundles passing along the posterior border of the scapula. Anteriorly, however, the separation is much clearer: Note how some of the vessels pass between the subscapularis and teres major. Carefully observe the muscles that cover the ventromedial region of the scapula. Most of the fibers belong to the subscapularis and converge toward the apex, but the small, strap-like **coracobrachialis**, about 0.5cm in width, crosses against the direction of the fibers. It also adducts the humerus. In this view, the coracobrachialis lies just posterior to the origin of the **biceps brachii** (see next paragraph).

Examine the medial surface of the brachium. Cut through the epitrochlearis and the pectoral musculature (mainly pectoantebrachialis) covering the brachium, and reflect the ends to expose the muscles on the brachium. Clear the connective tissue. A large vessel, the **brachial vein**, which is a continuation of the axillary vein, should be prominently exposed. Dissect carefully near it to reveal the **brachial artery** (continuing from the axillary artery) and several nerves from the brachial plexus. The large muscle anterior to the vessels and nerves is the biceps brachii, and the muscle that lies posteriorly is the long head of the triceps, which has already been observed in lateral view. Dissect dorsally to the vessels and nerves, and push them aside to expose the **medial head** of the triceps. Follow the biceps brachii proximally toward its origin on the scapula and reidentify the coracobrachialis. The biceps inserts distally on the radius and is a flexor of the antebrachium.

Key Terms: Muscles of the Forelimb

(See Table 7.2 for synonyms)

acromiodeltoid
acromiotrapezius
axillary artery
axillary vein
biceps brachii
brachial artery
brachial vein
brachialis
brachioradialis
cephalic vein
clavobrachialis
clavotrapezius
cleidomastoid
coracobrachialis
epitrochlearis
external jugular vein
external oblique
infraspinatus
latissimus dorsi
levator scapulae ventralis
lumbodorsal fascia
mandibular gland
parotid gland
pectoantebrachialis
pectoralis major
pectoralis minor
rhomboideus
rhomboideus capitis
serratus ventralis
spinodeltoid
spinotrapezius
splenius
sternomastoid
subscapularis
supraspinatus
teres major
teres minor
transverse scapular vein
triceps brachii: lateral head; long head; medial head
xiphihumeralis

TABLE 7.2   Muscles of the Forelimb.

Name	Origin	Insertion	Main Actions
Acromiodeltoid	Posterior margin of acromion of scapula	Lateral surface of spinodeltoid muscle	Flexes and rotates humerus laterally
Acromiotrapezius	Middorsal line from neural process of axis to 4th thoracic vertebra, by aponeurosis	Metacromion process and anterior half of spine of scapula,	Adducts and stabilizes scapula
Biceps brachii	Small tubercle near dorsal margin of glenoid fossa of scapula, by tendon	Bicipital tuberosity of radius, by tendon	Flexes antebrachium

Continued

TABLE 7.2   Muscles of the Forelimb.—cont'd

Name	Origin	Insertion	Main Actions
Brachialis	Lateral surface of humerus	Lateral surface of ulna, just distal to semilunar notch	Flexes antebrachium
Brachioradialis	Midshaft of humerus	Styloid process of radius	Supinates manus
Clavobrachialis	Clavicle and raphe shared with clavotrapezius	Medial surface of ulna, just distal to semilunar notch	Flexes antebrachium
Clavotrapezius	Medial half of nuchal crest of skull and middorsal line up to neural process of axis	Clavicle and raphe shared with clavobrachialis	Draws scapula anterodorsally
Cleidomastoid	Mastoid process of temporal	Clavicle	Turns head when clavicle stabilized; draws clavicle anteriorly when head stabilized
Coracobrachialis	Coracoid process of scapula	Medial surface of proximal end of humerus	Adducts humerus
Epitrochlearis	Surface of latissimus dorsi	Olecranon process of ulna, by fascia	Extends antebrachium
Infraspinatus	Infraspinous fossa of scapula	Greater tuberosity of humerus	Rotates humerus laterally
Latissimus dorsi	Lumbodorsal fascia	Medial surface of proximal diaphysis of humerus	Draws humerus posterodorsally
Levator scapulae ventralis	Occipital bone and transverse process of atlas	Ventrally on metacromion and infraspinous fossa of scapula	Draws scapula anteriorly
Pectoantebrachialis	Manubrium of sternum	Fascia covering proximal surface of antebrachium	Adducts humerus
Pectoralis major	Anterior sternebrae	Pectoral ridge of humerus	Adducts humerus
Pectoralis minor	Body of sternum	Pectoral ridge of humerus	Adducts humerus
Rhomboideus	Posterior cervical and anterior thoracic vertebrae	Posterior part of dorsal border of scapula	Draws scapula toward vertebral column
Rhomboideus capitis	Medial portion of nuchal crest	Anterior part of dorsal border of scapula	Rotates and draws scapula anteriorly
Serratus ventralis cervicis	Transverse processes of 3rd to 7th cervical vertebrae	Medial surface of scapula, near dorsal border	Draws scapula anteroventrally
Serratus ventralis thoracis	Lateral surface of first 9 or 10 ribs	Medial edge of scapula, near dorsal border	Draws scapula ventrally, helps support trunk on forelimb
Spinodeltoid	Middle third of spine of scapula	Deltoid ridge of humerus	Flexes and laterally rotates humerus
Spinotrapezius	Middorsal line from neural processes of most thoracic vertebrae	Tuberosity of spine of scapula; fascia of supraspinatus and infraspinatus mm	Draws scapula posterodorsally
Subscapularis	Subscapular fossa of scapula	Lesser tuberosity of humerus	Adducts humerus
Supraspinatus	Supraspinous fossa of scapula	Greater tuberosity of humerus	Extends humerus
Teres major	Dorsal third of posterior border of scapula	Medial surface of humerus, by tendon in common with latissimus dorsi	Flexes and medially rotates humerus
Teres minor	Posterior border of scapula, just distal to glenoid fossa	Greater tuberosity of humerus	Flexes and laterally rotates humerus
Triceps brachii Lateral head Long head Medial head	Deltoid ridge of humerus Posterior border of scapula, near glenoid fossa Shaft of humerus	All 3 heads of the triceps brachii insert on the olecranon process of ulna by a common tendon	All 3 heads of the triceps brachii act to extend antebrachium
Xiphihumeralis	Xiphoid process of sternum	Near distal border of bicipital groove of humerus	Adducts forelimb

Muscles of the Hind Limb

Table 7.3 lists the hind limb muscles and indicates their origin, insertion, and main functions.

Superficial Hind Limb Muscles: Lateral View (Figure 7.31)

The skin of the hind limb has already been removed. Carefully pick away the connective tissue and fat on the lateral surface of the hind limb and the base of the tail, but take care not to injure the fascia that covers the musculature (Figure 7.31a). Considerable fat is present in the popliteal fossa, the depression posterior to the knee joint. Carefully clear this region. You may notice the oval popliteal lymph node, which is embedded within the fat. The fat and lymph node are illustrated in Figure 7.31b.

Examine the musculature. The most conspicuous muscle of the thigh is the **biceps femoris**. This large muscle covers much of the posterior half of the thigh and is a main adductor and extensor of the thigh. It has a narrow origin dorsally from the ischial tuberosity but widens distally to insert by an aponeurosis on the patella and the proximal portion the tibia. The posterior portions of two deeper muscles, the **semitendinosus** and **semimembranosus**, are exposed posterior to the dorsal part of the biceps femoris.

Next, identify the **sartorius**, which lies along the anterior edge of the thigh. In lateral view this muscle appears to be elongated and narrow, but its more extensive medial portion will be seen shortly. Posterior to the dorsal portion of the sartorius is the **gluteus medius**, which is covered by fascia. Carefully remove the fascia to uncover the muscle, as shown in Figure 7.31a. Posterior to it is the smaller **gluteus superficialis**, followed by the **gluteofemoralis**. The latter is a strap-like muscle that continues distally, deep to the biceps femoris. The gluteal musculature generally functions in abducting the thigh. Extending between the origin of the biceps femoris and the posterior margin of the gluteofemoralis, identify the **obturatorius internus**. Dorsal to it, extending from beneath the posterodorsal margin of the gluteofemoralis to the ventral surface of the tail, is the **coccygeus**.

The **tensor fasciae latae**, a flexor of the thigh, is ventral to the gluteal musculature noted earlier. From its narrow dorsal origin, it fans out distally and inserts on the **fascia lata**, the tough membrane that extends toward the knee joint, anterior to the biceps femoris. The tensor fasciae latae is subdivided into a longer anterior portion and a triangular posterior portion. Trace the anterior portion as it extends deep to the sartorius and onto the anterior surface of the thigh.

The **tenuissimus** is a ribbon-like muscle that lies almost entirely deep to the biceps femoris. It will be exposed presently. Its very distal end, however, is exposed in lateral view along the posterodistal edge of the biceps femoris. Look for this exposed portion. It inserts by a short tendon that merges with that of the biceps femoris. When you have identified it, lift the posterior edge of the biceps femoris and gently scrape the loose connective tissue adhering to its deep surface to locate the rest of the tenuissimus. It extends dorsally and passes deep to the very dorsal portion of the gluteofemoralis (see Figure 7.33). It may serve as an extensor of the thigh or flexor of the crus, but its composition (a high number of muscle spindles) and size suggest that it may function as a tension sensor.

Superficial Hind Limb Muscles: Medial View (Figure 7.32)

Clear away connective tissue from the medial surface of the thigh, but leave the crural fascia for the time being. Avoid damaging vessels and nerves. Note again the wad of fat that lies between the body wall and the proximal surface of the thigh. In males the spermatic cords extend through this wad as they pass from the body wall to the testes. Dissect carefully to avoid injuring the spermatic cords. They are shown as having been dissected out in Figure 7.33a, but you do not need to expose them to the same extent.

Once the surface of the thigh has been cleared of connective tissue, the full extent of the sartorius, already observed in lateral view, may be appreciated. It covers the anteromedial surface of the thigh. The large muscle on the posteromedial surface is the **gracilis** muscle, which flexes the crus and adducts the thigh. The distal portion of the semitendinosus is exposed just posterior to the gracilis. Distally the sartorius and gracilis converge and lie side by side, with the former partially covering the latter, as they extend toward the knee. Examine the triangular region formed between their proximal portions and the body wall. The **femoral artery** and **femoral vein** are the conspicuous vessels emerging from within the body cavity and extending distally. The small muscle just anterior (keep in mind that the limb is essentially parasagittal in position) to the very proximal end of these vessels is the **iliopsoas**, which flexes and rotates the thigh. The **femoral nerve** lies anterior to the femoral vessels and passes over the iliopsoas. Probe between the iliopsoas and sartorius to identify the **vastus medialis**, which will be exposed shortly. Posterior to the vessels is the **pectineus**, a small, triangular adductor of the thigh, followed by the larger **adductor longus**, and the much larger **adductor femoris**, much of which lies deep to the gracilis. The last two muscles are also adductors of the thigh (but see also Table 7.3).

Examine the vessels more closely. They will be discussed in more detail later, but it is worth noting the main branching patterns at this time. The femoral artery and vein are the main continuations of the **external iliac artery** and **external iliac vein**, which lie within the body cavity (see pages 263 and 264). The femoral artery and vein are therefore the most conspicuous vessels on the

7

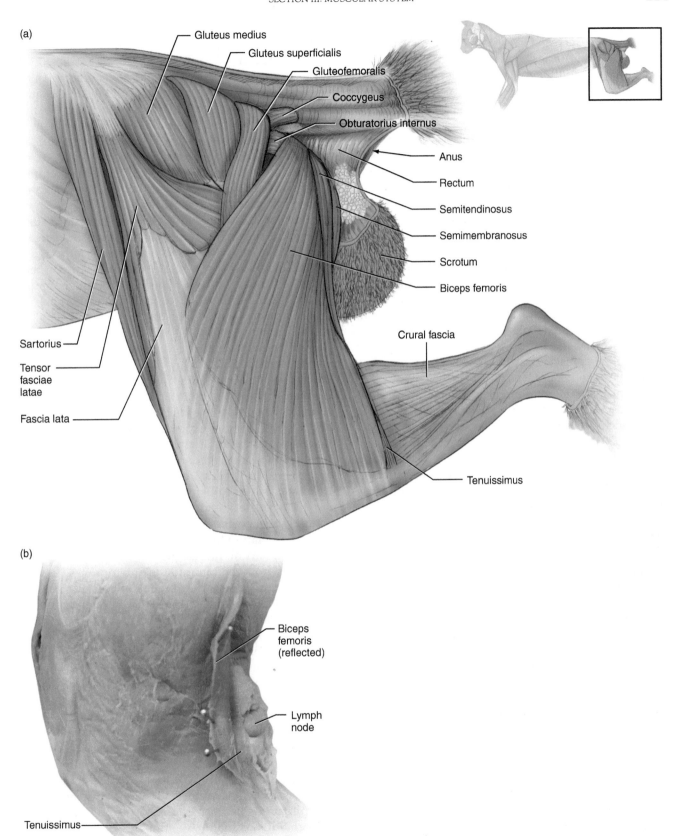

(a)

Gluteus medius

Gluteus superficialis

Gluteofemoralis

Coccygeus

Obturatorius internus

Anus

Rectum

Semitendinosus

Semimembranosus

Scrotum

Biceps femoris

Crural fascia

Sartorius

Tensor fasciae latae

Fascia lata

Tenuissimus

(b)

Biceps femoris (reflected)

Lymph node

Tenuissimus

FIGURE 7.31   Superficial muscles of the left hind limb of the cat in (a) lateral view and (b) detail of knee region.

medial surface of the thigh. However, the external iliac vessels branch just as they approach the body wall. The external iliac artery gives rise to the femoral artery and the **deep femoral artery**, and the external iliac vein is formed by the confluence of the femoral vein and the **deep femoral vein**. It is these four vessels that perforate the body wall onto the thigh. As noted, the femoral vessels are easily observed, but the deep femoral vessels may also be identified. Pick away the body musculature adjacent to the femoral artery, working your way toward the abdominal midline. The much smaller deep femoral artery, extending posteromedially, will be exposed. In many individuals, one of its branches is the **external pudendal artery**, which extends medially into the wad of fat in the groin region to supply the external genitalia. The **external pudendal vein** follows the artery (Figure 7.32a). In males the artery and vein pass dorsal to the spermatic cord. Follow the deep femoral artery, after the origin of the external pudendal artery, as it passes into the musculature. The deep femoral vein follows the artery but lies more deeply. In Figure 7.32b the femoral vein has been pulled anteriorly and the adductor longus and pectineus muscles have been pulled posteriorly to expose the deep femoral vein.

Return to the femoral artery and femoral vein. The **lateral circumflex artery** and the **lateral circumflex vein** branch from these vessels a short distance distally (Figure 7.32a). Note the **saphenous nerve** that accompanies the vessels. More distally, nearer the knee, the **saphenous artery** and **saphenous vein** branch from the femoral vessels and extend distally on the surface of the crural fascia. The saphenous nerve accompanies them.

Deep Hind Limb Muscles: Lateral View
(Figures 7.33–7.35)

Examine the lateral surface of the thigh. Continue to scrape away the connective tissue and fat from between the deep surface of the biceps femoris (working from both its anterior and posterior margins) and the underlying musculature, and lift the biceps from the musculature. Make an anteroposterior cut through the central part of the biceps femoris and reflect its ends. The exposed surface must be cleared of connective tissue and fat to produce a dissection similar to those presented in Figure 7.33. Although this is tedious, it furnishes an excellent example of how connective tissue and fat are distributed between muscles, both to provide a plane along which muscles may move against each other and to provide protection for structures such as nerves. In clearing the region, preserve the conspicuous **ischiadic nerve** but remove or cut through blood vessels.

The ischiadic nerve extends nearly parallel to the tenuissimus muscle, illustrated in different positions in Figures 7.33a and b). Note the distal end of the gluteofemoralis. Its fibers insert into a narrow, tapered tendon,

which is nearly as long as the fleshy part of the muscle and extends toward the knee joint. The tendon, shown passing over the tip of the pin in Figure 7.33a, may be distinguished most easily distally, where it lies on the posterior surface of the fascia lata, and appears as a narrow, glistening ribbon. Separate it from the fascia lata.

Reflection of the biceps femoris also exposes two muscles already identified, the semitendinosus and, anterior to it, the semimembranosus. The **adductor femoris** lies anterior to the latter, deep to the ischiadic nerve. Just anterior to the origin of the biceps femoris and ventral to the obturatorius internus, identify the **gemellus caudalis** and the **quadratus femoris**.

Poke a small opening into the fascia lata, ventral to the level of the tensor fasciae latae. Insert a scissor blade into the opening, bisect the fascia anteroposteriorly, and lift it from the underlying muscle. Cut along the posterior edges of the fascia where it meets the adductor femoris (avoid cutting the tendon of the gluteofemoralis), and then reflect its flaps, as illustrated in Figure 7.33b. The tensor fasciae latae will be reflected with the proximal flap of the fascia lata. The large muscle exposed is the **vastus lateralis**. Clean away connective tissue along the anterodorsal margin of this muscle to expose a portion of the **rectus femoris**. The tendon of the **gluteus profundus** (see later) should also be visible. The **vastus intermedius** has a small, narrow exposure along the posterodistal margin of the vastus lateralis. Gently pull the tendon of the gluteofemoralis anteriorly, as shown in Figure 7.33b, to observe the vastus intermedius. The vastus lateralis, vastus intermedius, vastus medialis (identified earlier, but see later), and rectus femoris form the muscle complex termed the **quadriceps femoris**, which is the main musculature that extends the crus. The **capsularis** is a small, narrow muscle just anterior to the hip joint. It extends from the ilium to the anteroproximal surface of the femur. To locate it, probe between the origin of the vastus lateralis and the posteroventral margin of the gluteus profundus (Figure 7.34). Its small size and high proportion of muscle spindles suggest it functions as a tension sensor, although its position indicates a thigh flexor function.

Cut anteroposteriorly through the belly of the gluteofemoralis and reflect it, along with the tenuissimus. Free the gluteus superficialis and gluteus medius muscles from underlying musculature. Cut anteroposteriorly through the center of each muscle and reflect them, as shown in Figure 7.34, to expose the underlying muscles. The gluteus minimus (noted earlier), with its conspicuous insertion tendon, is the more anterior. The **piriformis** lies more posteriorly and largely covers the **gemellus cranialis**, two muscles that act mainly to abduct the thigh.

Reflect the gluteofemoralis, as shown in Figure 7.35. Carefully cut through the middle of the quadratus

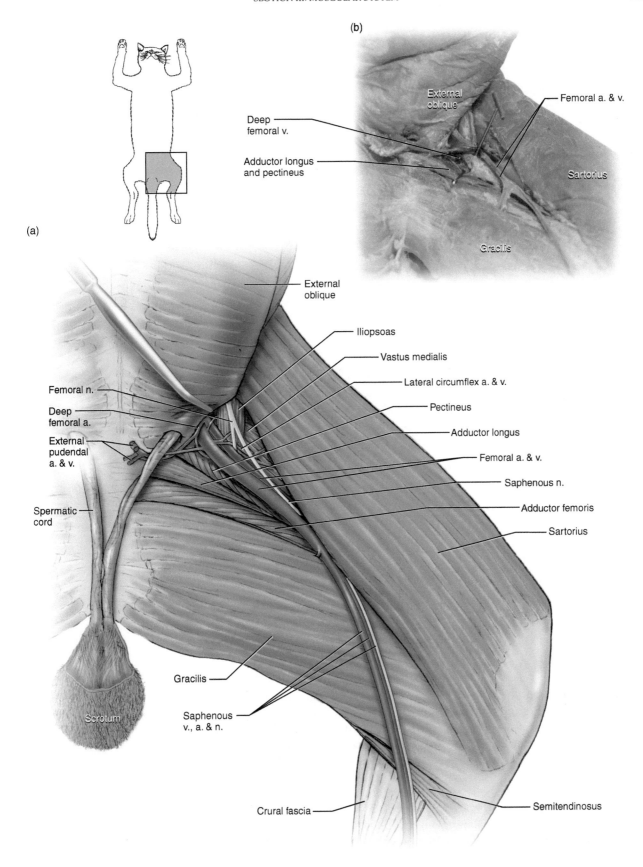

(b)

External
oblique

Deep
femoral v.

Femoral a. & v.

Adductor longus
and pectineus

Sartorius

Gracilis

(a)

External
oblique

Iliopsoas

Vastus medialis

Lateral circumflex a. & v.

Femoral n.

Pectineus

Deep
femoral a.

Adductor longus

External
pudendal
a. & v.

Femoral a. & v.

Saphenous n.

Adductor femoris

Spermatic
cord

Sartorius

Gracilis

Saphenous
v., a. & n.

Scrotum

Semitendinosus

Crural fascia

FIGURE 7.32 Superficial muscles of the left hind limb of the cat in (a) medial view and (b) showing detail of vessels.

(a)

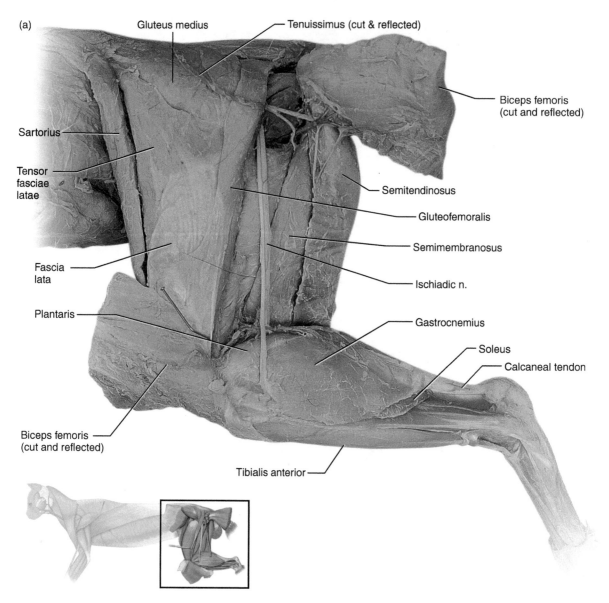

Gluteus medius

Tenuissimus (cut & reflected)

Biceps femoris
(cut and reflected)

Sartorius

Tensor
fasciae
latae

Semitendinosus

Gluteofemoralis

Semimembranosus

Fascia
lata

Ischiadic n.

Plantaris

Gastrocnemius

Soleus

Calcaneal tendon

Biceps femoris
(cut and reflected)

Tibialis anterior

FIGURE 7.33   Deep muscles of the left hind limb of the cat in lateral view, with (a) and (b) showing sequential reflection of musculature.

femoris, but do not injure underlying musculature. Continue cutting through the gemellus caudalis and obturatorius internus. Refer back to Figure 7.34 for the position of these muscles. Reflect the ends of the muscles you have cut through to expose the **obturatorius externus**, which extends anteroposteriorly (Figure 7.35).

Examine the muscles on the lateral surface of the crus by removing the crural fascia (Figure 7.31). There are actually several layers of fascia in the region, and they have been separated and partially cut away to produce the dissection in Figure 7.33 (as well as Figures 7.34 and 7.36), which preserves the insertions of the biceps femoris and tenuissimus. The most conspicuous muscle on the crus is the **gastrocnemius**, which lies posteriorly and has lateral

and medial heads. Its proximal portion is very thick, but it tapers distally into its tendon, the gastrocnemius tendon. A smaller muscle, the **soleus**, lies mainly deep to the gastrocnemius, but a small portion is exposed anterior to it (the soleus will also be noted later in medial view). It is also thick proximally and tapered distally. The tendons of the gastrocnemius and soleus muscles pass together through a sheath formed by the fascia as the **calcaneal tendon**, which inserts on the proximal end of the calcaneum. As their tendons converge, the soleus and gastrocnemius, which are the main extensors of the pes, are collectively termed the **triceps surae** muscle. A portion of the **plantaris** is exposed proximal to the gastrocnemius (Figure 7.33). This muscle will be observed later in medial view.

(b)

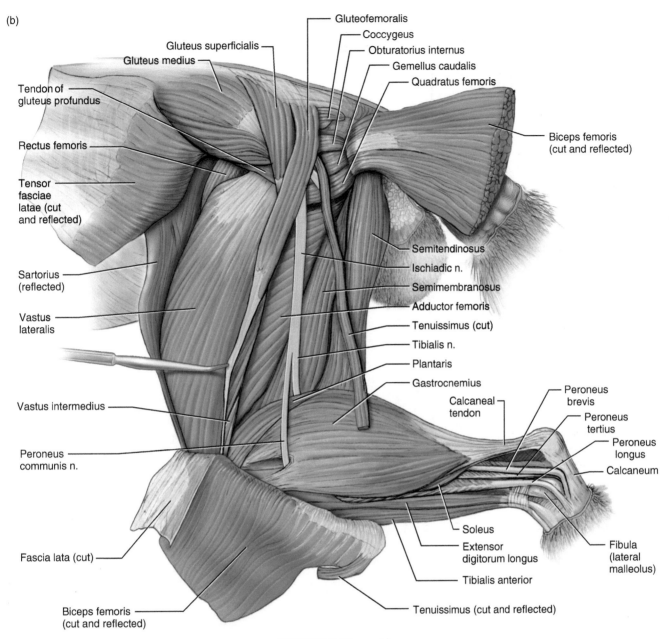

FIGURE 7.33, cont'd

The **tibialis anterior** lies on the anterior surface of the crus (Figures 7.33 and 7.36). Posterior to it is the slender, elongated **extensor digitorum longus** (Figure 7.33b). Follow the latter distally into its thick tendon on the dorsal surface of the pes. Between the extensor digitorum longus and the soleus is the peroneus musculature. You may note the prominent peroneus superficialis nerve (which is not illustrated but is a branch of the **peroneus communis nerve**; see Figures 7.33b and 7.34) that lies between the extensor digitorum longus and the peroneus muscles. Three peroneus muscles may be recognized. The **peroneus longus** is an elongated muscle lying posterior to the extensor digitorum longus. Follow it distally as it tapers into its prominent tendon, which passes along a groove on the lateral surface of the lateral malleolus of the fibula (see page 206). The **peroneus tertius** is a smaller muscle posterior to the peroneus longus. It may be difficult to isolate, but its narrow tendon lies posterior to the tendon of the peroneus longus. The **peroneus brevis** is larger and exposed posterior to the peroneus tertius. Its tendon is much larger and easily identified. The tendons of the peroneus tertius and peroneus brevis extend distally adjacent to each other, and both pass along a groove on the posterior surface of the lateral malleolus. As they pass distally, the tendon of the peroneus brevis obscures that of the peroneus tertius.

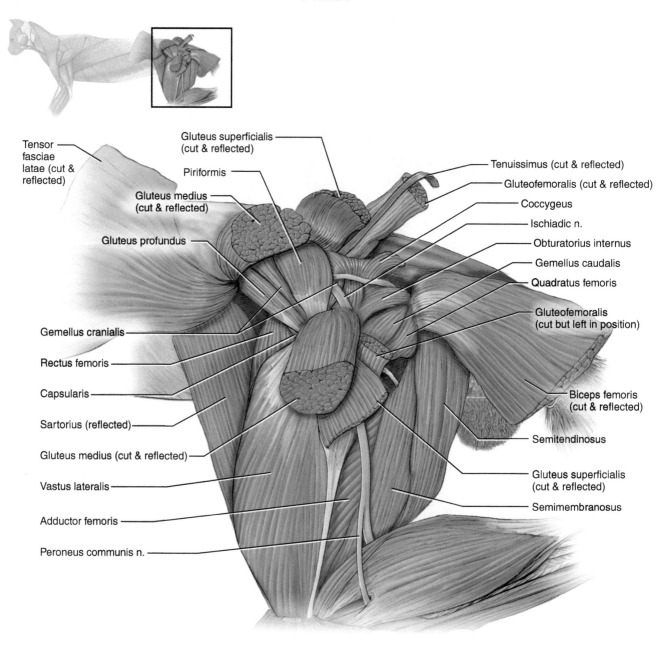

FIGURE 7.34   Deep muscles of the left hind limb of the cat in lateral view.

Deep Hind Limb Muscles: Medial View (Figure 7.36)

Return to the medial surface of the hind limb. Lift the sartorius by separating it from underlying musculature. Cut across its central portion and reflect its ends. Do likewise for the gracilis. This will expose much of the musculature that has already been identified. Anteriorly, identify the vastus medialis. Much of its anterior margin is bordered by the rectus femoris. Recall that the tensor fasciae latae and fascia lata, which was cut earlier (as shown in Figures 7.34 and 7.36), lie on the vastus lateralis. More posteriorly, note the adductor femoris and semimembranosus, both now uncovered. The semitendinosus inserts

on the tibia just distal to the insertion of the semimembranosus. The lateral-to-medial orientation of these last two muscles and their contribution to the medial wall of the popliteal fossa is easily appreciated.

Remove the crural fascia from the medial surface of the crus. Clear away connective tissue. The medial head of the gastrocnemius is the conspicuous muscle posteriorly. Probe deep to its anterior margin, where the gastrocnemius begins to taper distally, to locate the soleus, already observed in lateral view, and the plantaris (a portion of which was noted earlier in lateral view), lying between the soleus and gastrocnemius. The tendon of

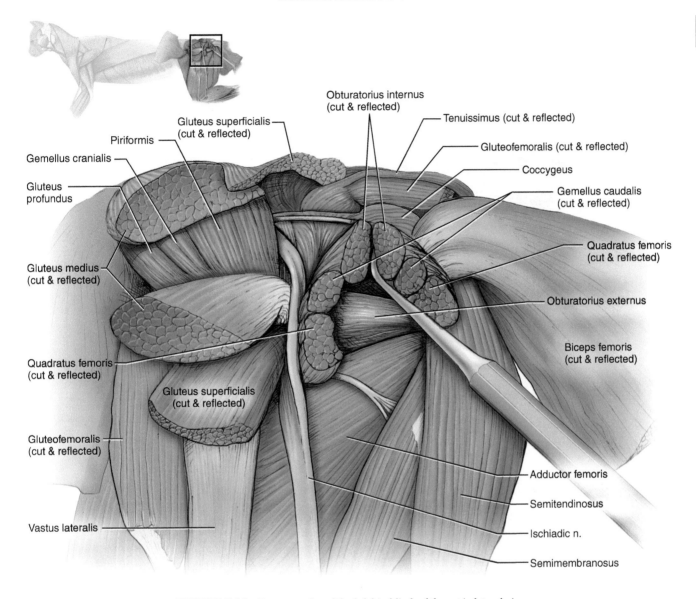

FIGURE 7.35   Deep muscles of the left hind limb of the cat in lateral view.

the plantaris extends through the crural sheath with the calcaneal tendon. It continues around the proximal end of the calcaneum onto the plantar surface of the pes, and extends distally onto the digits. Here, the small flexor digitorum brevis muscle (not illustrated) surrounds the tendon and contributes to flexion of the digits.

The large **tibialis nerve** emerges from beneath the musculature. Carefully dissect around the nerve to free it from surrounding structures, and follow it as it extends toward and then around the posterior margin of the medial malleolus of the tibia (see page 206). The medial surface of the tibia is the white strip of bone extending proximodistally near the anterior margin of the crus. On its anterior surface is the tibialis anterior, identified earlier. Follow it distally into its tendon and note that it extends medially across the proximal surface of the pes. Examine the musculature

posterior to the tibia. Near the insertion of the semitendinosus, locate the popliteus, consisting of a faint, thin band of fibers extending posterodorsally from the tibia. The popliteus, which flexes the crus and rotates the tibia medially, curves laterally and crosses the posterior side of the knee joint to its origin on the lateral epicondyle of the femur. The **flexor digitorum longus** is the larger mass of musculature just posterior to the tibia. Follow it distally. Its tendon is nearly parallel and anterior to the tibialis nerve, and is of similar thickness. The tendon curves around the posterior margin of the medial malleolus onto the medial surface of the pes. Anterior to the tendon of the flexor digitorum longus is the thicker tendon of the **tibialis posterior** muscle. As the tendon of the latter muscle extends proximally, it passes deep to that of the flexor digitorum longus. Follow the tendon of the tibialis posteriorly, pushing the

(a)

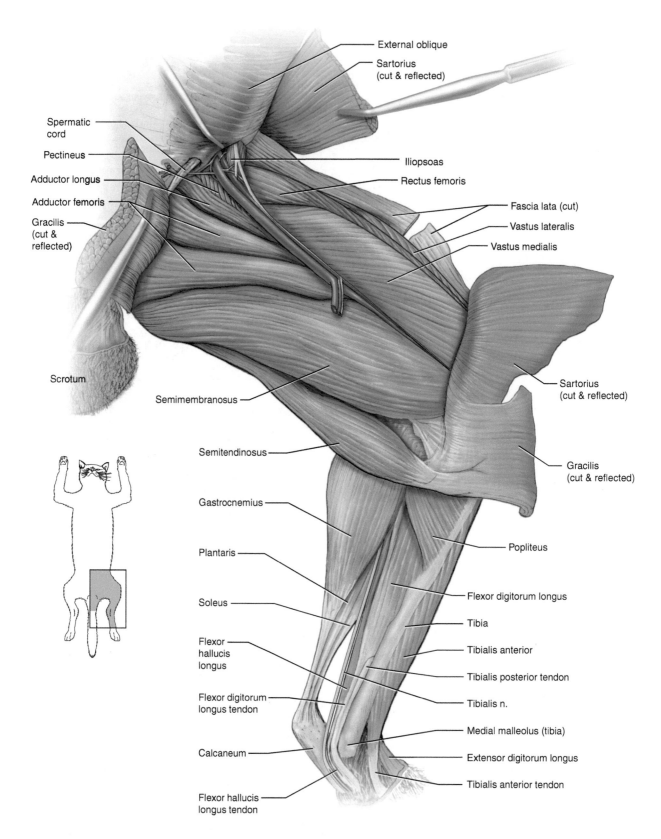

FIGURE 7.36 Deep muscles of the left hind limb of the cat in medial view, with (a) showing the thigh and crural musculature, and (b) showing detail of the crural musculature.

(b)

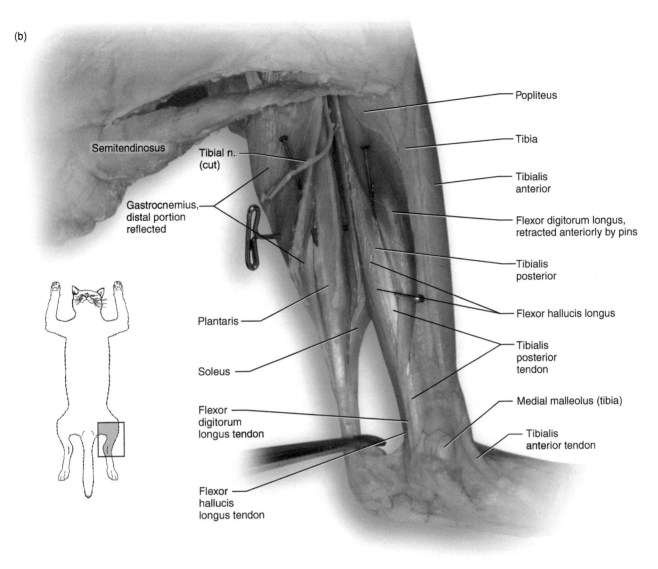

Popliteus

Tibia

Tibialis anterior

Flexor digitorum longus, retracted anteriorly by pins

Tibialis posterior

Flexor hallucis longus

Tibialis posterior tendon

Medial malleolus (tibia)

Tibialis anterior tendon

Semitendinosus

Tibial n. (cut)

Gastrocnemius, distal portion reflected

Plantaris

Soleus

Flexor digitorum longus tendon

Flexor hallucis longus tendon

FIGURE 7.36, cont'd

belly of the flexor digitorum longus anteriorly, to reveal more completely the tibialis posterior, as shown in Figure 7.36b. As you may now appreciate, the fibers of the tibialis posterior lie almost entirely deep to the flexor digitorum longus. The tendon of the tibialis posterior also curves around the malleolus onto the medial surface of the pes. Posterior to the tibialis posterior is the **flexor hallucis longus**. Follow it distally as it passes posterior, still largely fleshy, to the tendon of the flexor digitorum longus. Its tendon becomes apparent more distally.

Key Terms: Muscles of the Hind Limb

adductor femoris
adductor longus (adductor femoris longus)
biceps femoris
calcaneal tendon (Achilles tendon)

capsularis (articularis coxae)
coccygeus (abductor caudae internus)
deep femoral artery
deep femoral vein
extensor digitorum longus
external iliac artery
external iliac vein
external pudendal artery
external pudendal vein
fascia lata
femoral artery
femoral nerve
femoral vein
flexor digitorum longus
flexor hallucis longus
gastrocnemius
gemellus caudalis (gemellus inferior)

gemellus cranialis (gemellus superior)
gluteofemoralis (caudofemoralis, coccygeofemoralis, gluteobiceps)
gluteus medius
gluteus profundus (gluteus minimus)
gluteus superficialis (gluteus maximus)
gracilis
iliopsoas
ischiadic nerve (ischiatic nerve, sciatic nerve)
lateral circumflex artery
lateral circumflex vein
obturatorius externus
obturatorius internus
pectineus
peroneus brevis (fibularis brevis)
peroneus communis nerve (common peroneal nerve, fibular nerve)
peroneus longus (fibularis longus)
peroneus tertius (fibularis tertius)
piriformis (pyriformis)
plantaris (flexor digitorum brevis superficialis)

popliteus
quadratus femoris
quadriceps femoris (= rectus femoris + vastus intermedius + vastus lateralis + vastus medialis)
rectus femoris
saphenous artery
saphenous nerve
saphenous vein
sartorius
semimembranosus
semitendinosus
soleus
tensor fasciae latae
tenuissimus (abductor cruris caudalis)
tibialis anterior (tibialis cranialis)
tibialis nerve (tibial nerve)
tibialis posterior (tibialis caudalis)
triceps surae (= gastrocnemius + soleus)
vastus intermedius
vastus lateralis
vastus medialis

TABLE 7.3	Muscles of the Hind Limb.

Name	Origin	Insertion	Main Actions
Adductor femoris	Pubis and ischium	Shaft of femur (linea aspera)	Adducts and extends thigh
Adductor longus/adductor femoris longus	Anterior margin of pubis	Shaft of femur (linea aspera)	Adducts and flexes thigh
Articularis coxae/capsularis	Ilium	Anterior surface of femur	Flexes thigh
Biceps femoris	Ischial tuberosity	By broad aponeurosis to patella and proximal half of tibia	Flexes crus, abducts thigh
Coccygeus/abductor caudae internus	Dorsal margin of ilium and ischium	2nd to 5th caudal vertebrae	Flexes tail
Extensor digitorum longus	Lateral epicondyle of femur	Distal phalanges of digits 2 to 5, by tendon that branches into four portions	Extends digits 2 to 5 and flexes pes
Flexor digitorum longus	Proximal portion of fibula and tibia	Joins tendon of flexor hallucis longus to distal phalanges of digits 2 to 5	Flexes digits 2 to 5 and extends pes
Flexor hallucis longus	Proximal portion of fibula and tibia	Joins tendon of flexor digitorum longus to distal phalanges of digits 2 to 5	Flexes digits 2 to 5 and extends pes
Gastrocnemius	Lateral and medial epicondyles of femur and patella	By tendon onto proximal end of calcaneum, together with tendon of soleus muscle	Extends pes; flexes crus
Gemellus caudalis/inferior	Dorsolateral surface of ischium, anterior to ischial tuberosity	Tendon of obturatorius internus	Abducts thigh
Gemellus cranialis/superior	Dorsal margin of ischium and ilium	Greater trochanter of femur	Abducts and outwardly rotates thigh
Gluteofemoralis/ caudofemoralis/ coccygeofemoralis/ gluteobiceps	Anterior caudal vertebrae	Patella	Abducts thigh and extends crus
Gluteus superficialis/ maximus	Posterior sacral and anterior caudal vertebrae, and gluteal fascia	Distal part of greater trochanter of femur	Abducts thigh
Gluteus medius	Posterior sacral and anterior caudal vertebrae, dorsolateral surface of ilium, and sacral fascia	Proximal part of greater trochanter of femur	Abducts thigh

Continued

TABLE 7.3   Muscles of the Hind Limb.—cont'd

Name	Origin	Insertion	Main Actions
Gluteus profundus/minimus	Ventrolateral surface of ilium	Lateral part of greater trochanter of femur	Abducts and outwardly rotates thigh
Gracilis	Pubic and ischial symphyses	Proximomedial surface of tibia, crural fascia	Adducts thigh, flexes crus
Iliopsoas	Lumbar and posterior thoracic vertebrae, and ilium	Lesser trochanter of femur	Flexes and rotates thigh
Obturatorius externus	Ischium and pubis	Trochanteric fossa of femur	Flexes thigh
Obturatorius internus	Ischium and connective tissue covering obturator foramen	Trochanteric fossa of femur	Abducts thigh
Pectineus	Anterior margin of pubis	Surface of femur, just distal to lesser trochanter	Adducts thigh
Peroneus brevis/fibularis brevis	Distal half of fibula	Lateral surface of metatarsal 5	Extends pes
Peroneus longus/fibularis longus	Proximal half of fibula	By tendon onto proximal part of metatarsals 2 to 4	Flexes and everts pes
Peroneus tertius/fibularis tertius	Lateral surface of fibula	Tendon of extensor digitorum longus	Extends and abducts 5th digit, flexes pes
Piriformis/pyriformis	Posterior sacral and anterior caudal vertebrae	Greater trochanter of femur	Abducts thigh
Plantaris/flexor digitorum brevis superficialis	Lateral epicondyle of femur and patella	By tendon extending past proximal end of calcaneum and onto tendon of flexor digitorum brevis	Extends pes (and through flexor digitorum brevis, flexes digits 2 to 5)
Popliteus	Lateral epicondyle of femur	Proximal third of medial surface of tibia	Flexes crus and rotates tibia medially
Quadratus femoris	Ischial tuberosity	Greater trochanter and lesser trochanter of femur	Extends and outwardly rotates thigh
Rectus femoris	Ventral margin of ilium anterior to acetabulum	Lateral surface of patella	Extends crus
Sartorius	Iliac crest and anteromedial margin of ilium	Proximomedial surface of tibia, and patella	Adducts thigh, contributes to extension of crus
Semimembranosus	Ischial tuberosity and posterior margin of ilium	Distomedial surface of femur	Extends thigh, flexes crus
Semitendinosus	Ischial tuberosity	Proximomedial surface of tibia	Flexes crus, extends thigh
Soleus	Proximal third of fibula	Proximal end of calcaneum (with tendon of gastrocnemius)	Extends pes
Tensor fasciae latae	Anteroventral surface of ilium	Fascia lata, which merges with proximal part of aponeurosis of biceps femoris (see above)	Flexes thigh
Tenuissimus/abductor cruris caudalis	Anterior caudal vertebra	Crural fascia and tibia, with aponeurosis of biceps femoris	May serve as a tension sensor; effectiveness as extensor of thigh or flexor of crus is probably minimal
Tibialis anterior/cranialis	Proximolateral surface of tibia and proximomedial surface of fibula	Metatarsal 1	Flexes pes
Tibialis posterior/caudalis	Proximal end of tibia and fibula	Navicular and intermediate cuneiform (tarsals)	Extends pes
Vastus intermedius	Anterior surface of femur	Patella (capsule of knee joint)	Extends crus
Vastus lateralis	Greater trochanter and dorsolateral surface of femur	Lateral surface of patella	Extends crus
Vastus medialis	Shaft of femur	Medial surface of patella (and patellar ligament)	Extends crus

Muscles of the Head and Trunk

Table 7.4 lists the head and trunk muscles, and indicates their origin, insertion, and main functions.

Muscles of the Trunk (Figures 7.28–7.30)

The musculature forming the abdominal wall is arranged mainly as three thin, but extensive, muscular sheets or layers (Figure 7.28). The most superficial layer is the **external oblique** (which you noted previously), the middle layer is the **internal oblique**, and the deep layer is the **transversus abdominis**. A fourth muscle, the **rectus abdominis**, also contributes to the abdominal wall. It forms a ventral muscular band extending anteroposteriorly between the sternum and pelvis (Figure 7.30). These four muscles act to constrict the abdomen. The rectus abdominis can also draw the ribs and sternum posteriorly, flexing the trunk.

The fibers of the external oblique extend posteroventrally. They do not extend all the way to the midventral line but attach by way of an aponeurosis, which covers the deeper musculature of the ventral surface of the abdomen. To see the deeper muscles, cut a small flap in the external oblique, in the central part of the abdomen, as in Figure 7.28. Reflect the muscle and aponeurosis. You can now see the middle layer, the internal oblique. Its fibers extend anteroventrally, nearly at right angles to those of the external oblique. Note, though, that the internal oblique also does not extend to the midventral line, but is continued by an aponeurosis. The muscle lying deep to the aponeuroses is the rectus abdominis, and it will be seen clearly when the internal oblique is cut and reflected. To do so, cut a similar flap in the internal oblique and reflect the muscle and aponeurosis. The muscle exposed is the transversus abdominis, the deepest layer. In this region of the abdomen the ventral end of this muscle dips deep to the rectus abdominis. Lift the lateral margin of the rectus abdominis. The transversus abdominis is also continued to the midventral line by an aponeurosis. The relative positions of the transversus abdominis and rectus abdominis muscles change more posteriorly. There, the aponeurosis of the former divides so that part of it also covers the rectus abdominis.

Next examine muscles associated with the thorax. The **scalenus** (Figure 7.30) lies laterally on the thorax, along the ventral parts of the serratus ventralis. Posteriorly, it is subdivided into three narrow bands, with the middle part usually extending farthest posteriorly. Anteriorly, they unite into a single band that passes along the neck. The **transversus costarum** is a small, thin sheet. Its fibers extend anterodorsally from the sternum toward the ventral band of the scalenus, where it inserts on the first rib. The **longus colli** is a long, narrow muscle situated ventral to the cervical vertebrae. It extends anteroposteriorly between the anterior part of the scalenus and the external jugular vein. Note also the longitudinal muscle

ventral to the posterior part of the external jugular. It is the posterior end of the **sternohyoid muscle**, which will be considered again later.

The thoracic wall is formed mainly by musculature that extends between successive ribs. As in the abdominal region, there are three main muscles. Most superficially are the **external intercostals** (Figure 7.29), followed by the **internal intercostals**, and most medially, the **transversus thoracis**. Of these, only the internal intercostals extend from middorsally to midventrally. To see the first two muscles, examine the thoracic wall posterior to the serratus ventralis. The muscle fibers extending slightly posteroventrally between successive ribs are external intercostals. Carefully cut perpendicularly through the fibers of one set of external intercostals to expose the internal intercostals, the fibers of which extend steeply posterodorsally. The external intercostals do not extend all the way to the midventral line, but end at about the level of the lateral margin of the rectus abdominis. Locate the latter muscle and lift its lateral portion away from the thoracic wall. Near the midventral line, note that the internal intercostals are exposed between the ribs. The transversus thoracis is present only near the midventral line, deep to the internal intercostals. Its fibers extend almost transversely, with a slight posteroventral inclination. It is best viewed when the thoracic cavity is opened (see Figures 7.45 and 7.51).

Muscles of the Back and Neck (Figure 7.37)

Two thin muscular sheets, the **serratus dorsalis cranialis** and **serratus dorsalis caudalis**, lie over the back, deep to the latissimus dorsi and rhomboideus (Figure 7.30). The serratus dorsalis muscles are composed of slips extending from the middorsal line to the lateral surfaces of the ribs. The fibers do not extend all the way middorsally, but attach by aponeuroses. The fibers of the serratus dorsalis cranialis extend posteroventrally, and those of the serratus dorsalis caudalis extend anteroventrally.

Next examine the back or epaxial muscles (Figure 7.37), which lie deep to the musculature so far observed and act mainly to extend the vertebral column. This musculature is extensive, lying between the neural spines and transverse processes and proximal ends of the ribs, from the ilium to the cervical region. It is complex largely because its individual muscles extend between various parts of more posterior vertebrae to more anterior vertebrae, and fusion among its different parts occurs, particularly posteriorly.

Begin you examination posteriorly with the muscles of the lumbar and posterior thoracic regions. Make a longitudinal cut through the fleshy part of the serratus dorsalis muscles and reflect them. To examine the back musculature here, you must also cut through the lumbodorsal fascia, as follows. Poke a hole through the fascia about midway along its length, 0.5 cm to one side of the neural spines. There are two main fascial layers. Separate them

7

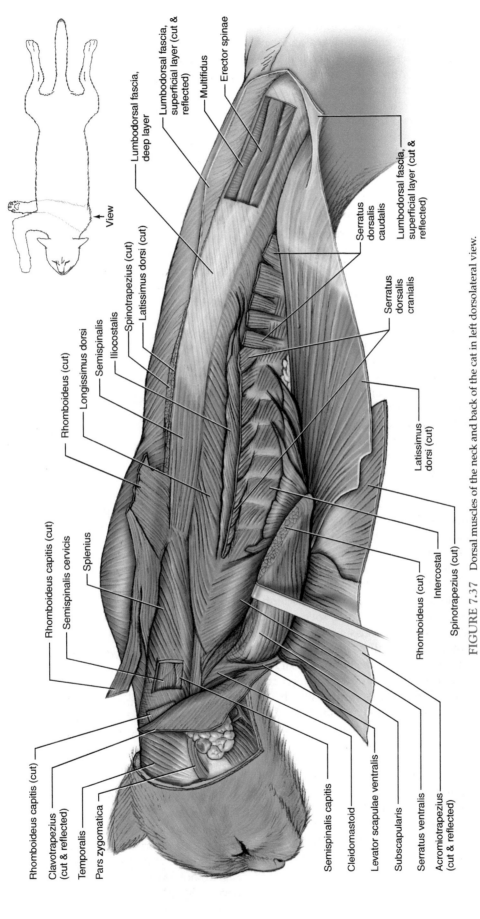

FIGURE 7.37 Dorsal muscles of the neck and back of the cat in left dorsolateral view.

Lumbodorsal fascia, deep layer

Lumbodorsal fascia, superficial layer (cut & reflected)

Multifidus

Erector spinae

Serratus dorsalis caudalis

Lumbodorsal fascia, superficial layer (cut & reflected)

Serratus dorsalis cranialis

Latissimus dorsi (cut)

Spinotrapezius (cut)

Latissimus dorsi (cut)

Iliocostalis

Semispinalis

Longissimus dorsi

Rhomboideus (cut)

Rhomboideus capitis (cut)

Semispinalis cervicis

Splenius

Rhomboideus (cut)

Intercostal

Spinotrapezius (cut)

Rhomboideus capitis (cut)

Clavotrapezius (cut & reflected)

Temporalis

Pars zygomatica

Semispinalis capitis

Cleidomastoid

Levator scapulae ventralis

Subscapularis

Serratus ventralis

Acromiotrapezius (cut & reflected)

THE DISSECTION OF VERTEBRATES

by passing a blunt probe between them. Cut through the superficial layer longitudinally, going forward as far as its anterior limit and back to about the level of the anterior part of the hind limb. Continue to remove the connective tissue and fat deep to this fascial layer—note that the latissimus dorsi arises from it laterally.

Clean the surface of the deep layer of fascia to determine its limits. Lateral to the sheet is a longitudinal and relatively narrow muscle bundle—the most lateral portion of the **longissimus dorsi** musculature. You should also be able to faintly see muscle bundles through the fascia. Cut through this deeper fascia, using the same method as with the superficial layer. In this case, dissection is essentially confined to the lumbar region, because you will not be able to go as far anteriorly, as muscle fibers adhere strongly to the deep surface of the fascia. Reflect the fascia to uncover three more longitudinally arranged muscle bundles. The most medial, as well as the narrowest, is the **multifidus**. It is an extensive muscle that is best seen in the lumbar region, lying next to the neural processes. It passes anteriorly deep to other muscles of the back into the cervical region. The two lateral bundles are part of the longissimus dorsi, which is therefore subdivided into three bundles in this region. Its two lateral bundles, separated by the lumbodorsal fascia, are considered to form the lateral division of the longissimus dorsi. The medial bundle is considered the medial division of this muscle.

Farther anteriorly, in the thoracic region, the longissimus musculature continues forward as three subdivisions, but these do not correspond entirely to those observed in the lumbar region. To see the more anterior muscles, reflect the anterior part of the latissimus dorsi (along its middorsal origin), as well as the serratus dorsalis. The three subdivisions of the musculature, arranged as longitudinal bundles and derived from the more posterior portion of the longissimus, may now be identified. The **spinalis dorsi** is the most medial, followed by the anterior continuation of the longissimus dorsi, and then, most laterally, by the **iliocostalis**. The spinalis dorsi arises mainly from the deep layer of the lumbodorsal fascia. Follow the longissimus dorsi anteriorly into its cervical extension, the narrow **longissimus capitis**, which lies along the ventral margin of the splenius on the neck. The latter was identified earlier as being deep to the rhomboideus capitis. The splenius and longissimus capitis tend to fuse, but they should be clearly demarcated. Separate these muscles, cut through the middle of the splenius, and reflect its ends. The two muscles thus exposed are the **semispinalis capitis** and **semispinalis cervicis** (which is actually the anterior portion of the multifidus, noted earlier), which pass forward from the vertebrae to insert on the skull. The longus colli, passing ventrally along the neck between the anterior part of the scalenus and the external jugular vein, was identified earlier.

Muscles of the Throat and Jaw (Figures 7.28, 7.38, and 7.57)

Examine the muscles associated with the jaws, throat, and tongue, many of which are small, thin, strap-like muscles. It is useful to be familiar with the meaning of the main roots used in the names of these muscles (*hyo* refers to the hyoid, *thyro* to the thyroid cartilage, *genio* to the chin, and *glossus* to the tongue). Reidentify the sternomastoid, and examine it in ventral view. The sternomastoid from each side passes ventrally and posteriorly from the occiput of the skull. Each follows the anterior border of the clavotrapezius dorsally. As the sternomastoid muscles approach the throat, they veer medially and meet in a V at the base of the throat. The muscles of the throat lie mainly between the V. Gently spread apart the sternomastoid muscles to expose this musculature. Pick away the connective tissue from the surface of the musculature, being careful not to injure the veins passing through the region. Examine the external jugular veins and note that each receives a medial branch that together form the **transverse jugular vein** near the anterior end of the throat. Extending from this union is a small vessel that passes between the musculature.

The most superficial muscles covering the throat are the **sternohyoid** muscles, the posterior portions of which were seen earlier along the neck ventral to the external jugular vein. On the throat, these thin, narrow muscles pass anteroposteriorly next to each other at the midline. Separate them at the midline. Lift one of them from the underlying muscles, cut it, and reflect the ends. You will uncover another set of longitudinal fibers, dorsal and slightly lateral to the sternohyoid. These are parallel to those of the sternohyoid and appear at first glance to constitute a single muscle, but they are two muscles, lying end to end. The posterior and longer muscle is the **sternothyroid**, the more anterior and shorter is the **thyrohyoid** (Figure 7.38). Gently pass a blunt probe beneath them to determine their extent.

Examine the muscles anterior to the transverse jugular vein. This area is roughly triangular, with the sides of the triangle formed by the **digastric muscles**. They lie along the medial edges of the dentary and converge toward the mandibular symphysis. Between the digastric muscles is the **mylohyoid**, which actually consists of a pair of muscles connected by a raphe at the midline. It is easily recognizable by its transverse and slightly curving fibers. Extending laterally along the posterior edge of the mylohyoid from the lateral edge of the sternohyoid is the very small and narrow **stylohyoid**. The deeper muscles of the throat and tongue lie beneath the mylohyoid. To view them, cut and reflect the mylohyoid along the midline. The thin, narrow **geniohyoid** muscles (see also Figure 7.40) extend anteroposteriorly next to each other at the midline. These arise from the dentary on either side of the mandibular symphysis (the region of the chin in humans; cats do not have a true chin) and insert on the hyoid. The narrow **hyoglossus** muscle arises deep and lateral to the

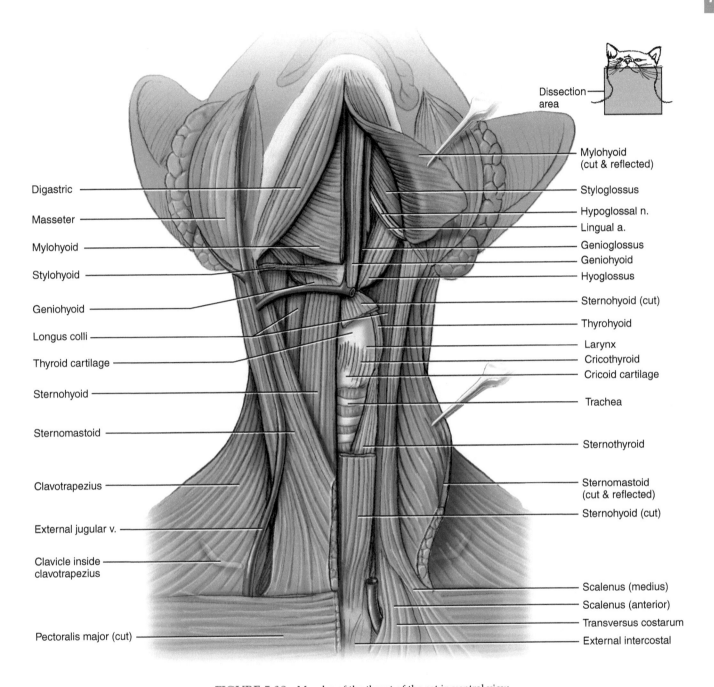

Dissection area

Digastric

Masseter

Mylohyoid

Stylohyoid

Geniohyoid

Longus colli

Thyroid cartilage

Sternohyoid

Sternomastoid

Clavotrapezius

External jugular v.

Clavicle inside clavotrapezius

Pectoralis major (cut)

Mylohyoid (cut & reflected)

Styloglossus

Hypoglossal n.

Lingual a.

Genioglossus

Geniohyoid

Hyoglossus

Sternohyoid (cut)

Thyrohyoid

Larynx

Cricothyroid

Cricoid cartilage

Trachea

Sternothyroid

Sternomastoid (cut & reflected)

Sternohyoid (cut)

Scalenus (medius)

Scalenus (anterior)

Transversus costarum

External intercostal

FIGURE 7.38 Muscles of the throat of the cat in ventral view.

insertion of the geniohyoid. Its fibers pass anteriorly into the tongue and its surface is crossed by the hypoglossal nerve and lingual artery. The **genioglossus** muscle arises anteriorly, lateral and deep to the origin of the geniohyoid, and passes posterodorsally into the tongue. The **styloglossus** is a long, narrow muscle that passes from the mastoid process of the skull into the tongue. It extends forward parallel to the medial border of the dentary. Push

the digastric laterally to help locate the styloglossus. It will cross the anterior part of the hyoglossus. The mass of the tongue is formed mainly by intrinsic fibers, termed the **lingualis proprius** within the tongue (Figure 7.40).

Examine the head. The larger muscles covering the posterolateral surface of the jaw is the **masseter**, one of the three main jaw-closing muscle groups (Figures 7.27, 7.29, 7.30, and 7.39). The **temporalis**, another jaw-closing muscle, lies on the lateral surface of the cranium and arises from the temporal fossa (see page 181). It is deep to the parotid and mandibular glands, large salivary glands that will be described presently, and covered by a thick fascia. A small slip of the temporalis, the pars zygomatica, curves over the zygomatic arch (Figure 7.37). The third and smallest jaw-closing muscle, the **pterygoid**, lies medial to the jaw and requires further dissection. It is not considered further.

Key Terms: Muscles of the Head and Trunk

digastric
external intercostals
external oblique
genioglossus
geniohyoid
hyoglossus
iliocostalis
internal intercostals
internal oblique
lingualis proprius
longissimus capitis

longissimus dorsi
longus colli
masseter
multifidus
mylohyoid
pterygoid
rectus abdominis
scalenus
semispinalis capitis
semispinalis cervicis
serratus dorsalis caudalis
serratus dorsalis cranialis
spinalis dorsi
sternohyoid
sternothyroid
styloglossus
stylohyoid
temporalis
thyrohyoid
transverse jugular vein
transversus abdominis
transversus costarum
transversus thoracis

TABLE 7.4 Muscles of the Head and Trunk.

Name	Origin	Insertion	Main Actions
Digastric	Mastoid process of temporal and jugular process of occipital	Ventromedial surface of dentary	Depresses mandible
External intercostals	Posterior margin of a rib	Anterior margin of the adjacent posterior rib	Protracts ribs, increasing diameter of thorax
External oblique	Posterior 9 or 10 ribs and lumbodorsal fascia	Mainly on linea alba from sternum to pubis, by aponeurosis	Constricts abdomen
Genioglossus	Medial surface of dentary	Tongue; posterior fibers insert on basihyoid and ceratohyoid	Depresses tongue; draws root of tongue anteriorly; curls tip of tongue ventrally
Geniohyoid	Ventromedial surface of dentary, just posterior to symphysis	Ventral surface of basihyoid	Draws hyoid anteriorly
Hyoglossus	Lateral part of ventral surface of basihyoid	Tongue	Depresses and retracts tongue
Iliocostalis	Lateral surface of ribs	Lateral surface of more anterior ribs	Draws ribs together
Internal intercostals	Anterior margin of a rib	Posterior margin of the adjacent anterior rib	Retracts ribs, decreasing diameter of thorax
Internal oblique	Lumbodorsal fascia and iliac crest	Linea alba, by aponeurosis	Constricts abdomen
Lingualis proprius	Intrinsic musculature of tongue consists of several bundles arranged mainly in longitudinal, transverse, and vertical groups. These attach to integument of tongue and insertion fascicles of extrinsic muscles of tongue (e.g., genioglossus, hyoglossus, styloglossus). Tongue musculature responsible for many complex movements, among others those during mastication and deglutition.		
Longissimus capitis	Prezygapophyses of 4th–7th cervical vertebrae	Mastoid process of temporal	Flexes head laterally
Longissimus dorsi Medial division Lateral division	Sacral and caudal vertebrae Ilium and deep layer of lumbodorsal fascia	More anterior lumbar, sacral, and caudal verterbrae More anterior lumbar and thoracic vertebrae	Extends vertebral column

Continued

TABLE 7.4   Muscles of the Head and Trunk.—cont'd

Name	Origin	Insertion	Main Actions
Longus colli	Ventral surface of first 6 thoracic vertebrae; slips from cervical vertebrae	Transverse processes of all cervical vertebrae	Flexes neck laterally and ventrally
Masseter	Zygomatic arch	Ventral part of masseteric fossa of dentary	Elevates mandible
Multifidus	Various parts of more posterior sacral, lumbar, thoracic, and cervical vertebrae	Neural process of more anterior vertebrae	Acting singly: Flexes vertebral column laterally; acting with other side multifidus: Extends vertebra; column
Mylohyoid	Medial surface of dentary	Midventral raphe; posterior fibers to basihyoid	Elevates floor of oral cavity; draws hyoid anteriorly
Pterygoid	Pterygoid blade of skull	Medial surface of angular region of dentary	Elevates mandible
Rectus abdominis	Pubis	Costal cartilages and sternum	Compresses abdomen; draws ribs and sternum posteriorly, flexing the trunk
Scalenus	Ribs	Transverse processes of all cervical vertebrae	Flexes neck; or draws ribs anteriorly
Semispinalis capitis	Prezygapophyses of 3rd–7th cervical and 1st–3rd thoracic vertebrae	Medial third of nuchal crest	Elevates head
Semispinalis cervicis	Neural processes of 7th cervical and 1st–3rd thoracic vertebrae, prezygapophyses of 2nd–5th thoracic vertebrae	Medial third of nuchal crest	Elevates head
Serratus dorsalis caudalis	Middorsally from neural processes of lumbar vertebrae	Posterior 4 or 5 ribs	Draws ribs posteriorly
Serratus dorsalis cranialis	Middorsal raphe between axis and 10th thoracic vertebrae, by aponeurosis	Lateral surface of first 9 ribs	Draws ribs anteriorly
Spinalis dorsi	Neural spines of 10th–13th thoracic vertebrae	Cervical and more anterior thoracic vertebrae	Extends vertebral column
Splenius	Anterior middorsal line	Nuchal crest	Acting singly: Flexes head laterally; acting with other side splenius: Elevates head
Sternohyoid	1st costal cartilage and manubrium of sternum	Basihyoid	Draws hyoid posteriorly
Sternomastoid	Anterior surface of manubrium of sternum	Lateral portion of nuchal crest and mastoid process of temporal	Acting singly: Flexes neck laterally; acting with other side sternomastoid: Depresses snout
Sternothyroid	1st costal cartilage	Posterolateral surface of thyroid cartilage of larynx	Draws larynx posteriorly
Styloglossus	Mastoid process of temporal and stylohyoid	Tongue	Elevates and retracts tongue
Stylohyoid	Lateral surface of stylohyoid	Ventral surface of basihyoid	Draws basihyoid dorsally
Temporalis	Fascia covering muscle, temporal fossa of skull	Coronoid process of dentary	Elevates mandible
Thyrohyoid	Lateral surface of thyroid cartilage of larynx	Thyrohyoid	Draws hyoid posteriorly and dorsally
Transversus abdominis	Costal cartilage of vertebrocostal and vertebral ribs, transverse processes of lumbar vertebrae, ventral margin of ilium	Linea alba	Constricts abdomen
Transversus costarum	Lateral margin of sternum	1st rib and costal cartilage	Draws ribs anteriorly
Transversus thoracis	Dorsolateral margin of sternum, between attachment of ribs 3–8	Costal cartilages near their attachment to ribs	Draws ventral portion of ribs posteriorly

SECTION IV: DIGESTIVE AND RESPIRATORY SYSTEMS

Salivary Glands

The cat has five pairs of salivary gland, two of which—the **parotid** and **mandibular glands**—are easily observed (Figure 7.39). They have already been noted in connection with the musculature of the throat. The parotid gland, lying almost directly ventral to the ear, is the largest. It is irregular, with obvious lobules and an elongated, tapered ventral portion. The smaller mandibular gland, smoother and nearly oval, lies just posterior to the ventral part of the parotid gland.

Note that three veins come together in this region to form the external jugular vein (Figure 7.39). The most dorsal tributary, the **maxillary vein**, turns dorsally, crosses the anterior part of the mandibular gland, and then passes deep to the parotid gland. **Lymph nodes** are also present in this region and may be confused with the salivary glands, but lymph nodes are typically smaller, and as they lack lobules, have a smoother surface. Usually, two are present ventral to the parotid and mandibular glands. Often there seems to be only a single

large node that is crossed by the **linguofacial vein**, but careful dissection will reveal two nodes, with the vein passing between them (Figure 7.39).

Examine the ducts of the parotid and mandibular glands. The **parotid duct** is the thick, whitish strand crossing over the middle part of the **masseter muscle** (noted earlier) and extending toward the upper lip. Also crossing the masseter are two branches of the **facial nerve**, the dorsal buccal and ventral buccal branches (Figure 7.39), which are thinner than the duct. One branch lies dorsal and the other ventral to the duct. If you were indelicate in removing the skin, you may have already taken off the duct and nerves.

The **mandibular duct** emerges from the anteroventral part of the mandibular gland, just dorsal to the lymph node. Dissect carefully in this region. The duct will appear as a whitish strand, thinner than the parotid duct, that passes ventral to a small glandular mass. This is the long and narrow **sublingual gland** (Figures 7.39 and 7.57). Its posterior end usually abuts against the mandibular gland and may initially be confused as an extension of that gland. Tugging the anterior end of the sublingual gland will reveal that it is more extensive than it at first appears. The duct of the sublingual gland is difficult to discern grossly.

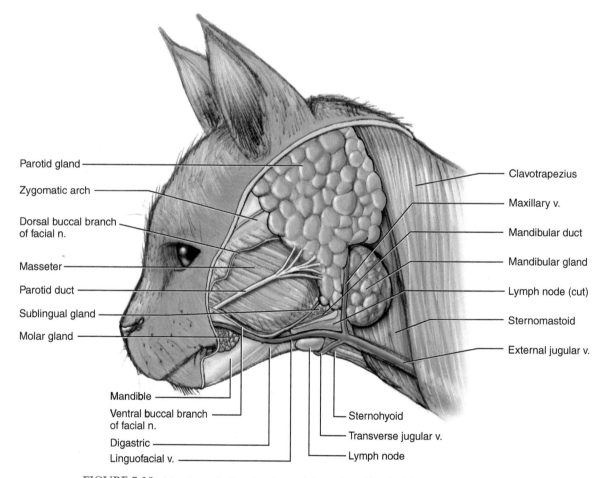

FIGURE 7.39   Muscles and other structures of the neck and head of the cat in left lateral view.

Return to the mandibular duct. Dissect the connective tissue binding it by gently pulling the duct taut with forceps and using a needle; a # 11 scalpel blade, worked backward (i.e., the noncutting edge) also works very well. The proximal part of the duct crosses the surface of the **digastric muscle** (observed earlier) as the latter turns dorsally toward the mastoid process of the skull. Continue dissecting anteriorly to reveal the duct and digastric. Note how the duct crosses the dorsal surface of the digastric and then passes medial to it. At this point, approach the digastric from the ventral surface of the jaw. Reflect and separate it from the mylohyoid. You should easily locate the mandibular duct as it passes deep to the edge of the mylohyoid. Reflect the mylohyoid (which was bisected earlier; see page 234 and Figure 7.38) to see the duct continuing anteromedially. The **lingual nerve** crosses the duct anteriorly (Figure 7.57). While studying this region, review the hypoglossal nerve (see also Figure 7.38). It is larger than the mandibular duct and extends almost parallel to it, but farther posteroventrally.

Oral Cavity and Pharynx

Once the salivary glands and their ducts have been examined, the **oral cavity** may be opened. To do this, the mandible must be spread apart and the tongue pulled ventrally. Begin by separating the jaws, as follows. Locate the position of the symphysis, which lies between the two middle incisor teeth of the mandible. Place the edge of a fresh scalpel blade between the incisors on the anteroventral margin of the mandible and rock it back and forth. The blade will soon pass into the symphysis, separating the dentaries. Do not push hard on the scalpel. A rocking motion with steady, gentle pressure and patience are all that are needed. Place a finger on the dull side of the blade to help guide the scalpel.

When you have passed through the symphysis, use your fingers (but be careful; the cat's teeth are sharp) to spread the dentaries apart. One of them will probably break. Cut through the musculature and other tissues attaching to the medial surface of each dentary as far back as about the level of the lymph nodes. Then pull the **tongue** down through the dentaries to give you a clear view into the oral cavity.

In its broad, everyday sense, the **mouth** includes the structures such as the tongue and the teeth. In a strict sense, however, these structures are part of the oral cavity. The mouth is only the part between the **labia** (sing., **labium**) or lips. The oral cavity is subdivided into the **vestibule**, between the lips and teeth, and the **oral cavity proper**, bounded by the teeth anteriorly and laterally (see Figure 7.40). Posteriorly it extends to just beyond the level of the **hard palate**. The epithelial covering of the hard palate has roughened transverse ridges, the **palatal rugae**.

The tongue almost completely fills the oral cavity when the mouth is closed. Pull the tongue dorsally and note the **lingual frenulum**, a vertical, median flap attaching the tongue to the cavity floor (see Figure 7.40). The surface of the tongue bears various projections or papillae (Figure 7.41). The **filiform papillae** are most numerous. Interspersed among them are the rounded **fungiform papillae**. The **vallate papillae** are set in the posterodorsal surface of the tongue, and the **foliate papillae** are along the side of the posterior end of the tongue.

The paired **palatine tonsils** are set partially within **tonsillar fossae**. Just anterior to the tonsillar fossae are the **palatoglossal arches**, lateral folds that may be made more prominent by pulling the tongue downward. In the adult these arches mark the end of the oral cavity. Thus, the oral cavity extends somewhat beyond the level of the hard palate (Figure 7.40). The bones of the hyoid apparatus that pass to the base of the skull are embedded within the palatoglossal arches, as are small muscles that assist in swallowing. Extend the cuts made to open the oral cavity farther posteriorly to the level of the palatoglossal arches, but do not cut through the hyoid apparatus. This provides a better view of the oral cavity and **pharynx**.

The pharynx is the passage common, in part, to both the digestive and respiratory systems. It extends from a line through the posterior edge of the hard palate to a line through the posterior boundary of the **larynx**. The anterior part of the pharynx is subdivided into dorsal and ventral portions by the **soft palate**, the fleshy posterior continuation of the hard palate. The region dorsal to the soft palate is the exclusively respiratory **nasopharynx**; that ventral to the soft palate is the dominantly digestive **oropharynx** (which therefore lies posterior to the oral cavity). Posterior to the soft palate these become continuous and extend posteriorly to the **laryngopharynx**, which communicates with the larynx posteroventrally and the esophagus posterodorsally (Figure 7.40). Make a median slit in the soft palate to expose the nasopharynx. Anteriorly, it leads to the **choanae**. The opening of the **auditory tube** (which leads to the middle ear) lies on the dorsolateral wall of the nasopharynx. However, it is often swollen with fluid and may be difficult to find.

The **esophagus** is the muscular tube of the digestive system that extends from the laryngopharynx to the stomach. The larynx is the chamber at the anterior end of the **trachea** or windpipe (see Figure 7.38) and is formed from several cartilages. The opening into the larynx is the **glottis**, which is guarded by the **epiglottis**, the plough-like, cartilaginous structure at the base of the tongue. The epiglottis, which is supported by an **epiglottal cartilage**, is flipped back to cover the glottis during swallowing, so that food passes posteriorly through the laryngopharynx and into the esophagus. Using a bent probe, flip the epiglottis forward and slide the probe

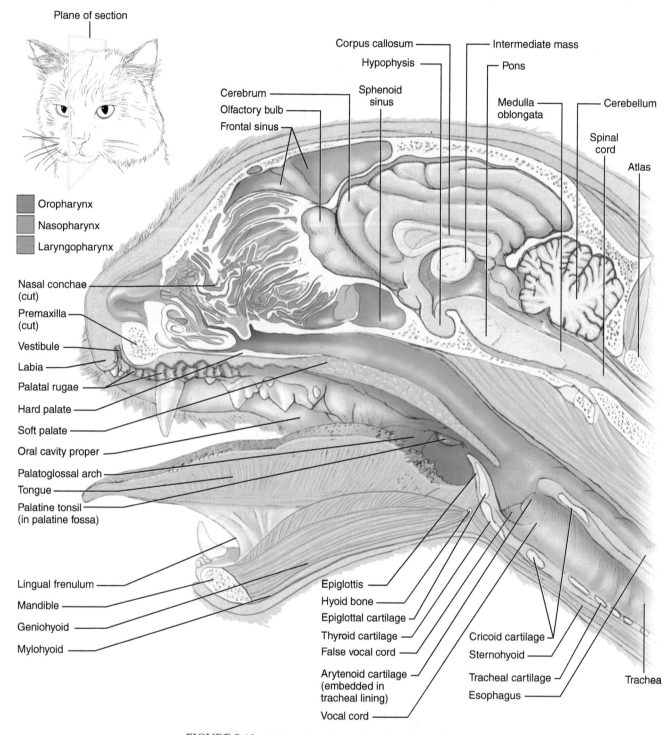

Plane of section

Oropharynx
Nasopharynx
Laryngopharynx

Cerebrum
Olfactory bulb
Frontal sinus

Corpus callosum
Hypophysis
Sphenoid sinus

Intermediate mass
Pons

Medulla oblongata

Cerebellum

Spinal cord

Atlas

Nasal conchae (cut)
Premaxilla (cut)
Vestibule
Labia
Palatal rugae
Hard palate
Soft palate
Oral cavity proper
Palatoglossal arch
Tongue
Palatine tonsil (in palatine fossa)

Lingual frenulum
Mandible
Geniohyoid
Mylohyoid

Epiglottis
Hyoid bone
Epiglottal cartilage
Thyroid cartilage
False vocal cord
Arytenoid cartilage (embedded in tracheal lining)
Vocal cord

Cricoid cartilage
Sternohyoid

Tracheal cartilage
Esophagus

Trachea

FIGURE 7.40   Right side of the head of the cat in sagittal section.

posteriorly. It will pass through the glottis and enter the trachea. Palpate the trachea to verify this.

Cartilages that contribute to the larynx include the **thyroid, cricoid**, a pair of **arytenoid cartilages**, and the epiglottal cartilage just mentioned. Examine them by observing the throat region in ventral view. The thyroid is the largest cartilage. It was noted in connection with the

thyrohyoid and sternothyroid muscles (see page 234 and Figures 7.38, 7.40, and 7.57). The thyroid is incomplete dorsally. Anteriorly, it contacts the thyrohyoid of the hyoid apparatus. The epiglottis rests against the antero-ventral part of the thyroid. The cricoid lies slightly distal to the thyroid and forms a complete ring, but is narrower ventrally than dorsally. The **cricothyroid muscle** extends

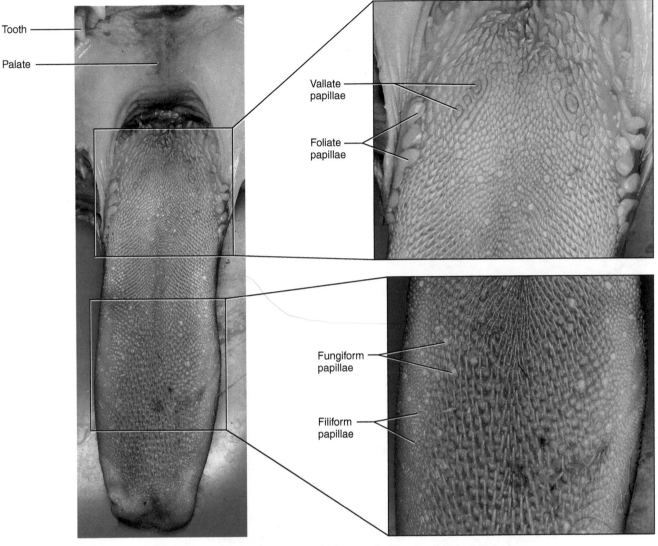

Tooth

Palate

Vallate papillae

Foliate papillae

Fungiform papillae

Filiform papillae

FIGURE 7.41   Tongue of the cat with blowups showing detail of the taste buds.

between these cartilages (Figure 7.38). Separate the larynx from surrounding tissue and observe its dorsal surface. The arytenoids (see Figure 7.40) are small cartilages that help complete the larynx dorsally, anterior to the cricoid cartilage.

Make a midventral slit through the larynx and spread it open. Two folds of tissue are present on each side of the larynx. The **vocal cords** are the more posterior folds, extending between the arytenoid and thyroid cartilages (Figure 7.40). The glottis lies between them. The **false vocal cords** are an accessory pair of folds, extending between the arytenoid and epiglottal cartilages. They are larger, looser and lie anterior to the vocal cords.

The **tracheal cartilages** are C-shaped, dorsally incomplete cartilaginous elements that keep the trachea open (Figures 7.38 and 7.40). The esophagus lies dorsal to the trachea. To examine the rest of the respiratory system, it will be necessary to open the **thorax**.

The thorax is the region that contains the **thoracic cavity**, the anterior part of the body cavity or coelom. The thoracic cavity and **abdominopelvic cavity** (the posterior part of the coelom) are separated by a muscular partition, the **diaphragm** (Figure 7.42). The thoracic cavity is subdivided into three cavities: left and right pleural cavities, each containing a **lung**, and a median space, the **mediastinum**, which contains most of the other structures that lie in or pass through the thorax (such as the heart, esophagus, trachea, and nearly all the vessels and nerves that pass through this region).

Subdivision of the thoracic cavity is due to the presence of thin epithelial membranes that line the inside wall of the cavity and cover the structures that lie within it. Enclosed body cavities (i.e., those that do not communicate with the exterior environment) such as the thoracic cavity are lined by serous epithelium, a membrane that produces a watery, lubricating secretion to help reduce

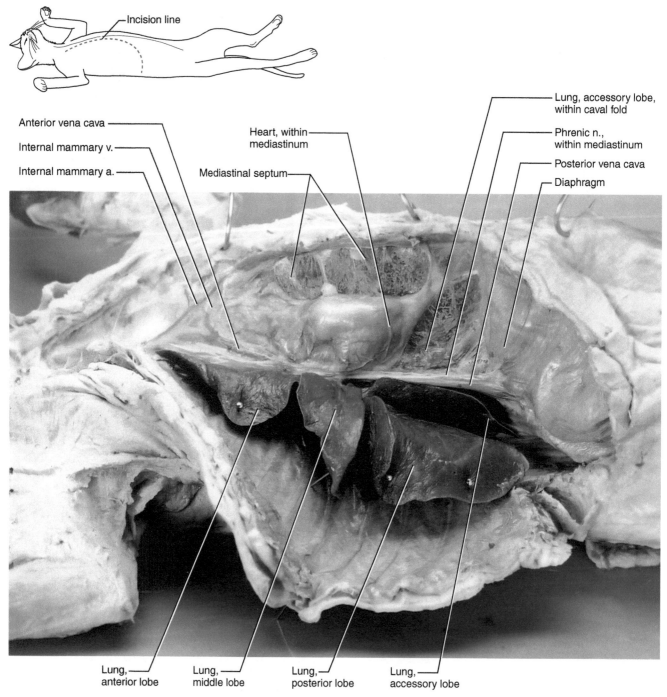

Incision line

Anterior vena cava

Internal mammary v.

Internal mammary a.

Heart, within
mediastinum

Mediastinal septum

Lung, accessory lobe,
within caval fold

Phrenic n.,
within mediastinum

Posterior vena cava

Diaphragm

Lung,
anterior lobe

Lung,
middle lobe

Lung,
posterior lobe

Lung,
accessory lobe

FIGURE 7.42   Right pleural cavity of the cat in lateral and slightly ventral view.

friction between structures lined or covered by the membrane. The serosa of the thoracic cavity is termed **pleura**. There are two sheets of pleura, one on the right side and the other on the left. Portions of the pleura are designated based on position. That portion that lines the inside of the cavity is **parietal pleura**, whereas the portion that envelops the lung is **visceral pleura**. As noted, these are actually formed by one continuous sheet. The parietal pleura lines the inside of the cavity, but near the sagittal midline

it reflects, so that it then passes to cover the lung. Where the right- and left-side parietal pleurae meet near the midline, they form a double layer termed the **mediastinal septum**. The mediastinum is the space or potential space between this double layer. Various structures, as noted earlier, may occupy this space, but in places the mediastinal septum remains as a double layered structure.

Open the right side of the thorax first by making a longitudinal cut about 1 cm to the right of the midventral

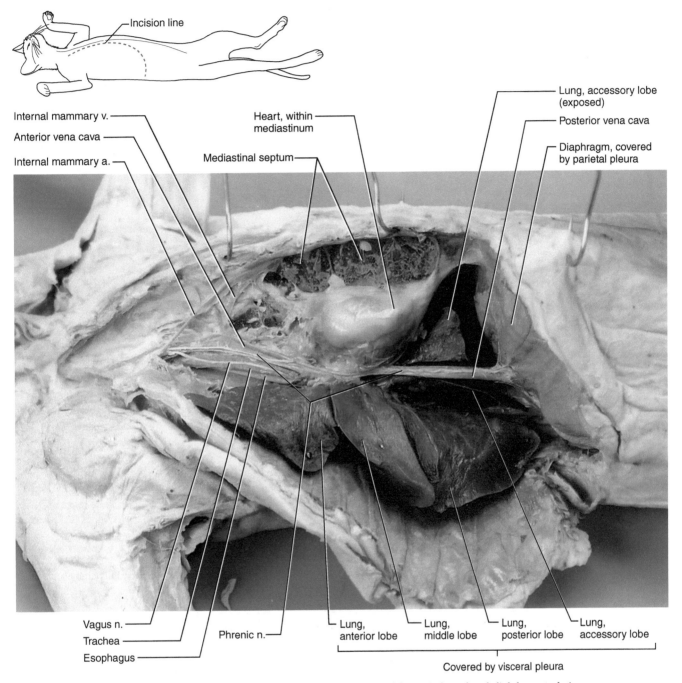

FIGURE 7.43   Deeper dissection of the right pleural cavity of the cat in lateral and slightly ventral view.

line. Begin posteriorly, at about the midpoint of the xiphihumeralis muscles. This position should be just anterior to the diaphragm. You will be cutting through costal cartilages and ribs, as well as musculature, so use a strong pair of scissors. Spread open the thorax and locate the diaphragm. Now cut laterally, following along the anterior surface of the diaphragm. Extend the cut as far dorsally as you can. Spread open the thorax further and snip the dorsal end of each rib, beginning posteriorly. This will allow you to push open the right side of

the thorax, exposing the right pleural cavity and lung (Figures 7.42–7.44).

Examine the right pleural cavity. It is really only potentially a cavity, because it is filled in life by the lung. Its walls are lined by parietal pleura, whereas the lung is covered by visceral pleura (Figure 7.44). The parietal pleura is clearly visible on the inside of the thorax, where it covers the musculature (including the diaphragm) and ribs. You cut through it in exposing the thorax. In addition, the parietal pleura forms the medial wall of the pleural cavity.

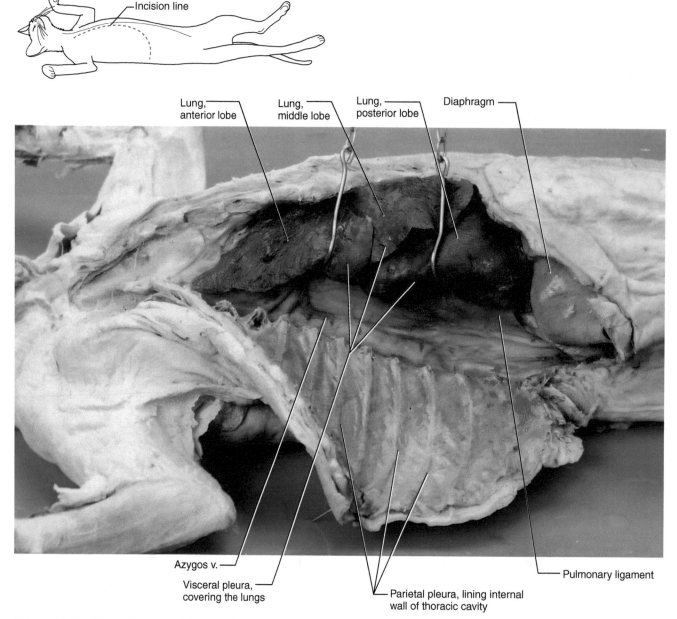

FIGURE 7.44 Deeper dissection of the right pleural cavity, with right lung lifted from dorsal wall, of the cat in lateral and slightly ventral view.

Between the medial walls of the right and left pleural cavities lies the mediastinum. The large bulge within it is the heart. Identify the mediastinal septum, which occurs where the pleurae of the right and left pleural cavities meet at the midline. Lift the sternum to observe the septum, but do not damage it (Figures 7.42 and 7.43).

The right lung consists of **anterior, middle, posterior,** and **accessory lobes**. The accessory lobe passes dorsal and then medial to a large vein, the **posterior vena cava**. This lobe may not be apparent at first because it lies in the **caval fold**, a pocket-like expansion of the mediastinal septum just posterolateral to the heart (Figures 7.42 and 7.43). Dissect the caval fold to expose the accessory lobe and

the vena cava (Figure 7.43). The lung is suspended in the pleural cavity by the **pulmonary ligament**, a flat, broad, sheet-like membrane that is the connection between the visceral pleura and parietal pleura. To view it, pull the posterior lobe of the lung ventrally, and examine in lateral view the region dorsal to the lobe (Figure 7.44).

Next, push the lung laterally and look between the lung and mediastinal wall. At about the middle of its medial surface, a fold of pleura will be seen passing from the lung to the mediastinal wall. This fold is part of the pulmonary ligament. Various structures (such as the pulmonary vessels and the bronchus), collectively forming the **root of the lung**, pass through it.

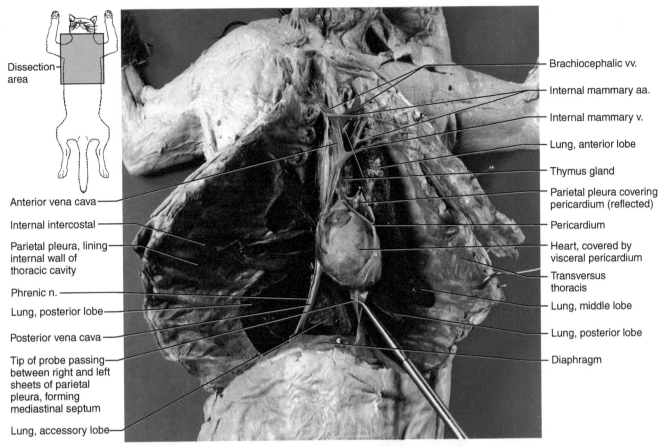

Dissection area

Anterior vena cava

Internal intercostal

Parietal pleura, lining internal wall of thoracic cavity

Phrenic n.

Lung, posterior lobe

Posterior vena cava

Tip of probe passing between right and left sheets of parietal pleura, forming mediastinal septum

Lung, accessory lobe

Brachiocephalic vv.

Internal mammary aa.

Internal mammary v.

Lung, anterior lobe

Thymus gland

Parietal pleura covering pericardium (reflected)

Pericardium

Heart, covered by visceral pericardium

Transversus thoracis

Lung, middle lobe

Lung, posterior lobe

Diaphragm

FIGURE 7.45 Thoracic cavity of the cat in ventral view.

Examine the mediastinal wall just ventral to the root of the lung. You should discern a thin, whitish strand extending anteroposteriorly. This is the **phrenic nerve**, which lies in the mediastinum and passes to the diaphragm (Figure 7.43). Using a needle, expose the phrenic nerve and follow it posteriorly as it passes along the ventral surface of the posterior vena cava. Also follow the nerve anteriorly as it passes along the mediastinum anterior to the heart. Break through into the mediastinum as you do so. Very delicately, clear away fat and connective tissue. The large vessel injected with blue latex is the **anterior vena cava**. Dorsal to it is the trachea; its cartilaginous elements should be easily recognizable. Note the **vagus nerve**, extending along the lateral surface of the trachea. Dorsal and slightly to the left of the trachea is the esophagus (Figure 7.43). Farther posteriorly, the esophagus passes through the diaphragm and joins the stomach. By contrast, the trachea remains within the thorax and bifurcates near the level of the sixth rib into right and left **primary bronchi**. Postpone their study until the heart has been examined. Each bronchus in turn branches into **secondary** and **tertiary bronchi**, which further branch into bronchioles that end in the tiny respiratory structures of the lungs, the alveoli. The bronchioles and alveoli, however, cannot be easily dissected and observed grossly.

Open the left pleural cavity by making a longitudinal cut to the left of the midventral line and by repeating the procedures described earlier for the right pleural cavity. This method produces a median flap that leaves the sternum intact. Examine the left lung, and identify the anterior, middle, and posterior lobes (Figure 7.45). The left lung lacks an accessory lobe. Find the left phrenic nerve, just ventral to the root of the lung.

Pericardial Cavity

Cut the median strip ventrally and break through the pleura ventral to the heart so you can lift the sternum (Figure 7.45). Be careful not to injure the internal mammary vessels (which will be considered later) passing to the inside of the strip. Peel the pleura away from the heart—there may be a considerable layer of fat deep to the pleura. Carefully clear the fat to reveal that the heart sits in its own space, the **pericardial cavity**, which is covered by a tough layer of **pericardium**. Note the great vessels of the heart anteriorly by cleaning away fat, but do not examine them at this time. Lift the apex of the heart (its posterior end) to better appreciate the extent of the accessory lobe of the right lung. With a pair of scissors, pierce the pericardium and cut it longitudinally to expose the surface of the heart, itself covered by **visceral pericardium** (Figure 7.45), whereas the inner surface of the pericardium is lined by **parietal pericardium**.

Abdominopelvic Cavity

The abdominopelvic cavity forms the rest of the coelom posterior to the diaphragm and contains most of the structures of the digestive and urogenital systems. Open the abdominopelvic cavity by making a longitudinal incision, using scissors, about 1 cm to the right of the midventral line. Then cut laterally along the posterior margin of the diaphragm and spread open the muscular walls of the abdomen (Figure 7.46).

The abdominopelvic cavity and its contained structures are lined or covered with serous epithelium termed **peritoneum**. That portion lining the walls of the cavity is **parietal peritoneum** and that covering structures within the cavity is **visceral peritoneum**. The peritoneum actually represents two separate sheets, each of which encloses separate cavities in the embryo. These cavities are the right coelom and left coelom, and the embryonic gut lies between them. Near the midsagittal plane, dorsal and ventral to the embryonic gut, the peritoneum of the right and left coela lie adjacent to each other, forming a sheet-like, double layer (similar to the pleurae of the thoracic cavity) of visceral peritoneum that is termed a *mesentery*. Other organs, as they arise and enlarge in

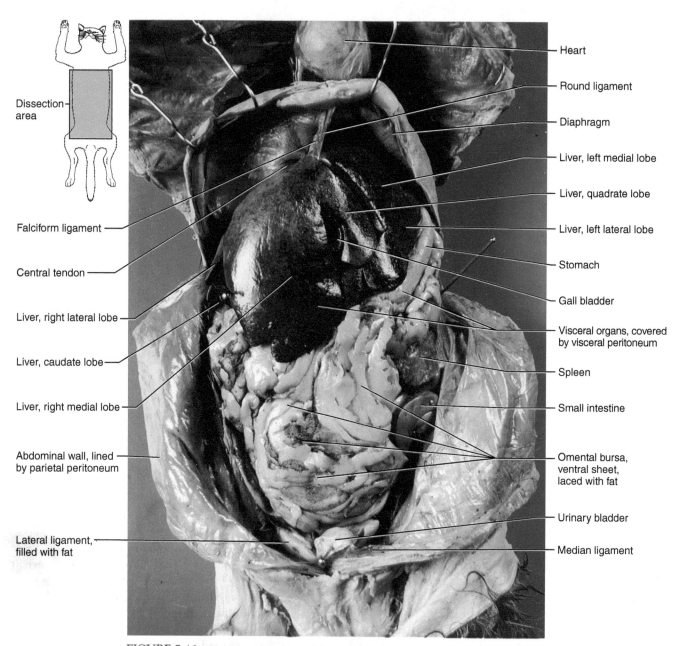

FIGURE 7.46 Abdominopelvic cavity of the cat in ventral view, with diaphragm reflected.

the embryo, come to occupy a position between the two layers and demarcate separate sections of mesentery. Subsequently in development, most of the ventral mesentery breaks down, creating a single space from the two originally separate coela. Whereas adults retain only small remnants of the ventral mesentery, nearly all of the dorsal mesentery remains. Mesenteries serve mainly to support structures in the abdominopelvic cavity by anchoring them to the body wall (either directly or by way of other mesenteries) and to provide a channel for the passage of nerves and vessels. Examine an inside wall of the abdominopelvic cavity to observe parietal peritoneum.

Pressed tightly to the posterior surface of the diaphragm is the **liver**. It is covered by visceral peritoneum. Spread apart the liver and diaphragm. This will pull taut the **falciform ligament** (an example of ventral mesentery), which passes from the diaphragm to the liver and separates the liver into left and right halves. In overweight cats the falciform ligament may be hard to distinguish because it is very fatty. The free ventral edge of the falciform ligament may contain a slight expansion, the **round ligament**. The fibers of the diaphragm converge toward its center and insert on the **central tendon**, which may be difficult to discern. To see it, separate the liver and diaphragm and observe the posterior surface of the diaphragm. Run a finger along the anterior surface of the diaphragm, toward its center, to the left of the heart. You should be able to make out your finger through the translucent central tendon. The **coronary ligament** is a very short structure between the diaphragm and liver on either side of the falciform ligament, and binds the liver tightly to the diaphragm and lateral body wall.

As noted earlier, the falciform ligament separates the liver into right and left halves, each of which is further subdivided (Figures 7.46 and 7.47). The **right medial lobe** is considerably larger than the **left medial lobe**. Between these lobes is the small **quadrate lobe**. It is separated from the right medial lobe by the greenish, sac-like **gallbladder**. The quadrate and right medial lobes are connected anterior to the gallbladder, and in some specimens posterior to it as well. The **left lateral lobe** is larger than the **right lateral lobe**, which is partly hidden by the right medial lobe. Lift the right lateral lobe to reveal the **caudate lobe**. The caudate lobe extends medially to the left, passing dorsal to the gastrohepatoduodenal ligament and hepatic portal vein (both are considered later), as a smaller component. This portion and its relationship to the main part of the caudate lobe is best viewed once the remaining structures have been examined, as shown in Figure 7.58. Some authors consider the main portion of the caudate lobe as part of the right lateral lobe, and thus subdivide the latter into anterior and posterior portions, and restrict the caudate lobe only to the smaller medial extension described earlier. However, our designations here follow the scheme of Homberger and

Walker (2003) for the cat (and Hermanson, de Lahunta, and Evans (2019) for the dog).

The **stomach** is the large, light-colored, sac-like organ partially exposed posterior to the left lateral lobe of the liver (Figure 7.46). Dorsal to the edge of this lobe, near the midline, find the esophagus passing through the diaphragm and into the stomach. The stomach's long, convex surface is the **greater curvature**; its shorter, concave surface is the **lesser curvature**. Distally, toward the right, the stomach constricts into the **pyloric sphincter**, the muscular, valve-like separation between the stomach and the duodenum (the first part of the intestine that will be considered presently). Make a slit in the ventral wall of the stomach and note the folds or **rugae** that line its inner walls. The large dark organ to the left of the stomach is the **spleen**.

Posterior to the stomach, the abdominopelvic cavity is covered ventrally by the **omental bursa**, a large, double-layered, fat-laced mesentery that covers the **intestines** like an apron (Figure 7.46). The bursa is a sac-like structure formed from the **mesogaster** or greater omentum, which is part of the dorsal mesentery. Its structure will be discussed shortly. The omental bursa extends posteriorly to the **urinary bladder**, the light-colored, median, sac-like organ lying ventrally. If it is empty, it resembles a collapsed balloon. The mesentery passing from the bladder to the midventral wall, just to the left of the incision made to open the abdominopelvic cavity, is the **median ligament** (Figure 7.46). The bladder is also supported by **lateral ligaments**, one on either side, that are often filled by wads of fat.

Return to the anterior part of the abdominopelvic cavity and spread apart the stomach and liver (Figure 7.48). The mesentery extending from the lesser curvature of the stomach and duodenum to the liver is the **gastrohepatoduodenal ligament** or lesser omentum (another example of ventral mesentery), which is divided into two portions. One part, the **hepatogastric ligament**, passes from the lesser curvature to the liver. The other, passing from the proximal part of the duodenum to the liver, is the **hepatoduodenal ligament**, which appears to head toward the gallbladder. Various structures pass through the gastrohepatoduodenal ligament, including the **common bile duct** and the **hepatic artery**. These will be considered later.

The remaining mesenteries associated with the digestive system are mainly part of the dorsal mesentery. Similar subdivisions as were noted for the dogfish may be recognized, but they are more complex in the cat. The most conspicuous part is the mesogaster, which as previously mentioned forms a double-layered, sac-like omental bursa that is draped over the intestines. The mesogaster consists mainly of two sheets. The ventral sheet attaches to the greater curvature of the stomach and extends posteriorly (Figure 7.46). Near the urinary

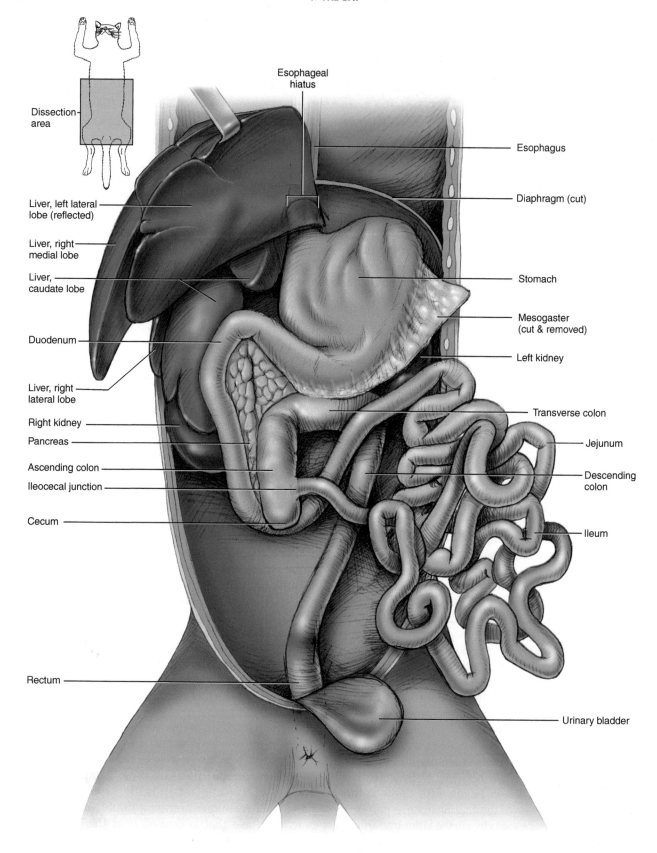

FIGURE 7.47 Diagrammatic illustration of the viscera of the cat in ventral view, with liver reflected to show arrangement of intestines.

bladder the sheet curves dorsally and extends back toward the stomach as the dorsal sheet. The omental bursa is delicate and should be handled carefully. Slowly lift its posterior end away from the intestines (you will also lift the spleen) to see the dorsal sheet, which passes dorsal to the stomach (Figure 7.49). Just posterior to the stomach, the dorsal sheet contains glandular material, the **tail of the pancreas**, but do not examine it now. The omental bursa is this sac-like part of the greater omentum. Within it is the potential space, normally closed (because the two sheets are pressed together), known as the **lesser peritoneal cavity**. It communicates with the abdominopelvic cavity via the **epiploic foramen**, which

will be described shortly. Part of the mesogaster, the **gastrosplenic ligament**, stretches between the stomach and the spleen (Figure 7.46).

Return to the pyloric sphincter and trace the digestive tract distally without tearing through any mesenteries. Use Figure 7.47 to locate the structures described. The **small intestine** fills nearly all the abdominopelvic cavity posterior to the liver and stomach. It is subdivided into three parts. The **duodenum** is the short initial or proximal segment, comprising the first loop of the small intestine. The rest of the small intestine is subdivided into the middle **jejunum** and the distal **ileum**. The small intestine is tightly coiled and ends toward the right posterior part of

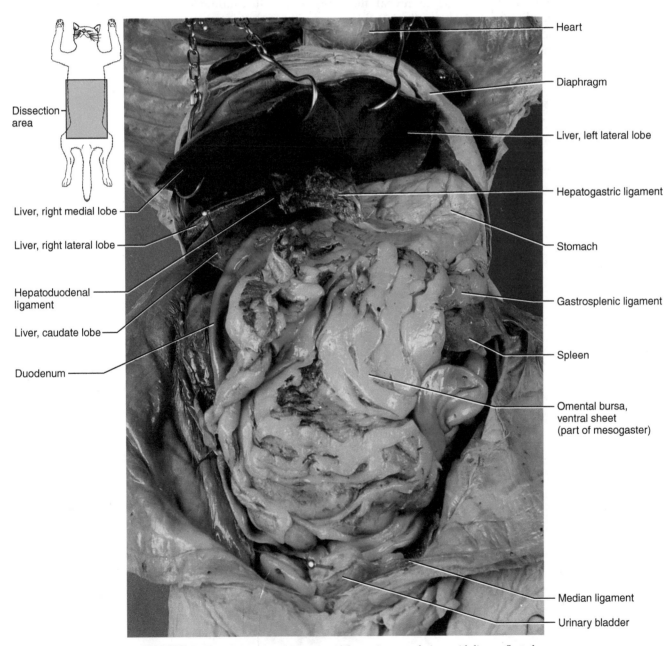

FIGURE 7.48   Abdominopelvic cavity of the cat in ventral view, with liver reflected.

the abdominopelvic cavity, where it passes into the **large intestine** or **colon**. A very short segment of the colon continues posterior to its union with the small intestine as the blind-ended **cecum**. The main part of the colon includes the **ascending colon**, which extends anteriorly on the right side; the **transverse colon**, which passes from right to left; and the **descending colon**, which extends posteriorly on the left side. Note that the duodenum loops dorsal to the colon and that the jejunum and rest of the small intestine pass ventrally to the rest of the colon. This position is due to rotation of the intestine during embryonic development. The colon continues as the **rectum**, which lies in the pelvic canal and leads to the **anus**.

Lift the coils of the small intestine to reveal the descending colon, extending posteriorly, on the left side of the abdominopelvic cavity (Figure 7.50). Pull it gently to see that it is suspended by the **mesocolon**, which has

a broad attachment to the middorsal body wall. The rectum is supported by the **mesorectum**.

The duodenum extends posteriorly along the right side of the abdominopelvic cavity and then turns anteromedially. Lift the duodenum to reveal the caudate lobe of the liver. Gently tug the duodenum ventrally and note that it is supported by the **mesoduodenum** (Figure 7.50), within which is the **head of the pancreas**. The **duodenocolic ligament**, a small, triangular mesentery with a free posterior margin, extends between the mesoduodenum and the mesocolon.

Examine the caudate lobe of the liver. The large oval swelling posterior to it is the right **kidney**. Lift the posterior end of the caudate lobe. The **hepatorenal ligament**, a small triangular mesentery, extends between the posteromedial end of the caudate lobe and the peritoneum covering the kidney (Figure 7.50).

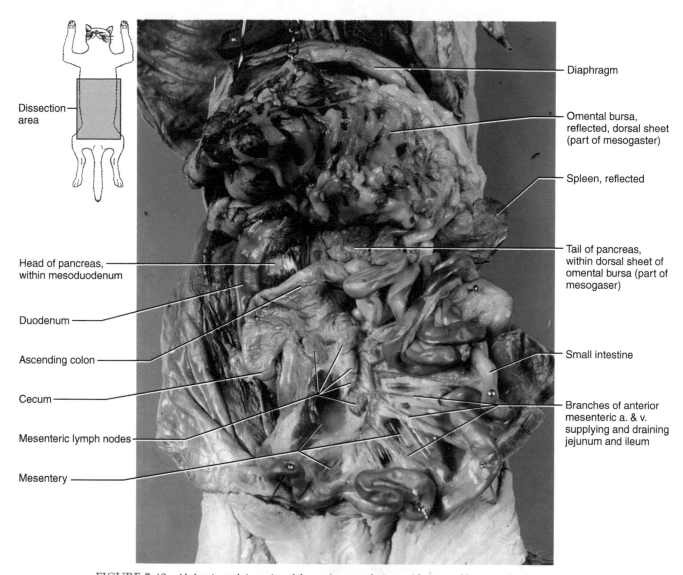

Dissection area

Head of pancreas, within mesoduodenum

Duodenum

Ascending colon

Cecum

Mesenteric lymph nodes

Mesentery

Diaphragm

Omental bursa, reflected, dorsal sheet (part of mesogaster)

Spleen, reflected

Tail of pancreas, within dorsal sheet of omental bursa (part of mesogaser)

Small intestine

Branches of anterior mesenteric a. & v. supplying and draining jejunum and ileum

FIGURE 7.49   Abdominopelvic cavity of the cat in ventral view, with omental bursa and spleen reflected.

The jejunum and ileum, the middle and distal parts of the small intestine, are supported by the **mesentery proper** (Figure 7.49). The mesentery has only a restricted attachment to the middorsal body wall, but it fans out to support all but the duodenal portion of the small intestine. Gently spread the coils of the intestine to see the mesentery clearly. Note the small **mesenteric lymph nodes** and the vessels within it.

The gallbladder is a sac used for bile storage. The ducts associated with the gallbladder and liver are greenish, like the gallbladder itself. The ducts pass to the duodenum through the gastrohepatoduodenal ligament. To view these structures, carefully dissect the gastrohepatoduodenal ligament near the caudate lobe of the liver and the gallbladder. Do not injure the vessels that are also present in the gastrohepatoduodenal ligament. The

cystic duct (Figure 7.58) is the duct of the gallbladder. It extends a short distance toward the duodenum before it is joined by usually two **hepatic ducts** from the liver. The confluence of the cystic and hepatic ducts forms the **common bile duct**, which continues to the duodenum. Follow the common bile duct to its entrance into the duodenum.

Return to the ventral sheet of the omental bursa and cut across its attachment to the stomach. Examine the dorsal sheet and distinguish the tail of the pancreas (Figure 7.49). Follow the tail toward the right, noting that it is partly tucked behind the stomach and then curves ventrally along the medial side of the duodenum, within the mesoduodenum, as the head of the pancreas. The pancreas consists of endocrine and exocrine glandular tissue. Of the two, the exocrine portion makes by far the

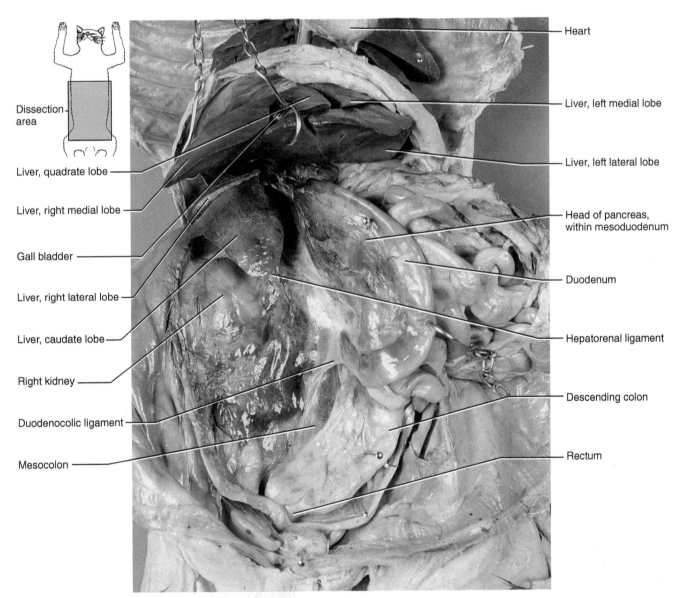

FIGURE 7.50 Abdominopelvic cavity of the cat in ventral view, with liver, intestines, and pancreas reflected.

larger part (the endocrine portion consists of small pockets of tissue termed *islets of Langerhans*, which cannot be distinguished grossly). The pancreas has two ducts that carry its exocrine secretion to the duodenum. The main **pancreatic duct** passes, together with the bile duct, to the duodenum. Find it in this region and trace it for a short distance into the head of the pancreas by carefully scraping away pancreatic tissue.

Key Terms: Digestive and Respiratory Systems

abdominopelvic cavity
anterior vena cava
anus
arytenoid cartilages
ascending colon
auditory tube (Eustachian tube, pharyngotympanic tube)
bronchus: primary, secondary, tertiary (pl., bronchi)
caval fold
cecum
central tendon
choanae
colon
common bile duct
coronary ligament
cricoid cartilage
cricothyroid muscle
cystic duct
descending colon
diaphragm
digastric muscle
duodenocolic ligament
duodenum
epiglottal cartilage
epiglottis
epiploic foramen
esophagus
facial nerve
falciform ligament
false vocal cords
filiform papillae
foliate papillae
fungiform papillae
gallbladder
gastrohepatoduodenal ligament (lesser omentum)
gastrosplenic ligament
glottis
greater curvature
hard palate
hepatic artery
hepatic ducts
hepatoduodenal ligament
hepatogastric ligament

hepatorenal ligament
ileum
intestines
jejunum
kidney
labia (sing., labium; lips)
large intestine
laryngopharynx
larynx
lateral ligaments
lesser curvature
lesser peritoneal cavity
lingual frenulum
lingual nerve
linguofacial vein (anterior facial vein)
liver: caudate lobe, left lateral lobe, right lateral lobe, left medial lobe, right medial lobe, quadrate lobe
lung: anterior, middle, posterior, accessory lobes
lymph nodes
mandibular duct
mandibular gland
maxillary vein (posterior facial vein)
median ligament
mediastinal septum
mediastinum
mesenteric lymph nodes
mesentery
mesocolon
mesogaster (greater omentum)
mesoduodenum
mesorectum
mouth
nasopharynx
omental bursa
oral cavity (= oral cavity proper + vestibule)
oral cavity proper
oropharynx
palatal rugae
palatine tonsils
palatoglossal arches
pancreas: head, tail
pancreatic duct
parietal pericardium
parietal peritoneum
parietal pleura
parotid duct
parotid gland
pericardial cavity
pericardium
peritoneum
pharynx
phrenic nerve
pleura
posterior vena cava

pulmonary ligament
pyloric sphincter
rectum
root of the lung
round ligament
ruga (pl., rugae)
small intestine
soft palate
spleen
stomach
sublingual gland
thoracic cavity
thorax
thyroid cartilage
tongue
tonsillar fossae (tonsillar pits)
trachea
tracheal cartilages
transverse colon
urinary bladder
vagus nerve
vallate papillae
vestibule
visceral pericardium
visceral peritoneum
visceral pleura
vocal cords

SECTION V: CARDIOVASCULAR SYSTEM

Heart

Here, the heart of the cat is described. However, its small size (compared to that of a larger mammal, such as the sheep) and the likelihood that it is filled with latex limit observation of its detailed anatomy. For these reasons a dissection and description of the sheep heart is provided for further study at the end of Section V. For now, proceed to examine the cat heart based on the following directions and descriptions, although many of the structures will be more clearly observed on the sheep heart.

The **heart** lies within the pericardial cavity, which is isolated from the mediastinum by the **pericardium** (see page 245). To complete the study of the heart and its vessels, remove the pericardium carefully, especially near the roots of the lungs and where the great vessels leave the heart (see Figure 7.45). The heart has a flattened, anterior **base** and a pointed, posterior **apex**. Within the heart are the left and right **atria** (sing., **atrium**) anteriorly and the left and right **ventricles** posteriorly. The small arteries and veins on the surface of the heart are the **coronary arteries** (red) and **veins** (blue), which respectively supply and drain the heart (see Figure 7.45).

On the ventral surface of the heart, the separation of the ventricles is marked by the **interventricular paraconal sulcus**, which passes anteriorly on the left to posteriorly on the right and is often filled with fat. A similar, though less obliquely oriented groove, the **interventricular subsinuosal sulcus**, is present on the dorsal surface of the heart. The separation between the atria and ventricles is represented by the **coronary sulcus**, which is also usually filled with fat. The lateral extension of each atrium is the **auricle** (Figures 7.52 and 7.53), which has scalloped margins and is usually darker than the rest of the heart tissue. The great vessels that leave the heart anteriorly pass between the left and right auricles.

Before examining the other structures of the cardiovascular system, it is useful to review the basic flow of blood. Oxygen-depleted blood from the body returns to the right atrium through the **anterior** and **posterior venae cavae** (sing., **vena cava**). From there, it is pumped to the right ventricle, which pumps it through the pulmonary arteries to the lungs for reoxygenation. From the lungs, oxygenated blood returns to the heart via the pulmonary veins to the left atrium. The blood then enters the left ventricle, which pumps it through the aorta and arteries to the rest of the body. Note that a vessel is defined based on the direction of blood flow within it (toward or away from the heart), not on the type of blood it carries (oxygenated or oxygen-depleted). A *vein* carries blood toward the heart, and an *artery* carries blood away from the heart.

Vessels

Main Vessels Associated with the Heart

The great vessels leave the base of the heart (Figures 7.45 and 7.55). The **pulmonary trunk** leads from the right ventricle. It is located on the middle part of the ventral surface and angled toward the left. The **aorta**, carrying blood from the left ventricle, extends anteriorly just dorsal to the pulmonary trunk and medial to the right auricle. It curves sharply to the left as the **aortic arch**, and then posteriorly, passing dorsal to the root of the left lung, as the **thoracic aorta** (Figure 7.51). Dissect carefully between the aortic arch and pulmonary trunk to find a tough ligamentous band connecting these vessels. This is the **ligamentum arteriosum**, a remnant of the ductus arteriosus, through which blood was shunted in the fetus.

The pulmonary trunk is very short and divides almost immediately into left and right **pulmonary arteries** (injected with blue latex). The left pulmonary artery is clearly discernible as it passes laterally to the left lung. The right pulmonary artery passes under the aortic arch to reach the right lung. Blood from the lungs returns to the left atrium through left and right **pulmonary veins** (injected with red latex), which are easily seen ventrally on the roots of the lungs (Figure 7.51). Blood from the body returns to the heart via the venae cavae. The **anterior vena cava**, draining the region anterior to the diaphragm has already been noted during dissection of the mediastinum. The **posterior vena cava** drains the regions posterior to the diaphragm. Lift the right side of the heart and note these vessels entering the right atrium (Figures 7.42, 7.43 and 7.53).

The peripheral distribution, including the main branches and tributaries, of vessels that leave the heart is treated in the following sections. Diagrams that summarize the patterns are provided at the end of this Section in Figures 7.64 through 7.67.

Vessels Anterior to the Diaphragm

The vessels described here are branches of the aortic arch and thoracic aorta and tributaries of the anterior vena cava, which, respectively, mainly supply and drain the regions of the body anterior to the diaphragm. The summit of the aortic arch gives rise to two main vessels (Figures 7.55 and 7.56). The larger, on the right, is the **brachiocephalic artery**; the other is the **left subclavian artery**. The anterior vena cava lies just to the right of the brachiocephalic artery. To trace these vessels anteriorly, remove the connective tissue, including the **thymus gland** (Figure 5.45) and fat associated with them, but avoid injuring the nerves. The thymus is a lymphoid organ and is especially important in development of the immune system: T lymphocytes mature and proliferate in the thymus. It is especially prominent in young mammals. Indeed, it reaches its maximum size just before sexual maturity, but thereafter decreases in size and may be difficult to find in old individuals. It lies in the mediastinum anterior to the heart, and may extend anteriorly beyond the first rib onto the ventral surface of the neck.

Begin by tracing the main tributaries of the anterior vena cava (Figures 7.53 and 7.54). One or two small branches arise from the anterior vena cava, but its most posterior branch is the **azygos vein** (Figures 7.44, 7.52–7.54) from the dorsal surface of the vena cava; lift the heart to observe it. The azygos passes a very short distance dorsally, turns abruptly posteriorly, and then passes dorsal to the root of the right lung along the right side of the body. It mainly drains the intercostal areas posterior to the heart through the **intercostal veins** from both sides of the body. The anterior intercostal areas are drained by the **highest intercostal veins**. That of the

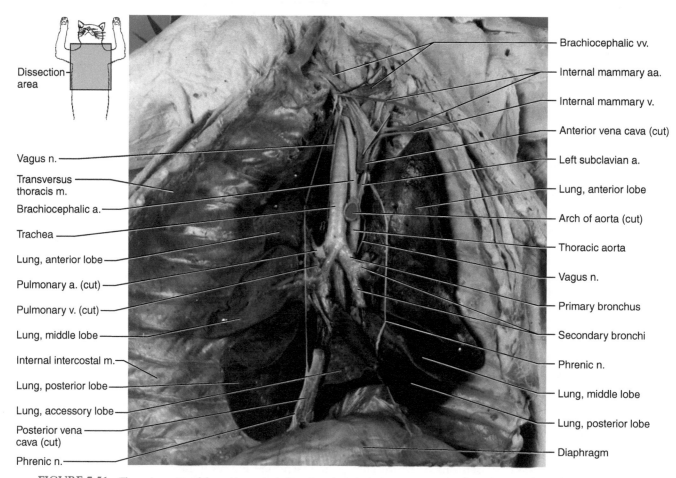

Dissection area

Vagus n.

Transversus thoracis m.

Brachiocephalic a.

Trachea

Lung, anterior lobe

Pulmonary a. (cut)

Pulmonary v. (cut)

Lung, middle lobe

Internal intercostal m.

Lung, posterior lobe

Lung, accessory lobe

Posterior vena cava (cut)

Phrenic n.

Brachiocephalic vv.

Internal mammary aa.

Internal mammary v.

Anterior vena cava (cut)

Left subclavian a.

Lung, anterior lobe

Arch of aorta (cut)

Thoracic aorta

Vagus n.

Primary bronchus

Secondary bronchi

Phrenic n.

Lung, middle lobe

Lung, posterior lobe

Diaphragm

FIGURE 7.51   Thoracic cavity of the cat in ventral view, showing respiratory system, vessels, and nerves; heart has been removed.

right side often enters the anterior part of the azygos, but may enter the vena cava directly, a short distance anterior to the azygos. The left highest intercostal vein typically enters the left side of the vena cava between the second and third ribs. The **intercostal arteries** and **highest intercostal arteries** that supply these areas extend closely in parallel with the veins; their origins are considered later.

The next large branch, the **internal mammary vein**, enters the ventral surface of the anterior vena cava. It is a short trunk formed from the confluence of **left** and **right internal mammary veins**, which pass along the inner wall of the chest on either side of the sternum (Figure 7.45). The **left** and **right internal mammary arteries** extend, for most of their length, parallel to the veins. A short distance anterior to the internal mammary vein, the anterior vena cava is formed by the confluence of the **left** and **right brachiocephalic veins**. The tributaries of the brachiocephalic veins are similar but usually not symmetric. The **costocervical + vertebral trunk,** a large

vein formed by the confluence of the **costocervical vein** and **vertebral vein,** commonly enters the right brachiocephalic dorsolaterally just before or very near the union of the right and left brachiocephalic veins. On the left side, the costocervical + vertebral trunk usually enters the left brachiocephalic slightly more anteriorly. In either case, the trunk curves dorsally. Follow the trunk to observe its tributaries. The vertebral vein heads almost directly anteriorly, whereas the costocervical heads mainly laterally and then posteriorly. In some specimens the right costocervical and vertebral veins have separate entrances, as is shown in the right side of Figure 7.54.

The brachiocephalic vein is formed by the **subclavian vein** and **bijugular trunk,** large vessels that unite just medial to the first rib. The subclavian passes laterally and helps drain the forelimb. The bijugular trunk is a very short vessel formed by the large **external jugular vein** and the much smaller **internal jugular vein.** The external jugular vein passes anteriorly along the lateroventral surface of the neck, whereas the internal jugular

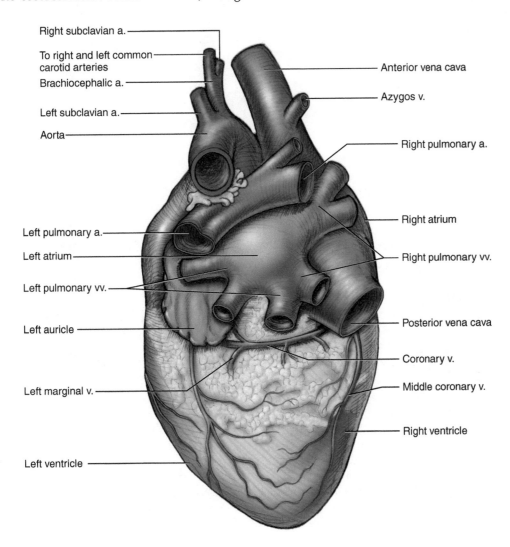

FIGURE 7.52 Heart and vessels of the cat in dorsal view.

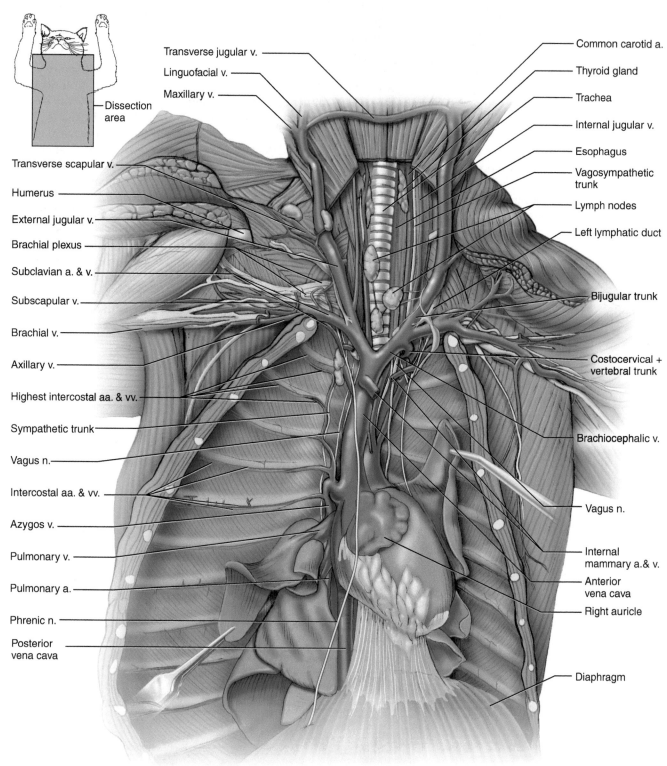

Dissection area

Transverse jugular v.

Linguofacial v.

Maxillary v.

Common carotid a.

Thyroid gland

Trachea

Internal jugular v.

Esophagus

Vagosympathetic trunk

Lymph nodes

Left lymphatic duct

Transverse scapular v.

Humerus

External jugular v.

Brachial plexus

Subclavian a. & v.

Subscapular v.

Brachial v.

Axillary v.

Highest intercostal aa. & vv.

Sympathetic trunk

Vagus n.

Intercostal aa. & vv.

Azygos v.

Pulmonary v.

Pulmonary a.

Phrenic n.

Posterior vena cava

Bijugular trunk

Costocervical + vertebral trunk

Brachiocephalic v.

Vagus n.

Internal mammary a. & v.

Anterior vena cava

Right auricle

Diaphragm

FIGURE 7.53 Vessels and nerves anterior to the heart of the cat in ventral view.

vein passes more medially along the neck. The internal jugular primarily drains the palate, pharynx, and inside of the skull, and may not be injected. In some individuals, the internal and external jugular veins unite at about the same level as the subclavian; in others, the internal jugular may even join the brachiocephalic vein. Thus, you will not necessarily be able to identify vessels based only on branching patterns; it is good practice to follow them distally toward the structures they serve. Do not confuse the internal jugular veins with the very small ventral thyroid vein. Although the latter is usually injected, it is a median unpaired vessel in the throat region. You will find an internal jugular vein lying just lateral to a common carotid artery (see later), whereas the ventral thyroid vein lies between the right and left common carotid arteries and empties into one of the brachiocephalic veins.

The external jugular vein helps drain two main areas, the forelimb and the external structures of the head. Drainage of the forelimb is accomplished through its first main tributary, the **transverse scapular vein**. As was observed earlier (page 217), this vessel enters the external jugular vein at about the level of the front of the shoulder. A main tributary of the transverse scapular vein is the **cephalic vein**, also observed earlier passing along the lateral surface of the forelimb (pages 211 and 213). Farther anteriorly the external jugular vein has three main tributaries, but their branching patterns vary (Figures 7.39).

Often, they converge close together to form the external jugular. The **transverse jugular vein** forms a ventral link between the left and right external jugular veins, and helps drain the neck. The **linguofacial vein** passes anteriorly to the face and lower jaw. The **maxillary vein** extends anterodorsally, crosses the mandibular gland, and passes deep to the parotid gland (see page 238).

Return to the aortic arch, and reidentify the brachiocephalic and left subclavian arteries (Figures 7.55 and 7.56). Just to the left of the subclavian artery, two slender **intercostal arteries** usually arise from the arch to help supply the more anterior intercostal regions. Follow the aorta as it curves posteriorly and dorsally. Lift the left lung to observe the remaining intercostal arteries that arise from the thoracic aorta and are accompanied by the intercostal veins, which were described earlier. The left subclavian artery eventually supplies the left forearm, but before doing so gives off (as does the right subclavian artery) several substantial branches that supply other structures in the trunk and neck. Tracing of these branches is deferred until the branches of the brachiocephalic artery have been studied.

The brachiocephalic artery heads anteriorly, just medial to the anterior vena cava. At about the level of the internal thoracic vein, it gives rise to three branches, the right **subclavian artery**, and the **right** and **left common carotid arteries**. The subclavian artery lies most laterally, while the common carotid arteries pass anteriorly on either side of the trachea. The arteries pass deep to the anterior vena cava and brachiocephalic veins. The left common carotid sometimes arises independently, with the right common carotid and right subclavian arising farther distally from a very short common trunk. All three arteries, however, may arise very close together.

As the common carotid arteries ascend the neck, they send out several small branches to supply the **thyroid gland**, muscles, and **lymph nodes** (Figure 7.52). The thyroid gland is a bilobed endocrine gland. Its two lobes lie laterally on the anterior end of the trachea near the larynx and are connected by a thin isthmus that is often destroyed during dissection. This gland produces, stores, and releases two iodine-containing thyroid hormones, tetraiodothyronine (T4 or thyroxine) and triiodothyronine (T3), that are important in regulating metabolic rate, growth, and reproduction. Incorporated in the thyroid are scattered parafollicular cells that produce calcitonin, a hormone that functions to lower blood calcium levels. In most other vertebrates, such cells occur as usually paired masses, termed ultimobranchial bodies, in the throat region. Their embryonic origin is distinct from that of the thyroid (see Kardong, 2018, p. 790). A small pair of parathyroid glands are embedded in each lobe of the thyroid; their identification requires magnification. The parathyroid glands produce parathyroid hormone (or parathormone), which acts to raise calcium blood levels.

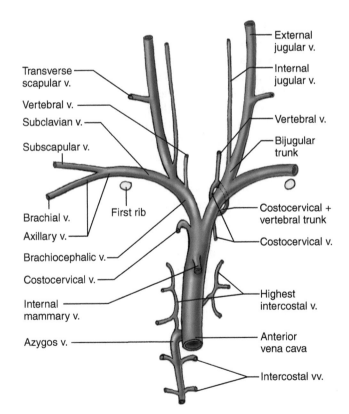

Transverse scapular v.
Vertebral v.
Subclavian v.
Subscapular v.
Brachial v.
First rib
Axillary v.
Brachiocephalic v.
Costocervical v.
Internal mammary v.
Azygos v.

External jugular v.
Internal jugular v.
Vertebral v.
Bijugular trunk
Costocervical + vertebral trunk
Costocervical v.
Highest intercostal v.
Anterior vena cava
Intercostal vv.

FIGURE 7.54 Schematic illustration of the veins anterior to the heart of the cat in ventral view.

Note the nerves present in this region. The **phrenic** and **vagus nerves** were observed passing through the thorax (Figures 7.42 and 7.43). In the cervical region, the vagus and the anterior extension of the **sympathetic trunk** (Figure 7.53) form the **vagosympathetic trunk**, which is bound together with the common carotid artery and internal jugular vein by a tough connective sheath. The vagus and sympathetic trunk separate near the branching of the brachiocephalic artery. The sympathetic trunk extends anteroposteriorly along the dorsal wall of the thorax, crossing ventral to the intercostal arteries and veins. Carefully dissect it out, and note the series of small ganglia along its course.

Follow one of the common carotid arteries as it approaches the skull, near which it divides. Its main branch is the **external carotid artery**, which is as wide as the common carotid artery, so the division is not readily apparent (Figure 7.57). Find the branching point as follows. A short distance posterior to the masseter muscle, locate the **hypoglossal nerve**, which was observed earlier (page 235). The nerve crosses the very proximal part of the external carotid artery, so the division of the common carotid artery occurs just posterior to the nerve. Carefully dissect along the dorsal surface of the common carotid artery, just posterior to the hypoglossal nerve. In most mammals three small arteries usually arise in very close proximity. There are the internal carotid artery and the **occipital** and **ascending pharyngeal arteries**; the last two usually arise by a short, common trunk. These arteries can be dissected, but they may be difficult to find because they are small, are closely bound by connective tissue, and in the cat one of the arteries, the internal carotid artery, is largely occluded and represented by a whitish strand, the **internal carotid ligament**. (Only the proximal portion of the internal carotid artery is occluded. Its distal part, within the skull, remains functional, receiving blood via anastomoses with other arteries, such as the maxillary and ascending pharyngeal arteries.) The arteries and ligament pass anterodorsally with the ligament being most posterior of the three. Thus, even though they arise almost together with the internal carotid ligament, the occipital and ascending pharyngeal arteries are actually the first branches of the external carotid artery. The ligament extends nearly dorsally, and so is not visible in Figure 7.57. These three small vessels may be observed, but this requires time and delicate dissection. If you cannot find them, use the hypoglossal nerve as a landmark to indicate the beginning of the external carotid artery.

The distribution of the main superficial vessels on the skull may be examined next. Anterior to the hypoglossal nerve, the external carotid artery gives rise to the **lingual artery**, which passes anteromedially toward the base of the tongue. The lingual artery is also crossed by the hypoglossal nerve. The external carotid artery continues as a large vessel and soon gives rise to the **facial artery**, which passes anteriorly along the ventral margin of the masseter muscle and subdivides into various branches to supply the jaws and facial structures. The external carotid continues dorsally, deep to the mandibular gland. Follow it by dissecting between the ventral parts of the mandibular and parotid glands. It gives rise first to the **posterior auricular artery**, a fairly large branch that extends dorsally posterior to the ear, and then the **superior temporal artery**, a smaller branch, extending dorsally anterior to the ear. The superficial temporal divides into the **anterior auricular artery** and **transverse facial artery**. After it gives off the superficial temporal, the external carotid continues as the **maxillary artery**, a large vessel that passes into the dorsal part of the masseter muscle to supply orbital and palatal regions.

The regions supplied by the arteries discussed in the preceding paragraph are drained by veins that extend mainly parallel to the arteries but with slightly different branching patterns. Return to the external jugular vein and trace it forward toward the skull. Three large vessels were noted in connection with the beginning of the external jugular vein on page 238 and Figure 7.39. These branches will now be considered in detail. The external jugular is formed by the confluence of the **maxillary vein** and **linguofacial vein**. The latter extends anteriorly between the two prominent lymph nodes lying ventral to the salivary glands (see page 238 and Figure 7.39). The **transverse jugular vein** enters the very proximal end of the linguofacial vein, although in some individuals the transverse facial, linguofacial, and maxillary veins appear nearly to branch from the same point. Follow the linguofacial anteriorly. It is formed by the union of the **facial vein**, which extends along the ventral margin of the masseter muscle, and the **lingual vein**, which extends ventromedially in draining the mandible, lower lip, and tongue. The maxillary vein extends dorsally deep to the parotid gland. Follow it, dissecting away potions of the gland as necessary, to see the **posterior auricular vein**, from behind the ear, entering the maxillary vein. The latter continues a short distance anterodorsally to receive the **superficial temporal vein** and then continues deep to the posterodorsal margin of the masseter muscle. The superficial temporal vein receives the **anterior auricular vein** from in front of the ear and continues anteriorly toward the eye as the **transverse facial vein**.

Finally, consider the branches of the subclavian arteries and veins (Figures 7.53–7.56). These vessels help supply and drain the forelimbs, as well as regions of the neck and thorax. Trace them on the side opposite that on which the forelimb muscles were dissected. For the most part, it is easier to find the arteries and then note the corresponding veins, which extend beside them but are not injected.

The subclavian artery and subclavian vein pass toward the forelimb laterally, just in front of the first rib. Follow the subclavian artery. Just before it passes out of the thorax, it

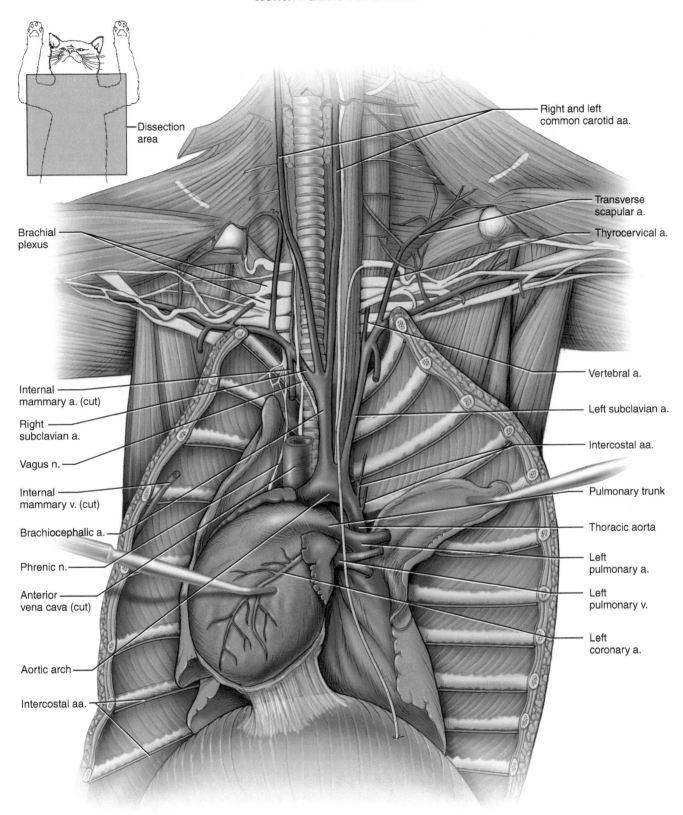

Dissection area

Right and left common carotid aa.

Transverse scapular a.

Thyrocervical a.

Brachial plexus

Vertebral a.

Internal mammary a. (cut)

Left subclavian a.

Right subclavian a.

Intercostal aa.

Vagus n.

Internal mammary v. (cut)

Pulmonary trunk

Brachiocephalic a.

Thoracic aorta

Phrenic n.

Left pulmonary a.

Anterior vena cava (cut)

Left pulmonary v.

Left coronary a.

Aortic arch

Intercostal aa.

FIGURE 7.55 Arterial and partial venous systems and nerves anterior to the heart of the cat in ventral view.

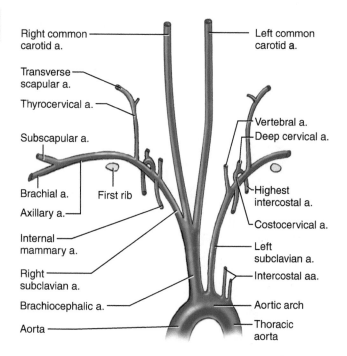

Right common carotid a.

Left common carotid a.

Transverse scapular a.

Thyrocervical a.

Subscapular a.

Vertebral a.

Deep cervical a.

Brachial a.   First rib

Axillary a.

Highest intercostal a.

Costocervical a.

Internal mammary a.

Left subclavian a.

Right subclavian a.

Intercostal aa.

Brachiocephalic a.

Aortic arch

Aorta

Thoracic aorta

FIGURE 7.56 Schematic illustration of the arteries anterior to the heart of the cat in ventral view.

gives rise to four arteries. Dissect around the artery to find these branches. The **internal mammary artery** arises from the ventral surface of the subclavian artery and extends posteriorly. The left and right internal mammary arteries converge toward the ventral midline, and pass along either side of the sternum in parallel with the internal mammary veins, noted earlier (see page 255). If you have trouble locating its origin, follow the artery proximally to the subclavian artery from the midventral strip of the thorax. Note that the internal mammary vein empties into the anterior vena cava and not the subclavian vein. The **costocervical artery** arises from the posterodorsal surface of the subclavian artery almost opposite to the internal mammary artery. The costocervical is a short trunk that soon branches into two vessels. The **highest intercostal artery** is the smaller branch. It passes back over the medial surface of the first rib onto the anterior part of the thorax. The other, larger branch is the **deep cervical artery**, which extends almost directly dorsally to supply deep neck muscles.

The other two arteries, the **vertebral** and **thyrocervical arteries**, can be exposed by gently pulling the subclavian posteriorly and carefully dissecting along its anterior surface. The vertebral artery is more medial and passes anteriorly and dorsally. The thyrocervical artery is more lateral. It passes anteriorly deep to the subclavian vein and then for a short distance follows along the dorsolateral side of the external jugular vein. It gives off a small medial branch toward the thyroid and then passes laterally as the **transverse scapular artery**, in company with the transverse scapular vein, at about the level of the anterior part of the shoulder.

Just anterior to the first rib, the subclavian artery and vein pass from the thorax into the axilla (armpit) as the **axillary artery** and **vein**. These vessels and their branches extend toward the forelimb in company with the complex of nerves termed the *brachial plexus*, which was examined briefly in Section III (see pages 217 and 218), so the vessels require careful dissection. The axillary artery sends out several branches, mainly to surrounding pectoral musculature, before it branches into two large vessels, the **brachial artery**, which continues onto the medial surface of the brachium, and the **subscapular artery**, which extends into the musculature toward the anterior part of the shoulder. The axillary vein receives the **subscapular vein**. Distally it follows the subscapular artery, but it enters the axillary vein more proximally than the origin of the subscapular artery from the axillary artery. The axillary vein continues distally, receiving tributaries from the surrounding musculature, and passes onto the medial surface of the brachium as the **brachial vein**.

Vessels Posterior to the Diaphragm

The vessels considered here mainly supply and drain the musculature and organs in the posterior part of the body. Trace the **posterior vena cava** posteriorly from the heart (Figures 7.42, 7.43, 7.45, 7.46, 7.53, and 7.54). It passes through the diaphragm and enters the liver. It receives several **hepatic veins** that drain the liver. Scrape away some of the liver, just anterior to its quadrate lobe, to expose the veins. The posterior vena cava continues through the liver. It emerges into the abdominopelvic cavity from beneath the caudate lobe of the liver, passes between the left and right kidneys, and continues posteriorly (Figures 7.58 and 7.59). Return to the thoracic aorta and, lifting the left lung, trace it posteriorly (Figure 7.51). It passes through the diaphragm and into the abdominopelvic cavity as the **abdominal aorta**, and extends posteriorly to the left of the posterior vena cava (Figure 7.58).

Push the digestive tract and spleen to the right, so you can view the abdominal aorta from the left side. Immediately on passing through the diaphragm, it gives rise to the **celiac artery** and then the **anterior mesenteric artery**, both large, unpaired vessels that mainly supply the viscera (Figure 7.58). With the viscera pushed to the right, these vessels pass directly laterally toward the right. They will be traced shortly. The veins that drain these structures are part of the **hepatic portal system** and are also discussed later. Careful dissection will reveal several ganglia along the ventral surface of the aorta between the origins of these arteries. Continue to follow the aorta posteriorly. The next large branches of the aorta are the paired **renal arteries**. Remove the peritoneum covering the kidneys and follow the right and left renal arteries as they pass toward the central part of the medial surface of each kidney. As the right kidney lies slightly anterior to the left kidney, the right renal artery is slightly anterior to the left renal artery. The **renal veins**

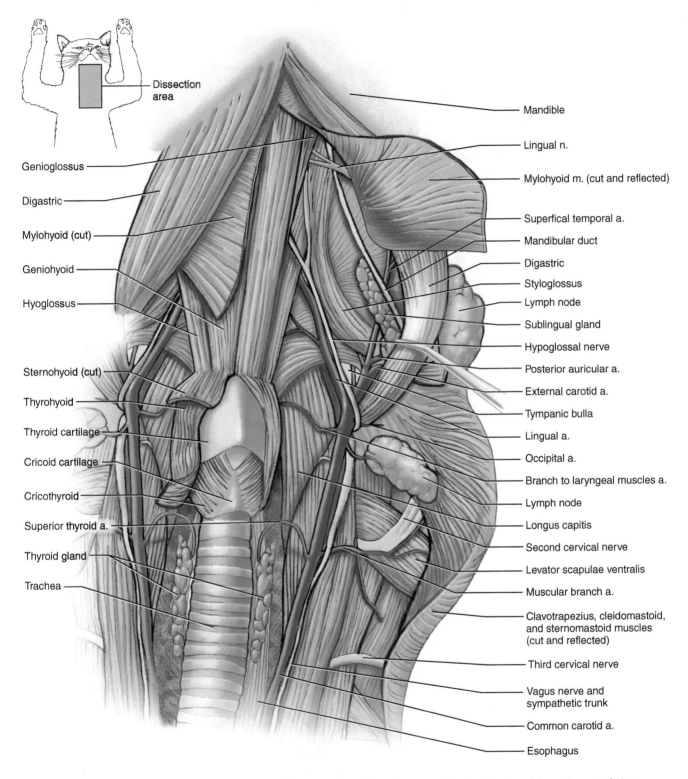

Dissection area

Genioglossus

Digastric

Mylohyoid (cut)

Geniohyoid

Hyoglossus

Sternohyoid (cut)

Thyrohyoid

Thyroid cartilage

Cricoid cartilage

Cricothyroid

Superior thyroid a.

Thyroid gland

Trachea

Mandible

Lingual n.

Mylohyoid m. (cut and reflected)

Superfical temporal a.

Mandibular duct

Digastric

Styloglossus

Lymph node

Sublingual gland

Hypoglossal nerve

Posterior auricular a.

External carotid a.

Tympanic bulla

Lingual a.

Occipital a.

Branch to laryngeal muscles a.

Lymph node

Longus capitis

Second cervical nerve

Levator scapulae ventralis

Muscular branch a.

Clavotrapezius, cleidomastoid, and sternomastoid muscles (cut and reflected)

Third cervical nerve

Vagus nerve and sympathetic trunk

Common carotid a.

Esophagus

FIGURE 7.57 Branches of the common carotid artery, and muscles and nerves of the throat region of the cat in ventral view.

(Figures 7.58 and 7.59), of which the left is usually larger, follow the renal arteries.

Between the anterior mesenteric and renal arteries, the aorta usually gives rise to slender and paired **adrenolumbar arteries**, which may also arise from the renal arteries (Figures 7.58 and 7.60). On each side, the adrenolumbar artery supplies and passes the small, nodular **adrenal gland**. The right adrenal gland lies between the anterior end of the kidney and the vena cava, and the left adrenal gland lies between the kidney and the aorta. Remove the fat and tissue from around the glands to expose them. Lifting the kidneys, follow the adrenolumbar arteries.

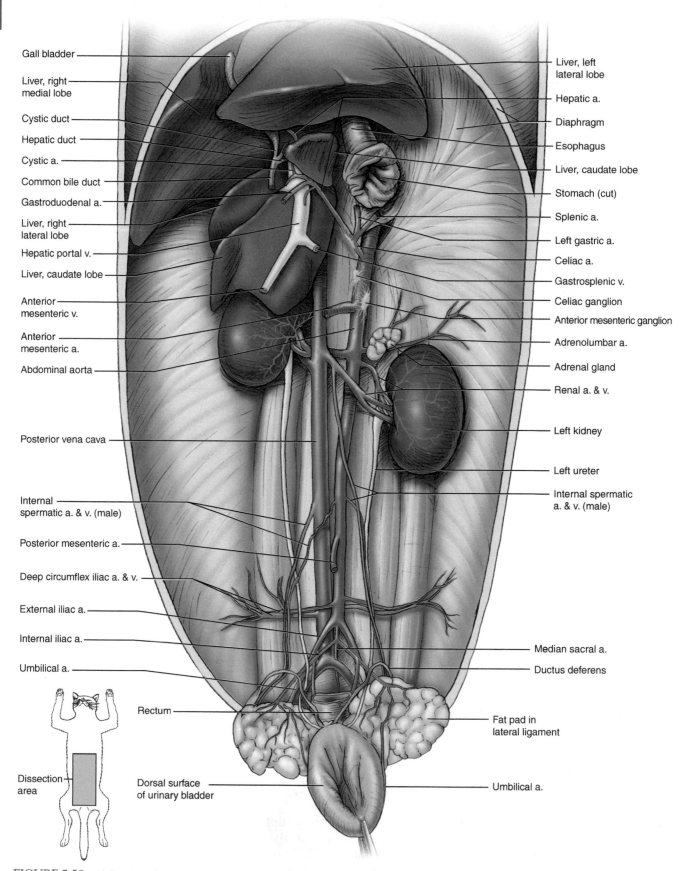

FIGURE 7.58 Abdominopelvic cavity of the cat showing arteries and veins posterior to the heart, with most of digestive system removed and urinary bladder reflected, in ventral view.

They continue onto the dorsal wall of the abdominal cavity. The **adrenolumbar veins** accompany the arteries. They may empty into the posterior vena cava or the renal veins (Figures 7.58 and 7.59).

Posterior to the renal arteries, the aorta gives rise to paired gonadal arteries, **internal spermatic arteries** in the male and **ovarian arteries** in the female (Figures 7.58 and 7.60). In the female they pass almost directly laterally to the ovaries, which lie at the anterior end of each uterine horn, the paired proximal extensions of the uterus (Figure 7.71). In the male, however, the internal spermatic arteries extend posteriorly, through the abdominal wall (by way of the inguinal canal) to the scrotum, which contains the testes (Figures 7.58 and 7.70). The right **ovarian** and **internal spermatic veins** closely follow the corresponding arteries to enter the posterior vena cava. On the left side, however, the proximal portions of the artery and vein are farther apart because the vein enters the renal vein.

The next aortic branch, the **posterior mesenteric artery**, is an unpaired vessel arising from the ventral surface of the aorta that helps supply the viscera. Its peripheral distribution is discussed later. Almost immediately posterior to the origin of the posterior mesenteric artery, the aorta gives rise to the paired **deep circumflex iliac arteries**. The **deep circumflex iliac veins** closely follow the arteries to enter the posterior vena cava.

Just beyond these vessels, the aorta gives rise to two large branches, the **external iliac arteries**, which extend posterolaterally into each hind limb. The aorta continues as a narrower vessel for a very short distance before giving off the **internal iliac arteries**, which also extend posterolaterally but are smaller than the external iliac arteries. Follow the internal iliac arteries for a short distance to see that they give off branches to the pelvic viscera. Past the origin of the internal iliac arteries, the aorta continues as the small **median sacral artery** and then enters the tail as the **caudal artery** (but you will not look for it).

In contrast to the similar branching patterns of most of the paired arteries and veins of the aorta and posterior vena cava, the patterns of the iliac arteries and veins differ. Near the origin of the external iliac arteries, the posterior vena cava begins with the confluence of two large tributaries,

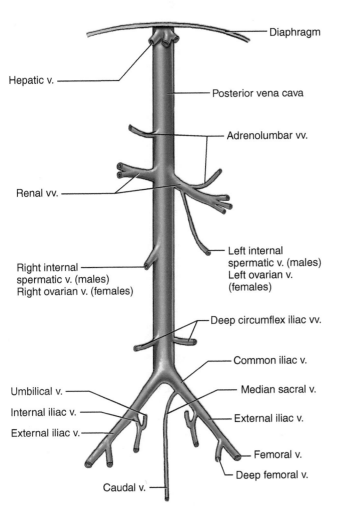

FIGURE 7.59 Schematic illustration of the veins posterior to the heart of the cat in ventral view.

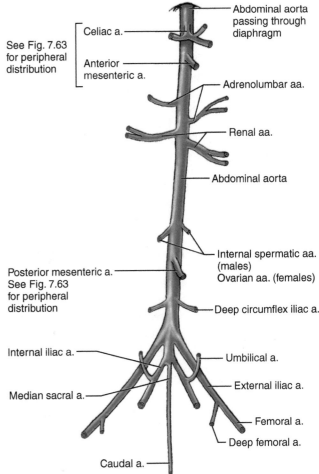

FIGURE 7.60 Schematic illustration of the arteries posterior to the heart of the cat in ventral view.

the **common iliac veins**. Each common iliac vein, in turn, is formed by confluence of the **internal** and **external iliac veins**. The **median sacral vein**, which accompanies the median sacral artery, usually joins one of the common iliac veins, and continues in the tail as the **caudal vein**.

Return to the external iliac arteries and follow one of them distally. It passes through the abdominal wall and enters the hind limb as the **femoral artery**. Just before it passes through the body wall, it gives rise from its medial surface to the **deep femoral artery**. The latter gives rise to several branches and then passes deep into the medial side of the thigh. These arteries and veins were considered during the dissection of the musculature of the hind limb (see pages 220 and 222 and Figure 7.32). The internal iliac artery mainly supplies muscular and visceral structures in the pelvic canal. Its first branch, the **umbilical artery**, supplies the urinary bladder. The **femoral vein** and **deep femoral vein** empty into the external iliac vein. The **umbilical vein** usually empties into the internal iliac vein.

Vessels Associated with the Viscera

In this section, the vessels that supply and drain the viscera are examined. The arterial system consists of three vessels, the celiac, anterior mesenteric, and posterior mesenteric arteries, and their branches. The origins of the three main arteries from the aorta have already been identified. The structures supplied by these arteries are drained by veins that ultimately form the hepatic portal vein, which returns blood to the liver (Figures 7.58, 7.61, and 7.62). The hepatic portal vein breaks up into capillaries within the liver. These recollect into several hepatic veins, which then enter the posterior vena cava.

Reflect the viscera to the right, and locate the origins of the celiac and anterior mesenteric arteries from the aorta, as it emerges through the diaphragm, and extend toward the right. The celiac artery is a short trunk that divides into three branches. Two of these, the **hepatic** and **left gastric arteries**, slant anteriorly toward the right in this view. The third, the **splenic artery**, is the largest and extends almost directly toward the right. You can easily follow the hepatic artery for about 2 cm in this view, but change perspective to follow it more distally.

Reflect the viscera (including the duodenum and head of the pancreas) to the left, and locate the hepatic portal vein, sitting primarily on the caudate lobe of the liver, and the common bile duct as it extends toward the duodenum; these structures were exposed earlier (page 247; see also Figure 7.58). The hepatic portal vein is difficult to see if it is not injected. Carefully dissect along the

hepatic portal vein, freeing it as much as possible from connective tissue without destroying any of its branches. (In some specimens the vein is injected with yellow latex, which greatly facilitates dissection.) In doing so, you should reveal the hepatic artery (Figures 7.61 and 7.63). Ascertain that this is indeed the hepatic artery by reflecting the viscera to the right again, gently but firmly grasping the hepatic artery with blunt forceps, and then reflecting the viscera to the left again, thus verifying that the gripped vessel is the one you saw from the other view. This method of using forceps to identify a vessel can be used to advantage in many situations.

Follow the hepatic artery anteriorly. Near the anterior margin of the caudate lobe, a prominent branch (the hepatic artery proper) heads toward the liver. The hepatic artery then gives rise to two final branches, a larger **gastroduodenal artery**, which follows the common bile duct as it heads toward the duodenum, and a smaller branch, the **right gastric artery**. The latter supplies the pylorus, gastrohepatoduodenal ligament, and lesser curvature of the stomach, and then anastomoses with the left gastric artery, which is larger. The **right gastric vein**, which accompanies right gastric artery, enters the hepatic portal vein. Trace the gastroduodenal artery until it divides into posteriorly curving and anteriorly curving branches dorsal to the pylorus. The posterior branch, the **anterior pancreaticoduodenal artery**, passes between the duodenum and pancreas. The anterior branch, the **right gastroepiploic artery**, passes to the distal part of the greater curvature of the stomach. Follow the right gastroepiploic artery with the duodenum in natural position but with the stomach reflected anteriorly (i.e., so you view its dorsal surface). The arteries are followed closely by corresponding veins. Trace the **right gastroepiploic vein** and **anterior pancreaticoduodenal vein**; their confluence forms the **gastroduodenal vein**, which enters the hepatic portal vein.

With the stomach reflected anteriorly, gently pull the tail of the pancreas posteriorly. Dissect away the connective tissue and fat between the pancreas and stomach to reveal a system of arteries and veins. Then, reflect the viscera to the right and grasp the left gastric artery with forceps, and again reflect the stomach anteriorly. Trace the left gastric artery as it passes onto the lesser curvature of the stomach. The **left gastric vein** accompanies the artery.

The splenic artery passes to the left toward the spleen (verify this by reflecting the viscera to the right and grasping the vessel with forceps). The **gastrosplenic vein**,[2] which accompanies the splenic artery (note that there

[2] Gastrosplenic and splenic vessels may also be referred to as lienogastric and lienic; for example lienogastric vein is used for the shark and mudpuppy. However, the pattern of the branches of the celiac artery and hepatic portal vein differ in these vertebrates and the cat, and usage of gastrosplenic vein in the cat follows Crouch's (1969) usage, and splenic (rather than lienic or lienal) is used here for associated veins and arteries for consistency with the term gastrosplenic. Hermanson et al. (2019) follow these designations for the vessels in the dog.

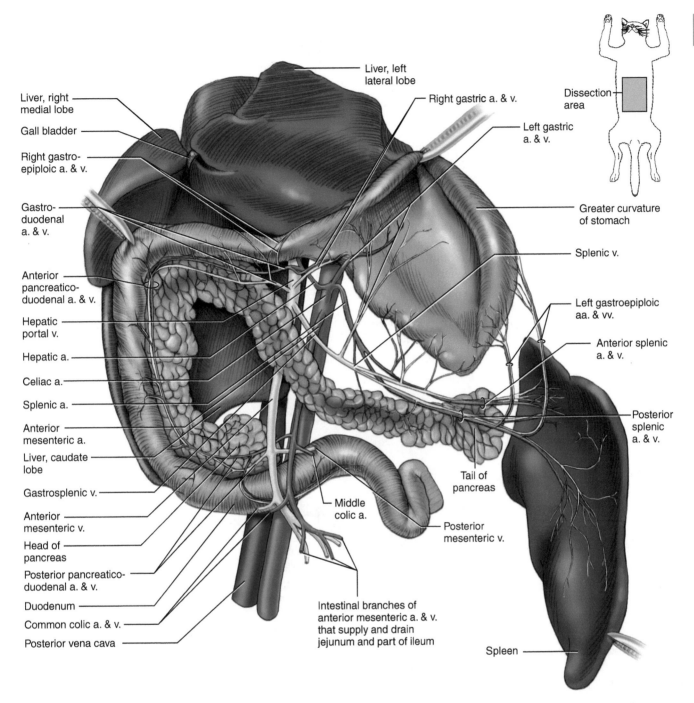

Liver, left
lateral lobe

Right gastric a. & v.

Left gastric
a. & v.

Dissection
area

Liver, right
medial lobe

Gall bladder

Right gastro-
epiploic a. & v.

Gastro-
duodenal
a. & v.

Anterior
pancreatico-
duodenal a. & v.

Hepatic
portal v.

Hepatic a.

Celiac a.

Splenic a.

Anterior
mesenteric a.

Liver, caudate
lobe

Gastrosplenic v.

Anterior
mesenteric v.

Head of
pancreas

Posterior pancreatico-
duodenal a. & v.

Duodenum

Common colic a. & v.

Posterior vena cava

Greater curvature
of stomach

Splenic v.

Left gastroepiploic
aa. & vv.

Anterior splenic
a. & v.

Posterior
splenic
a. & v.

Tail of
pancreas

Middle
colic a.

Posterior
mesenteric v.

Intestinal branches of
anterior mesenteric a. & v.
that supply and drain
jejunum and part of ileum

Spleen

FIGURE 7.61 Detail of vessels of the viscera, showing liver (reflected), stomach, duodenum, pancreas, and spleen (reflected) of the cat in ventral view.

is no gastrosplenic artery), is conspicuous even in unin-jected specimens. It is formed by the confluence of the left gastric vein and the **splenic vein**. Follow the splenic artery and vein. After one or two small branches to the pancreas, the vein and artery divide. Follow the vein first, with both the stomach and spleen reflected anteriorly. In this view, one tributary, the **anterior splenic vein**, heads almost directly laterally and drains the left side of the spleen; the other, the **posterior splenic vein**, passes posterolaterally

and drains the right side of the spleen. The **anterior** and **posterior splenic arteries** also follow this pattern, but the anterior splenic artery lies deep to the vein and is not apparent at first. Trace the posterior splenic artery and vein into the spleen. Some of their branches will be seen to extend across to the greater curvature of the stomach. These are the **left gastroepiploic arteries** and **veins**.

Work your way back proximally along the gastro-splenic vein. Its confluence with the **anterior mesenteric**

vein, oriented nearly anteroposteriorly, forms the hepatic portal vein. Dissect along the anterior mesenteric vein for about 3 cm. With the viscera reflected to the right, trace the anterior mesenteric artery as it angles posteriorly to the right. Note the anterior mesenteric vein converging toward the artery. Continue to expose the anterior mesenteric artery. You will need to move through the coils of the intestines individually and tear the mesentery binding the coils together. As the artery and vein extend distally, they give off numerous branches, which mainly supply and drain the jejunum, ileum, cecum, and proximal portion of the colon, as described later.

Two proximal branches of the anterior mesenteric artery should be traced. They often arise in close proximity, and either may have the more proximal origin (see Figures 7.61 and 7.63). The **posterior pancreaticoduodenal artery** extends to the distal part of the duodenum and head of the pancreas. It passes anteriorly and anastomoses with the anterior pancreaticoduodenal artery. If you have trouble tracing the posterior pancreaticoduodenal

artery from the anterior mesenteric artery, begin by tracing the anterior pancreaticoduodenal artery posteriorly, and then follow the posterior pancreaticoduodenal artery to the anterior mesenteric artery. The artery is accompanied by the **posterior pancreaticoduodenal vein**, which empties into the anterior mesenteric vein.

The other branch of the anterior mesenteric artery is the **middle colic artery** (Figures 7.61 and 7.63). It gives off several branches, mainly supplying the distal part of the transverse colon and proximal part of the descending colon. Its main part continues posteriorly along the medial side of the descending colon and anastomoses with a branch of the posterior mesenteric artery (see later). The vein that accompanies the middle colic artery proximally is the **posterior mesenteric vein**, which empties into the anterior mesenteric vein near the entrance of the posterior pancreaticoduodenal vein.

Another readily recognizable branch of the anterior mesenteric artery arises just distal to the middle colic. This is the **ileocolic artery** (Figures 7.61 and 7.63), which

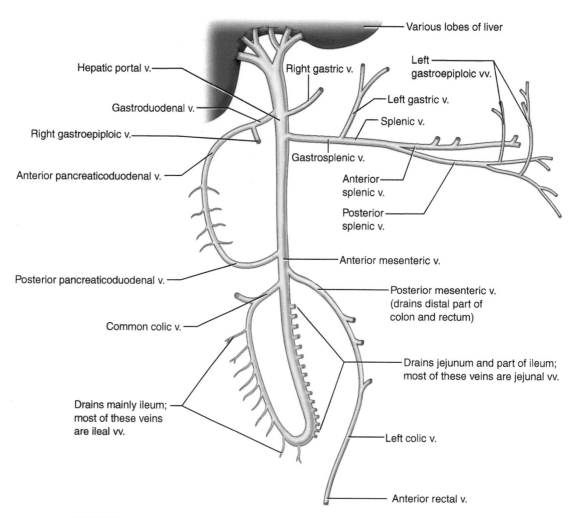

FIGURE 7.62   Schematic illustration of the hepatic portal system of the cat in ventral view.

extends initially toward the right. It is accompanied by the **ileocolic vein**, which empties into the anterior mesenteric vein. The artery's proximal branches mainly supply the ascending colon and proximal part of the transverse colon, but the vessel then turns posteriorly to supply the distal part of the ileum and eventually anastomoses with the distal end of the anterior mesenteric artery. Once the ileocolic artery has branched off, the anterior mesenteric artery continues posteriorly, giving rise to numerous branches that supply the jejunum and ileum (as shown in Figure 7.49). These vessels anastomose with each other, forming loops or arcades.

Next, trace the **posterior mesenteric artery**. It is the smallest and least complexly branched of the arteries that supply the viscera. Usually, the posterior mesenteric artery forms a short trunk that branches into the **left colic artery** and the **anterior rectal artery**. The left colic passes anteriorly on the surface of the descending colon and anastomoses with the middle colic artery. The anterior rectal artery passes posteriorly to supply the distal end of the colon and the rectum. These arteries are accompanied by the posterior continuation of the posterior mesenteric vein, which sequentially becomes the **left colic vein** and the **anterior rectal vein**.

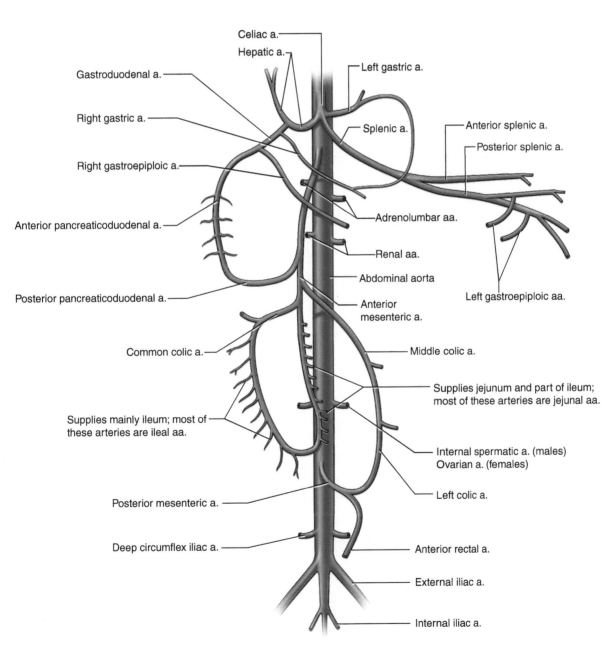

Celiac a.

Hepatic a.

Left gastric a.

Gastroduodenal a.

Right gastric a.

Splenic a.

Anterior splenic a.

Posterior splenic a.

Right gastroepiploic a.

Anterior pancreaticoduodenal a.

Adrenolumbar aa.

Renal aa.

Abdominal aorta

Left gastroepiploic aa.

Posterior pancreaticoduodenal a.

Anterior mesenteric a.

Common colic a.

Middle colic a.

Supplies jejunum and part of ileum; most of these arteries are jejunal aa.

Supplies mainly ileum; most of these arteries are ileal aa.

Internal spermatic a. (males)
Ovarian a. (females)

Left colic a.

Posterior mesenteric a.

Deep circumflex iliac a.

Anterior rectal a.

External iliac a.

Internal iliac a.

FIGURE 7.63   Schematic illustration of the visceral arterial system of the cat in ventral view.

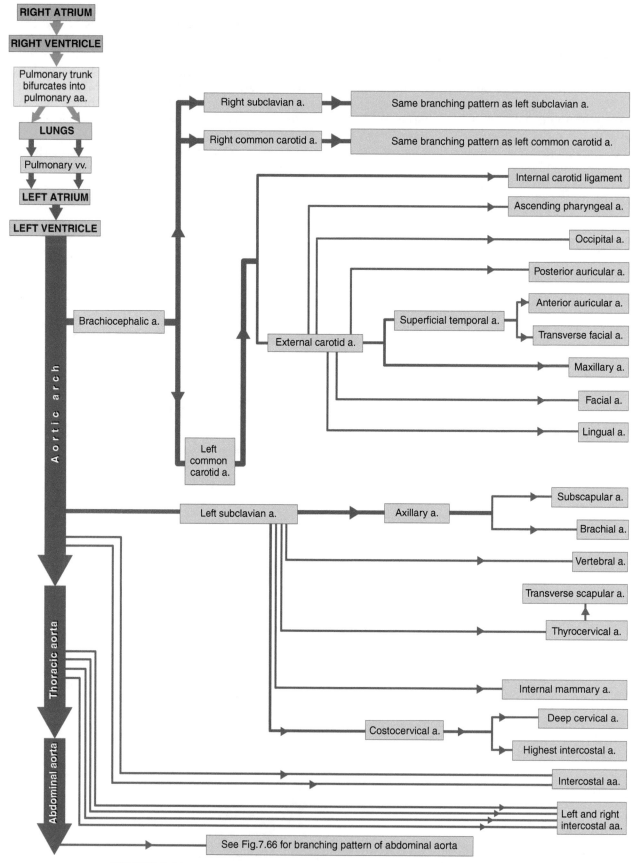

FIGURE 7.64   Flow chart diagram of the arterial system anterior to the heart of the cat.

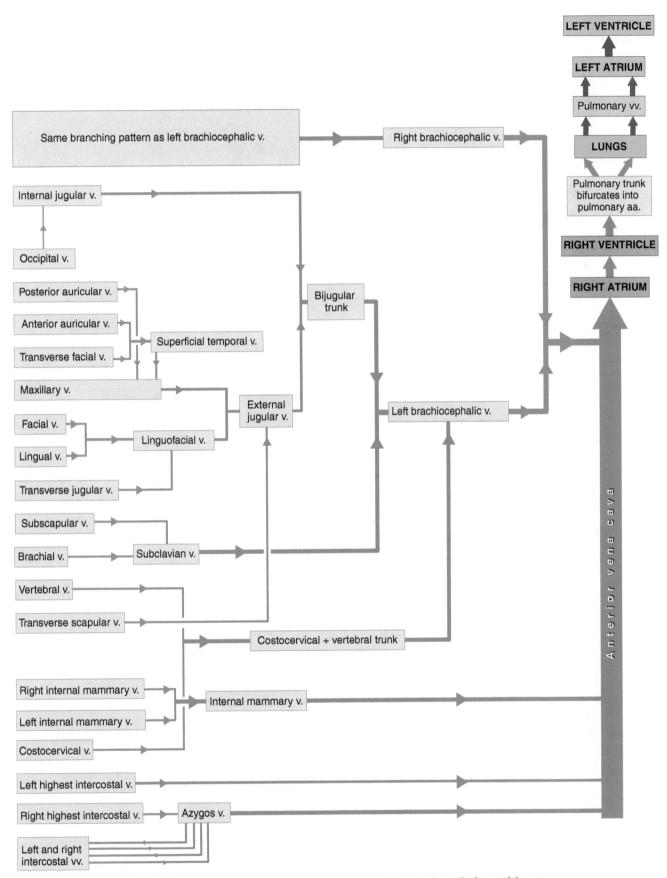

FIGURE 7.65   Flow chart diagram of the venous system anterior to the heart of the cat.

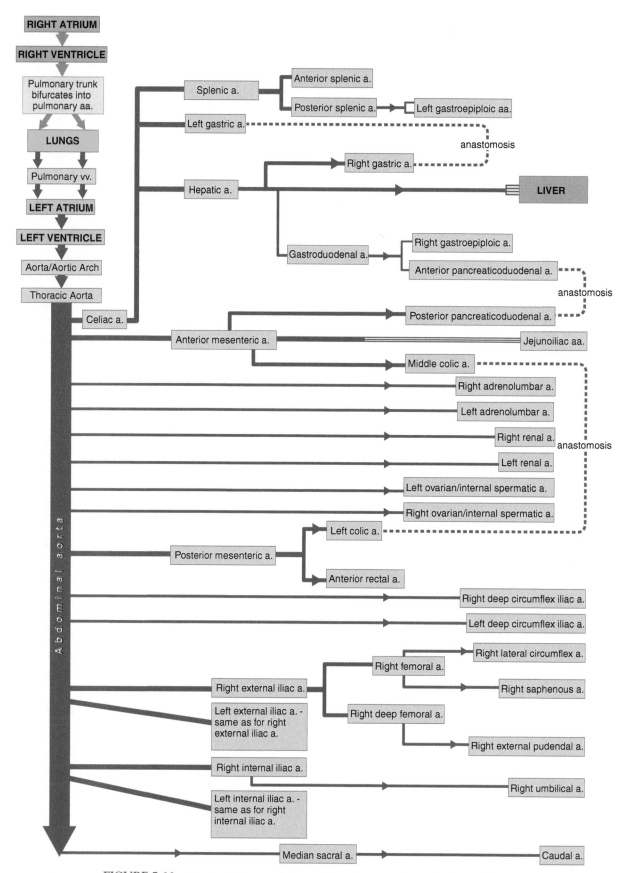

FIGURE 7.66   Flow chart diagram of the arterial system posterior to the heart of the cat.

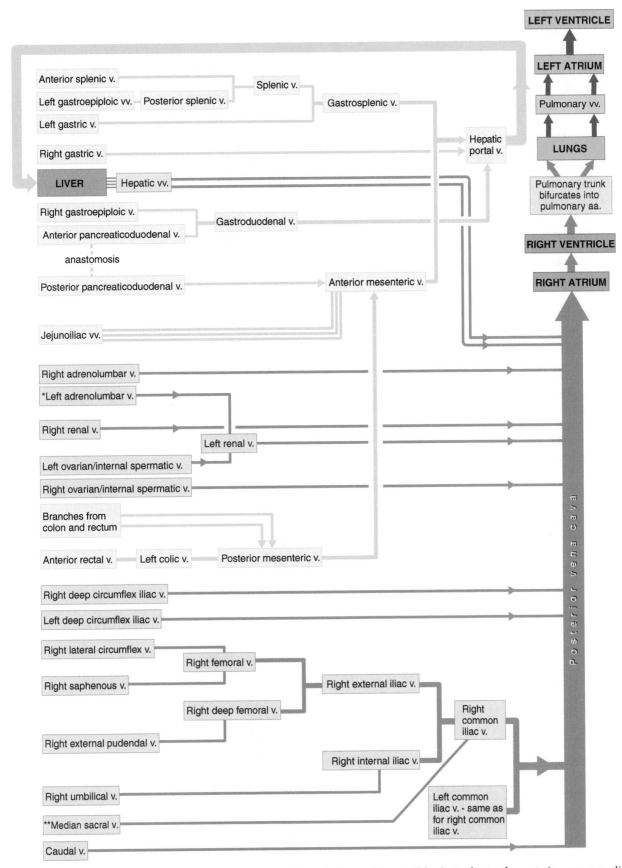

FIGURE 7.67 Flow chart diagram of the venous system posterior to the heart of the cat. *May instead enter the posterior vena cava directly. **May instead enter the left common iliac vein.

Key Terms: Cardiovascular System

abdominal aorta
adrenal gland
adrenolumbar artery
adrenolumbar vein
anterior auricular artery
anterior auricular vein
anterior mesenteric artery
anterior mesenteric vein
anterior pancreaticoduodenal artery
anterior pancreaticoduodenal vein
anterior rectal artery
anterior rectal vein
anterior splenic artery
anterior splenic vein
anterior vena cava
aorta
aortic arch
apex of heart
ascending pharyngeal artery
atrium (pl., atria)
auricle (of right atrium and of left atrium)
axillary artery
axillary vein
azygos vein
base of heart
brachial artery
brachial vein
brachiocephalic artery
brachiocephalic vein
cardiac vein
caudal artery
caudal vein
celiac artery
cephalic vein
common carotid artery
common iliac vein
coronary artery
coronary sulcus
coronary vein (cardiac vein)
costocervical artery
costocervical vein
costocervical + vertebral trunk
deep cervical artery
deep circumflex iliac artery
deep circumflex iliac vein
deep femoral artery
deep femoral vein
external carotid artery
external iliac artery
external iliac vein
external jugular vein
facial artery
facial vein

femoral artery
femoral vein
gastroduodenal artery
gastroduodenal vein
gastrosplenic vein (lienogastric vein)
heart
hepatic artery
hepatic portal system
hepatic portal vein
hepatic vein
highest intercostal arteries
highest intercostal veins
hypoglossal nerve
ileocolic artery
ileocolic vein
intercostal artery
intercostal vein
internal carotid artery
internal carotid ligament
internal iliac artery
internal iliac vein
internal jugular vein
internal mammary artery
internal mammary vein
internal spermatic artery
internal spermatic vein
interventricular paraconal sulcus
interventricular subsinuosal sulcus
left colic artery
left colic vein
left gastric artery
left gastric vein
left gastroepiploic artery
left gastroepiploic vein
ligamentum arteriosum
lingual artery
lingual vein
linguofacial vein (anterior facial vein)
lymph nodes
maxillary artery
maxillary vein (posterior facial vein)
median sacral artery
median sacral vein
middle colic artery
occipital artery
ovarian artery
ovarian vein
pericardial cavity
pericardium
phrenic nerve
posterior auricular artery
posterior mesenteric artery
posterior mesenteric vein
posterior pancreaticoduodenal artery
posterior pancreaticoduodenal vein

posterior splenic artery
posterior splenic vein
posterior vena cava
pulmonary artery
pulmonary trunk
pulmonary vein
renal artery
renal vein
right gastric artery
right gastric vein
right gastroepiploic artery
right gastroepiploic vein
splenic artery
splenic vein
subclavian artery
subclavian vein
subscapular artery
subscapular vein
superior temporal artery
superior temporal vein
sympathetic trunk
thoracic aorta
thymus gland
thyrocervical artery
thyroid gland
transverse facial artery
transverse facial vein
transverse jugular vein (hyoid venous arch)
transverse scapular artery
transverse scapular vein
umbilical artery
umbilical vein
vagosympathetic trunk
vagus nerve
ventricle
vertebral artery
vertebral vein

Cardiovascular System Supplement— Sheep Heart

The **heart** was examined briefly in the cat, but a deeper and more detailed understanding of its form and function is afforded by dissection of a sheep heart. There are advantages and disadvantages to dissecting a sheep heart. One advantage is that it is larger, which facilitates dissection and observation of some structures. The disadvantages are that the sheep heart and its vessels are not usually injected with latex and in many specimens several of the vessels are cut very close to the heart, making it harder to identify and follow them. On the other hand, removal of the latex in injected cat hearts is very tedious and often results in destruction of some finer structures. In any event, review the basic anatomy of the heart as noted for the cat, as the terms (nonetheless repeated in

this section) will facilitate dissection of the sheep heart, which is accompanied by Figures 7.H1 through 7.H18. As the degree of detail and quantity of heart structures is considerable greater than that for the cat a Key Terms section for the sheep heart is provided at the end of this section.

Given that the heart available to you will have several of its vessels cut short, the following illustrations and accompanying text of a detailed dissection is provided. Reading this section provides an orientation that will facilitate interpreting structures of the heart (particularly its great vessels) that you will dissect. The specimen illustrated here is almost certainly more complete than the specimen available to you. It retains parts of the respiratory system for context, but nearly all of the **pericardium** (which is considered below) has been removed. The dissection of the sheep heart in this section may be replicated by beginning with a heart and lung pluck, but it requires prior knowledge of structure and takes considerable time.

The heart is somewhat conical (Figure 7.H1). In the sheep it sits in the thoracic cavity with its flattened **base** facing nearly dorsally and tapered **apex** facing nearly ventrally, so that the surface observed in Figure 7.H1 is its anterior surface. By contrast, in the cat the apex is oriented nearly posteriorly and the base anteriorly, so that the homologous surface of the heart faces nearly ventrally (see Figure 7.45).

Two large vessels, the **aorta** and the **pulmonary trunk**, arise from the central part of the base. The aorta is a large vessel. It extends initially anterodorsally as the **ascending aorta** and then curves sharply posterodorsally as the **arch of the aorta** (May 1964, p. 369), as is evident in Figure 7.H2. Note the **right auricle**, an extension of the **right atrium**, to the right of the base of the aorta (Figures 7.H1 and 7.H4). From the aorta arises the **brachiocephalic trunk**, which extends anterodorsally (Figures 7.H1 and 7.H2). In the sheep this vessel gives rise to four main arteries, the **right subclavian** and **left subclavian arteries** and the **right common carotid** and **left common carotid arteries**, which supply blood to the head, neck, thoracic limbs, and much of the thoracic wall (May 1964, p. 369). This branching pattern differs from that of the cat, in which the brachiocephalic trunk usually gives rise to the right subclavian and the right and left common carotid arteries, whereas the left subclavian artery arises separately form the aorta. Typically, as is the case for the specimen illustrated in Figures 7.H1 through 7.H6, the brachiocephalic trunk is usually cut proximal to the origin of these vessels, but the occasional specimen does retain them (see page 284 and Figure 7.H10).

The pulmonary trunk arises from the **right ventricle** and extends toward the left, crossing the base of the aorta (this portion is not shown in Figure 7.H1, but

(a)

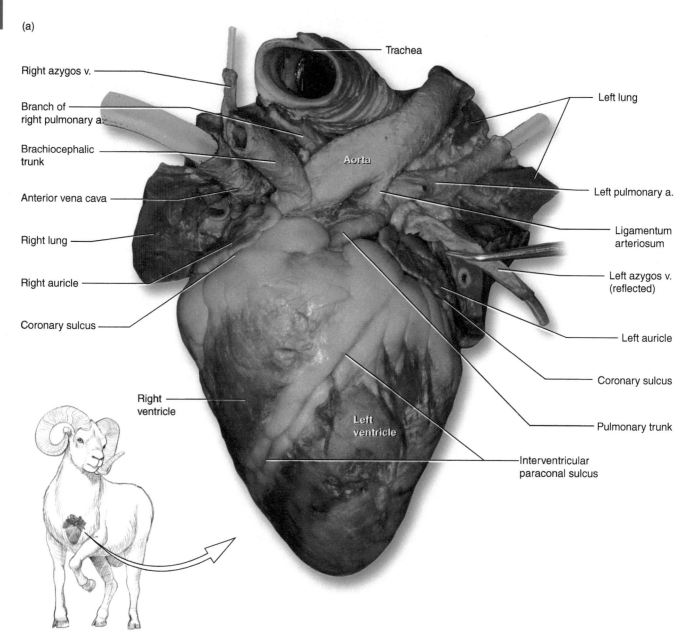

Right azygos v.

Branch of
right pulmonary a.

Brachiocephalic
trunk

Anterior vena cava

Right lung

Right auricle

Coronary sulcus

Right
ventricle

Trachea

Aorta

Left
ventricle

Left lung

Left pulmonary a.

Ligamentum
arteriosum

Left azygos v.
(reflected)

Left auricle

Coronary sulcus

Pulmonary trunk

Interventricular
paraconal sulcus

FIGURE 7.H1   Heart of the sheep in ventral view, with (a) a photograph and (b) a color-coded illustration of the same specimen.

will be noted again later). It extends obliquely ventral to the aorta, passing dorsally and posteriorly (Figures 7.H1 and 7.H2), and divides into the **left** and **right pulmonary arteries** (Figure 7.H3). Just before doing so, the pulmonary trunk is connected to the arch of the aorta by the **ligamentum arteriosum** (Figures 7.H1–7.H3), which is a fibrous band that is the remnant of the fetal ductus arteriosus. In the fetus, the pulmonary circulation is not functional in terms of gas exchange and most of the blood passing through the pulmonary trunk is diverted to the aorta through the ductus arteriosus.

Note the **coronary sulcus,** a deep groove that marks the division of the atria dorsally from the ventricles ventrally and is usually filled with considerable fat (Figures 7.H1 through 7.H6). The sulcus almost entirely encircles the base, being interrupted only anteriorly by the pulmonary trunk. The right auricle and the **left auricle** are readily observable dorsal to the coronary sulcus. As in the cat, the auricles extend the atria and have scalloped margins. The anterior and posterior surfaces of the heart each bear an interventricular sulcus (Figure 7.H1 and Figures 7.H3 through 7.H5); that on the anterior surface

(b)

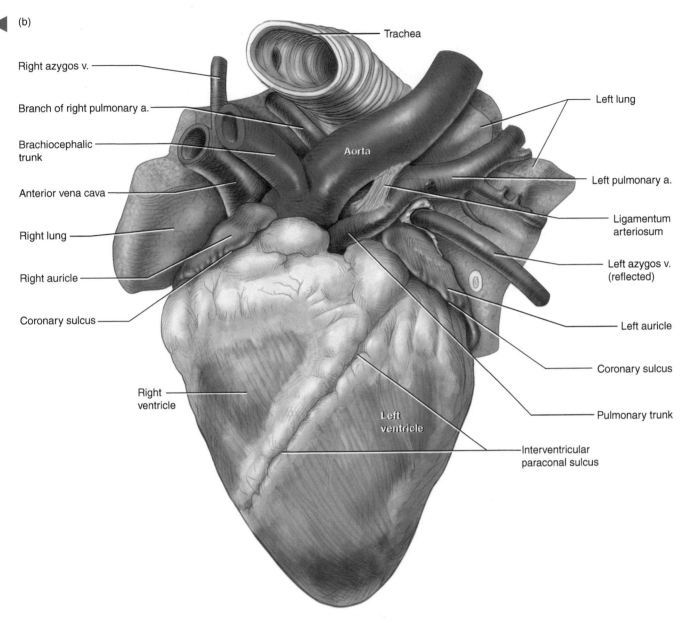

Trachea

Right azygos v.

Branch of right pulmonary a.

Brachiocephalic trunk

Anterior vena cava

Right lung

Right auricle

Coronary sulcus

Aorta

Left lung

Left pulmonary a.

Ligamentum arteriosum

Left azygos v. (reflected)

Left auricle

Coronary sulcus

Pulmonary trunk

Right ventricle

Left ventricle

Interventricular paraconal sulcus

FIGURE 7.H1, cont'd

is the **interventricular paraconal sulcus**, whereas that on the posterior surface is the **interventricular subsinuosal sulcus**. The coronary sulcus and interventricular sulci mark, externally, the positions of the four chambers of the heart. Thus, as may be observed in Figure 7.H1, the apex of the heart is part of the **left ventricle**. The sulci generally contain fat and mark the main path of vessels, the **coronary arteries** and **coronary veins** (the veins are also commonly referred to as cardiac veins, and Hermanson et al., 2019, use coronary and cardiac interchangeably for the veins in the dog) that respectively supply and drain the **myocardium**, the muscular portion of the heart. These vessels are easily observable

on the cat heart, as the vessels are injected, but they are not as clear in the uninjected sheep heart; however, several of the vessels are indicated.

The interventricular paraconal sulcus begins at the coronary sulcus and extends obliquely from left to right toward the apex. Along it passes the descending branch of the **left coronary artery** (labeled in the cat heart, Figure 7.55) and the **great coronary vein**. The **circumflex branch** of the left coronary artery (noted again later) extends to the left along the coronary sulcus and reaches the posterior surface of the heart. The great coronary vein follows the circumflex branch along the coronary sulcus. The **right coronary artery** extends to

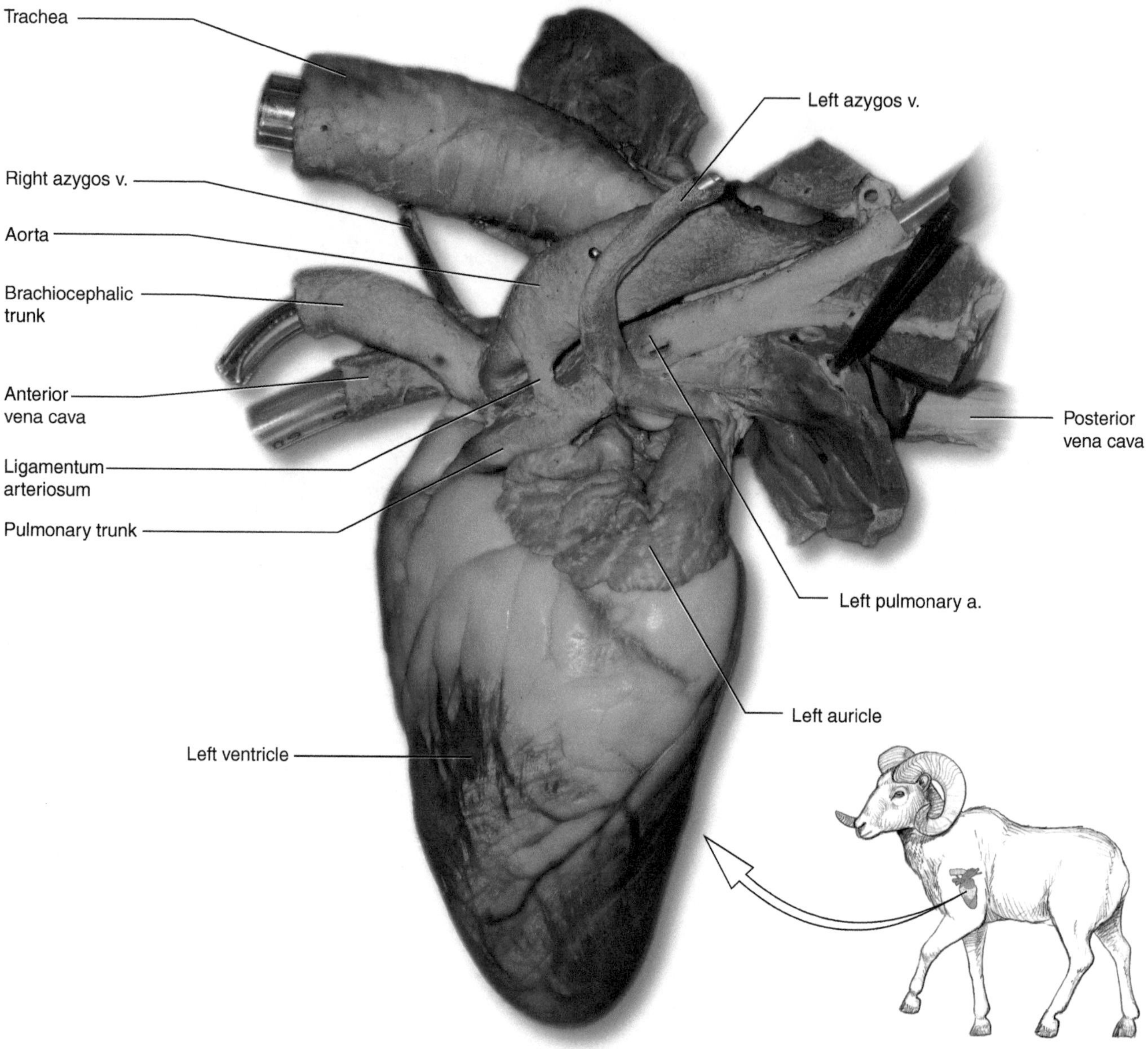

Trachea

Left azygos v.

Right azygos v.

Aorta

Brachiocephalic
trunk

Anterior
vena cava

Posterior
vena cava

Ligamentum
arteriosum

Pulmonary trunk

Left pulmonary a.

Left auricle

Left ventricle

FIGURE 7.H2   Heart of the sheep in left lateral view.

the right along the coronary sulcus and also reaches the posterior surface of the heart. The origins of the coronary arteries from the aorta are considered later. The interventricular subsinuosal sulcus is less obliquely oriented. The **middle coronary vein** (Figure 7.H4b) and the **subsinuosal branch** of the left coronary artery pass along this sulcus.

Consider next the **venae cavae** (sing., **vena cava**; Figures. 7.H1 through 7.H6), the vessels that return nearly all of the deoxygenated blood from the tissues of the body back into the heart. The **anterior vena cava** lies to the right of and posteriorly to the aorta (and brachiocephalic trunk, as seen in Figure 7.H1). It extends nearly

anteroposteriorly and curves sharply ventrally to enter the right atrium; the **posterior vena cava** also extends anteroposteriorly and curves ventrally to enter the right atrium (Figure. 7.H6).

A main tributary of the anterior vena cava, the **right azygos vein**, is illustrated in Figures 7.H1 through 7.H6. The more distal part of this vessel (which is not represented in the illustrations) extends anteriorly along the dorsal thoracic wall. Anterior to the level of the apical bronchus of the trachea, it curves ventrally, crossing the trachea on the right side, and joins the anterior vena cava (Figure 7.H6). The right azygos vein receives tributaries mainly from the right side intercostal veins and drains

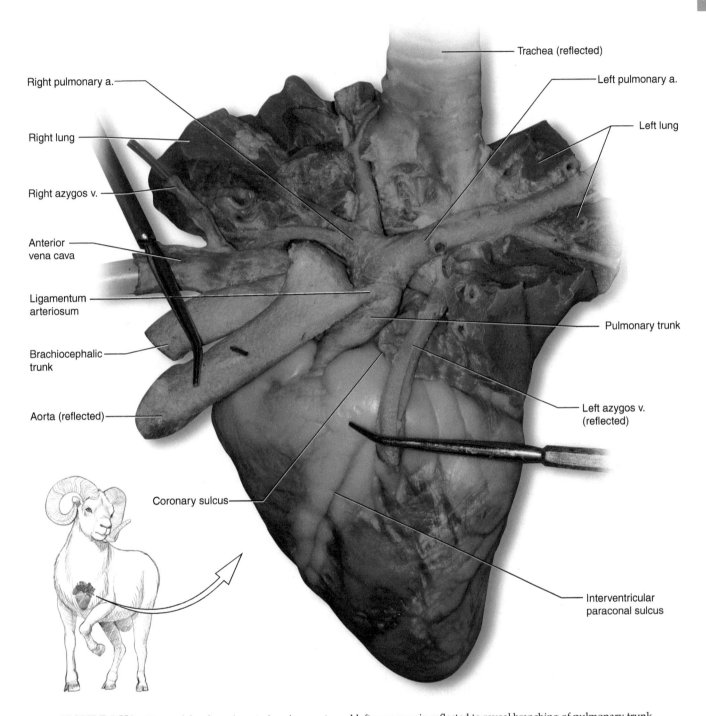

FIGURE 7.H3   Heart of the sheep in anterior view, aorta and left azygos vein reflected to reveal branching of pulmonary trunk.

most of the dorsal thoracic wall from about the 10th thoracic vertebra.

The **left azygos vein** drains the left side of the dorsal thoracic wall (and part of the right side, posterior to the 11th thoracic vertebra) and also receives a small portion of the bronchial arterial blood (see Charan, Turk, & Dhand, 1984; 2007). It passes along the dorsal thoracic wall and curves ventrally, as indicated in Figure 7.H2, crossing the aorta and pulmonary trunk (very close to where the latter bifurcates into the left and right pulmonary arteries). It then passes to the posterior surface of the heart, where it joins the great coronary vein (Figure 7.H4).

(a)

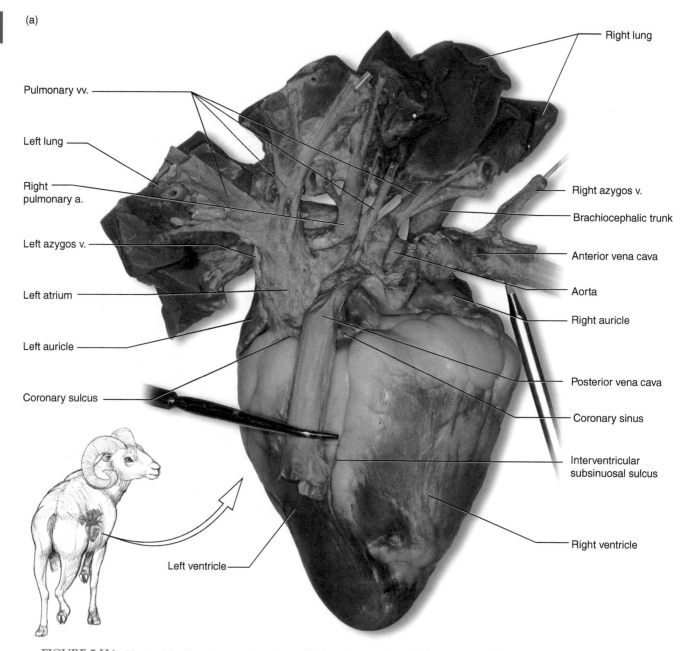

FIGURE 7.H4   Heart of the sheep in posterior view, with (a) a photograph and (b) a color-coded illustration of the same specimen.

The pattern and terminology of the vessels that mainly drain the right side and left side thoracic walls—that is, the right and the left azygos veins—differ from those described in the cat, in which one vessel, the azygos (rather than right azygos) vein, mainly receives the left and right intervertebral veins, and thus drains the intercostal regions of both sides of the body posterior to the heart (page 254, Figures 7.44, 7.52–7.54). In some mammals, the sheep included, a separate vein drains most of the left side dorsal thoracic wall (in the cat this vessel atrophies during fetal development). This vein is generally termed the hemiazygos vein in those mammals (such as humans) in which it joins the vessel draining the right side dorsal thoracic wall—that is, the azygos vein. However, in the sheep this does not occur—the left side (i.e., left azygos) vein does not join the right side (i.e., right azygos) vein. Although this left side vessel has been termed the hemiazygos vein in the sheep, May's (1964, p. 369) suggestion that the left side vessel be recognized as the left azygos vein and the right side vessel as the right azygos vein is followed here and elsewhere (e.g., Charan, Thompson, & Carvalho, 2007, 1984; Hill & Iaizzo, 2009).

(b)

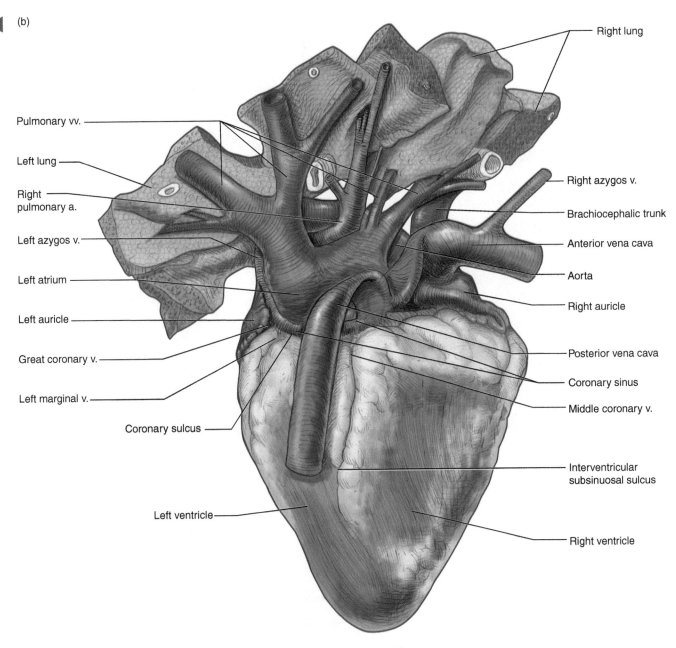

Pulmonary vv.

Left lung

Right
pulmonary a.

Left azygos v.

Left atrium

Left auricle

Great coronary v.

Left marginal v.

Coronary sulcus

Left ventricle

Right lung

Right azygos v.

Brachiocephalic trunk

Anterior vena cava

Aorta

Right auricle

Posterior vena cava

Coronary sinus

Middle coronary v.

Interventricular
subsinuosal sulcus

Right ventricle

FIGURE 7.H4, cont'd

The union of the left azygos vein and the great coronary vein form a common vessel, the **coronary sinus**. It passes from left to right along the coronary sulcus on the posterior surface of the heart (Figure 7.H4 through 7.H6) and opens into the right atrium very near the entrance of the posterior vena cava (see later). Another vein, the **left marginal vein**, joins the sinus just beyond the union of the left azygos and great coronary veins. The middle coronary vein, mentioned earlier, does not enter the sinus, but returns blood directly to the right atrium (see later). Besoluk and Tipirdamaz (2001) provide further detail on the pattern of these vessels and the coronary sinus.

Consider next the vessels of the pulmonary circulation in more detail. The pulmonary trunk and arteries have already been identified (Figure 7.H3). The right and left pulmonary arteries branch repeatedly as they conduct deoxygenated blood to the right and left lungs, respectively. These branches may be followed especially in Figures 7.S3 through 7.H5. Once gaseous exchange has occurred between the alveoli of the lungs and the cardiovascular capillaries, the oxygenated blood is returned to the **left atrium** of the heart through several **pulmonary veins**. The pattern of the pulmonary veins is well displayed in Figures 7.H4 and 7.H5.

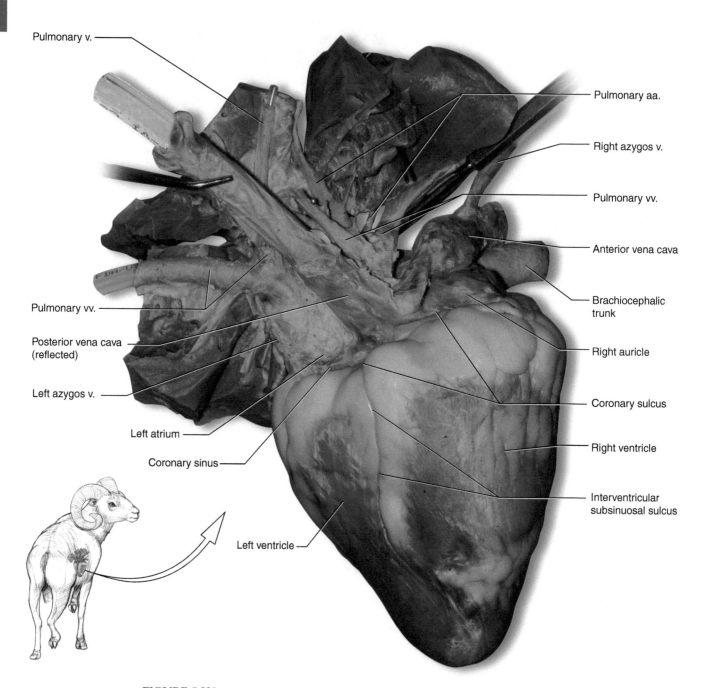

FIGURE 7.H5 Heart of the sheep in posterior view, with posterior vena cava reflected.

Once you are familiar with the structures indicated above, proceed with your dissection of a sheep heart. The specimen available to you may or may not have an intact pericardium. The specimen illustrated in Figures 7.H7 through 7.H9 retains the pericardium, whereas the pericardium is almost entirely missing from the specimen illustrated in Figures 7.H1 through 7.H6. If the pericardium is present, you must first orient your specimen correctly before you begin. To do this, you may rely on several landmarks. You may be able to identify the position of the pulmonary trunk, for example (as indicated in Figures 7.H7a, 7.H8, and 7.H9), thus identifying the anterior surface. As well, examine your specimen for the cut edges of several of the great vessels by inspecting the wider end of your specimen. You should be able to identify those of the aorta and the brachiocephalic trunk (Figure 7.H7) without much difficulty. This will also help you orient the heart (see

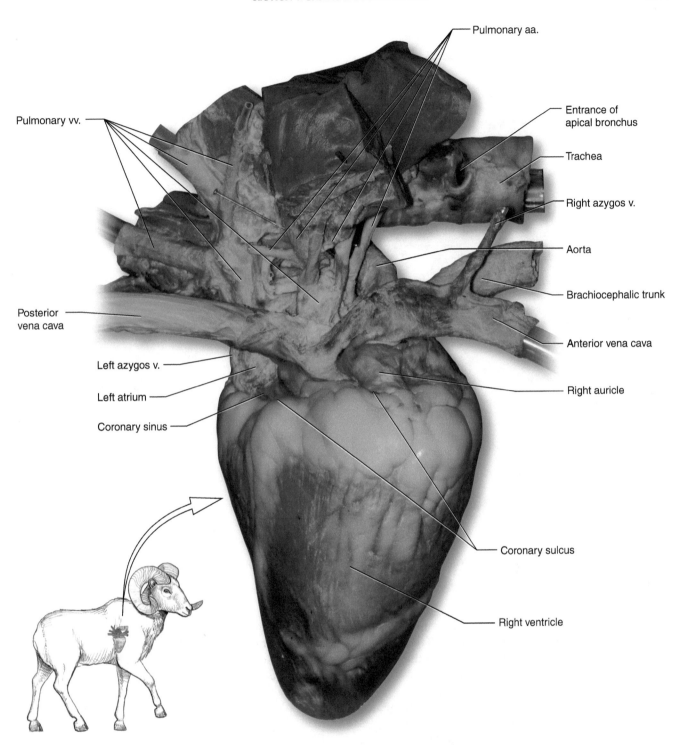

Pulmonary aa.

Pulmonary vv.

Entrance of
apical bronchus

Trachea

Right azygos v.

Aorta

Brachiocephalic trunk

Posterior
vena cava

Anterior vena cava

Left azygos v.

Left atrium

Coronary sinus

Right auricle

Coronary sulcus

Right ventricle

FIGURE 7.H6   Heart of the sheep in right lateral view.

later). Although the aorta is much the larger, both vessels have a thick wall and remain open. The venae cavae
may also be identified, but they have a thinner wall and
so are usually collapsed. The anterior vena cava lies to
the right of the base of the aorta (Figures 7.H7 through
7.H10a). Examine the posterior surface of the heart

for the posterior vena cava, particularly in the region
indicated in Figure 7.H7b. The venae cavae may not be
located precisely as indicated in these figures, depending on how the heart and pericardium were cut from
their surrounding structures, but by probing carefully
you should be able to reveal their lumen. Once you

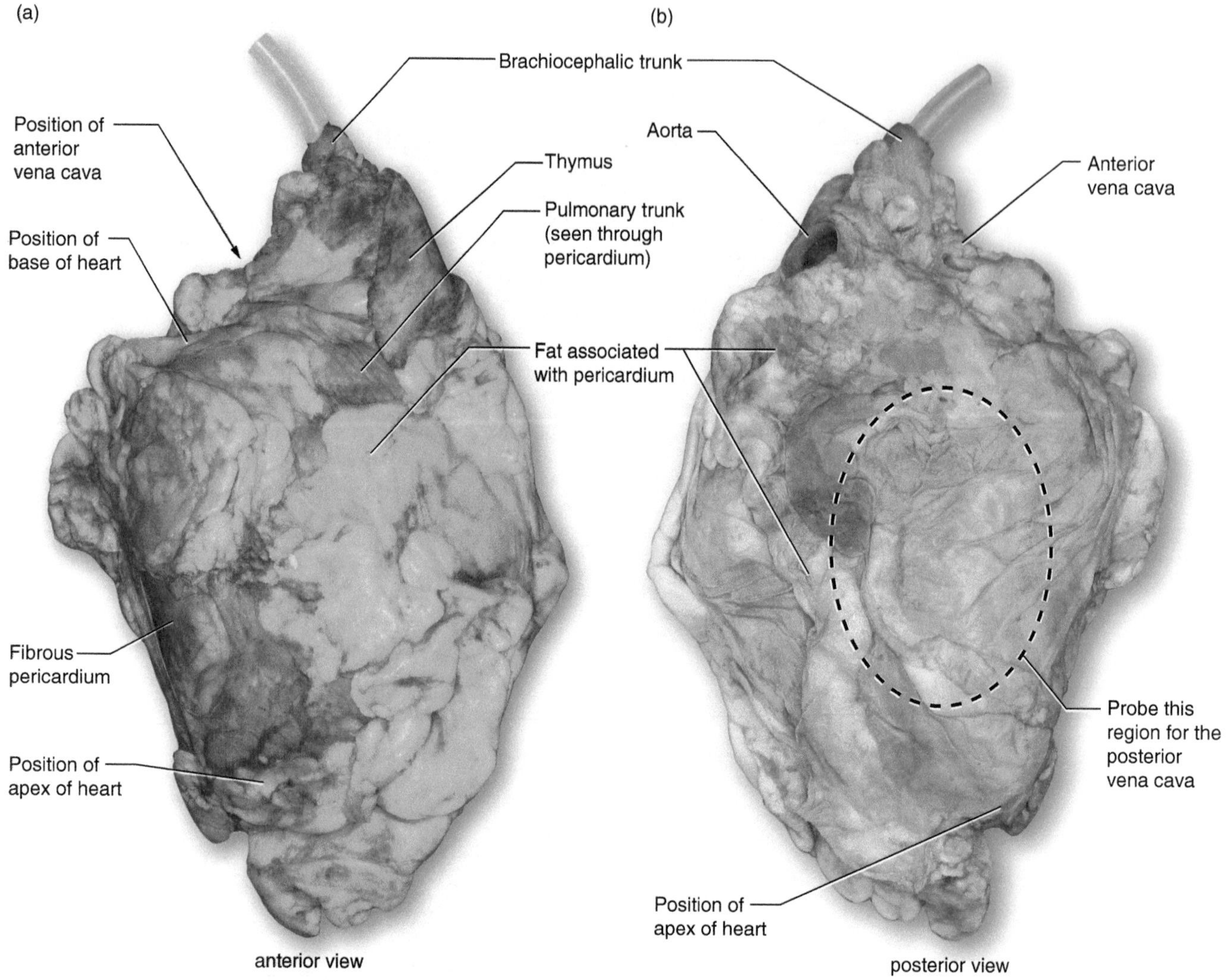

(a)

(b)

Brachiocephalic trunk

Position of anterior vena cava

Aorta

Thymus

Anterior vena cava

Pulmonary trunk (seen through pericardium)

Position of base of heart

Fat associated with pericardium

Fibrous pericardium

Probe this region for the posterior vena cava

Position of apex of heart

Position of apex of heart

anterior view

posterior view

FIGURE 7.H7   Heart of the sheep within pericardial sac in (a) anterior and (b) posterior views.

have identified these vessels, it is useful to insert plastic tubing into them (you will need several different sizes). Note that the brachiocephalic trunk veers to the right in anterior view and to the left in posterior view (compare Figure 7.H7a and 7.H7b).

The pericardium forms the **pericardial sac**, which encloses the **pericardial cavity**. The heart occupies nearly the entire pericardial cavity. The pericardium was examined in the cat, and that of the sheep is very similar. Considerable fat is usually associated with it but it is not worth trying to remove it—it is best taken off with the pericardium, as described presently. You may also notice grayish or pinkish-yellow patches of glandular material associated with the pericardium covering the anterodorsal surface of heart (Figures 7.H7a and 7.H8). These are parts of the thoracic portion of the **thymus**, a gland that varies considerably in size, being larger in young individuals (as noted for the cat; see page 254). The

pericardium is composed mainly of a superficial fibrous layer, the **fibrous pericardium**, which continues and attaches to several of the large vessels associated with the heart, such as the aorta (see, e.g., Figure 7.H10a) and venae cavae, among others (May 1964, p. 369). A layer of parietal pleura, which forms the mediastinal walls, may in places adhere to its outer surface.

Place your specimen with its anterior surface facing you, as shown in Figure 7.H7a. Near the middle of the heart, make a dorsoventral incision through the pericardium as shown in Figure 7.H8. Do not at this time extend your incision farther dorsally than shown in this figure (note the position of the pulmonary trunk). Reflect the pericardium to observe the **parietal pericardium**, which is the serosal epithelium that lines the pericardial cavity, forming the internal layer of the pericardial sac. The parietal pericardium reflects onto the great vessels at the base of the heart and onto the surface

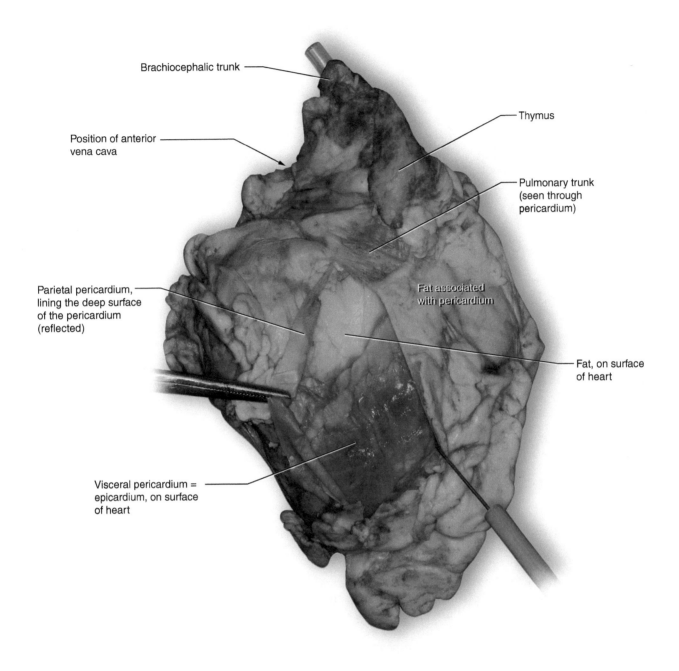

Brachiocephalic trunk

Thymus

Position of anterior
vena cava

Pulmonary trunk
(seen through
pericardium)

Parietal pericardium,
lining the deep surface
of the pericardium
(reflected)

Fat associated
with pericardium

Fat, on surface
of heart

Visceral pericardium =
epicardium, on surface
of heart

FIGURE 7.H8   Heart of the sheep in anterior view with pericardium partially opened.

of the heart as the **visceral pericardium**, which is also known as the **epicardium**, for it forms the outermost layer of the heart itself. The middle layer of the heart, by far the thickest, is the myocardium, noted earlier as composing the musculature of the heart. The internal layer of the heart, also serosal epithelium, is the **endocardium**, which is continuous with the internal lining, termed **endothelium**, of the cardiovascular vessels. The

myocardium, endocardium, and endothelium will be observed later.

Extend your incision through the pericardium ventrally around the apex of the heart, and then dorsally on the posterior surface to about the middle of the heart. Veer slightly to the left as you cut dorsally to avoid the posterior vena cava (for its approximate location review Figure 7.H7b and also see Figure 7.H11). Turn the heart

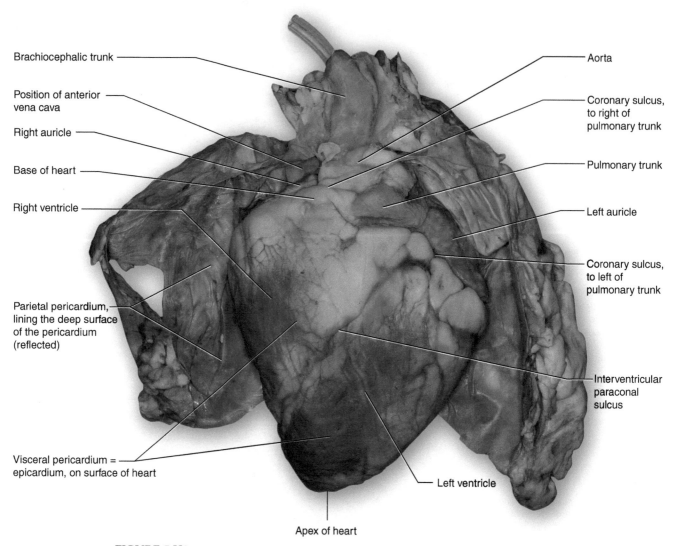

Brachiocephalic trunk

Position of anterior
vena cava

Right auricle

Base of heart

Right ventricle

Parietal pericardium,
lining the deep surface
of the pericardium
(reflected)

Visceral pericardium =
epicardium, on surface of heart

Aorta

Coronary sulcus,
to right of
pulmonary trunk

Pulmonary trunk

Left auricle

Coronary sulcus,
to left of
pulmonary trunk

Interventricular
paraconal
sulcus

Left ventricle

Apex of heart

FIGURE 7.H9   Heart of the sheep in anterior view, with pericardial sac opened to expose heart.

over so you observe it in anterior view again. Beginning from the cut edge of the brachiocephalic trunk, use scissors and a probe to snip the pericardium and separate it from the vessel. Work your way ventrally until you reach the first incision you made through the pericardium. You should now be able to spread the pericardium as shown in Figure 7.H9. Note the position of the several structures already noted, such as the pulmonary trunk, coronary sulcus, left and right auricles, and interventricular paraconal groove.

Clean away fat from the base of the brachiocephalic trunk until you expose the aorta and then separate the pericardium from the latter vessel. Figure 7.H10a indicates the line of attachment of the pericardium to the aorta. Once the aorta and brachiocephalic trunk are well exposed, the path of the pulmonary trunk as it crosses the base of the aorta should be easily appreciated (Figure 7.H10b). Dissect, using needle, forceps,

and scissors, between the aorta and the pulmonary trunk, clearing fat and pericardium until you expose the ligamentum arteriosum, as shown in Figure 7.H10b. The pericardium has not been entirely removed from around the heart in this figure, but has simply been tucked in behind it. Figure 7.H10c is of a specimen that includes the proximal portions of the branches of the brachiocephalic trunk. The first large branch is the left subclavian artery, beyond which the vessel is termed the **brachiocephalic artery**. The latter gives rise to the very short **bicarotid trunk**, which splits almost immediately into the right and left common carotid arteries, and the right subclavian artery.

Examine the heart in posterior view and locate the main vessels already noted (Figure 7.H11). Find the left azygos vein near the left margin of the heart—use the opening of the aorta as a guide. It will not be obvious but inspect the pericardium for an opening.

7

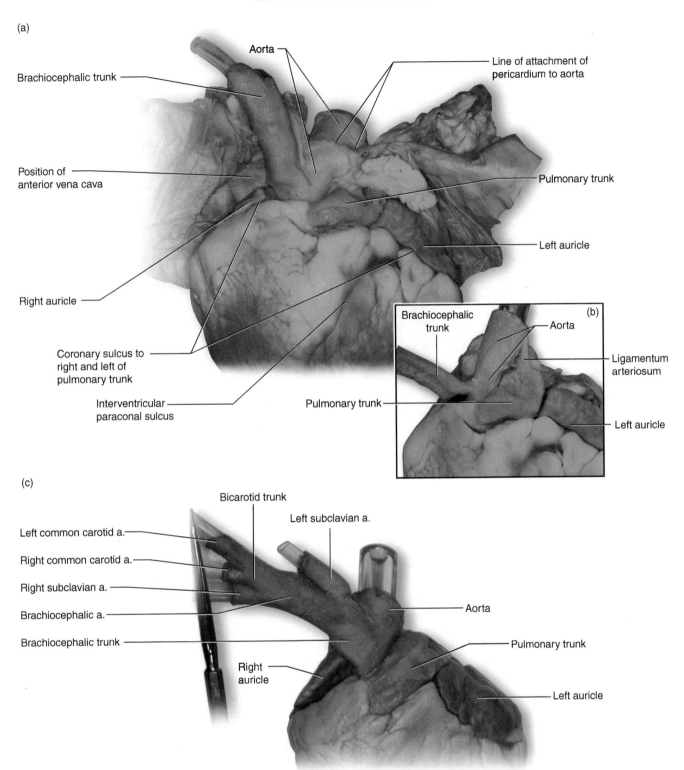

(a)

Aorta

Line of attachment of
pericardium to aorta

Brachiocephalic trunk

Position of
anterior vena cava

Pulmonary trunk

Left auricle

Right auricle

(b)

Brachiocephalic
trunk

Aorta

Ligamentum
arteriosum

Coronary sulcus to
right and left of
pulmonary trunk

Pulmonary trunk

Left auricle

Interventricular
paraconal sulcus

(c)

Bicarotid trunk

Left subclavian a.

Left common carotid a.

Right common carotid a.

Right subclavian a.

Aorta

Brachiocephalic a.

Brachiocephalic trunk

Pulmonary trunk

Right
auricle

Left auricle

FIGURE 7.H10   Dorsal portion of the heart of the sheep in anterior view, showing (a) great vessels, (b) detail indicating the ligamentum arteriosum, and (c) branches of the brachiocephalic trunk.

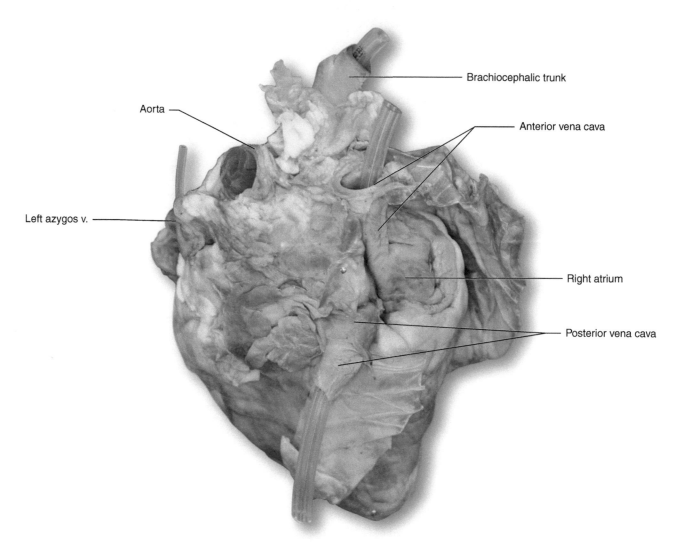

Brachiocephalic trunk

Aorta

Anterior vena cava

Left azygos v.

Right atrium

Posterior vena cava

FIGURE 7.H11   Heart of the sheep in posterior view indicating vessels.

Near the position where the left azygos extends onto the posterior surface of the heart, you may note several small openings, but do not examine them at this time—they are labeled as pulmonary vessels in Figure 7.H12a. Instead, concentrate first on the much larger venae cavae (Figure 7.H11). Begin by exposing the anterior vena cava, using the same technique as for the brachiocephalic trunk, and then proceed to do the same for the posterior vena cava. You should produce a dissection similar to that shown in Figure 7.H11.

Return to the pulmonary vessels near the left azygos vein (Figure 7.H12a). In the specimen illustrated in this figure, two openings are initially observable. Examine them closely. It will be evident that the wall of one of the openings is thicker than that of the other. This is therefore an artery and the vessel is a pulmonary artery. The other opening is a pulmonary vein. These vessels may be exposed by using the same technique described

for other vessels, but the pulmonary arteries, and especially the pulmonary veins, are more delicate; therefore, greater care and patience is required. Begin dissection of these vessels near the left azygos vein and work your way toward the right. Eventually you should produce dissections similar to those illustrated in Figure 7.H12b and c. You may, if time permits, follow the pulmonary arteries to their origin at the bifurcation of the pulmonary trunk.

Once the heart has been examined externally, its internal structure may be considered. The heart may be sectioned in several ways to expose its internal structure, but the easiest is to make a nearly transverse section, which subdivides a sheep heart into nearly equal anterior and posterior halves. However, before proceeding to this step, first examine structures associated with the pulmonary trunk and right ventricle. Cut lengthwise along the anterior surface of the pulmonary trunk and

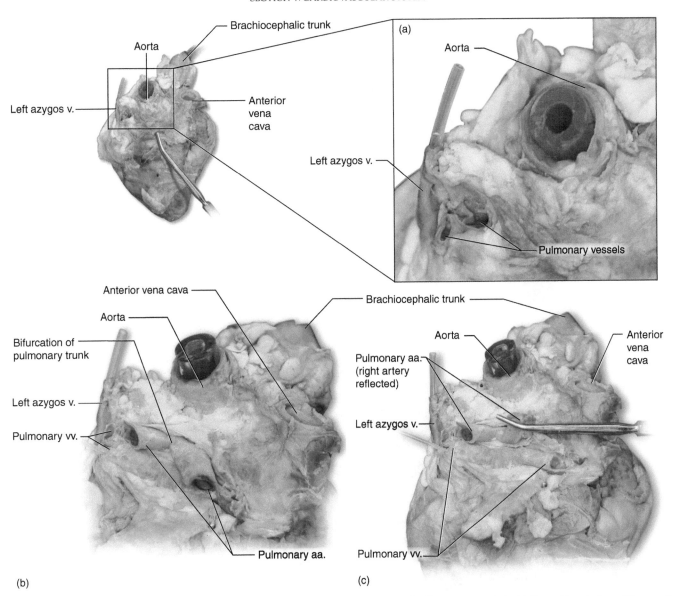

FIGURE 7.H12   Dorsal portion of heart of the sheep in posterior view, with detail of pulmonary vessels in (a) the undissected condition, and (b) and (c), sequential dissections to expose the vessels.

extend the incision into the wall of the right ventricle, as shown in Figure 7.H13a and b, stopping about half way along the dorsoventral height of the heart. Spread open the walls of the pulmonary trunk and right ventricle (Figure 7.H13c). The interior of the ventricle is now clearly evident. At the base of the pulmonary trunk, identify the three flaps or cusps of the **pulmonary semilunar valve**. This valve allows blood to pass from the right ventricle into the pulmonary trunk, but prevents blood, once in the pulmonary trunk, from flowing back into the right ventricle.

Next section the heart into anterior and posterior halves. Begin by cutting the aorta into anterior and posterior halves and continuing through the heart along the plane of these incisions.

View the interior of the anterior half of the heart (you will be observing it in posterior view), as shown in Figures 7.H14 and 7.H15. Identify the aorta and the opening of the brachiocephalic trunk. To the left of the aorta, find the pulmonary trunk as it bifurcates into the left and right pulmonary arteries. Return to the aorta and examine its base for the cusps of the **aortic semilunar valve**. This valve prevents blood from flowing back into the left ventricle once it has entered the aorta. Two of the three cups that comprise the valve should be observable in this view. On the anterior wall of the base of the aorta, just past the aortic semilunar valve, are the openings of the coronary arteries (Figure 7.H14b).

The internal lining of the heart, the endocardium, may be observed in any of it chambers, whereas the

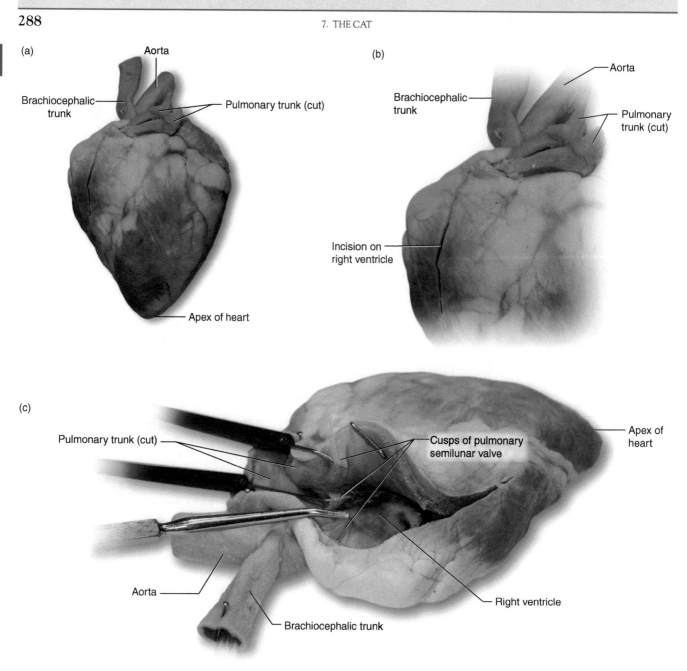

FIGURE 7.H13   Heart of the sheep in (a) anterior view, indicating the course of the incision to expose the right ventricle, (b) anterior view indicating detail of the incision, and (c) anterior and right lateral view, exposing the right ventricle and pulmonary semilunar valve.

endothelium, the internal lining of the blood vessels, is observable on the inner surface of the aorta (Figure 7.H14a). In this view the depth of the coronary sulcus, filled with fat, is clearly discernible, and the position of the circumflex branch of the left coronary artery is apparent (Figure 7.H14a). The atria lie dorsal to the level of the sulcus and the ventricles lie ventrally, as noted earlier, but in this view the right atrium is not visible. Locate the left atrium and its auricle to the left of the aorta. Its walls are relatively thin compared to the much thicker ventricular walls, but note that the wall of the right ventricle is not as thick as that of the left ventricle. The atria are

separated by the **interatrial septum** (see later), and the ventricles by a much thicker **interventricular septum**. The walls that define the various chambers of the heart are composed almost entirely of cardiac musculature, or the myocardium; the epicardium and endocardium are very thin serosal sheets.

Examine the left atrium and left ventricle and identify the valve between them, the **left atrioventricular valve**, which is also known as the mitral or bicuspid valve; as implied by the latter term, two cusps or flaps comprise this valve, although smaller accessory cusps are generally present between them. One main cusp is

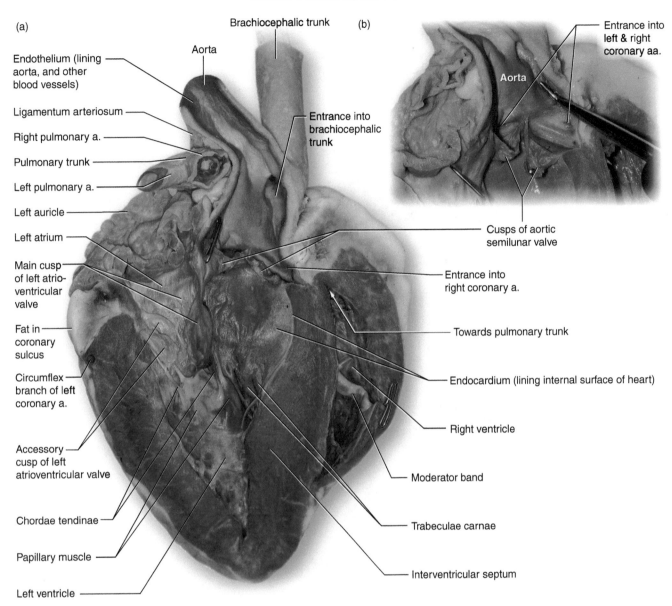

(a)

Brachiocephalic trunk

Aorta

Endothelium (lining aorta, and other blood vessels)

Ligamentum arteriosum

Right pulmonary a.

Pulmonary trunk

Left pulmonary a.

Left auricle

Left atrium

Main cusp of left atrio-ventricular valve

Fat in coronary sulcus

Circumflex branch of left coronary a.

Accessory cusp of left atrioventricular valve

Chordae tendinae

Papillary muscle

Left ventricle

Entrance into brachiocephalic trunk

(b)

Aorta

Entrance into left & right coronary aa.

Cusps of aortic semilunar valve

Entrance into right coronary a.

Towards pulmonary trunk

Endocardium (lining internal surface of heart)

Right ventricle

Moderator band

Trabeculae carnae

Interventricular septum

FIGURE 7.H14   Anterior half of the heart of the sheep in posterior view, indicating (a) internal structures and (b) detail indicating aortic semilunar valve and entrances into coronary arteries.

observable in this view (Figures 7.H14a and 7.H15); the other will be seen when the posterior half of the heart is examined. The cusps project (or "hang down") into the ventricles. During ventricular filling (the atrium undergoes both diastole and systole in this phase), blood flows easily over the cusps to enter the left ventricle, but during ventricular contraction the left atrioventricular valve is forced shut, preventing blood from reentering the atrium; the blood therefore passes through the aortic semilunar valve and into the aorta. The great pressures generated by a contracting ventricle can force the cusps to evert ("flip backwards") into the atria. To prevent this, the narrow and string-like **chordae tendinae** extend from the edges of the cusps to the internal wall of a ventricle,

thereby anchoring the cusps during ventricular contraction (see especially Figure 7.H15b). Many of them may be seen attaching to finger-like projections, **papillary muscles**, of the ventricular musculature in Figures 7.H14 and 7.H15. The internal ventricular walls also bear irregular fleshy ridges termed **trabeculae carnae**, as well as cordlike bands, termed **moderator bands**, that extend across the ventricular cavity (Figures 7.H14a and 7.H15a). An especially prominent moderator band is present in the right ventricle (Figure 7.H14a). Such bands may serve to prevent overexpansion of the ventricle (Fishbeck & Sebastiani, 2015, p. 566).

Examine the posterior half of the heart in anterior view (Figure 7.H16). Many of the structures (e.g.,

(a)

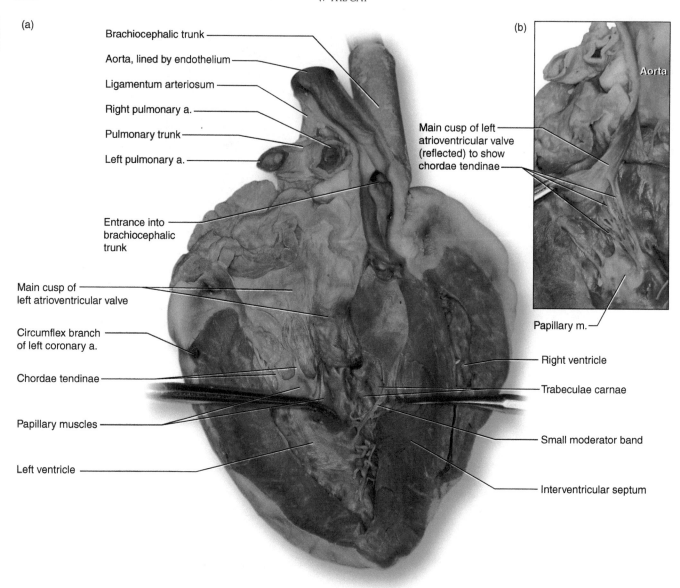

Brachiocephalic trunk

Aorta, lined by endothelium

Ligamentum arteriosum

Right pulmonary a.

Pulmonary trunk

Left pulmonary a.

Entrance into brachiocephalic trunk

Main cusp of left atrioventricular valve

Circumflex branch of left coronary a.

Chordae tendinae

Papillary muscles

Left ventricle

(b)

Aorta

Main cusp of left atrioventricular valve (reflected) to show chordae tendinae

Papillary m.

Right ventricle

Trabeculae carnae

Small moderator band

Interventricular septum

FIGURE 7.H15   Anterior half of the sheep heart in posterior view, with (a) interventricular septum retracted to the right, indicating detail of trabeculae carnae and (b) detail of left atrioventricular valve and chordae tendinae.

papillary muscles, left and right ventricles, interventricular septum, fat in the coronary sulcus) just noted in the preceding paragraphs are also observable in this view. In addition, note the other cusps that form the aortic semilunar valve at the base of the aorta and the left atrioventricular valve between the left atrium and left ventricle. Moreover, the right ventricle has been spread in Figure 7.H16a to reveal the cusps of the **right atrioventricular valve** (also known as the tricuspid valve, and thus composed of three main cusps or flaps) between the right atrium and right ventricle, and the right atrium is visible beyond them. Reflect the dorsal margin of the left auricle, as shown in Figure 7.H16b, to expose the left atrium more completely and reveal the entrances of pulmonary veins.

Several authors have noted the presence in sheep of an **os cordis**, a small and variable piece of cartilage and bone associated with and likely reinforcing the base of the aorta (see e.g., Frink & Merrick, 1974; Getty, 1975; Gopalakrishnan, Blevins, & Van Alstine, 2007; May 1964, p. 369). It is considered a regular feature of the heart of larger ruminants (and some other mammals; see e.g., Ghonimi, Balah, Bareedy, & Abuel-Atta, 2014), although its precise location may be somewhat variable. In the specimen dissected here, an os cordis was noted by probing for a small, hard structure near the attached end of the cusp of the aortic semilunar valve. Careful dissection revealed the element, as illustrated in Figure 7.H16c.

Lastly, examine the right atrium by making three small incisions through the myocardium, as indicated

(a)

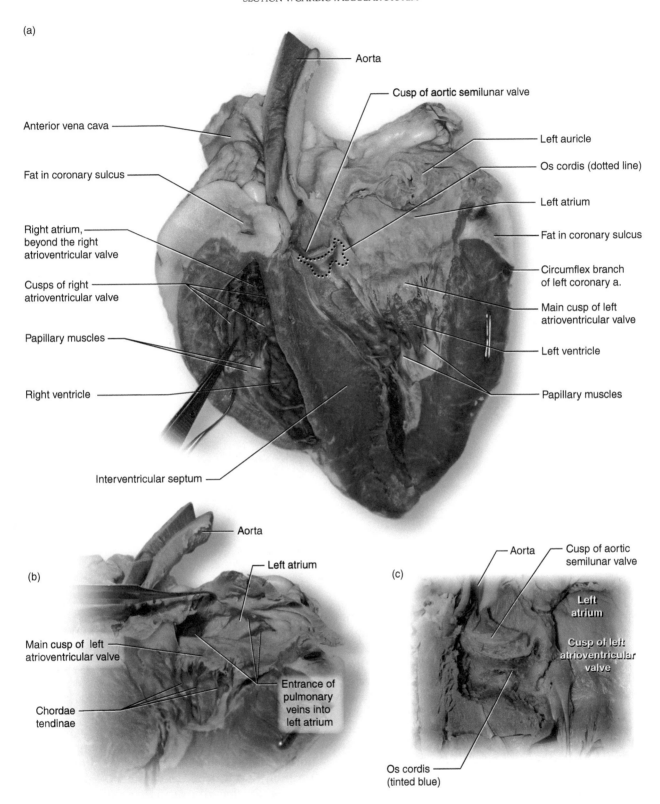

Aorta

Cusp of aortic semilunar valve

Anterior vena cava

Left auricle

Os cordis (dotted line)

Fat in coronary sulcus

Left atrium

Right atrium,
beyond the right
atrioventricular valve

Fat in coronary sulcus

Circumflex branch
of left coronary a.

Cusps of right
atrioventricular valve

Main cusp of left
atrioventricular valve

Papillary muscles

Left ventricle

Right ventricle

Papillary muscles

Interventricular septum

Aorta

(b)

Left atrium

(c)

Aorta

Cusp of aortic
semilunar valve

Left
atrium

Main cusp of left
atrioventricular valve

Cusp of left
atrioventricular
valve

Entrance of
pulmonary
veins into
left atrium

Chordae
tendinae

Os cordis
(tinted blue)

FIGURE 7.H16 Posterior half of the sheep heart in posterior view, with (a) showing detail of internal structures, (b) detail of left atrium, and (c) detail of os cordis.

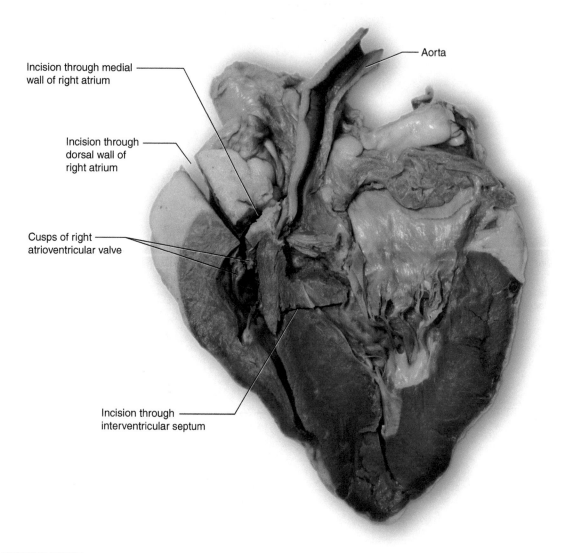

Aorta

Incision through medial
wall of right atrium

Incision through
dorsal wall of
right atrium

Cusps of right
atrioventricular valve

Incision through
interventricular septum

FIGURE 7.H17   Posterior half of the sheep heart in anterior view, indicating incisions required for exposing the right atrium.

in Figure 7.H17. This will facilitate the observation of the interior of the atrium. The incisions through the interventricular septum, just ventral to the level of the tricuspid valve, and the medial wall of the atrium (the interatrial septum) just to the right of the aorta, should be short; see Figure 7.H18 to gauge the depth of these incisions. That through the dorsal atrial wall can be longer, but keep to the right of the position of the anterior vena cava. Once you have made these incisions, reflect the walls of the atrium and observe the heart in right anterolateral view, as shown in Figure 7.H18. Three large opening will be evident. These are the entrances of the anterior vena cava, posterior vena cava, and coronary sinus into the right atrium. These three veins deliver nearly all of the deoxygenated blood from the systemic circulation into the heart. In Figure 7.H18a and b, plastic

tubes may be seen emerging from these openings, but the tubes have been removed in Figure 7.H18c to allow the observation of the entrance of the middle coronary vein and the **fossa ovalis**. The middle coronary vein was noted earlier as passing along the interventricular subsinuosal sulcus. The fossa ovalis is a shallow depression in the interatrial septum, very near the entrance of the posterior vena cava. It represents the position of the fenestra ovalis in the fetus, an opening that allows shunting of blood from the right atrium directly into the left atrium during fetal development. Like the ductus arteriosus, the fenestra ovalis becomes nonfunctional once the lungs begin to work. The fenestra ovalis closes soon after birth, but the tissue is thinner than the rest of the interatrial septum and so appears as a shallow depression.

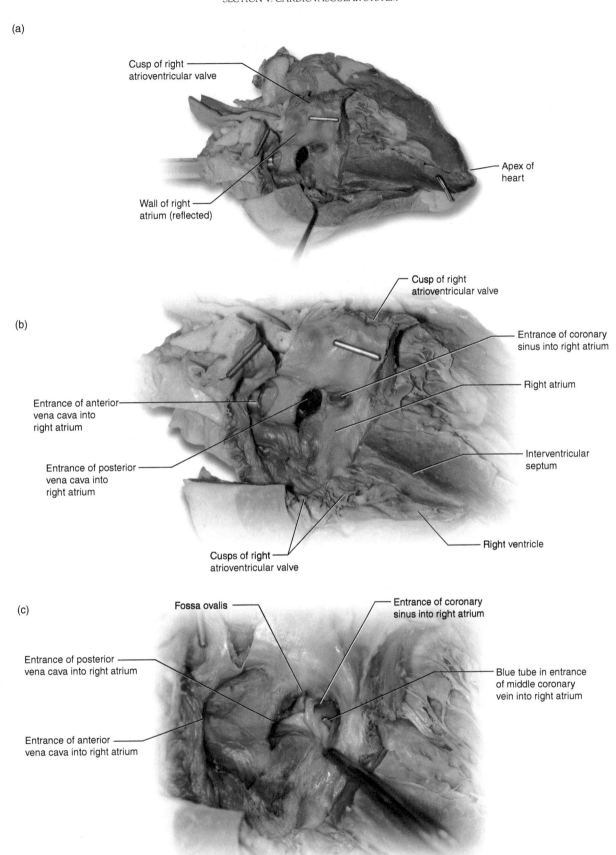

(a)

Cusp of right
atrioventricular valve

Apex of
heart

Wall of right
atrium (reflected)

(b)

Cusp of right
atrioventricular valve

Entrance of coronary
sinus into right atrium

Right atrium

Entrance of anterior
vena cava into
right atrium

Interventricular
septum

Entrance of posterior
vena cava into
right atrium

Right ventricle

Cusps of right
atrioventricular valve

(c)

Fossa ovalis

Entrance of coronary
sinus into right atrium

Entrance of posterior
vena cava into right atrium

Blue tube in entrance
of middle coronary
vein into right atrium

Entrance of anterior
vena cava into right atrium

FIGURE 7.H18   Posterior half of the sheep heart in right lateral view, with (a) right atrium opened, (b) detail indicating entrances of venae cave and coronary sinus into right atrium, and (c) detail indicating fossa ovalis and entrance of middle coronary vein into right atrium.

Key Terms: Sheep Heart

anterior vena cava
aorta
aortic semilunar valve
apex of heart
ascending aorta
base of heart
bicarotid trunk
brachiocephalic artery
brachiocephalic trunk
chordae tendinae
circumflex branch of left coronary artery
coronary arteries
coronary sinus
coronary sulcus
coronary veins (cardiac veins)
endocardium
epicardium
fibrous pericardium
fossa ovalis
great coronary vein (great cardiac vein)
heart
interatrial septum
interventricular paraconal sulcus
interventricular septum
interventricular subsinuosal sulcus
left atrioventricular valve (mitral valve, bicuspid valve)
left atrium
left auricle
left azygos vein
left common carotid artery
left coronary artery
left marginal vein
left pulmonary artery
left subclavian artery
left ventricle
ligamentum arteriosum
middle coronary vein (middle cardiac vein)
moderator bands
myocardium
os cordis
papillary muscles
parietal pericardium
pericardial cavity
pericardial sac
pericardium
posterior vena cava
pulmonary semilunar valve
pulmonary trunk
pulmonary veins
right atrioventricular valve (tricuspid valve)
right atrium
right auricle

right azygos vein
right common carotid artery
right coronary artery
right pulmonary artery
right subclavian artery
right ventricle
subsinuosal branch of left coronary artery
thymus
trabeculae carnae
venae cavae (sing., vena cava)
visceral pericardium

SECTION VI: UROGENITAL SYSTEM

The urogenital system includes the excretory and reproductive systems, which perform distinct roles. The excretory system functions mainly in helping to maintain homeostasis by maintaining water balance and ridding the body of nitrogenous waste products. The reproductive system functions in producing gametes (sperm or ova), meeting all the needs of the fetuses during gestation (in the female), and maintaining sex-specific features, mainly through the production of hormones. Despite the fact that the two systems are distinct, their accessory structures, particularly in the male, become intimately associated. Thus, it is convenient to consider them together as the *urogenital system*.

Excretory System

The **kidneys** (Figures 7.58, 7.68, 7.69, and 7.71), the main organs of the excretory system, were exposed during the dissection of the cardiovascular system. They are retroperitoneal, lying dorsal to the parietal peritoneum in the lumbar region. The kidneys are surrounded by considerable fat that protects them from mechanical injury. Note that the left kidney lies slightly posterior to the right kidney, which abuts anteriorly against the caudate lobe of the liver. Note again the **adrenal glands**, which were described earlier (see page 261 and Figures 7.58 and 7.71).

Each kidney is bean-shaped and covered by a thin, tough, fibrous **renal capsule**. The **hilus** is a medial indentation through which the **ureter** and blood vessels pass. Follow one of the ureters posteriorly toward the **urinary bladder**. It passes dorsal to the **ductus deferens** in the male (Figure 7.69) and **uterine horn** in the female (Figure 7.71; see later), and then through the fat of the lateral ligament to enter the dorsal surface of the urinary bladder. Dissect carefully through the fat; recall that the umbilical artery was also traced to the bladder (page 264). The urinary bladder is a sac-like reservoir for urine. Its broad anterior portion gradually narrows posteriorly into the **urethra**, a narrow tube that passes through the pelvic canal (see later).

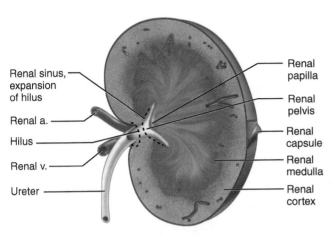

Renal sinus, expansion of hilus

Renal a.

Hilus

Renal v.

Ureter

Renal papilla

Renal pelvis

Renal capsule

Renal medulla

Renal cortex

FIGURE 7.68   Frontal section through left kidney of the cat.

Return to one of the kidneys, preferably the left. Remove the renal capsule, cutting carefully around the structures passing through the hilus. Section the kidney in the frontal plane (i.e., so that you have dorsal and ventral halves) and examine one of the sectioned surfaces (Figure 7.68). The substance of the kidney is subdivided into a lighter peripheral zone, the **renal cortex**, and the darker **renal medulla**. The medullary portion constitutes a **renal pyramid**, which tapers toward the hilus into the **renal papilla**.

The ureter and renal vessels enter the hilus, which expands within the kidney into a space termed the **renal sinus**. The expanded proximal portion of the ureter within the kidney is the **renal pelvis**. If it is not evident in the section you are studying, examine the other half of the kidney.

Key Terms: Excretory System

adrenal gland
ductus deferens
hilus
kidneys
renal capsule
renal cortex
renal medulla
renal papilla
renal pelvis
renal pyramid
renal sinus
ureter
urethra
urinary bladder
uterine horn

Male Reproductive System

The paired **testes** (Figure 7.69) are the male gonads and lie in the **scrotum**, which is a skin-covered sac just ventral to the anus. Externally, the **penis**, the copulatory organ, lies midventrally with respect to the scrotum. Skin the scrotum by making a small incision toward the ventrolateral side of one of the testes. Extend the incision longitudinally and peel the ventral skin of the scrotum. This exposes the left and right **cremasteric pouches**, sac-like extensions of the abdominopelvic cavity in which the testes descend from their abdominal position in the embryo.

The testes, as well as other structures, are contained within the pouches. The posterior end of each pouch contains a testis and so is expanded. Follow the pouches anteriorly. Each narrows into a thin tube that passes toward the abdominal wall along the ventral surface of the pelvis, just to one side of the pubic symphysis, and passes through the wad of fat present in the groin. You were cautioned against removing this fat during skinning of the cat precisely because the cremasteric pouches pass through it. Now, follow the pouches, teasing away the fat as needed.

The penis will still have its sheath of skin attached. Remove it by making a cut through the sheath and picking through the connective tissue with needle and forceps. Work your way anteriorly until you have exposed the penis and can appreciate that it passes toward the pelvic canal.

Return to the posterior end of one of the pouches, slit its ventral surface longitudinally, and continue the cut onto its narrowed portion. The space within the pouch, the **vaginal cavity**, is analogous to the space of the abdominopelvic cavity. It is lined by **parietal tunica vaginalis**. The structures within the pouch are covered by **visceral tunica vaginalis**. The **mesorchium** is the mesentery supporting these structures. The most notable structure within the pouch is the testis. Pull the posterior end of the cremasteric pouch. The short connective tissue extending to the testis is the **gubernaculum**.

The **epididymis** is a thin, flattened band of tissue on the dorsomedial part of the testis. It consists of expanded head and tail regions and a central, narrower body. As it may be difficult to discern the epididymis in ventral view, turn the testis over by reflecting its lateral surface dorsally. The concave margin of the epididymis is easily apparent. Grasp it with a forceps and return the testis to its anatomical position.

Spermatozoa, the male gametes, from the testis pass through the head, body, and tail of the epididymis and then enter the strand-like **ductus deferens**, which passes mainly along the medial side of the testis. This portion of the ductus deferens is highly convoluted. A thicker strand-like structure passes to the anterior end of the testis. This is the **pampiniform plexus**, formed by the intertwining of the convoluted distal ends of the internal spermatic artery and vein. Follow the artery, vein, and ductus deferens anteriorly by slitting the narrow part of the cremasteric pouch. These structures, bound together

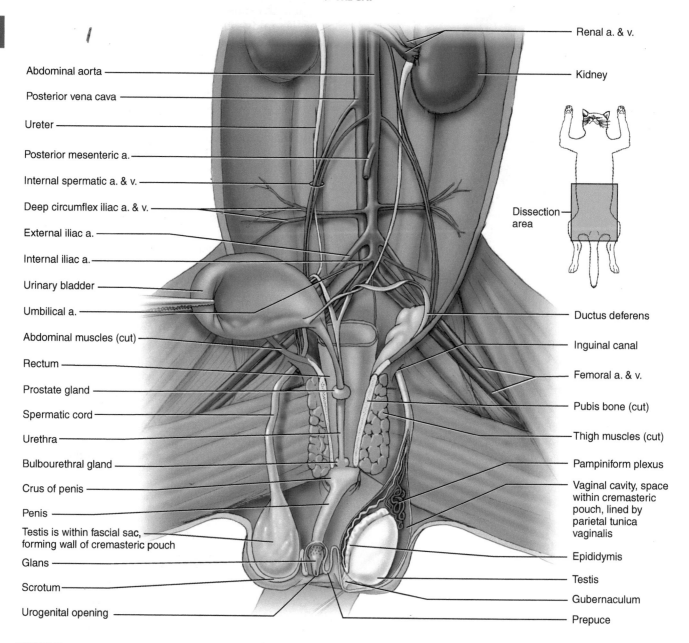

Renal a. & v.

Kidney

Abdominal aorta

Posterior vena cava

Ureter

Posterior mesenteric a.

Internal spermatic a. & v.

Deep circumflex iliac a. & v.

External iliac a.

Internal iliac a.

Urinary bladder

Umbilical a.

Abdominal muscles (cut)

Rectum

Prostate gland

Spermatic cord

Urethra

Bulbourethral gland

Crus of penis

Penis

Testis is within fascial sac, forming wall of cremasteric pouch

Glans

Scrotum

Urogenital opening

Dissection area

Ductus deferens

Inguinal canal

Femoral a. & v.

Pubis bone (cut)

Thigh muscles (cut)

Pampiniform plexus

Vaginal cavity, space within cremasteric pouch, lined by parietal tunica vaginalis

Epididymis

Testis

Gubernaculum

Prepuce

FIGURE 7.69 Abdominopelvic cavity of the male cat with the viscera removed and urinary bladder reflected to the right to show vessels and urogenital structures, in ventral view. Structures within the vaginal cavity, such as the left side testis, epididymis, and vessels, are covered by visceral tunica vaginalis.

by the visceral tunica vaginalis, constitute the **spermatic cord**. The cord passes through the **inguinal canal**, a short passageway in the abdominal wall, into the abdominopelvic cavity. Within the abdominopelvic cavity, the vessels and ductus deferens go their separate ways. The vessels extend anteromedially. The ductus deferens loops around the ureter and then extends posteromedially, passing through the fat within the lateral ligament of the urinary bladder. Locate the ductus on the other side so that both ducti deferentes may be traced as they extend into the pelvic canal and meet the urethra (see later).

Opening the Pelvic Canal

The pelvic canal must be opened in order to continue following the structures of the urogenital system. Doing so involves cutting through the pelvic symphysis, which is fairly easy, as the symphysis is not fused. Finding the symphysis, however, can be tricky. The most direct way is to continue the incision cut through the body wall posteriorly, all the way to the symphysis. In doing so, try to cut along the midsagittal plane (recall that originally you cut to one side of the midsagittal plane). Exposing the bladder and urethra may help, as the latter, which lies more or less midsagittally, can serve as a guide. When you find the anterior end of the pelvis, attempt to push through the symphysis with a scalpel. (Scraping the pubes a little can help show the symphysis as distinct from the bones.) If you are not precisely on the symphysis, the scalpel will meet with resistance. Do not attempt to force the scalpel through. Instead, move it over slightly to one side or the other and try again; it will pass through fairly easily if it is on the symphysis. Keep trying until you cut through the symphysis. Spread the symphysis by grasping the thighs and twisting them dorsally.

Once the pelvic canal has been opened, follow the urethra posteriorly, clearing away connective tissue and fat as you do so. Just within the canal, the urethra seems to expand laterally. Careful dissection reveals that the expansion is actually the **prostate gland**. Return to the ducti deferentes and trace them through the canal. They converge and pass along the dorsal surface of the urethra. The prostate gland surrounds their entrance into the urethra. The urethra continues posteriorly and enters the penis very near the end of the pelvic canal.

The penis is formed by three columns of erectile tissue that, when filled with blood, cause its erection (Figure 7.70). The **corpus spongiosum** is the middorsal column; the urethra is embedded within it. Posteriorly the corpus forms

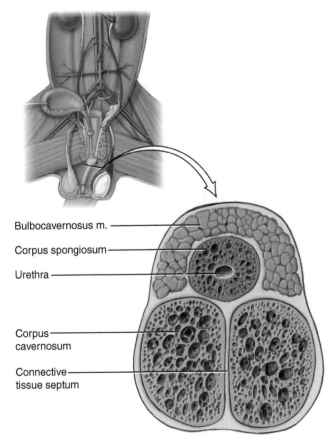

Bulbocavernosus m. ———

Corpus spongiosum ———

Urethra ———

Corpus ———
cavernosum

Connective ———
tissue septum

FIGURE 7.70   Cross-section through the penis of the cat.

the free end of the penis, the **glans penis**, from which the urethra emerges. The other two columns, the **corpora cavernosa** (sing., **corpus cavernosum**), lie side by side on the ventral surface. The dorsal position of the urethra and corpus spongiosum is a peculiarity of the posterior-pointing penis of the cat. In most other mammals the urethra and corpus spongiosum lie ventrally, and the corpora cavernosa are dorsal. Posteriorly, the corpora cavernosa diverge laterally as the **crura** (sing., **crus**) of the **penis**, which anchor the organ to the ischia. The small, paired **bulbourethral glands** lie just dorsal to the crura. Cut and reflect a crus to see them. Then section the penis near its center to view the three columns of spongy tissue. The **os penis** is a small bone (three to seven mm long) lying in the glans penis.

Key Terms: Male Reproductive System

bulbourethral glands
corpora cavernosa (sing., **corpus cavernosum)**
corpus spongiosum
cremasteric pouches
crus (pl., **crura**) of **penis**
ductus deferens
epididymis
glans penis
gubernaculum
inguinal canal
mesorchium
os penis (baculum)
pampiniform plexus
parietal tunica vaginalis
penis
prostate gland
scrotum
spermatic cord
testis (pl., **testes**)
vaginal cavity
visceral tunica vaginalis

Female Reproductive System

The small, oval **ovaries** are the female gonads (Figure 7.71) and lie in the abdominopelvic cavity just posterior to the kidneys. Follow the ovarian artery and vein of one side to an ovary. As in most vertebrates, ova produced by the ovaries pass posteriorly through paired tubes that have become longitudinally differentiated into several distinct regions in mammals. The **uterine tube**, the most anterior portion, is a thin, convoluted tube lateral to the ovary. Its proximal end enlarges into the **infundibulum**, which forms a hood-like expansion over the anterior end of the ovary. Its margin bears **fimbriae**, frill-like projections that help ensure the ovum passes through the **ostium tubae**, which is the opening of the infundibulum, and into the uterine tube.

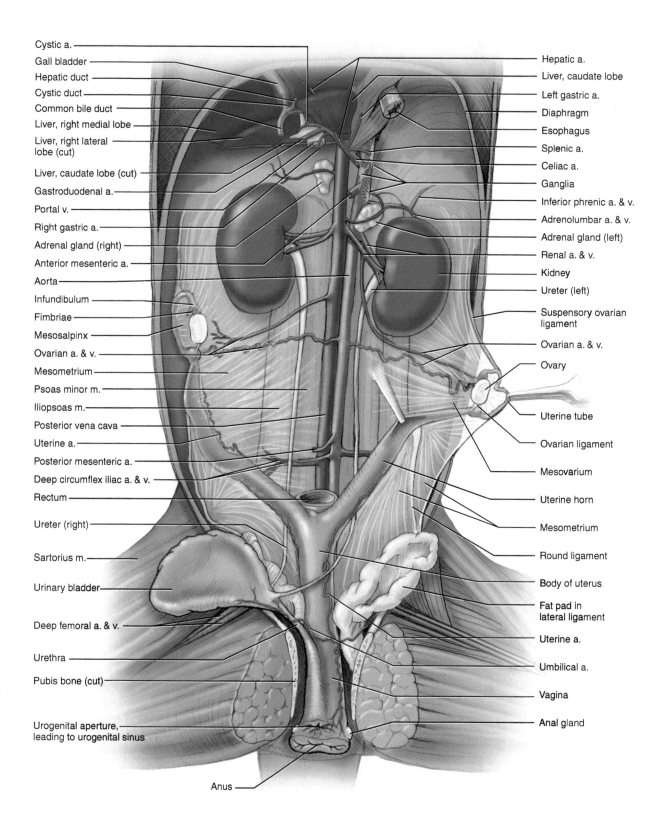

Cystic a.
Gall bladder
Hepatic duct
Cystic duct
Common bile duct
Liver, right medial lobe
Liver, right lateral lobe (cut)
Liver, caudate lobe (cut)
Gastroduodenal a.
Portal v.
Right gastric a.
Adrenal gland (right)
Anterior mesenteric a.
Aorta
Infundibulum
Fimbriae
Mesosalpinx
Ovarian a. & v.
Mesometrium
Psoas minor m.
Iliopsoas m.
Posterior vena cava
Uterine a.
Posterior mesenteric a.
Deep circumflex iliac a. & v.
Rectum
Ureter (right)
Sartorius m.
Urinary bladder
Deep femoral a. & v.
Urethra
Pubis bone (cut)
Urogenital aperture, leading to urogenital sinus
Anus

Hepatic a.
Liver, caudate lobe
Left gastric a.
Diaphragm
Esophagus
Splenic a.
Celiac a.
Ganglia
Inferior phrenic a. & v.
Adrenolumbar a. & v.
Adrenal gland (left)
Renal a. & v.
Kidney
Ureter (left)
Suspensory ovarian ligament
Ovarian a. & v.
Ovary
Uterine tube
Ovarian ligament
Mesovarium
Uterine horn
Mesometrium
Round ligament
Body of uterus
Fat pad in lateral ligament
Uterine a.
Umbilical a.
Vagina
Anal gland

FIGURE 7.71 Abdominopelvic cavity of the female cat with many of the viscera removed and urinary bladder reflected to the right to show vessels and urogenital structures, in ventral view.

Each uterine tube leads posteriorly into the larger **uterine horn**. The uterine horns will be narrow unless the female is gravid or has previously carried a litter. The horns of each side pass posteromedially and merge dorsal to the urinary bladder. The short, tough **ovarian ligament** anchors the ovary and the anterior end of the horn. At first glance, however, it appears to form a connection between the uterine horn and ovary. Thus, carefully follow the uterine horn into the uterine tube.

The union of the uterine horns posteriorly form the **body** of the **uterus**, a wider canal that then leads into the **vagina**. As there is no external division of these structures, their separation will be seen when they are opened. Note that the ureters pass dorsal to the uterine horns.

The reproductive tract is supported by several mesenteries. The main supporting structure is the **broad ligament**, of which several portions are recognized. Lift the reproductive structures to clearly observe the following mesenteries. The portion of the broad ligament supporting the ovary is the **mesovarium**. The **mesosalpinx** attaches to the uterine tube, and the **mesometrium** attaches to the uterus, including the uterine horn. Lift the uterine horn near its center and tug it medially, as shown in Figure 7.71. The **round ligament** is the fibrous band in the mesometrium that extends diagonally from the uterine horn posterolaterally toward the body wall. Also, note the **suspensory ovarian ligament**, which supports the ovary anteriorly, as it extends to the dorsal body wall just lateral to the kidney.

Open the pelvic canal by cutting through the pubic symphysis following the instructions given for the male on pages 296 and 297. Pick away connective tissue to reveal and separate the structures passing through the canal (Figure 7.71). The urethra and vagina unite toward the posterior end of the pelvic canal to form a common passageway, the **urogenital sinus**, which opens to the outside through the **urogenital aperture**. Dissect dorsal to the vagina and urogenital sinus to separate them from the rectum. On each side, dissect the tissue lateral to rectum, very near the anus, to reveal the **anal glands**. Then completely separate the urogenital sinus from the rectum so that it can be turned over. Make a slit along one side of the body of the uterus, vagina, and urogenital sinus, and spread the flaps apart (Figure 7.72). Find the **urethral orifice**, the entrance of the urethra. The vagina extends anteriorly from this point to the **cervix** of the uterus, which is the distally tapered, sphincter-like portion of the uterus. Directly posterior to the urethral orifice, and just before the urogenital aperture, lies the small **clitoris**, which may be difficult to see. It is the female homologue of the penis and generally considerably smaller, but may be quite large in a few mammals; in spotted hyenas, for example, the clitoris may be as large as the flaccid penis (Cunha et al., 2014).

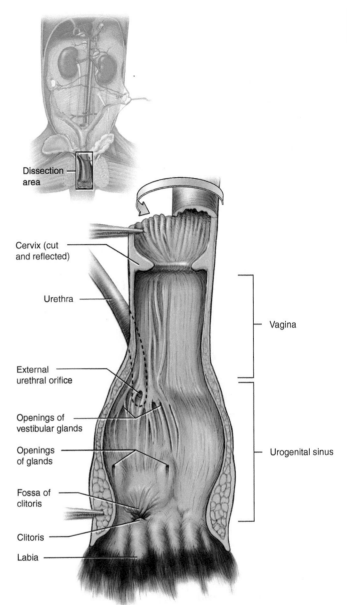

FIGURE 7.72   Distal portion of female urogenital tract of the cat in ventral view. The tract has been cut along its lateral margin and opened to reveal the internal anatomy.

The **mammary glands** are important reproductive structures in mammals. They were observed beneath the skin of the thorax and abdomen during skinning of the cat (see page 211).

Key Terms: Female Reproductive System

anal glands
body of the **uterus**
broad ligament
cervix of the **uterus**
clitoris
fimbriae

infundibulum
mammary glands
mesometrium
mesosalpinx
mesovarium
ostium tubae
ovarian ligament
ovary
round ligament
suspensory ovarian ligament
urethral orifice
urogenital aperture (vulva)
urogenital sinus (vaginal vestibule, urogenital canal)
uterine horn
uterine tube (fallopian tube)
vagina

SECTION VII: BRAIN AND CRANIAL NERVES

In comparative anatomy courses, the brain and cranial nerves of a sheep, rather than of a cat, usually are studied. There are two main reasons for this. First, the cat brain is small and many of its structures, particularly the nerves, are extremely difficult to find. Second, the brain of the cat would have to be carefully removed from the skull, which requires patience and the proper tools. To avoid these problems, the mammalian brain should be studied using a prepared sheep brain. Illustrations of the sheep brain (Figures 7.73–7.77) are provided here, as well as for the cat (Figures 7.78 and 7.79). Compare the structures in the cat in these illustrations with those you observe in the sheep brain.

Meninges

The brain is surrounded by three membranes, or **meninges** (sing., **meninx**), which protect and isolate it (Figure 7.73). The thickest is the dura mater. It fuses with the periosteum of the bones enclosing the cranial cavity and is usually left on the skull when the brain is removed, although a ventral portion of the dura mater, covering the hypophysis, is present in most specimens. If your specimen still has the dura mater adhering to it, carefully remove it, but keep its ventral and posteroventral portions. Some specimens that preserve the dura mater also preserve the branching of the trigeminal nerve (see later). In lateral view, this is a large, Y-shaped, horizontal structure lateral to the hypophysis. If present, sever the nerve just before it branches.

The other meningeal layers are the pia mater, a thin layer adhering to the surface of the brain, and the **arachnoid mater**, lying between the dura and pia. The arachnoid largely remains with the dura mater in the cranial cavity of the skull.

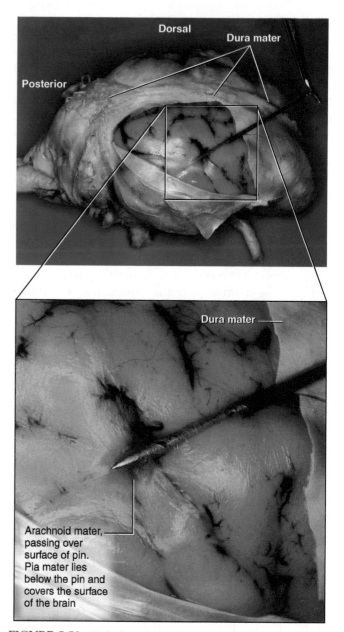

FIGURE 7.73   Right lateral view (anterior to right of page) of the sheep brain showing meninges. The dura mater has been partially cut and reflected to reveal deeper meninges. Blowup shows the pin passing deep to the arachnoid mater.

Telencephalon

The **telencephalon** includes the **cerebrum** and **olfactory bulbs** (Figures 7.74 and 7.75). It is greatly expanded in mammals, due mainly to the enormous size of the cerebrum. Indeed, the largest parts of the brain are the **cerebral hemispheres**, which lie dorsally and together comprise the cerebrum. The **cerebellum** (part of the **metencephalon**; see later) is another prominent part of the brain, and lies posterior to the cerebrum (Figure 7.74). Separating the cerebellum from the cerebrum is a deep cleft, the **transverse fissure**. The surface of the cerebrum

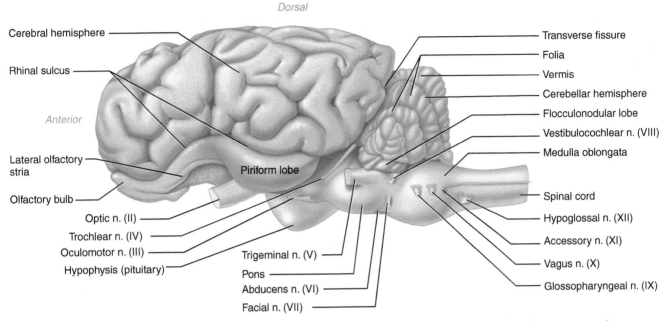

FIGURE 7.74 Left lateral view of the sheep brain and cranial nerves, with dura mater and arachnoid mater removed.

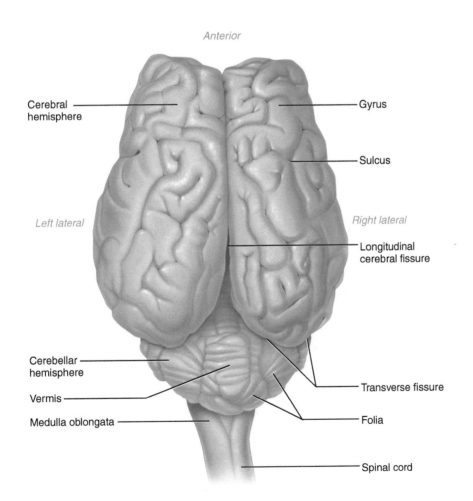

FIGURE 7.75 Dorsal view of the sheep brain, with dura mater and arachnoid mater removed.

bears numerous folds, the **gyri** (sing., **gyrus**), separated by grooves, the **sulci** (sing., **sulcus**). Right and left cerebral hemispheres are separated by the deep **longitudinal cerebral fissure** (Figure 7.75). Gently spread the hemispheres to reveal the **corpus callosum**, a white structure consisting of fibers that connect the hemispheres.

The elongated, flattened olfactory bulbs lie anteroventrally on the telencephalon (Figures 7.74 and 7.76). A band of fibers, the **olfactory tract**, continues posteriorly from the olfactory bulb and almost immediately separates into **lateral** and **medial olfactory striae**. The medial olfactory stria extends posteromedially, whereas the lateral olfactory stria extends posterolaterally. Each stria is accompanied by a gyrus, and only the lateral olfactory stria is easily discernible. It continues posteriorly into the **piriform lobe**, which is separated from the rest of the cerebrum by the **rhinal sulcus** (Figure 7.74).

Diencephalon

The greatly expanded cerebrum covers the roof of the **diencephalon**, so that without dissection only the **hypothalamus**, or floor of the diencephalon, may be clearly observed (Figure 7.76). The **optic chiasm**, at the anterior end of the hypothalamus, represents a partial decussation of the **optic nerves**. Just posterior to the optic chiasm, a thin, delicate stalk, the **infundibulum**, suspends the **hypophysis** (Figure 7.77). In prepared specimens, the hypophysis and infundibulum may be missing. Remove them if present in your specimen. The opening for the infundibulum is a continuation of the **third ventricle** (see later). The area of the hypothalamus adjacent to the opening is the **tuber cinereum** (Figure 7.76). Immediately posterior to the tuber are the paired **mammillary bodies** (which may appear as a single rounded structure), which mark the posterior end of the hypothalamus.

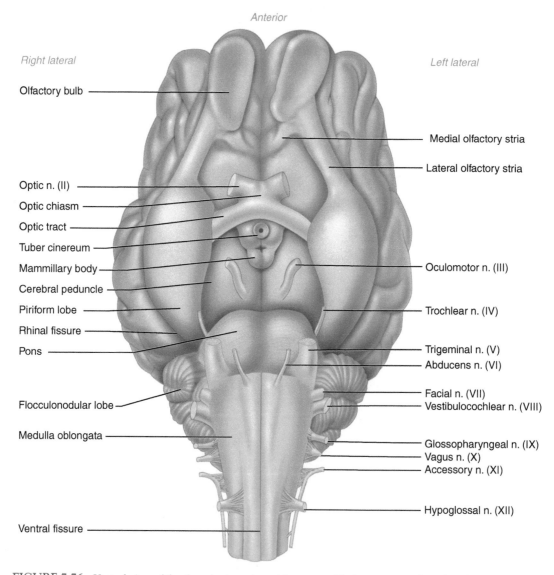

FIGURE 7.76 Ventral view of the sheep brain and cranial nerves, with dura mater and arachnoid mater removed.

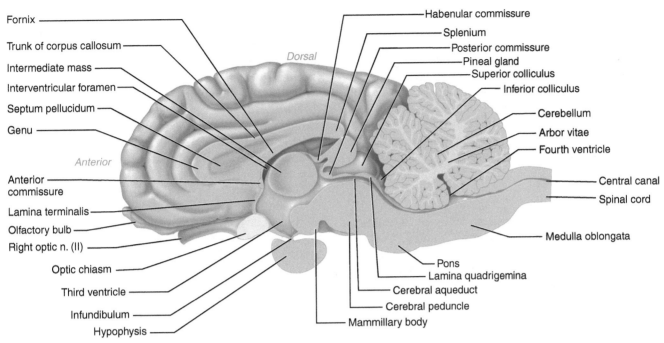

FIGURE 7.77 Sagittal section of sheep brain.

The dorsal part of the diencephalon is formed by the **epithalamus**. To see it, observe the brain in dorsal view, spread apart the cerebral hemispheres, and make a longitudinal cut through the corpus callosum. The epithalamus lies mainly ventral to the latter's posterior portion. Remove its roof, the **tela choroidea**, to reveal the third ventricle, the narrow cavity of the diencephalon. The slightly thickened posterolateral margin of the ventricle forms the **habenula** on each side. These margins converge posteriorly toward the midline and form the **habenular commissure**, which is more readily discernible in sagittal section (Figure 7.77). The rounded **pineal gland** lies posterior to the commissure.

The walls of the **thalamus**, the lateral portions of the diencephalon, are mainly lateral to the dorsal margins of the third ventricle, but their extent is difficult to appreciate. Part of the thalamus, the **intermediate mass** (see later), extends across the third ventricle to connect the left thalamus and right thalamus (Figure 7.77). You may have cut through it during exposure of the third ventricle. The positional relationships of these structures will be examined again in sagittal view.

Mesencephalon

The **tectum**, or roof, of the **mesencephalon** was partially exposed during dissection of the third ventricle. To see it more completely, spread the cerebral hemispheres from each other, as well as from the cerebellum. The tectum is characterized by two paired, prominent swellings that together form the **corpora quadrigemina**, which sit

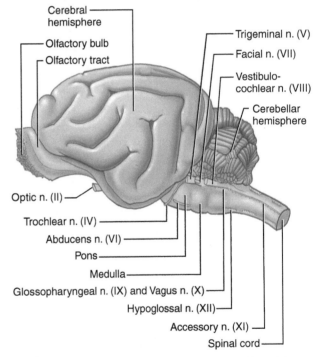

FIGURE 7.78 Left lateral view of the cat brain and cranial nerves, with dura mater and arachnoid mater removed.

on the **lamina quadrigemina** (the latter will be seen in sagittal section). The two anterior swellings, the **superior colliculi**, are the larger, well-rounded structures that lie just posterior to the pineal gland. The **inferior colliculi** are the smaller, flattened structures that protrude from the posterior end of the superior colliculi. The **trochlear nerves** arise from the dorsal surface of the mesencephalon just posterior to the inferior colliculi.

Examine the brain in ventral view (Figure 7.76). The floor of the mesencephalon is formed by the cerebral peduncles, slightly elevated regions lateral and posterior to the mammillary bodies. The relatively wide and flattened **oculomotor nerve** arises from the surface of each cerebral peduncle.

Metencephalon

The **metencephalon** consists mainly of two regions. Dorsally is the cerebellum (Figures 7.74 and 7.75). Its surface is highly folded into **folia**, which are separated by **sulci**. The median part of the cerebellum is the **vermis**. The **cerebellar hemispheres** lie on either side of the vermis. A **flocculonodular lobe** may be distinguished on the ventrolateral part of each hemisphere. On the ventral surface, the metencephalon consists of the **pons**, a rounded region of transverse fibers posterior to the cerebral peduncles. On each side, the **trigeminal nerve** arises laterally from the posterolateral part of the pons. Distally the trigeminal has three main branches: the ophthalmic, maxillary, and mandibular nerves, which are often respectively abbreviated as V_1, V_2, and V_3 (Table 7.5).

Myelencephalon

The **myelencephalon** consists of the **medulla oblongata**, which forms the brain posterior to the metencephalon and connects to the **spinal cord** (Figures 7.76 and 7.77). The **ventral fissure** is the median ventral groove of

the medulla. To either side are narrow longitudinal bands termed **pyramids**. Lateral to the anterior part of each pyramid, and just posterior to the pons, lies a **trapezoid body**.

The remaining cranial nerves are mainly associated with the medulla oblongata, but some of them may be difficult to discern if the meninges have been stripped. Thus, identify the stumps of these nerves before removing the meninges. The nerves to be identified are the **abducens, facial, vestibulocochlear, glossopharyngeal, vagus, accessory**, and **hypoglossal nerves** (Figures 7.74, 7.76, and 7.80).

An abducens nerve arises from the anterior part of the medulla, between the trapezoid body and pyramid. The facial nerve arises lateral to the trapezoid body just posterior to the trigeminal nerve, and the vestibulocochlear nerve arises slightly more dorsally, from beneath the flocculonodular lobe. The glossopharyngeal, vagus, and accessory nerves arise in sequence and more posteriorly from the lateral surface of the medulla. The hypoglossal nerve arises farther posteriorly and ventrally. If the meninges have been stripped from the medulla, however, it will be extremely difficult to identify these nerves.

One method of exposing the nerves is to make a midventral cut through the meninges using fine scissors. Begin posteriorly and work your way forward. When you reach the level of the abducens nerves, carefully reflect the meninges. As you peel the meninges back, look for a series of fine rootlets arising from the ventrolateral surface of the posterior part of the medulla. These rootlets merge to form the hypoglossal nerve. Separate it from the meninges and continue peeling the latter. Soon you will note the glossopharyngeal, vagus, and accessory nerves. The accessory will probably be the most prominent at first, because it forms a longitudinal nerve adjacent to the surface of the medulla. Follow it forward, and you will find the stumps of the three nerves arising very close to each other. The glossopharyngeal and vagus arise by a series of small rootlets, so they may not be readily discernible. The accessory nerve arises as a series of very fine rootlets along the posterior part of the medulla. Posterior to it, your specimen may preserve the first spinal nerve. Its origin appears similar to that of the hypoglossal nerve.

Examine the dorsal surface of the medulla. A good part of its roof is covered by a **tela choroidea**. Remove it to expose the **fourth ventricle**, the cavity within the medulla that continues forward under the cerebellum. This part of the roof of the fourth ventricle is covered by a separate membranous structure, the **medullary velum**.

Sagittal Section of the Brain

With a new scalpel blade, make a sagittal section through the brain by extending the section made earlier to expose the third ventricle. Carefully continue the

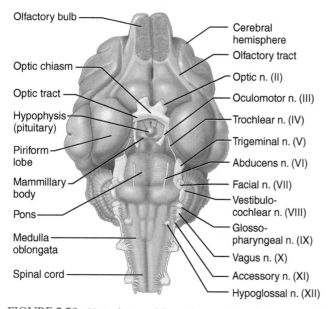

Olfactory bulb
Optic chiasm
Optic tract
Hypophysis (pituitary)
Piriform lobe
Mammillary body
Pons
Medulla oblongata
Spinal cord

Cerebral hemisphere
Olfactory tract
Optic n. (II)
Oculomotor n. (III)
Trochlear n. (IV)
Trigeminal n. (V)
Abducens n. (VI)
Facial n. (VII)
Vestibulo-cochlear n. (VIII)
Glosso-pharyngeal n. (IX)
Vagus n. (X)
Accessory n. (XI)
Hypoglossal n. (XII)

FIGURE 7.79 Ventral view of the cat brain and cranial nerves, with dura mater and arachnoid mater removed.

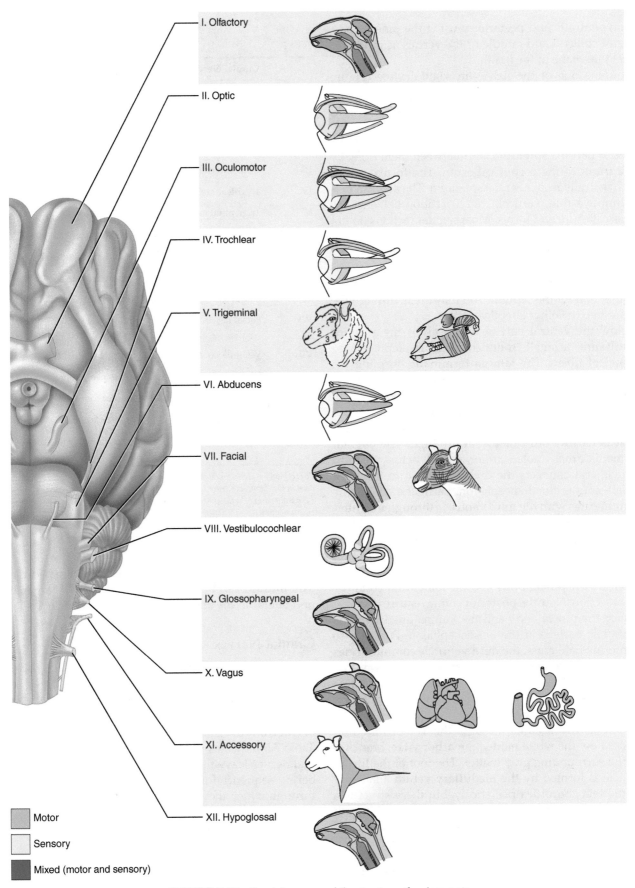

I. Olfactory

II. Optic

III. Oculomotor

IV. Trochlear

V. Trigeminal

VI. Abducens

VII. Facial

VIII. Vestibulocochlear

IX. Glossopharyngeal

X. Vagus

XI. Accessory

XII. Hypoglossal

Motor

Sensory

Mixed (motor and sensory)

FIGURE 7.80   Cranial nerves and the structures they innervate.

THE DISSECTION OF VERTEBRATES

section ventrally and posteriorly using the pineal body, superior colliculi and middle of the vermis as guides to the sagittal plane of the brain.

Examine one of the halves in sagittal view (Figure 7.77) and note the great extension of the cerebrum. The **corpus callosum** (see also earlier), the structure allowing communication between the cerebral hemispheres, is also clearly revealed. Its anterior, curved part is the **genu**, its posterior part the **splenium**, and between them dorsally is the **trunk** of the corpus callosum. The **fornix** curves anteroventrally from near the splenium. The **septum pellucidum** is a thin, vertical, shiny partition between the corpus callosum and fornix. Most sagittal sections do not coincide precisely with the midsagittal plane, and so the septum remains on one or the other half of the brain. If it is absent on the half you are examining, you will see a cavity, the **lateral ventricle** of the cerebral hemisphere; check the other half for the septum. You may break through the septum into the other lateral ventricle.

Below the ventral part of the fornix is the **anterior commissure**, a small, rounded structure representing a group of fibers. The **lamina terminalis**, forming the anterior wall of the third ventricle, extends ventrally from the anterior commissure to the **optic chiasm**, which is oval in section. The third ventricle lies mainly ventral to the fornix and splenium of the corpus callosum. Note that it is narrow but dorsoventrally high. The circular intermediate mass, noted earlier, extends across the third ventricle and connects the thalami that form the right and left sides of the diencephalon. Each lateral ventricle communicates with the third ventricle through the **interventricular foramen**, located just anterior to the intermediate mass. Identify the **mammillary body** ventrally on the hypothalamus, and more posteriorly, the **cerebral peduncle** and **pons**.

Identify the **pineal body** and the **habenular commissure** in section. Note the **posterior commissure** extending between the pineal body and the lamina quadrigemina (mentioned earlier) of the mesencephalon. Posterior to the intermediate mass, the third ventricle communicates with the **cerebral aqueduct**, a narrow passageway that extends posteriorly beneath the lamina quadrigemina and then into the fourth ventricle. The latter, in turn, extends back into the **central canal** of the spinal cord.

Examine the section through the cerebellum and observe how the white matter, the **arbor vitae**, branches into the surrounding gray matter. The roof of the fourth ventricle is formed by the **medullary velum** anteriorly and the **tela choroidea** posteriorly, but these structures were probably removed during exposure of the fourth ventricle. Note how the cerebellum attaches to the rest of the brain on each side of the fourth ventricle via a series of fiber tracts or peduncles. Make a frontal section through the peduncles to remove the cerebellum. The posterior curved part of the section consists of the **posterior**

TABLE 7.5 Cranial Nerves and Associated Foramina of the Sheep.

Cranial Nerve		
Number	Nerve	Foramen
I	Olfactory nerve	Cribriform foramina of the cribriform plate
II	Optic nerve	Optic canal
III	Oculomotor nerve	Orbital fissure[1]
IV	Trochlear nerve	Orbital fissure[1]
V	Trigeminal nerve	
V_1	Ophthalmic nerve	Orbital fissure[1]
V_2	Maxillary nerve	Foramen rotundum[1]
V_3	Mandibular nerve	Foramen ovale
VI	Abducens nerve	Orbital fissure[1]
VII	Facial nerve	Internal acoustic meatus, then stylomastoid foramen[2]
VIII	Vestibulocochlear nerve	Internal acoustic meatus[3]
IX	Glossopharyngeal nerve	Jugular foramen
X	Vagus nerve	Jugular foramen
XI	Accessory nerve	Jugular foramen
XII	Hypoglossal nerve	Hypoglossal canal, then jugular foramen

[1]The orbital fissure and foramen rotundum fuse to form a single foramen orbito-rotundum in the sheep; see Figure 7.S1.
[2]The main part of the facial nerve passes through the inner ear and middle ear before emerging from the stylomastoid foramen.
[3]As the vestibulocochlear supplies structures of the inner ear, it does not leave the skull.

cerebellar peduncle. Anteriorly there are two peduncles. The **middle cerebellar peduncle** is slightly larger than and lateral to the **anterior cerebellar peduncle**.

Cranial Nerves

The stumps of the 12 cranial nerves have already been identified during the dissection of the brain, but their peripheral distribution cannot easily be followed. Figure 7.80 provides a summary of the cranial nerves, their origin from the brain, and their peripheral distribution. Table 7.5 indicates the foramina through which the cranial nerves leave the skull. Review the names and numbering sequence for the nerves, as well as the foramina through which they pass.

Key Terms: Brain and Cranial Nerves

　　abducens nerve
　　accessory nerve
　　anterior commissure

arachnoid mater
arbor vitae
central canal
cerebellar hemispheres
cerebellar peduncle, anterior, middle and posterior
cerebellum
cerebral aqueduct (aqueduct of Sylvius)
cerebral hemispheres
cerebral peduncle
cerebrum
corpora quadrigemina
corpus callosum
diencephalon
dura mater
epithalamus
facial nerve
flocculonodular lobe
folia
fornix
fourth ventricle
genu
glossopharyngeal nerve
gyrus (pl., gyri)
habenula
habenular commissure
hypoglossal nerve
hypophysis
hypothalamus
inferior colliculi
infundibulum
intermediate mass (massa intermedia)
interventricular foramen (foramen of Munro)
lamina quadrigemina
lamina terminalis
lateral olfactory stria
lateral ventricle
longitudinal cerebral fissure
mammillary body
medial olfactory stria
medulla oblongata

medullary velum
meninx (pl., meninges)
mesencephalon
metencephalon
myelencephalon
oculomotor nerve
olfactory bulbs
olfactory tract
optic chiasm
optic nerves
pia mater
pineal body
pineal gland
piriform lobe
pons
posterior commissure
pyramids
rhinal sulcus
septum pellucidum
spinal cord
splenium
sulcus (pl., sulci)
superior colliculi
tectum
tela choroidea of diencephalon
tela choroidea of myelencephalon
telencephalon
thalamus
third ventricle
transverse fissure
trapezoid body
trigeminal nerve
trochlear nerve
trunk of corpus callosum
tuber cinereum
vagus nerve
ventral fissure
vermis
vestibulocochlear nerve (auditory nerve, octaval nerve)

8

Reptile Skulls and Mandibles

INTRODUCTION

The great taxonomic diversity of reptiles—including such forms as turtles, alligators, lizards, snakes, and birds—is reflected in the enormous anatomical variation present in this clade. This chapter treats the cranial skeleton, including the skull and mandible, of representative forms of the major reptilian clades that are not covered in a separate chapter: a turtle, lizard, snake, alligator, and a tyrannosaurid dinosaur. The skull and mandible of a bird are discussed in Chapter 9. The taxa covered represent the main subdivisions of Reptilia. Although the specific examples discussed here may not be available for study, a closely related form is likely at your disposal and sufficient for appreciating at least the more general aspects of the cranial skeletal anatomy. In most cases, it will also be possible to follow more detailed anatomical features, such as conformation of individual bones and foramina. The exception is the tyrannosaurid skull, which is unlikely available for direct study.[1] Still, the exercise of identifying the main bones and features of the skull using accompanying figures allows an understanding of the basic pattern and some of the specific features that characterize a main reptilian clade.

SECTION I: TURTLE SKULL AND MANDIBLE

Turtles (Testudines) were traditionally considered representative of the anapsid condition, in which the lateral temporal region of the skull is unfenestrated, reflecting the ancestral amniote condition, and were thus included in Parareptilia. However, as noted in Chapter 1 (see also Figure 1.15), the anapsid skull pattern present in all but the most basal turtles is now understood to be a modification of an ancestrally diapsid condition. The main

taxon discussed here is *Chelydra serpentina*, the common snapping turtle. As in most turtles, its temporal region is derived, so the skull of *Chelonia mydas*, the green sea turtle, is used as an example of the more stereotypical anapsid condition as well as to supplement aspects of anatomical detail. The nomenclature of the skeletal elements of the turtle largely follow Gaffney (1972); Further images may also be found in Olori (2004).

Observe the skull of *C. serpentina* and identify the structures described below. Many of the sutures between the bones are difficult to discern, but the position of the bones will be apparent from the following descriptions and figures. The sutural patterns of most other turtle species you are likely to have at your disposal will differ slightly, but most of the bones should still be identifiable. Another consideration is that there may be considerable tissue covering and thus obscuring some of the features noted below. If so, you will need to clear it away.

In lateral view (Figures 8.1a and 8.4a), note the relatively short **rostrum** and anteriorly placed **orbits**. Anteriorly, the **naris** (in life the opening is separated by soft tissue into left and right nares) is bounded by the **maxillae** laterally, the **premaxillae** ventrally, and the **prefrontals** dorsally. The **nasals**, present in most other tetrapods, are usually absent in turtles (nasals are present, for example, in the earliest fossil turtles such as *Odontochelys*; see Li et al., 2008). Also absent in *Chelydra*, as in nearly all turtles, are teeth. Instead, these structures are replaced functionally by a beak covered by the keratinous **rhamphotheca**, anchored to the ventral margin of the maxillae and premaxillae as well as to the lower jaw (Figure 8.2). Teeth are present in early turtles—for example, in the roof of the oral cavity of *Proganochelys* (see Reisz & Head, 2008) and as a marginal series and within the oral cavity of *Odontochelys* (Li et al., 2008).

The margin of the orbit (Figures 8.1a and 8.3a) is formed by the maxilla and prefrontal as well as the **postorbital** and **jugal**; in some species the **frontal** may make a small contribution to the orbit. The lacrimal, a bone commonly bordering the orbit in other amniote skulls,

---

[1]Casts, though expensive, are available. A more affordable option would be to 3D print a scaled down model of the many dinosaur skulls available free online.

is missing in turtle skulls (except possibly *Proganochelys*; see Gaffney, 1979; Schoch & Sues, 2017, and note that there is no definitive evidence for the presence of a lacrimal in even the stem turtle *Pappochelys*). A large opening, the **interorbital foramen**, is present on the medial wall of the orbit. The foramen is paired with another on the medial wall of the orbit of the other side of the skull. In life, much of the foramen is filled with cartilage (some may still be present in your specimen) through which several structures pass (e.g., the optic nerve; portions of the eye muscles as well as oculomotor and trochlear nerves; a branch of the trigeminal nerve). A smaller opening, the **orbitonasal foramen**, is visible on the floor of the orbit for passage of an artery between the orbital and nasal regions. Look posteriorly into the orbit to view the braincase, which supports the brain and covers it ventrally, posteriorly, and partially laterally with bony elements. The space within the braincase is the

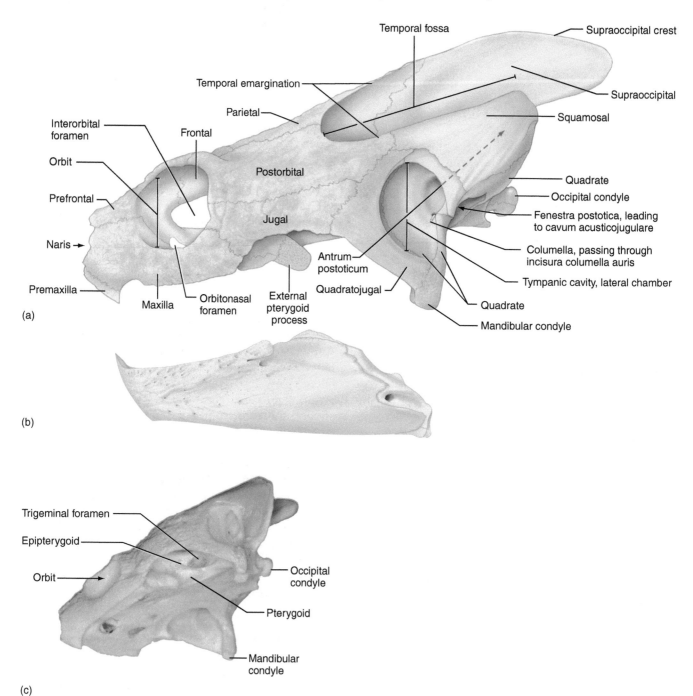

FIGURE 8.1   Skull and mandible of the snapping turtle: (a) skull in lateral view; (b) mandible in lateral view (see Figure 8.6 for labels); and (c) skull in ventrolateral view.

cranial cavity (Figure 8.4b). Parts of the braincase are not ossified, with the sphenoid and ethmoid regions being mainly cartilaginous. The rest of the brain is covered dorsally by the **epipterygoid** (see later) and bones of the skull roof (Figures 8.1, 8.3a, and 8.4). The **parietal** forms the roof of the skull posterior to the frontal and extends into the prominent, blade-like **supraoccipital crest** of

the **supraoccipital**. The parietal and supraoccipital also extend lateroventrally to help cover the brain.

The temporal region lies posterodorsally. The skull of turtles is commonly considered an example of an anapsid skull because there is no definite opening, or **temporal fenestra**, in the lateral skull roof posterior to the orbit, except in basal turtles, as noted earlier in this section and in Chapter 1. This anapsid condition is best observed in a sea turtle (Figure 8.4). In most turtles, however, the posterior margin of the skull is emarginated, forming a large notch, the **temporal emargination** (Figures 8.1a and 8.3a). The large jaw-closing or adductor musculature is located here, and emargination is related to the attachment and bulging of this musculature (see Rieppel, 1990). The space occupied by the musculature is the **temporal fossa**, and the walls forming the fossa provide much of the origin for the musculature. Compared to the sea turtle, in which the fossa is almost entirely covered over (Figure 8.4a), it is apparent that emargination has occurred mainly in the parietal. In *Chelydra* the margin is formed by the parietal and postorbital anteriorly and the **squamosal** posteriorly (but this may differ in other turtles). Note that the squamosal forms the posterodorsal corner of the skull. The depressor mandibulae muscle, which opens the jaw, takes origin from the posterolateral and posterior surfaces of the squamosal.

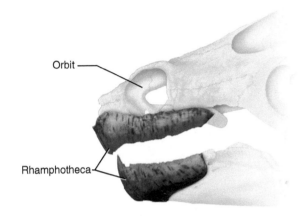

FIGURE 8.2 Anterior part of skull and mandible of the snapping turtle showing rhamphotheca.

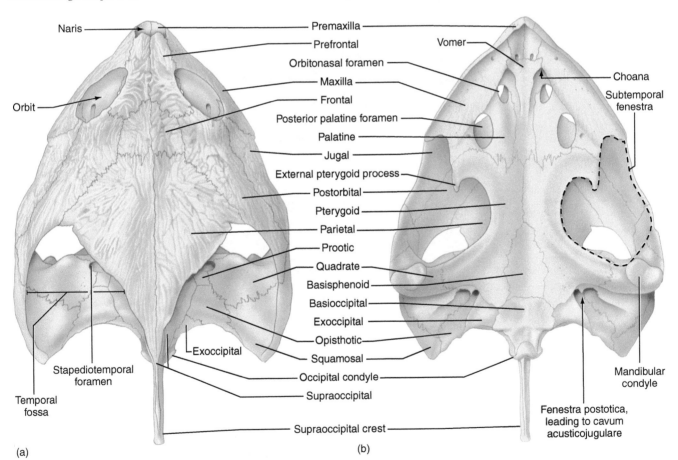

(a)   (b)

FIGURE 8.3 Skull of the snapping turtle in (a) dorsal and (b) ventral views.

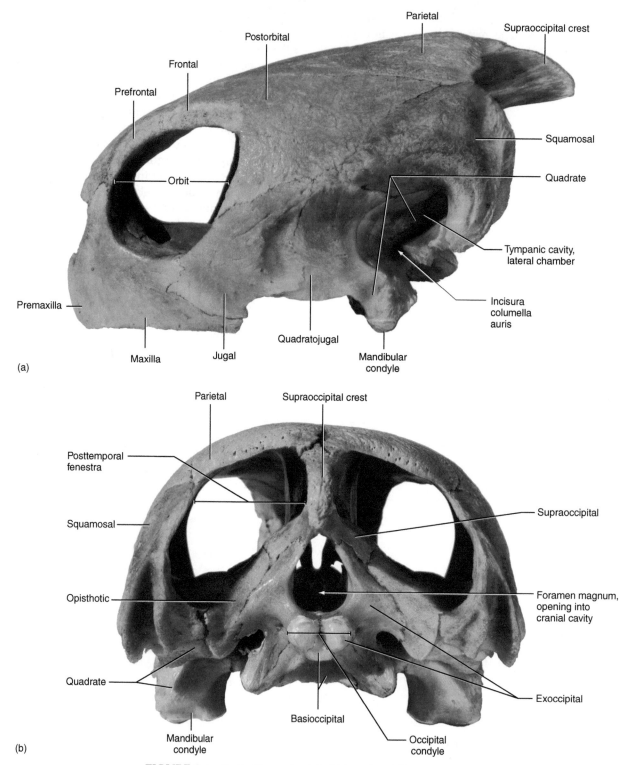

(a)

(b)

FIGURE 8.4   Skull of the sea turtle in (a) lateral and (b) posterior views.

Ventral to the anterior part of the squamosal is a large, somewhat oval or kidney-shaped depression, notched posteriorly. This is the **tympanic cavity**, which is formed almost entirely by the **quadrate** and contains the middle ear (Figures 8.1a and 8.4a). Posteriorly, a small and delicate rod-like bone (it may be absent in your specimen) projects into the cavity. This is the most distal portion of the **columella** (or stapes; Figures 8.1a and 8.5) and is continued laterally by a cartilaginous portion, the **extracolumella** (which will almost certainly not be

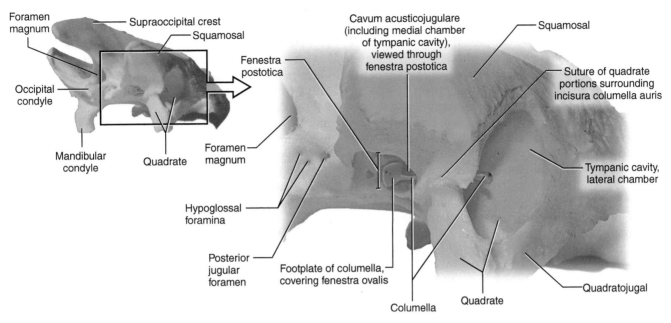

FIGURE 8.5 Skull of the snapping turtle in posterolateral view.

preserved in your specimen), to the eardrum or tympanum (but see later for *Chelonia*), which is attached to the margin of the tympanic cavity. To provide greater stability for the placement of the columella, the quadrate constricts around (as in sea turtles) and in some cases (as in *Chelydra*) completely surrounds it, thus forming a passageway, the **incisura columella auris**. The suture indicating where the ends of the quadrate meet in forming the incisura can be seen just posterior to the columella in *Chelydra* (Figure 8.5). In surrounding the columella, the quadrate subdivides the middle ear into a lateral and a medial chamber. This is distinct from the typical condition of the middle ear in other tetrapod skulls. We have already noted the lateral chamber in the *Chelydra* skull. The medial chamber is traversed by the medial end of the columella, once it has passed through the incisura, to reach the inner ear (Figure 8.5). The medial end of the columella expands into a circular footplate that covers the **fenestra ovalis**. The medial chamber is best viewed in posterior view through the **fenestra postotica**, which is, however, probably covered by tissue and may have to be prepared (Figure 8.5).

Return to a lateral view (Figure 8.1a). Posterodorsally the tympanic cavity is continuous with a second cavity, the **antrum postoticum**, which is formed mainly by the squamosal. To see it, look posteriorly into the tympanic cavity. The antrum serves to enlarge the middle ear cavity and thus to control the range of sounds that can be heard. In general, the more terrestrial a turtle, the larger its antrum. Given that the antrum is filled with air, all turtles that dive deeply need to reduce its size and thicken its walls—in marine turtles the antrum is quite small. As well, *Chelonia* lacks a tympanic membrane, and

the tympanic cavity is mainly filled with a fatty tissue, which according to Rieppel (1993) apparently compensates for the relative vacuum resulting in the middle ear during diving (and it may also be related to solving the problem of acoustic impedance in aquatic tetrapods).

The quadrate continues ventrally below the tympanic cavity to form the **mandibular condyle**, which articulates with the lower jaw (Figures 8.1a, 8.3b, and 8.4). The **quadratojugal** articulates with the quadrate here and extends dorsally, curving around the margin of the tympanic cavity, to articulate with the jugal, postorbital, and squamosal. The ventral margin of the skull posterior to the orbit is formed by the quadratojugal and jugal and is concave. Tilt the skull to observe the braincase ventral to this margin (Figure 8.1c). A large oval opening, probably covered by tissue, will be visible. This is the **trigeminal nerve foramen**, which communicates between the cranial cavity and temporal fossa and transmits the abducens nerve, the mandibular artery, and two branches of the trigeminal nerve. The small epipterygoid, noted earlier, helps form the braincase. It is located anterior to the foramen and contributes to its anterior margin. The epipterygoid lies between the parietal and pterygoid, but is fused in adult individuals of *Chelydra* (but not in all turtles) and not apparent as an individual element.

Observe the skull in dorsal view and note the position and form of the bones already described (Figure 8.3a). Consider the edges of the temporal emarginations. These define the **posttemporal fenestrae**, paired openings that face posteriorly and are best appreciated in posterior view of a sea turtle skull (Figure 8.4b). They are less apparent, though still present, in *Chelydra*. In dorsal view (Figure 8.3a), note the medial extensions of the quadrates and squamosals and the lateral extensions of the supraoccipital. Between these extensions, at the back of the skull, note the **opisthotic** posteriorly and the smaller **prootic** anterior to it. These bones contain the inner ear. The small triangular wedge between the posterior parts of the opisthotic and supraoccipital is the **exoccipital**. The **stapediotemporal foramen**, through which the stapedial artery passes, lies between the quadrate and prootic.

In ventral view, note again the edentulous maxillae anteriorly, each with a sharp-edged ventral margin (Figure 8.3b). They help form the palate, along with the premaxillae, **vomer**, **palatines**, and **pterygoids**. There is usually considerable fibrous tissue adhering to the anterior part of the palate. The nasal cavities emerge onto the palate through the **choanae** (the internal opening of the nasal cavities). Although the bony naris is a single opening, the nasal cavities are paired, separated mainly by cartilage (which may be observed by looking through the choana). The ventral opening of the orbitonasal foramen is located just posterolateral to the choana. A larger opening, the **posterior palatine foramen**, lies posterior to the orbitonasal foramen and transmits the inframaxillary artery. The pterygoids are clearly the largest palatal elements. They curve posteriorly to meet the quadrates laterally and the basicranial elements more medially. Along with the maxilla, jugal, quadratojugal, and quadrate, the pterygoid helps define the **subtemporal fenestra**, the ventral opening of the temporal fossa through which the adductor musculature (in addition to the mandibular artery and a branch of the trigeminal nerve) descends to insert on the lower jaw. Note the **external pterygoid process** (also visible in lateral view, Figure 8.1a) of the pterygoid. It forms a nearly vertical plate with a flat, lateral surface that acts as a guide for the lower jaw during jaw closing.

The **basisphenoid** and **basioccipital** are the midline elements between and posterior to, respectively, the posterior ends of the pterygoid (Figure 8.3a). These bones help support the brain ventrally. The basisphenoid includes the **parasphenoid** in most turtles. We note the parasphenoid (a dermal bone of the braincase) even though it is not present as a separate element, because it is generally a conspicuous element in some tetrapod skulls. The basioccipital forms the ventromedial part of the **occipital condyle** (Figures 8.3 and 8.4b), which articulates with the first cervical vertebra (or atlas). The condyle is completed by the paired exoccipitals (Figure 8.4b). The **foramen magnum**, the large opening through which the spinal cord passes, lies dorsal to the condyle (Figures 8.4b and 8.5). The exoccipital also contributes to the basicranium ventrally, extending laterally from the basioccipital, and the posterior surface of the skull, which is pierced (Figure 8.5) by three openings, two

small and medial **hypoglossal foramina** and a larger and lateral **posterior jugular foramen**. The large opening on each side of the basicranium is the fenestra postotica, mentioned earlier with regard to the medial part of the columella. The fenestra leads into a spacious cavity, the **cavum acusticojugulare**. In addition to the columella, and thus the medial portion of the middle ear, it contains several vessels and nerves such as the glossopharyngeal, hyomandibular, and vagus (which exits through the fenestra postotica).

The palatal and braincase bones on the underside of the skull (Figure 8.3b) are all solidly fused and are unable to move relative to each other and the skull roof. Thus, the turtle skull is termed akinetic. Movement between components of the skull is possible in other reptiles. Such skulls are regarded as kinetic and will be considered in Section II and in Chapter 9.

The lower jaw of reptiles is composed of left and right rami, each formed by several bones rather than the single bone that occurs in mammals. As for the skull, many of the sutures are fused in the adult, so the limits of the individual bones are not easily discernible. Examine a lower jaw ramus of *Chelydra* in lateral view and identify the **dentary** anteriorly (Figure 8.6a). It is the largest bone and occupies nearly the entire lateral surface. Anteriorly the left and right dentaries are fused at the **mandibular symphysis** (Figure 8.6b). The **coronoid** has a small lateral exposure, forming the dorsalmost tip of the jaw ramus, the **coronoid process**. The process serves as an insertion site for some of the mandibular adductor musculature, mainly by way of a strong central tendon termed the bodenaponeurosis. The **surangular** is exposed along the posterodorsal margin of the dentary. Posteriorly it contributes a small, lateral portion to the **Meckelian fossa** (see later) and bears a conspicuous opening, the **auriculotemporal nerve foramen**, which communicates medially with the fossa and transmits a branch of the mandibular nerve. Posteroventrally, the **angular** and **articular** have small lateral exposures.

Examine the ramus in medial view (Figure 8.6b). The dentary forms approximately the anterior half of the ramus in this view. The **Meckelian sulcus** is the longitudinal groove on the medial surface of the dentary. It contains the embryonic **Meckel's cartilage**, remnants of which may be present in your specimen. The posterior end of Meckel's cartilage ossifies as the articular, which bears a smooth surface forming most of the articulation (the surangular contributes a small lateral portion) with the quadrate of the skull. The area posterior to the articular surface serves for insertion of the depressor mandibulae muscle, which opens the jaw. The Meckelian fossa, noted earlier, is the large space between the medial and lateral walls of the posterior part of the mandible and serves for attachment of adductor musculature.

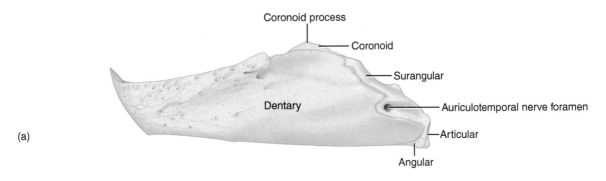

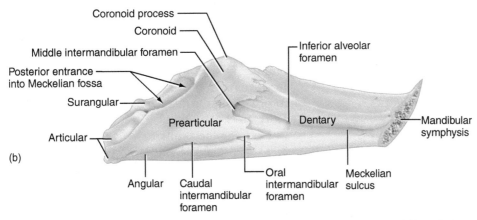

FIGURE 8.6 Mandible of the snapping turtle in (a) lateral and (b) medial views.

On the dentary, follow the Meckelian sulcus posteriorly where it becomes continuous with the Meckelian fossa, which is covered medially mainly by the flat and sheet-like **prearticular**, the largest bone in medial view after the dentary. The posterior entrance into the fossa (Figure 8.6b) is a large oval opening on the posterodorsal surface of the mandible between the coronoid process and the quadrate. The anterior entrance into the fossa is the conspicuous **middle intermandibular foramen**. Through it passes a branch of the mandibular nerve as well as Meckel's cartilage. Just anterior to this foramen along the dorsal margin of the Meckelian sulcus is a smaller opening, the **inferior alveolar foramen**, which is the entrance into the inferior alveolar canal through which passes a branch of the mandibular nerve.

The coronoid and angular are dorsal and ventral, respectively, to the prearticular. These bones were seen in lateral view, but they have much broader medial exposures. The surangular, also noted as having a small exposure in lateral view, is a large sheet-like bone, most of which extends medial to the dentary and forms much of the lateral wall of the Meckelian fossa. The surangular is largely covered in medial view by the prearticular. Two small foramina are usually apparent along or near the suture between the prearticular and angular. The **oral intermandibular foramen** for a branch of the mandibular nerve is the more anterior. The other is the **caudal intermandibular foramen** for another branch of the same nerve.

Key Terms: Turtle Skull and Mandible

angular
antrum postoticum
articular
auriculotemporal nerve foramen
basioccipital
basisphenoid
caudal intermandibular foramen
cavum acusticojugulare
choana (pl., choanae)
columella
coronoid
coronoid process
cranial cavity
dentary
epipterygoid
exoccipital
external pterygoid process
fenestra ovalis
fenestra postotica
foramen magnum
frontal
hypoglossal foramina
incisura columella auris
inferior alveolar foramen
interorbital foramen
jugal

mandibular condyle
mandibular symphysis
maxilla
Meckelian fossa
Meckelian sulcus
Meckel's cartilage
middle intermandibular foramen
naris (pl., nares)
nasal
occipital condyle
opisthotic
oral intermandibular foramen
orbit
orbitonasal foramen
palatine
parasphenoid
parietal
posterior jugular foramen
posterior palatine foramen
postorbital
posttemporal fenestrae
prearticular
prefrontal
premaxilla
prootic
pterygoid
quadrate
quadratojugal
rhamphotheca
rostrum
squamosal
stapediotemporal foramen
subtemporal fenestra
supraoccipital
supraoccipital crest
surangular
symphysis
temporal emargination
temporal fenestra
temporal fossa
trigeminal nerve foramen
tympanic cavity
vomer

SECTION II: IGUANA SKULL AND MANDIBLE

The skull of the squamate *Iguana iguana* (Figures 8.7 and 8.8), the green iguana, appears much less solid than those of the turtle and alligator. Like these last two, squamates are also diapsids, but belong to Lepidosauromorpha (see Figure 1.16). The orbits are large openings, as are the supratemporal and infratemporal fenestrae. As noted later, the latter is unbounded ventrally due to loss of the lower temporal bar. As well, the openings into the nasal cavities and on the palatal surfaces are conspicuous, and the braincase is less ossified. These features are present generally in lizards, but in the burrowing and largely legless lizards the skull appears more solid, which reflects the use of the skull in such activities.

The loss of the lower temporal bar in lizards (and snakes) is generally associated with increased movement among elements of the skull. Skulls with the capacity to undergo such movements are said to be **kinetic**. The skull of the iguana described here, however, is incapable of such movements and so is **akinetic**. Nonetheless, it is worth noting the types of kinesis possible in lizards and identifying the positions of the joints allowing kinesis where appropriate. First the skull is described, to allow familiarity with the various bones involved in kinesis. In addition to the literature cited in this section, further information is available in the works of Condrad (2008), Fairman (1999), and Oelrich (1956).

Examine the skull of *I. iguana* in lateral view (Figure 8.7a). The **maxilla** is a large bone that forms much of the rostrum. The **premaxilla** is a single bone (formed by fusion of paired premaxillae) lying anterior to it. These bones bear a marginal row of teeth. The premaxilla sends a median process posteriorly between the **nares**, the anterior openings of the nasal cavities. The maxilla forms the lateral margin of each naris, with the midline **nasal** forming the posterior margin. A small **septomaxilla** may be seen through the naris, within the nasal cavity, and supports the cartilaginous nasal capsule.

The **orbit** is bounded anteriorly by the **lacrimal**. The **lacrimal foramen** lies just medial to the orbit's margin, which is completed by the **prefrontal**, **frontal**, **postfrontal**, **postorbital**, and **jugal**. Note the extensive contact between the jugal and maxilla. The small postfrontal is wedged between the frontal and postorbital. The large interorbital opening is filled in life by a network of cartilages and membranous septa that form the interorbital septum and the anterior, nonosseous part of the braincase.

The frontal and **parietal**, both single elements, meet at the **parietal foramen** (Figure 8.8a). It is along the suture between the frontal and parietal that mesokinetic movement (see the section "Kinesis in Lizard Skulls") occurs in several other kinds of lizards. The parietal forms much of the roof of the braincase and of the medial margin of the **supratemporal fenestra** (Figures 8.7a and 8.8). The fenestra is bounded ventrally mainly by the postorbital and **squamosal**, with the **supratemporal** contributing a slender process. Posteriorly, the parietal curves ventrally and laterally to form the posterior margin of the fenestra, and its lateral surface provides origin sites for some of the adductor musculature that closes the mouth.

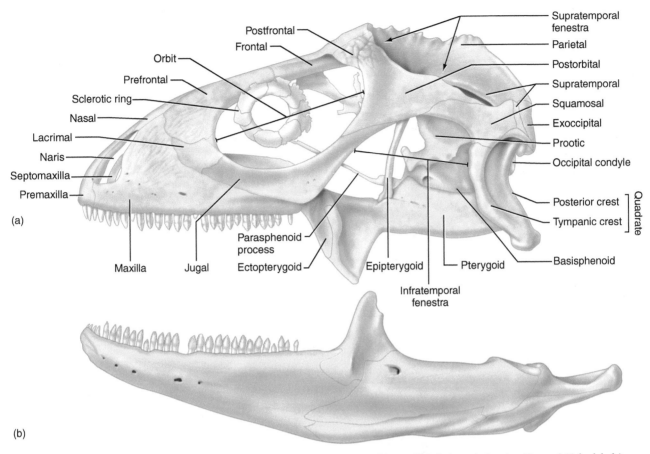

FIGURE 8.7 Skull and mandible of the iguana: (a) skull in lateral view; (b) mandible in lateral view (see Figure 8.11 for labels).

The parietal contacts the squamosal (Figure 8.7a). Posterior to this contact, the squamosal meets the supratemporal, which also continues anteriorly, partly covered by the squamosal, to contact the parietal. The **exoccipital** lies posterior to the supratemporal. This portion of the exoccipital is the **paraoccipital process** of the exoccipital (see Figure 8.10). Ventral to these bones (Figure 8.7a) lies the **quadrate**, a large, curved, and posteriorly concave bone. It is at this point that streptostylic movement (see "Kinesis in Lizard Skulls") between the quadrate and the other skull bones is possible in several other lizards. The anterior margin of the quadrate bears a curved, laterally projecting **tympanic crest** (see also Figure 8.10) to which the anterior part of the tympanic membrane is attached. The broad anterior surface of the quadrate is a site of origin for adductor musculature. Its posterior surface, subdivided by the **posterior crest** (see also Figure 8.10), helps bound the cavity of the middle ear. The lateral end of the slender **columella** (which may not be preserved in your specimen) is visible just posterior to the posterior crest (Figure 8.7a). The ventral surface of the quadrate bears the **mandibular condyle** for articulation with the lower jaw.

Anterior to the quadrate is the **infratemporal fenestra**. It is bounded dorsally by the squamosal and postorbital and anteriorly by the jugal. It is unbounded ventrally, as the quadratojugal has been lost (in an evolutionary sense). In lateral view, particularly when figured, it may appear that there is a ventral bar, but this is not the diapsid lower temporal bar formed mainly by the jugal and squamosal (Figure 1.17), as in, for example, the alligator (Figure 8.12). Lizards (as well as snakes) have lost this bar. Rather, careful observation reveals that the structure that appears to enclose the fenestra ventrally is the **quadrate process** of the **pterygoid**, a bone of the palate (Figure 8.9). This part of the pterygoid extends between the conspicuous **descending flange** of the pterygoid, just posterior to the tooth row, and the medial surface of the quadrate.

The ossified **braincase**, enclosing the **cranial cavity**, is visible in lateral view through the supratemporal and infratemporal fenestrae (Figure 8.9). It covers the posterior part of the brain (which anteriorly is covered by the mainly nonosseous braincase). Dorsally in this region, the brain is covered by the parietal, which also sends a flange ventrally to cover the brain dorsolaterally. Ventral to this is a unit of bone, the braincase proper, which is composed of several elements, but these elements are almost always solidly fused to each other, so the sutures between them are very difficult to discern. The lateral

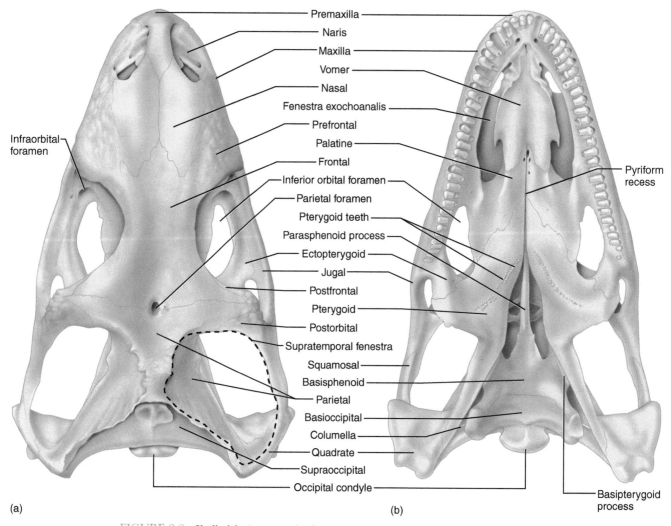

FIGURE 8.8   Skull of the iguana in (a) dorsal (sclerotic ring removed) and (b) ventral views.

surface of the unit provides origin sites for adductor musculature. Its posterodorsal portion is formed by the **supraoccipital** (which is more broadly exposed in posterior view; Figure 8.10), but is largely covered in lateral view by the parietal (Figure 8.9).

Most of the rest of the lateral wall, better exposed in lateral view, is formed by the **prootic**, which houses part of the inner ear (Figures 8.7a and 8.9). The medial portion of the exoccipital, which contains the **opisthotic** (also housing the inner ear and sometimes a separate element, as in the turtle), contributes a small portion to the lateral wall of the braincase. More ventrally, but largely hidden behind the quadrate process of the pterygoid (as well as the quadrate), the ventral part of the braincase is formed mainly by the **basioccipital** and **basisphenoid**, which are considered later. The prootic sits above these last two bones, and has three processes (Figure 8.9). The **alar process** extends dorsally and is bounded anteriorly by the **alar crest**. The **inferior process** lies anteroventrally and forms a hook-shaped process. The **trigeminal notch**

is formed where the alar and inferior processes meet. The trigeminal nerve passes along the ventral part of the notch. The **posterior process** of the prootic extends posterolaterally to meet the paraoccipital process of the exoccipital. The ventral margin along the inferior and posterior processes forms a sharp **prootic crest** formed by the prootic and, posteriorly, the exoccipital. Medial to the crest is a deep recess, the **jugular vein recess**, along which passes the internal jugular vein. The recess is part of the middle ear and contains the **fenestra ovalis**, which is formed by these bones and covered by the footplate of the columella. Just ventral to this region is the curved **interfenestral crest**, medial to which is another recess (the occipital recess) that is also part of the middle ear and dorsally contains the **foramen rotundum** (transmitting the glossopharyngeal nerve). This recess is bounded posteriorly by the **tuberal crest**, formed by the exoccipital and basioccipital.

The anteroventral part of the basisphenoid forms a short anterior process, from which the elongated and

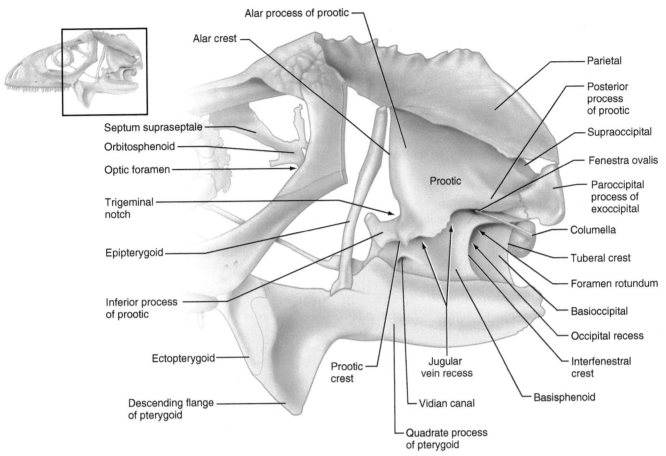

FIGURE 8.9   Skull of the iguana in lateral view showing detail of braincase region with the quadrate, squamosal, supratemporal, and parts of the postorbital and jugal removed.

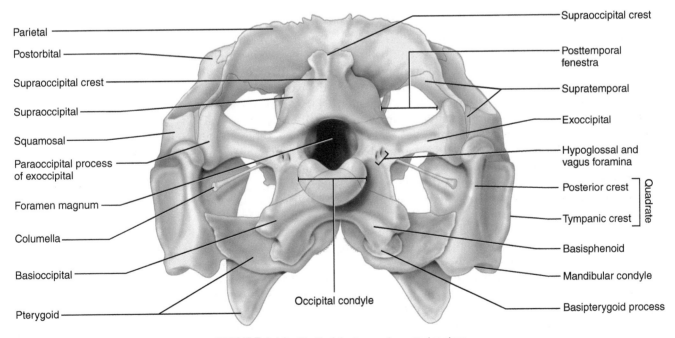

FIGURE 8.10   Skull of the iguana in posterior view.

slender **parasphenoid process** slopes anterodorsally and underlies the interorbital septum (Figure 8.8b). The posterior opening of the **vidian canal** lies on the lateral surface of the basisphenoid, just medial and slightly posterior to the dorsal ridge of the quadrate process of the pterygoid. The **epipterygoid**, a slender and nearly vertically oriented bar of bone, extends from a notch on the pterygoid to the ventrolateral margin of the parietal, forming a cylindrical, strut-like brace between these bones.

Within the posterior part of the orbit is the delicate **orbitosphenoid** (Figure 8.9), which bears several processes. Anterodorsally it meets a partially calcified **septum supraseptale**, an anterior plate-like expansion of a cartilaginous bar, which is not preserved, that extends along the ventral surface of the parietal and frontal. The septum is suspended from the frontal. The orbitosphenoid anteriorly bears a C-shaped notch that forms the posterior margin of the **optic foramen**. The planum and orbitosphenoid are usually the only preserved elements of a complex of interconnected cartilaginous bars. The openings formed by these bars are covered by membrane. The structures approximately posterior to the septum and orbitosphenoid constitute the anterior part of the braincase, which, as noted earlier, is mainly membranous rather than osseous. The structures ventral to the septum and anterior to the orbitosphenoid comprise the interorbital septum.

Examine the skull in dorsal view (Figure 8.8a). Many of the bones observable in this view have already been discussed, but examine their form and positional relationships. Among them are the premaxilla, maxilla, nasal, jugal, prefrontal, postfrontal, frontal, parietal, and squamosal. Note as well the openings in the skull that have already been mentioned: the nares, orbits, supratemporal fenestrae, and the parietal foramen. In those lizards that exhibit mesokinesis (at the frontoparietal suture; see "Kinesis in Lizard Skulls"), the parietal foramen is usually shifted away from the frontoparietal suture, and commonly lies entirely within the parietal. Identify the large **inferior orbital foramen**. It lies on the palatal surface of the skull, but is observable through the orbit. Just anterior and dorsal to the anterior margin of the inferior orbital foramen is the **infraorbital foramen**.

Examine the skull in ventral view (Figure 8.8b). The premaxilla and maxillae form the marginal rim of the rostrum and bear the marginal series of **teeth**. The teeth are borne by the medial surfaces of these bones, a condition termed *pleurodont*, rather than being set in sockets. Posterior to the premaxilla are the paired **vomers**, which form the anterior part of the palate. The vomers are concave ventrally and constricted anteriorly. Laterally the vomer forms most of the medial margin of the anteroposteriorly elongated **fenestra exochoanalis**. The maxilla bounds this fenestra laterally. The **palatine** lies posterior

to the vomer, forms the posterior margin of the fenestra, and has a lateral process that contacts the maxilla. Posterior to this contact, the palatine forms the anterolateral margin of the inferior orbital foramen, with the maxilla forming the anterolateral margin. Posteriorly the palatine contacts the pterygoid. It is in this region that hypokinetic movement is possible in kinetic skulls (see "Kinesis in Lizard Skulls"). The left and right palatines may approach each other closely and touch each other at the midline (perhaps with advancing age), but are generally separated from each other by the narrow **pyriform recess**, which widens posteriorly between the pterygoids (Figure 8.8b). The palatine's dorsal surface forms the posterior floor of the nasal cavity and anterior floor of the orbit. Laterally it meets the prefrontal, jugal, and lacrimal. The infraorbital foramen is located where these bones meet, with the palatine forming most of its margin.

The paired pterygoid bones are large elements (Figure 8.8b). Each extends posteriorly from the palatine, with its lateral margin forming the posteromedial margin of the inferior orbital foramen. More medially, the pterygoid bears a short, slightly curved row of **pterygoid teeth**. In this region, the left and right pterygoids become more widely separated at the midline, reflecting the increased size of the pyriform recess. Posterior to the inferior orbital foramen, the pterygoid extends laterally and curves ventrally to form a large descending flange. Here it meets the **ectopterygoid** (see also Figure 8.9), a relatively short bone that extends dorsally to meet the maxilla and jugal. The ectopterygoid helps form the rest of the margin of the inferior orbital foramen. Farther posteriorly, the pterygoid becomes mediolaterally compressed and extends posterolaterally to contact the quadrate. This is the quadrate process of the pterygoid and has been noted already.

Between the quadrate processes lie the unpaired elements that form the ventral part of the osseous braincase, the basisphenoid and, more posteriorly, the basioccipital (Figure 8.8b). The basisphenoid bears a pair of anterolaterally projecting **basipterygoid processes** that meet the pterygoid bones. Between them is an anterior projection from which the parasphenoid process extends. The latter structure may be observed though the pyriform recess. Posteriorly, the basisphenoid constricts (the degree of constriction may vary) and then widens again to contact the basioccipital along a mainly anteriorly convex suture. The basioccipital extends laterally to form a pair of stout processes. Between them is the **occipital condyle** for articulation with the atlas of the vertebral column. The basioccipital forms the midventral portion of the occipital condyle. The dorsal surface of each lateral projection of the basioccipital is deeply excavated to help form the occipital sinus, which was considered earlier. The posterior end of the basioccipital curves dorsally to form the ventral part of the occiput.

Observe the skull in posterior view (Figure 8.10). In this view, it becomes clear that the braincase is a central structure surrounded all but ventrally by a covering of other bony elements. The **foramen magnum** is the central and nearly circular opening through which the spinal cord passes to reach the brain. The occipital condyle lies ventral to the foramen. Its midventral portion is formed by the basioccipital (which thus contributes to the margin of the foramen magnum) and its lateral portions are formed by the exoccipitals (which form the lateral margins of the foramen). Several foramina open just lateral to the dorsal part of the occipital condyle. Usually there are four, a relatively large dorsal **vagus foramen**, for the vagus nerve, and three smaller **hypoglossal foramina** that transmit divisions of the hypoglossal nerve.

Laterally each exoccipital sends out a stout paraoccipital process that contacts the quadrate, supratemporal, and parietal (see also Figures 8.7a, in which it is the portion labeled as exoccipital, and 8.9). The supratemporal is a delicate, sliver-like bone adhering to the anteromedial surface of the parietal, but it is difficult to discern. Its ventral end passes anterior to the parietal and has a small exposure in lateral view (Figure 8.7a). The supraoccipital is a large median bone, bearing a raised median **supraoccipital crest**, forming the posterodorsal part of the osseous braincase (Figures 8.9 and 8.10). It contacts the parietal dorsally, where the sliding action of the metakinetic joint occurs in kinetic skulls (see "Kinesis in Lizard

Skulls"). The supraoccipital contacts the exoccipitals laterally. The axis of rotation in metakinetic skulls passes through the lateral ends of the exoccipitals. Ventrally the supraoccipital forms the margin of the foramen magnum. On either side of the supraoccipital is a **posttemporal fenestra**, bounded by the supraoccipital, parietal, supratemporal, and exoccipital (Figure 8.10).

As in other reptiles, the lower jaw is composed of several bones. Examine the lower jaw in lateral view (Figure 8.11a). The **dentary** is the large bone making up the anterior half of the lower jaw and bearing all the lower teeth. Posteriorly the dentary contacts the **coronoid** dorsally, the **surangular** centrally, and the **angular** ventrally. The coronoid is a relatively small bone with a dorsal projection, the **coronoid process**. The coronoid serves largely for insertion of the parts of the adductor musculature. Posteroventrally the coronoid contacts the surangular. The surangular and angular are relatively large and form much of the dorsal and ventral surfaces of the posterior half of the lower jaw. The conspicuous **anterior surangular foramen** lies on the anterodorsal surface of the surangular. Parts of the adductor musculature insert on the dorsal and lateral surfaces the surangular. Posterior to the surangular and angular is the **articular**, which bears the smooth and concave surface for articulation with the quadrate. The articular extends posteriorly as the **retroarticular process** that ends as a tubercle onto which the depressor mandibulae, the muscle that opens the jaws,

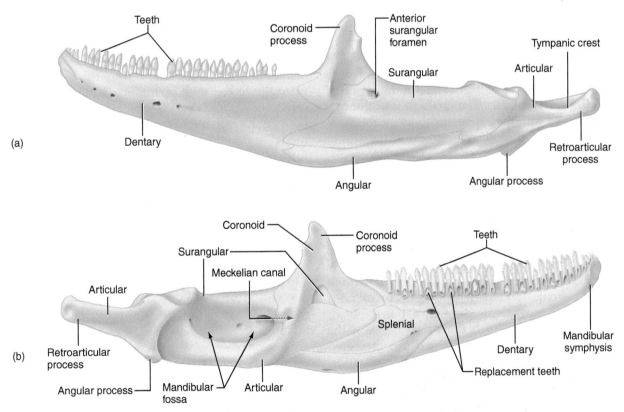

(a)

(b)

FIGURE 8.11 Mandible of the iguana in (a) lateral and (b) medial views.

inserts. The lateral border of the retroarticular process is the **tympanic crest**, to which is attached the skin covering the middle ear. Extending medially and slightly ventrally is a second, triangular projection, the **angular process**, which projects slightly ventral to the retroarticular process in lateral view. The pterygoid musculature inserts on much of the ventral surface of the articular, including its retroarticular and angular processes.

Examine the lower jaw in medial view (Figure 8.11b) and follow the angular ventrally as it curves onto the medial surface. Here it extends anteriorly, forming the rounded ventral margin of the **mandibular fossa**, and continues forward to contact the **splenial** and the two ventral processes of the coronoid. The surangular has a small exposure between the coronoid and articular. The angular also curves ventrally from the lateral to medial surfaces, where it has a small exposure ventrally. Anteriorly it meets the splenial. The latter, exposed only on the medial surface of the lower jaw, tapers anteriorly, where it contacts the dentary, which completes the rest of the lower jaw. Examine the base of the teeth—you should observe several smaller replacement teeth emerging from the dentary (Figure 8.11b).

Reexamine the mandibular fossa. It is an expansive and anteroposteriorly elongated opening leading into the **Meckelian canal**, a passage that extends anteriorly into and through the dentary. Much of the lower jaw may thus be viewed as a tubular structure, with its bones forming the walls surrounding the canal. The canal contains **Meckel's cartilage**, which extends through the length of the canal. Posteriorly, Meckel's cartilage lies ventrally within the mandibular fossa. It passes anteriorly through the canal, which opens onto the anteromedial surface of the dentary in the region of the **mandibular symphysis**. The cartilage exits through this opening and into the cutaneous tissue of the symphyseal region. The medial wall and dorsal margin of the mandibular fossa are formed by the surangular.

Kinesis in Lizard Skulls

The main types of kinesis described for lizards (see Metzger, 2002, for a detailed review) are mesokinesis, metakinesis, hypokinesis, and streptostyly. Mesokinesis involves movement of the rostrum dorsally and ventrally through a transverse axis passing through the suture between the frontal and parietal bones. Hypokinesis involves similar movements through a transverse or oblique axis passing through the palate between the pterygoid and both the palatine and ectopterygoid bones. The rostrum may thus be moved dorsally (dorsiflexion) and ventrally (ventroflexion) through the mesokinetic and hypokinetic joints. Metakinesis refers to a sliding movement of the supraoccipital and the parietal through a transverse axis passing through the

paraoccipital process of the exoccipital. Streptostyly is the anteroposterior swinging movement of the quadrate around its dorsal attachment to other skull elements (principally the squamosal and supratemporal).

The classic model of kinesis involves protraction (swinging forward) of the quadrate, pushing the basal unit (the pterygoid bone) anteriorly, which in turn causes dorsiflexion of the rostrum through the mesokinetic and hypokinetic joints. This movement causes the parietal bone to be depressed, which is made possible by the metakinetic joint. The combination of mesokinesis and metakinesis is termed amphikinesis. These movements are reversed by retraction of the quadrate, pulling the basal unit backward and causing ventroflexion of the rostrum.

This model (the quadratic-crank model), for many years considered as describing the typical kinetic abilities of lizards, is currently viewed as one of the possible conditions among lizards. One problem is that parts of the sequence of movements described earlier have been difficult to determine experimentally, either qualitatively or quantitatively. In many cases, the potential for kinesis is determined mainly through manipulation of skulls, but experimental evidence for particular movements is lacking or ambiguous. For example, hypokinesis has not been quantified experimentally, yet it must occur, in combination with mesokinesis, if the rostrum is to be dorsiflexed and ventroflexed as a unit. Studies have also failed to show metakinetic movements in some taxa that, based on skull manipulations, would be expected to possess this ability—and thus that mesokinetic and metakinetic movements are always correlated with each other. Other studies indicate that both mesokinesis and streptostyly may occur together but that they are not kinematically linked with each other; and that streptostyly can occur without causing mesokinesis. The picture that is emerging is that the classic model is not a generalized model widely applicable to lizards that display some form of kinesis. Rather, it seems restricted to some taxa, whereas others have different kinetic abilities.

The functional significance of kinesis is not entirely clear, although numerous hypotheses have been proposed. It seems that the function(s) may be specific to the kinds of movement possible in and the probable dietary habits of any particular group of taxa; in other words, it is inappropriate to generalize. Among the hypotheses advanced for streptostyly are increased gape, increased mechanical advantage for some jaw adductor musculature, efficiency of intraoral transport, cropping (in herbivorous lizards), and shearing of prey. The following are a sample of hypotheses advanced for mesokinesis: increased gape, decrease in duration of gape, prey subjugation, increased bite force, and control during jaw prehension. Metzger (2002) provided a comprehensive review of these and other aspects of kinesis, and Iordansky (2011) commented on its origin and evolution.

Key Terms: Iguana Skull and Mandible

alar crest
alar process of prootic
angular
angular process of articular
anterior surangular foramen
articular
basioccipital
basipterygoid processes
basisphenoid
braincase
columella
coronoid
coronoid process
cranial cavity
dentary
descending flange of pterygoid
ectopterygoid
epipterygoid
exoccipital
fenestra exochoanalis
fenestra ovalis
foramen magnum
foramen rotundum
frontal
hypoglossal foramina
inferior orbital foramen
inferior process of prootic
infraorbital foramen
infratemporal fenestra
jugal
jugular vein recess
lacrimal
lacrimal foramen
mandibular condyle
mandibular fossa
mandibular symphysis
maxilla
Meckelian canal (mandibular canal)
Meckel's cartilage
nares (sing., naris)
nasal
occipital condyle
opisthotic
optic foramen
orbit
orbitosphenoid
palatine
paraoccipital process of exoccipital
parasphenoid process
parietal
parietal foramen
posterior crest of quadrate
posterior process of prootic

postfrontal
postorbital
posttemporal fenestra
prefrontal
premaxilla
prootic
prootic crest
pterygoid
pterygoid teeth
pyriform recess
quadrate
quadrate process of pterygoid
retroarticular process of articular
septomaxilla
septum supraseptale
splenial
squamosal
supraoccipital
supraoccipital crest
supratemporal
supratemporal fenestra
surangular
teeth
trigeminal notch
tuberal crest
tympanic crest of articular
tympanic crest of quadrate
vagus foramen
vidian canal
vomer

SECTION III: SNAKE SKULL AND MANDIBLE

Snakes, as noted in Chapter 1 (see also Figure 1.16), are a group of lizards, and the snake skull represents a modified diapsid type. As is typical of lizards, the skull of snakes (Figure 8.12) appears less solid than in other reptiles considered here, with several large openings and spaces between the bony elements. In snakes, however, these features are taken to extremes, with the result that snake skulls, and their jaws in particular, exhibit extensive mobility. These adaptations allow snakes to feed on relatively large prey. The main features of the snake skull are described in this section, based on the skull of *Python molurus*, the Indian (or Burmese) python, but several of the features noted apply to snakes in general, at least in terms of the typical morphology of the bony elements and their relationship to each other. In contrast to the other skulls described in this and other chapters, the skeletal elements of the skull are usually quite readily discernible, because in snakes most of the bones are not solidly fused together. Students interested in further information on snake skulls may wish to consult Evans (2003) and Frazzetta (1959).

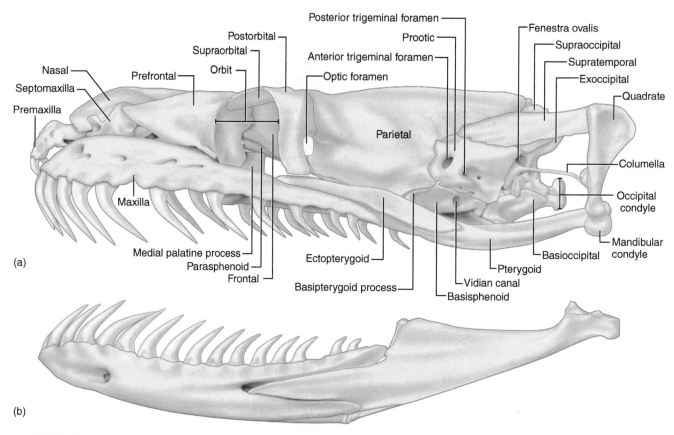

FIGURE 8.12   Skull and mandible of the python: (a) skull in lateral view; (b) mandible in lateral view (see Figure 8.15 for labels).

Examine the skull in lateral view (Figure 8.12a). In contrast to the other skulls discussed in Chapter 8, the braincase is readily observable, and it is apparent that several bones are absent. Indeed, many of the bones of the skull roof, as well as of other parts of the skull, have been lost in snakes, including the lacrimal, jugal, postfrontal, squamosal, and postparietal; the quadratojugal, a bone of the palatal region, is also lost (for some bones, however, it is unclear whether they are missing or indistinguishably fused to other elements; see Scanlon, 2005, for further discussion). In effect, the extensive mobility of the skull of snakes reflects the loss of many of these elements, particularly of elements in the temporal region. In snakes, both the upper and lower temporal bars have been lost, whereas in other diapsid skulls the temporal bars typically form struts across the temporal regions and tend to restrict movement. Mobility is also made possible by the types of joints present among skeletal elements. Rieppel (1993) noted that mobility was not achieved by further elaboration of synovial joints (the mandibular joint and the links in the ear region are the only synovial joints in the snake head) but by the presence of relatively loose connections held together by soft tissue (mainly ligamentous) attachments. The kinematic linkage system of elements permitting kinesis is more numerous and complex than in the lizard skull, and

the interested reader may refer to Frazzetta (1966) and Cundall (1987) for further details.

The **frontals** and (particularly) the **parietals**, originally bones of the skull roof, have expanded ventrally to help form the walls of the solid braincase, which houses the brain (Figures 8.12a and 8.13a). The **postorbital** extends laterally from between the contact of the frontal and parietal and then turns ventrally to form the posterior margin of the **orbit**. The ventral end of the postorbital has a ligamentous attachment to the bony elements ventral to it. The **supraorbital** (Figure 8.13a), a small, flattened bone, lies anterior to it. The frontal forms much of the medial wall of the orbit. The conspicuous **optic foramen**, for the optic nerve, lies between the frontal and parietal (Figure 8.12a).

The rostral end of the skull is formed laterally by the triangular (in dorsal view) **prefrontals** and the smaller **septomaxillae**. The **nasals** lie dorsally between these elements. The nasal is a wide and horizontally flat bone dorsally. Note its descending lamina, along its medial portion. The laminae of the nasals lie adjacent to each other at the midline, separating the nasal cavities, and extend farther anteriorly than the dorsal portions to contact the **premaxilla** (representing the two fused premaxillae), which bears **premaxillary teeth**. Ventrally (Figure 8.13b), the **vomers** (described in further detail later) underlie

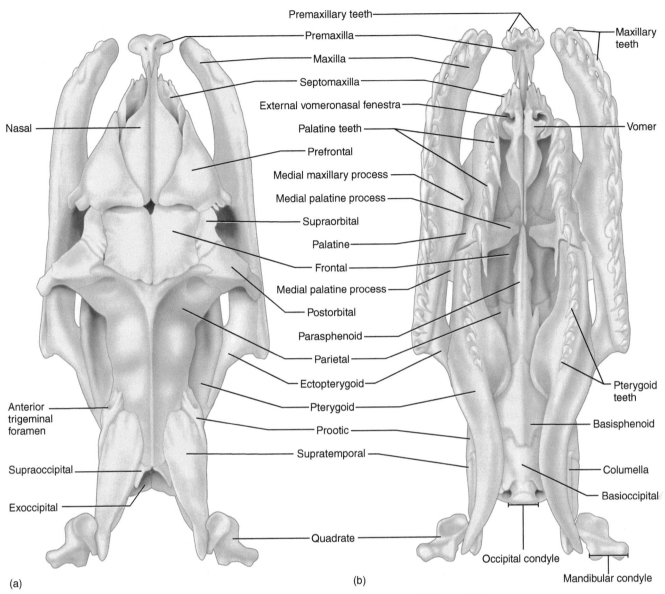

FIGURE 8.13   Skull of the python in (a) dorsal and (b) ventral views.

these bones. The unit formed by these bones is generally termed the snout, and it can move about the nasofrontal joints. As this movement occurs anterior to the orbit, the skull is considered to be **prokinetic**.

Consider the prefrontal again. Its dorsal portion is a curved triangular plate, but its posterior portion extends ventrally and forms the anterior wall of the orbit. A **lacrimal foramen** perforates the lateroventral surface of this posterior portion of the prefrontal. The other element forming the wall of the orbit, ventrally and posterior to the prefrontal, is the medial process of the **palatine**, which is discussed later.

Return to the parietal in lateral view (Figure 8.12a). Posteriorly, it contacts the **prootic**, which bears several openings. The large anterior opening is the **anterior trigeminal foramen**, for the maxillary nerve, and the large

posterior opening is the **posterior trigeminal foramen**, for the mandibular nerve. The slender region separating these openings was considered by Rieppel and Zaher (2001) to represent the laterosphenoid (but see comments in Scanlon, 2005). The much smaller opening posterior to the posterior trigeminal foramen is the **facial foramen**, for the nerve of the same name. Ventral to these foramina, the posterior opening for the **vidian canal** pierces the basisphenoid. The posterior edge of the prootic and the anterior edge of the **opisthotic** (which is fused with the **exoccipital**) form the **fenestra ovalis**, the entrance into the inner ear, which is covered by the footplate of the **columella** (or stapes). The distal end of the columella, however, contacts a process of the **quadrate** (Figure 8.14) by way of a cartilaginous extension. This is in marked contrast to the condition described for the other reptile skulls, in which the columella, usually via a different cartilaginous extension (the extracolumella), contacts a tympanic membrane or eardrum. Snakes lack a tympanic membrane and it has been suggested that their ears mainly detect ground vibrations, as well as low-frequency airborne vibrations.

This posterior part of the skull of snakes is also distinct in the suspension of the lower jaw. Note the **supratemporal**, mainly contacting the prootic and parietal but also the **supraoccipital** and exoccipital (which have small exposures in lateral view), and extending posteriorly to articulate with the quadrate (Figures 8.12, 8.13a, and 8.14). In the other reptile skulls described here, the quadrate is generally partially buttressed by the squamosal (as well as other elements; for examples, see Figures 8.1a, 8.7a, and 8.16a), which is absent in the snake. The supratemporal, if present as an independent element, generally has a small contact with the quadrate (compare, for example, the condition described for the iguana, Figures 8.7a and 8.10). In most snakes, however, the supratemporal is a long, flattened, and beam-like element that helps support the quadrate, which is streptostylic; that is, it is movable at its dorsal attachment (see "Kinesis in Lizard Skulls," earlier). Ventrally, the quadrate contacts the **pterygoid**, but via a looser articulation than in lizards. The ventral end of the quadrate bears the **mandibular condyle** for articulation with the lower jaw.

The posterior surface of the skull (Figure 8.14), or occiput, has a large central opening, the **foramen magnum**, through which the spinal cord passes to the brain. The **occipital condyle**, for articulation with the first vertebra (atlas) of the vertebral column, lies ventral to the foramen. The condyle is formed by the exoccipitals laterally and the **basioccipital** ventrally. The exoccipitals form most of the margin of the foramen magnum (the basioccipital contributes a small ventral portion) and they extend dorsal to the foramen, thereby excluding the supraoccipital, which forms the dorsal part of the

occiput. The **jugular foramen**, for the vagus nerve, lies lateral to the foramen magnum. Note how the supratemporal extends posteriorly, contacting the supraoccipital and exoccipital and, at its posterolateral surface, the quadrate.

In ventral view (Figure 8.13b), the basioccipital forms the posterior part of the basicranium. Anterior to it is the **basisphenoid**, which bears a sharp, midventral ridge that serves for muscular insertion. The palatal region of snakes is commonly considered as comprising palatomaxillary arches, on each side composed of a medial ramus and a lateral ramus. The medial ramus consists of the pterygoid, noted earlier, and **palatine**, and the lateral ramus includes the **maxilla** and **ectopterygoid**. Due to the absence of bony cross connections among left and right elements of the palatal region, the kinematic linkages can move independently. This allows the elements of the palate to specialize for prey capture and swallowing. Typically, the lateral palatomaxillary arches are used for capture, whereas the left and right medial palatomaxillary arches are moved alternatively to "walk" (often termed "pterygoid walk") over prey, sending it back for swallowing.

Examine the posterior end of the skull in ventral view. On either side of and ventral to the basicranium, the pterygoids bow medially from their articulation with the quadrates. Follow one of the pterygoids anteriorly. At about its midlength, it expands transversely and then narrows again into its anterior half. This latter portion bears a row of six to 10 **pterygoid teeth** and then contacts the palatine. Just medial and dorsal to the last teeth, the pterygoid articulates with the **basipterygoid process** (or basitrabecular process) of the basisphenoid (though it may not be homologous to the basipterygoid process of the iguana). The palatine continues anteriorly from the pterygoid. It also bears a row of usually six **palatine teeth** and is bound loosely to the vomer and septomaxilla, bones which were noted earlier. Just anterior to

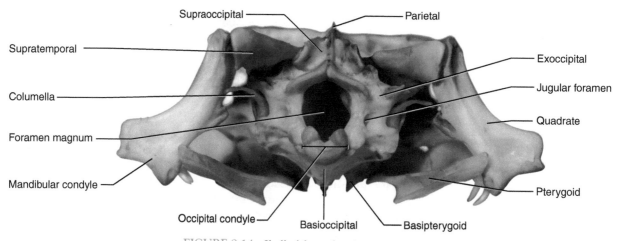

FIGURE 8.14   Skull of the python in posterior view.

the pterygopalatine joint, the palatine bears lateral and medial processes. The **lateral palatine process** is barely visible in ventral view because it passes dorsal to the **medial maxillary process** of the maxilla. These bones are loosely connected here and the joint permits sliding and rotation between these bones. The lateral palatine process also contacts the prefrontal in the ventral orbital wall. The **medial palatine process** extends dorsomedially toward the ventral surface of the braincase. The dorsal surface of this process was noted earlier in lateral view on the floor of the orbit.

The maxilla is an elongated bone, dorsoventrally high anteriorly and tapered posteriorly, that usually bears 18 **maxillary teeth**. The maxilla articulates with the ventral margin of the prefrontal anteriorly (Figure 8.12a). Its medial maxillary process, contacting the palatine, has already been described (Figure 8.13b). Posteriorly the maxilla articulates with the ectopterygoid along a fairly extensive surface (Figures 8.12a and 8.13b). Note that the ectopterygoid extends farther anteromedially than anterolaterally. The maxilla and ectopterygoid are tightly bound at the anteromedial extension of the ectopterygoid, but the maxilla may rotate about this point. Posteriorly, the ectopterygoid passes posteromedially to form an essentially immobile articulation with the pterygoid (Figures 8.12a and 8.13b).

Consider the region between the pterygoid and palatine in ventral view (Figure 8.13b). The **parasphenoid** is a slender, elongated, midventral element extending anteriorly from the basisphenoid. The parasphenoid tapers anteriorly and contacts the vomers. On either side, the parasphenoid contacts the frontal. Also note the parietal in this view, extending ventrally to contact the basisphenoid on either side of the skull. The vomer consists of a vertical plate posteriorly, where it contacts the parasphenoid. The plates of the right and left vomers are in contact throughout their length and articulate with the nasals. Farther forward, the vomer expands to form a ventral horizontal flange that extends laterally toward the palatine. This flange constricts anteriorly, but the vomer expands again to help form a conspicuous opening, the **external vomeronasal fenestra**. The septomaxilla, contacting the vomer laterally, contributes to the margin of the fenestra. The vomer then contacts the premaxilla. The external vomeronasal fenestrae allow access to the vomeronasal (or Jacobson's) organs. These are olfactory structures formed from the medioventral part of the olfactory epithelium of the nasal cavities, and are present in many tetrapods (though usually absent in aquatic forms). In lizards, including snakes, they usually form distinct, sac-like structures with their own entrance into the oral cavity. These are the structures that snakes

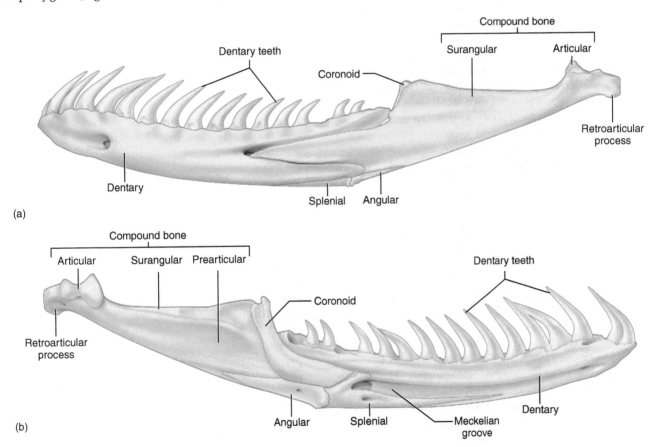

FIGURE 8.15   Mandible of the python in (a) lateral and (b) medial views.

use in combination with their forked tongue as it darts into and out of the oral cavity. Odorants adhere to the tongue, which is then brought into close proximity with the external vomeronasal fenestrae.

Examine the mandible. As usual in reptiles, it is formed from several bony elements: the **dentary, splenial, angular, coronoid,** and **compound bone.** The latter represents a fusion of the **surangular, prearticular,** and **articular,** although some of the sutures between these bones may be partly visible. In lateral view (Figure 8.15a), the anterior half of the mandible is formed mainly by the dentary, which bears all the lower or **dentary teeth.** Posteriorly, the dentary bifurcates into dorsal and ventral laminae, between which the dentary articulates mainly with the surangular portion of the compound bone. Ventrally, the dentary contacts the angular. In medial view (Figure 8.15b), the dentary forms most of the anterior part of the mandible. Firmly attached to its ventromedial surface is the splenial, which is a delicate, splint-like, and anteriorly tapering bone for much of its length, though its posterior end is expanded. The splenial helps form the margin of the **Meckelian groove,** which extends along the medial surface of the dentary. **Meckel's cartilage,** which is unlikely preserved, extends along the groove and passes posteriorly between the lateral and medial surfaces of the mandible. The dentary contacts the coronoid posteriorly, whereas the splenial contacts the angular. The coronoid, which has a small exposure in lateral view (Figure 8.15a), is more broadly exposed medially and consists of anterior and vertical portions that cover the compound bone (Figure 8.15b). The angular, already noted, has a small lateral exposure, but a larger medial exposure. Anteriorly, it articulates with the dentary, splenial, and coronoid; posteriorly, it articulates with the compound bone.

The compound bone forms much of the posterior half of the mandible. For much of its length, it consists of medial and lateral lamellae, representing the prearticular and surangular, respectively; these elements fuse dorsal and ventral to Meckel's cartilage during ontogeny and generally are referred to as the prearticular and surangular processes of the compound bone. The surangular process rises farther dorsally than the prearticular process. The articular represents the posterior part of the compound bone. It bears the articular surface forming the jaw joint with the quadrate of the skull. Posterior to it is a short **retroarticular process.**

The lower jaws are united anteriorly by soft tissue at the symphysis, rather than by bony fusion. The tips of the lower jaws may therefore move independently. Each lower jaw comprises two functional units, one including the dentary and splenial, and the other including the angular, compound, and coronoid. These components are functionally divided at the intramandibular joint, which permits movement between them: the dentary

and splenial may move dorsoventrally and lateromedially at this joint.

Key Terms: Snake Skull and Mandible

angular
anterior trigeminal foramen
articular
basioccipital
basipterygoid process
basisphenoid
columella
compound bone
coronoid
dentary
dentary teeth
ectopterygoid
exoccipital
external vomeronasal fenestra
facial foramen
fenestra ovalis
foramen magnum
frontal
jugular foramen
lacrimal foramen
lateral palatine process of **palatine**
mandibular condyle
maxilla
maxillary teeth
Meckelian groove
Meckel's cartilage
medial maxillary process of **maxilla**
medial palatine process of **palatine**
nasal
occipital condyle
opisthotic
optic foramen
orbit
palatine
palatine teeth
parasphenoid
parietal
posterior trigeminal foramen
postorbital
prearticular
prefrontal
premaxilla
premaxillary teeth
prokinetic
prootic
pterygoid
pterygoid teeth
quadrate
retroarticular process

septomaxilla
splenial
supraoccipital
supraorbital
supratemporal
surangular
vidian canal
vomer

SECTION IV: ALLIGATOR SKULL AND MANDIBLE

Alligators and crocodiles are crocodylians, which are the closest living relatives of birds (see Figure 1.15). The skull of *Alligator mississippiensis*, the American alligator, has an overall long and low appearance owing mainly to its elongated and flattened rostrum, as occurs generally in crocodylians. Posteriorly, the roof of the skull forms a nearly flat table. Although there are more openings in the temporal region compared to the turtle skull, the alligator skull appears more solid than in lizards and snakes. The dorsal surface of the skull has a sculpted texture, with many furrows interspersed among numerous irregular ridges. This pattern is the result of dermal ossifications or osteoderms that adhere to the bones covering the skull dorsally. In addition to the information presented here, the interested reader may consult Brochu (1999), Grigg and Gans (1993), Iordansky (1973), and Rowe et al. (2003) for further detail on morphology.

Examine the skull in lateral view (Figure 8.16a). Most of the rostrum consists, on either side, of the **maxilla**, with a small anterior contribution made by the **premaxilla**. These two bones bear all the **teeth** in a single marginal series. The teeth are set in sockets of the bone, a condition termed *thecodont*. The **naris** is positioned dorsally near the tip of the snout. More posteriorly, the **orbit** is dorsally placed as well. It faces dorsolaterally and so appears as an elongated opening. Internally, the left and right orbits appear to be continuous, but in life they are separated mainly by a cartilaginous and plate-like **interorbital septum** (Figure 8.17). A large osteoderm, the **palpebral**, ossifies in the eyelid (Figures 8.16a and 8.17); this bone is often lost in specimens. The dorsal position of the eye and nostril are adaptations to the crocodylian lifestyle, allowing the animal to remain almost completely underwater while continuing to see and breathe.

The **jugal** forms the orbit's ventral margin and shares an extensive contact with the maxilla (Figure 8.16a). Ventrally the jugal meets the **ectopterygoid**, which sends

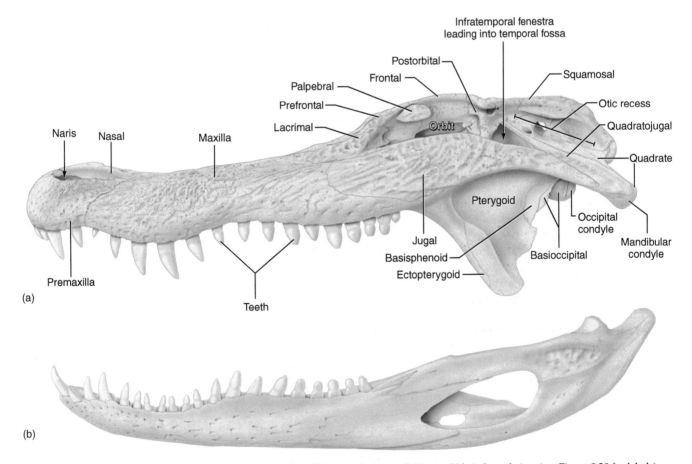

FIGURE 8.16   Skull and mandible of the alligator: (a) skull in lateral view and (b) mandible in lateral view (see Figure 8.20 for labels).

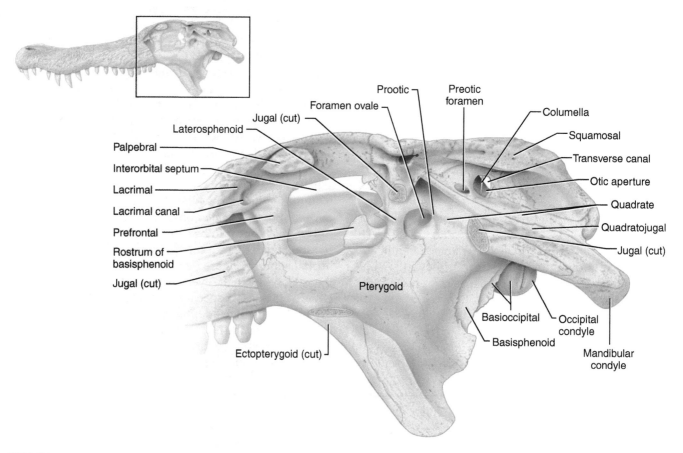

FIGURE 8.17   Skull of the alligator in lateral view, showing detail of braincase region with most of the jugal and small parts of the postorbital and maxilla removed.

a stout, conspicuous descending process ventrally to meet the transverse process of the **pterygoid** (see later). The **lacrimal** is a smaller bone medial to the jugal and forms the anterior margin of the orbit. The opening for the **lacrimal canal** is visible just inside the margin of the orbit (Figure 8.17). The dorsal margin of the orbit is formed by the **prefrontal**, **frontal**, and **postorbital** (Figure 8.16a). A cylindrical bony bar, formed by the descending process of the postorbital and ascending process of the jugal and slightly inset from the surrounding bony elements, defines the orbit posteriorly. A vascular foramen pierces this bar dorsally (Figures 8.16a and 8.17).

The **infratemporal fenestra**, approximately triangular in outline, lies posterior to this bar and leads into the **temporal fossa** (Figure 8.16a). The fenestra is bounded ventrally mainly by the jugal, which tapers posteriorly and nearly reaches the **mandibular condyle** (see later). The **quadratojugal** contributes to the ventral margin posteriorly and its slender ascending process forms the fenestra's posterior margin. The **supratemporal fenestra**, described later, is not visible in lateral view, but can be seen in dorsal view (Figure 8.18a). The quadratojugal (Figure 8.16a) shares an extensive suture with the **quadrate**, which slopes posteroventrally to form the

mandibular condyle for articulation with the lower jaw. The **occipital condyle**, which articulates with the atlas (the first element of the vertebral column), is clearly evident at the posterior midline of the skull.

Posterodorsally, the quadrate articulates with the **squamosal**, which continues onto the roof of the skull (Figures 8.16a–8.18a). The region of the skull bounded by the squamosal and quadrate forms a depression, the **otic recess**, across which in life is stretched the tympanic membrane or eardrum (Figure 8.16a). Medial to it is the middle ear. The conspicuous **otic aperture** is a large opening through which the **columella** (stapes) passes. The columella is a fine, needle-like element that extends to the inner ear (Figure 8.17). It may not be preserved in your specimen. The middle ear cavity is continuous with a complex series of interconnected passages and sinuses (or cavities) within many of the bones in the posterior part of the skull (e.g., the quadrate, **exoccipital**, **supraoccipital**) and even the **articular** of the lower jaw (see later). An example of such passages is the **preotic foramen**, the opening leading into the cavity within the quadrate. This foramen is the smaller circular opening just anterior to the otic aperture (Figure 8.17). Another opening into quadrate is the **foramen aereum** (Figure 8.19).

The latter lies on the posterodorsal surface of the quadrate, just dorsal to the mandibular condyle. The cavity of the quadrate communicates with the cavity within the articular of the lower jaw by way of a fibrous tube, the siphonium (which is not preserved). The system of cavities of the right and left sides of the skull are also interconnected through ventral passages in the **basioccipital** and **basisphenoid**, which are discussed later, and a dorsal passage, the **transverse canal**, between the two middles ear cavities (Figure 8.17). You should be able to see the transverse canal by peering into the dorsal part of the otic aperture. In addition, the system of cavities is connected to the pharynx by eustachian tubes. Although internally complex, these passages open externally by a median and a pair of lateral eustachian foramina, as noted later in ventral view.

Ventral to the otic region the quadrate forms a considerable portion of the wall of the temporal fossa and provides origin areas for the adductor musculature that closes the jaw. Crests and rugose areas that serve for tendons of this musculature may be observed on the quadrate. The quadrate forms the posterior margin of the large **foramen ovale** (Figure 8.17), through which the trigeminal nerve passes. The anterior margin is formed by the **laterosphenoid**, a bone that forms much of the anterior wall of the braincase. The anterior part of each laterosphenoid turns medially but the left and right laterosphenoids meet only briefly at the midline, so that a large opening is left between them. The opening dorsal to their meeting is roofed over by the frontal, forming a rounded passage for the olfactory tracts of the brain. Ventral to this, two notches are present along the anterior margin of the laterosphenoid (Figure 8.17). The optic nerve passes through the dorsal notch and the oculomotor nerve passes through the ventral notch. Medial to this notch is the vertical plate of the basisphenoid, termed the **rostrum** of the **basisphenoid**. The rostrum is actually an ossification of the **parasphenoid**, which fuses completely to the basisphenoid in adults. It helps form the interorbital septum between the orbits. The rest of the septum is cartilaginous and was noted earlier. Portions of the basisphenoid are also exposed posteriorly between the quadrate and pterygoid and ventrally posterior to the pterygoid. Although the basisphenoid is a fairly large bone, much of it is covered laterally by the laterosphenoid and pterygoid.

The pterygoid is a large bone. Ventral to the foramen ovale is forms a nearly vertical plate that curves laterally to meet the ectopterygoid. A process of the pterygoid extends anteriorly, past the rostrum of the basisphenoid, to contact the **palatine** and the prefrontal. The pterygoid also contributes significantly to the palate, which is considered later. The **prootic**, associated with the inner ear, is also largely covered laterally. A small exposure is present between the quadrate and pterygoid along the posteroventral margin of the foramen ovale, which it helps to form (Figure 8.17).

Examine the skull in dorsal view (Figure 8.18a). The premaxillae form the tip of the rostrum and almost completely encircle the nares. A thin, bony **internarial septum** is formed mainly by the narrow, elongated **nasals**, which follow the premaxillae posteriorly. Observe the **incisive foramen** through the naris. The foramen opens on the ventral surface of the palate and will be discussed again later. The maxillae lie lateral to the nasals. Note again the bones that encircle the orbit: the lacrimal, prefrontal, frontal, postorbital, and jugal. The frontal is a single element, narrowed between the orbits and sending an elongated anterior process to meet the nasal. Posteriorly, it is sutured to the **parietal**. The parietal, postorbital, and squamosal delimit the supratemporal fenestra, the dorsal opening of the diapsid skull, which in crocodylians characteristically lies in the horizontal plane and is visible only in dorsal view. The fenestra leads ventrally into the spacious temporal fossa. A second, much smaller opening, the **temporal canal**, emerges just within the posterior margin of the supratemporal fenestra. The infratemporal fenestra is also visible in dorsal view lateral to the supratemporal fenestra and posterior to the orbit. The large **palatal fenestra**, opening onto the ventral surface of the skull, is clearly seen through the orbit. Note again other bones described in lateral view (quadrate and quadratojugal), as well as the bar, between the orbit and infratemporal fenestra, formed by the jugal and postorbital.

In ventral view (Figure 8.18b), the most conspicuous feature is the length and form of the palate. This palate, forming the roof of the oral cavity, is not the original, or primary, palate present in such reptiles as lizards. In crocodylians the bones forming the primary palate have been modified to form a bony shelf, the secondary palate, ventral to the primary palate. This has been accomplished mainly through downfolding and inturning of the premaxillae, maxillae, and palatines. Together with the pterygoids, these bones form the secondary palate; the secondary palate of mammals is an independently derived, but only slightly differently constructed, analogous structure. The function of the secondary palate is to extend the **choanae**, the posterior openings of the nasal cavities, far posteriorly, thereby separating the oral cavity from the nasal passages, a feature that allows crocodylians to continue breathing while the oral cavity is engaged with prey. The incisive foramen, noted briefly in dorsal view of the skull, is a single opening between the premaxillae (Figure 8.18b). The much larger maxillae follow the premaxillae posteriorly. Both of these bones bear conical teeth, set in a single marginal series in deep sockets in the bone, the *thecodont* condition, as noted earlier.

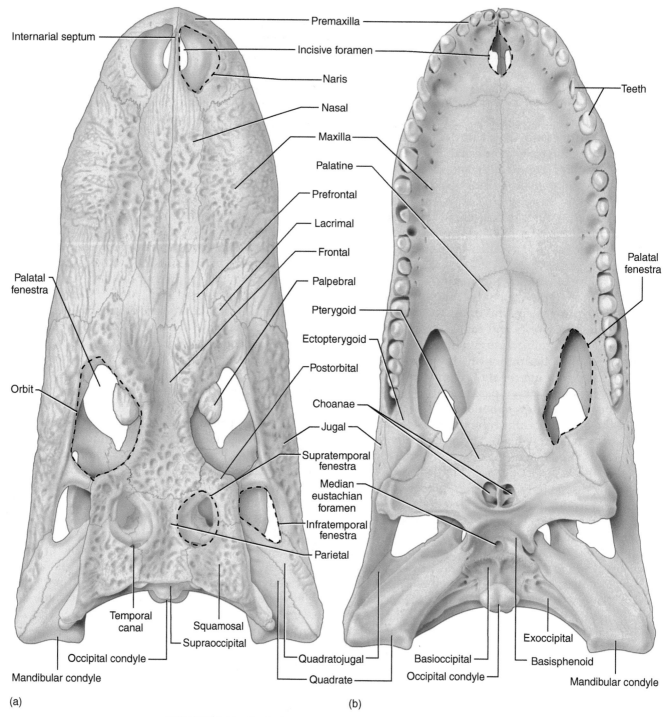

FIGURE 8.18　Skull of the alligator in (a) dorsal and (b) ventral views.

The palatines, narrower than the maxillae, form the medial margins of the palatal fenestrae and contact the pterygoids posteriorly (Figure 8.18b). The choanae lie far back between the pterygoids and are separated by a bony septum. The right and left pterygoids are solidly fused posterior to the choanae. The pterygoid expands laterally into a transverse flange that meets the ectopterygoid, which extends anterolaterally to meet the maxilla. These three bones form the rest of the margin of the palatal foramen. The basioccipital, forming the **occipital condyle**, which articulates with the atlas, can be seen posterior to the choanae. The eustachian foramina open in this region. The **median eustachian foramen** is large enough to be seen easily, opening between the basisphenoid and basioccipital (Figure 8.18b), but the two **lateral eustachian foramina**, which are narrow,

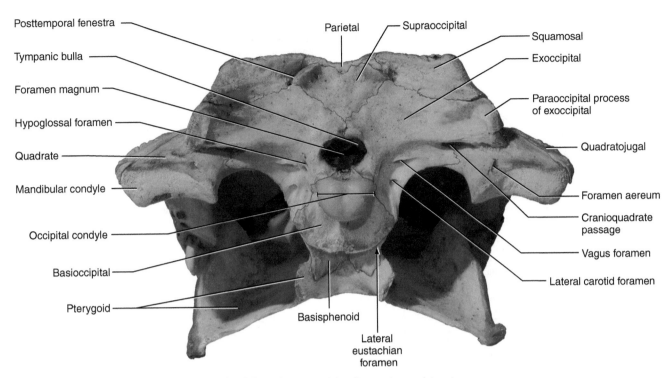

Posttemporal fenestra

Tympanic bulla

Foramen magnum

Hypoglossal foramen

Quadrate

Mandibular condyle

Occipital condyle

Basioccipital

Pterygoid

Parietal

Supraoccipital

Squamosal

Exoccipital

Paraoccipital process of exoccipital

Quadratojugal

Foramen aereum

Cranioquadrate passage

Vagus foramen

Lateral carotid foramen

Basisphenoid

Lateral eustachian foramen

FIGURE 8.19 Skull of the alligator in posterior view.

slit-like openings, are not (their position is indicated in Figure 8.19). Lateral to the occipital condyle the exoccipitals form narrow wedges contacting the quadrates. The mandibular condyles form the posterolateral corners of the skull. Note as well the quadratojugals.

Examine the occiput, the posterior part of the skull, which is formed by several bones (Figure 8.19). Locate the occipital condyle, formed solely by the basioccipital, which continues ventral to the condyle as a nearly flat plate with a curved ventral margin. The thin posterior portion of the basisphenoid lies immediately ventral to it, flattened against the posterior surface of the pterygoid and assuming its curved surface. The eustachian foramina, mentioned earlier, lie between this laminar portion of the basisphenoid and the basioccipital. The **foramen magnum** is the large, nearly circular opening dorsal to the occipital condyle. The paired exoccipitals lie to either side of and complete the margin of the foramen magnum (Figure 8.19).

The exoccipital includes the **opisthotic**, a bone helping to house the inner ear, that is present as an individual element in some other reptile skulls. It helps form the **tympanic bulla** (to which the prootic also contributes), which may be viewed though the foramen magnum. The exoccipital is a large bone that extends laterally over the mandibular condyle. Its lateral portion is the **paraoccipital process** (which also has a small exposure in lateral view). Several foramina piercing the exoccipital may be noted. A group of foramina lies just lateral to the condyle. The more

medial of the two small medial foramina is the **hypoglossal foramen** and transmits the hypoglossal nerve. Laterally is the large **vagus foramen**, for the vagus nerve. The **lateral carotid foramen**, nearly as large, is ventral to the vagus foramen and serves for passage of the internal carotid artery. Farther laterally, ventral to a prominent crest on the paraoccipital process, is the **cranioquadrate passage**. This lies between the exoccipital and quadrate and extends anteriorly into the middle ear. It transmits the orbitotemporal artery and lateral cephalic vein, as well as the main branch of the facial nerve. Just ventral and slightly lateral to the opening of the cranioquadrate passage is the small **foramen aereum**. This opening, entirely within the quadrate, was mentioned earlier in connection with the continuation of the cavity of the quadrate with that of the articular of the lower jaw.

Dorsally the occiput is completed by the median **supraoccipital** (which does not contribute to the margin of the foramen magnum) and laterally by the squamosal (Figure 8.19). The latter has an extensive suture with the exoccipital. A **posttemporal fenestra** lies where these three bones meet, high on each side of the occiput. The fenestra is small and effectively occluded by cartilage, in contrast to the spacious openings present in the turtle and iguana.

Examine the lower jaw (Figure 8.20). As in other reptiles, each half is formed by several bones. The **dentary**, as usual, is the large anterior bone. It bears all the teeth, conical and set in sockets like the upper teeth.

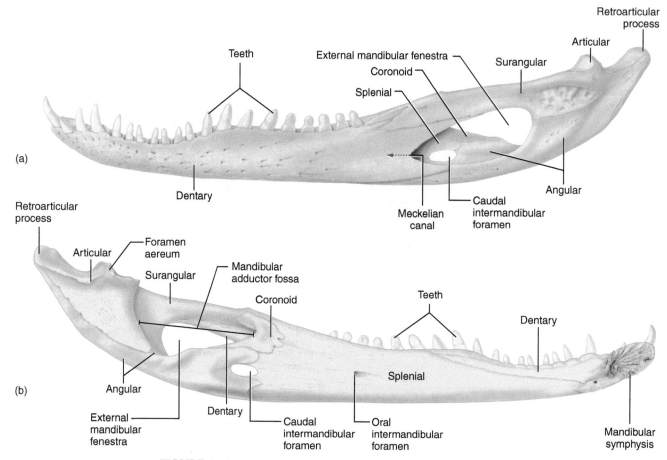

FIGURE 8.20   Mandible of the alligator in (a) lateral and (b) medial views.

In lateral view (Figure 8.20a), the dentary extends far posteriorly and bifurcates to form the anterior margin of the **external mandibular fenestra**. The **surangular** extends dorsal to the fenestra, whereas the **angular** lies ventral to it. These bones meet posterior to the fenestra. Posteriorly the surangular forms the lateral margin of the surface that articulates with the quadrate of the upper jaw. The **articular** has a small exposure in lateral view, forming the posterior margin of the articular surface for the quadrate and the dorsal surface of the **retroarticular process**. The angular extends farther back onto the retroarticular process of the lower jaw and forms its ventral margin. Anteriorly, the angular extends far forward as a narrow tapered process along the ventral edge of the lower jaw, wedged between the dentary and **splenial**. The latter, broadly exposed in medial view, forms a narrow strip medial and just posterior to the teeth. Several bones forming the medial surface of the lower jaw are visible through the external mandibular fenestra. The ventral bone is the angular, forming the ventral margin of the **caudal intermandibular foramen**. The splenial forms the dorsal margin of the foramen in lateral view. The **coronoid**, tapering posteriorly, lies dorsal to these bones.

Examine the lower jaw in medial view (Figure 8.20b). The dentary has a triangular exposure anteriorly, but much of the middle third of the lower jaw, including its dorsomedial edge, is covered by the splenial. The **oral intermandibular foramen** opens on the splenial ventral to the more posterior teeth. Posteriorly, the splenial meets the angular ventrally, forming with it the margin of the caudal intermandibular foramen; the splenial also meets the coronoid dorsally. The latter does not reach the dorsal margin of the lower jaw, but forms the anterodorsal margin of the opening of the **mandibular adductor fossa**. The angular is long and transversely wide ventrally but tapers dorsally, and contacts the surangular laterally. Its dorsal surface bears nearly all the mandibular adductor fossa. Posterior to the fossa, the angular contacts the articular and contributes to the retroarticular process. The articular forms a triangular, anteroventrally tapering, wedge-like bone. Its anterodorsal part forms most of the smooth articular surface for the quadrate of the skull, and its posterodorsal portion forms much of the retroarticular process, a narrow notch separating these two portions medially is pierced by the foramen aereum (recall that the quadrate also has a foramen aereum).

The spacious mandibular adductor fossa is a large trough that opens dorsally and medially from the posteromedial surface of the lower jaw. The margin of its opening is formed by the coronoid, angular, articular, and surangular. **Meckel's cartilage** (only a small posterior portion is usually preserved) extends anteriorly from the anteroventral tip of the articular, passing along a sulcus on the trough-like floor of the fossa. The fossa and cartilage continue forward into the spacious **Meckelian canal**, which is bounded laterally by the dentary and medially by the splenial. The canal is easily appreciated by observing anteriorly through the external mandibular fenestra. A stout process of the angular, projecting medially along the posteroventral margin of the opening of the fossa, provides an insertion site for the mandibular adductor musculature. The ventral portions of the retroarticular process provide insertion sites for the pterygoid musculature. The depressor mandibulae musculature, which opens the jaws, inserts more dorsally on the retroarticular process.

Key Terms: Alligator Skull and Mandible

angular
articular
basioccipital
basisphenoid
caudal intermandibular foramen
choanae
columella
coronoid
cranioquadrate passage
dentary
ectopterygoid
exoccipital
external mandibular fenestra
foramen aereum of articular
foramen aereum of quadrate
foramen magnum
foramen ovale
frontal
hypoglossal foramen
incisive foramen
infratemporal fenestra
internarial septum
interorbital septum
jugal
lacrimal
lacrimal canal
lateral carotid foramen
lateral eustachian foramina
laterosphenoid
mandibular adductor fossa
mandibular condyle
maxilla

Meckelian canal (mandibular canal)
Meckel's cartilage
median eustachian foramen
naris (pl., **nares**)
nasal
occipital condyle
opisthotic
oral intermandibular foramen
orbit
otic aperture
otic recess
palatal fenestra
palatine
palpebral
paraoccipital process of **exoccipital**
parasphenoid
parietal
postorbital
posttemporal fenestra
prefrontal
premaxilla
preotic foramen
prootic
pterygoid
quadrate
quadratojugal
retroarticular process
rostrum of **basisphenoid**
splenial
squamosal
supraoccipital
supratemporal fenestra
surangular
teeth
temporal canal
temporal fossa
transverse canal
tympanic bulla
vagus foramen

SECTION V: DINOSAUR SKULL AND MANDIBLE

Although it is unlikely that you will have the opportunity to examine a nonavian dinosaur skull directly, several of its more notable features may be appreciated on illustrations. Doing so will permit a broader scope in our consideration of reptile skulls, particularly of Archosauria. As noted in Chapter 1, dinosaurs belong to Archosauromorpha, one of the two main clades of Diapsida, the other being Lepidosauromorpha (see Figure 1.15). Living archosauromorphs, other than turtles and their kin (Pantestudines), belong to Archosauria and includes crocodylians and birds, the skulls of which are also described here (Section IV and Chapter 9, respectively), but the group has a rich fossil representation. The main archosaur divisions are the two lineages leading to crocodylians on the one hand and to the dinosaurs on the other. Several archosaur synapomorphies are better appreciated in some nonavian dinosaurs that in living archosaurs, and hence we present a description here of the cranial skeleton of two tyrannosaurid theropod dinosaurs, *Albertosaurus libratus* and *Tyrannosaurus rex*. Among the skeletal features characteristic of archosaurs are an antorbital fenestra, antorbital fossa, laterosphenoid, and external mandibular fenestra (the last two were noted in the alligator). Further information on these dinosaur skulls is available in Brochu (2002), Carr (1999), Currie (2003), Holtz (2004), Hurum and Sabath (2003), and Witmer (1997).

Using Figures 8.21 through 8.24, follow the descriptions for these theropods. Among the more notable features of the long, high, and narrow skull is its open appearance, due to the presence of many fenestrae, and elongated rostrum (Figure 8.21a, c). The diapsid nature of the skull is clearly evident: the **infratemporal** (Figure 8.21a) and **supratemporal** (Figure 8.21c) **fenestrae** are conspicuous openings.

Consider the anterior end of the skull. The **maxilla** is a large bone contacting the **premaxilla** anteriorly and the **nasal** dorsally. The premaxilla and nasal contribute to the margin of the **naris**, the external opening of the nasal cavity. The right and left nares are separated by the narial processes of these two bones. The maxilla and premaxilla bear a single marginal series of **teeth**, which are set in sockets within the bone, the *thecodont* condition as already noted for the alligator. Those implanted in the maxilla are elongated, curved, and dagger-like. Those in the premaxillae are smaller and incisiform (Figures 8.21a and 8.22).

The maxilla's posterior edge is concave, forming the anterior margin of the large **antorbital fenestra** (Figure 8.21a), which is a synapomorphy of archosaurs, as noted earlier and in Chapter 1. The rest of the margin is formed by the **jugal** and **lacrimal**. The fenestra lies in the **antorbital fossa**, a recessed region of the bony surface of the rostrum. The fossa is also developed early in the history of archosaurs, but the fossa and antorbital fenestra have been secondarily lost in later crocodylians, and thus was not noted in the description of the alligator. Much of the bony surface surrounding the fossa has a sculptured texture. The smaller **maxillary fenestra**, also within the antorbital fossa, lies anterior to the antorbital fenestra, but this feature is a synapomorphy of more derived theropod dinosaurs.

The lacrimal is a conspicuous element, contributing to the dorsal surface of the skull (Figure 8.21a, c) and the margin of the antorbital fossa, as already noted. Its posteroventral portion forms a bony bar between the antorbital fenestra and the **orbit**, the smaller, somewhat elliptical opening that housed the eyeball. The **lacrimal pneumatic recess** is visible near the posterodorsal edge of the antorbital fossa (Figure 8.21a). A prominent protuberance, the corneal process of the lacrimal, projects dorsolaterally, overhanging the region of the lacrimal pneumatic recess in *Albertosaurus* (but not *Tyrannosaurus*). Ventrally, the descending portion of the lacrimal meets the jugal, a large bone forming the cheek region and contributing to the margin of the orbit. Anteriorly, the jugal also contacts the maxilla. Posteriorly, it articulates with the quadratojugal. Posterodorsally, it sends a stout triangular flange that meets a similar descending flange of the **postorbital**. Together, these flanges form a stout partition between the orbit and the infratemporal fenestra.

The postorbital contributes to the dorsal surface of the skull, but does not meet the lacrimal in *Albertosaurus* (in contrast to adult *Tyrannosaurus*; not figured). These two bones are separated by a narrow space, the supraorbital sulcus. The small **prefrontal** forms the anterodorsal margin of the orbit, but has a larger dorsal exposure. Posteriorly the postorbital meets the **squamosal** (Figure 8.21a, c). The infratemporal fenestra is oddly shaped in these dinosaurs, as it is constricted near its middle. This is due to anterior extensions of the squamosal and **quadratojugal** that nearly bisect the infratemporal fenestra, thereby resulting in the constriction that produces dorsal and ventral openings.

The more anterior part of the **quadrate** is exposed through the ventral part of the fenestra, but much of the rest is obscured in lateral view by the quadratojugal and squamosal (Figure 8.21a). The quadrate extends posteroventrally to form the mandibular condyle for articulation with the lower jaw. It also has a small exposure just behind the posterodorsal part of the quadratojugal. Anteriorly the quadrate contacts the **pterygoid**.

The osseous portion of the braincase lies in the posterior part of the skull. It includes bones that surround and protect the dorsal, lateral, ventral, and posterior surfaces of the brain. A portion of this osseous braincase is visible through the dorsal part of the infratemporal fenestra. The main part of the bony mass visible is the **prootic**, which also houses part of the inner ear. The **laterosphenoid** has a small exposure anterior to the prootic (Figure 8.21a). The laterosphenoid is an ossification that helps form the anterolateral walls of the braincase. It was also noted in the alligator and is an early development in the history of archosaurs. The small **epipterygoid** contributes to the braincase in this region but is not visible due to the bar formed by the postorbital and the jugal (the position of some of these bones is more clearly observable in the skull of the alligator; see Figure 8.17).

The dorsal part of the braincase is formed by the **frontal** and **parietal** (Figure 8.21c). The parietal bears a

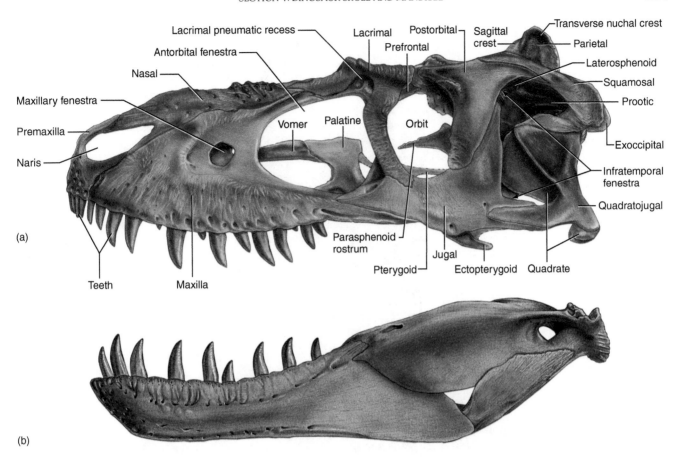

(a)

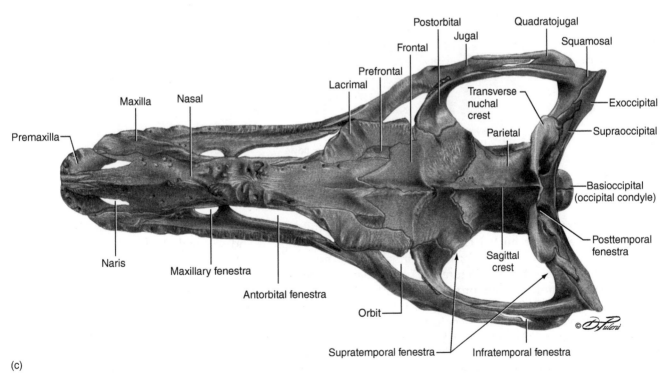

(b)

(c)

FIGURE 8.21   Skull of *Albertosaurus libratus* in (a) lateral and (c) dorsal views; and mandible in (b) lateral view (see Figure 8.24 for labels).

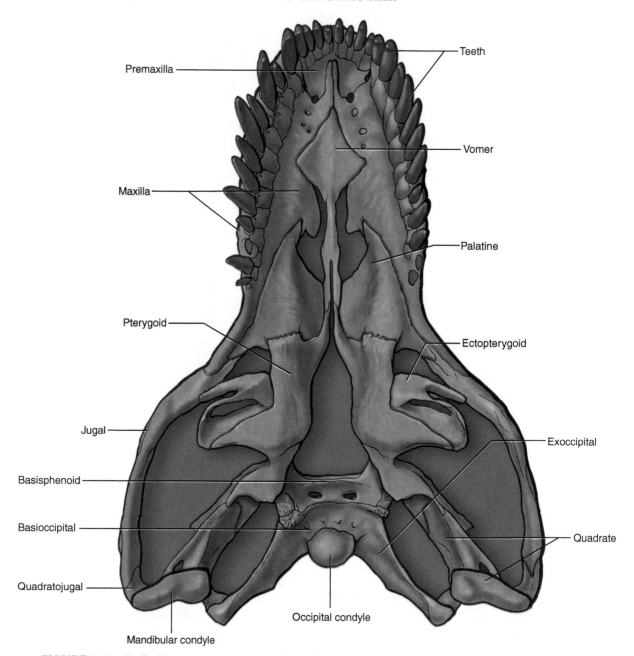

FIGURE 8.22   Skull of *Tyrannosaurus rex* in ventral view (line drawing © Dr. T. D. Carr; color rendering by D. Pulerà).

median **sagittal crest**, which extends posterodorsally and meets the tall **transverse nuchal crest** (Figure 8.21a,c). Ventrally the osseous braincase is formed mainly by the **basioccipital** and **basisphenoid** (Figure 8.22). A median blade-like **parasphenoid rostrum** projects anterior to the basisphenoid (Figure 8.22a), as similarly occurs in the alligator, but in which the parasphenoid is completely fused to the basisphenoid in adults, and was termed the rostrum of the basisphenoid.

Note how open the skull appears in dorsal view (Figure 8.21c). The fenestrae previously described are all clearly evident. In addition, note the supratemporal fenestrae, one on either side in the posterior region of the skull. As in the alligator, they are almost entirely in the frontal plane and so visible in dorsal view, but they are much larger. Anteriorly, the nasals are narrow, elongated bones that contact the frontal, lacrimal, and prefrontal. The nasals tend to fuse to each other, with only partial separation between them, whereas the parietals fuse solidly to each other with no trace of a suture. The frontal contacts the postorbital posterolaterally and the parietal posteriorly. A ridge occurs where the frontal and parietal bones meet. The prominence of the transverse nuchal crests is easily appreciated in this view. The margin of the supratemporal fenestra is formed by the postorbital, frontal, parietal, and squamosal.

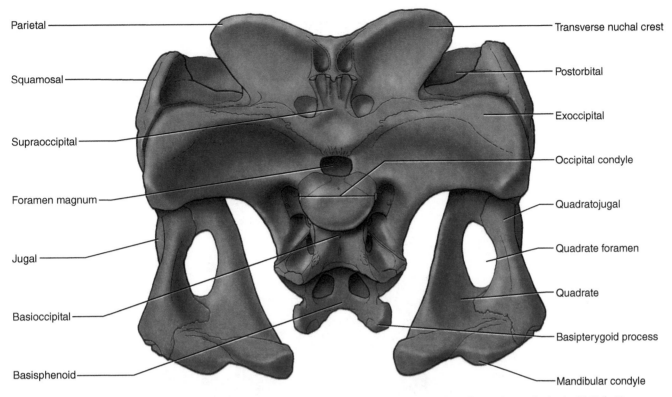

Parietal

Squamosal

Supraoccipital

Foramen magnum

Jugal

Basioccipital

Basisphenoid

Transverse nuchal crest

Postorbital

Exoccipital

Occipital condyle

Quadratojugal

Quadrate foramen

Quadrate

Basipterygoid process

Mandibular condyle

FIGURE 8.23   Skull of *Albertosaurus libratus* in posterior view (line drawing © Dr. T. D. Carr; color rendering by D. Pulerà).

The occiput or posterior surface of the skull, which covers the brain posteriorly, is partly visible in dorsal view (Figure 8.21c). The **posttemporal fenestra** is much smaller than in the turtle and iguana. The **supraoccipital** is the median bone dorsal to the **foramen magnum**, the opening though which the spinal cord joins the brain (Figure 8.23). This bone forms the dorsal margin of the foramen, and contacts the parietal dorsally and the **exoccipitals** laterally (Figure 8.23). The exoccipital is a combination of the fused exoccipital and **opisthotic**. These bones tend to remain separate in some reptiles, such as turtles, but are fused together in others. In the latter case, the unit is generally termed an exoccipital in the iguana and alligator. Anatomists differ on the term applied to this structure in dinosaurs, with some following the typical reptilian designation as exoccipital, as is done here. Others, however, follow the specific terminology applied to birds for dinosaurian anatomy, and term it the otoccipital. This term perhaps more accurately reflects the compound nature of the bone in referring to the otic (or ear, for the opisthotic, which helps house the inner ear) and occipital (for the exoccipital). The lateral portion of the exoccipital is the **paraoccipital process**. The basioccipital forms the ventral bone of the occiput, but also has a ventral exposure (see Figure 8.22). It forms most of the spherical **occipital condyle**, which articulates with the first vertebra of the vertebral column. The

basisphenoid, which lies anterior to the basioccipital, may also be observed in posterior view (Figure 8.23). The exoccipital contributes a small dorsolateral portion to the condyle. The quadrate extends ventrally and forms the **mandibular condyle**. The quadrate contacts the quadratojugal medially and the **quadrate foramen** lies between these bones.

In ventral view (Figure 8.22) the palate is formed anteriorly by the maxillae, premaxillae, and **vomer**, a narrow compound bone that has a diamond-like expansion anteriorly. The triangular **palatines** follow posteriorly and help enclose the **choanae**, the internal openings of the nasal cavities. The choanae are farther back than in the turtle and iguana, but not as far as in the alligator (compare with Figure 8.18b). Farther posteriorly, each pterygoid, a large bone, contacts the **ectopterygoid** laterally, and then constricts. At this point, the pterygoid articulates medially with the basisphenoid, which helps form the basicranium with the basioccipital. The exoccipitals, extending posterolaterally from either side of the basioccipital, contribute to the basicranium and form much of the occiput, as noted earlier. The posterior end of the pterygoid articulates with the quadrate.

Examine the mandible. Three bones have prominent exposures in lateral view (Figure 8.24a). The **dentary** is the large anterior bone forming at least half of the mandible, and bears the teeth. The **surangular**, also large,

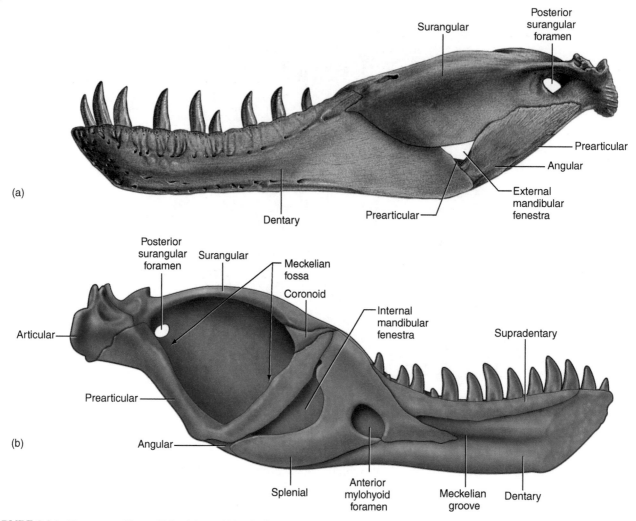

FIGURE 8.24   Tyrannosaurid mandibles: (a) mandible of *Albertosaurus libratus* in lateral view and (b) mandible of *Tyrannosaurus rex* in medial view.

lies posterodorsally, and the smaller **angular** lies posteroventrally. Between these three bones is the **external mandibular fenestra**, a characteristic feature of archosauromorphs that was also noted in the alligator. Another prominent opening, the **posterior surangular foramen**, pierces the posterior part of the surangular. The **prearticular**, which lies mainly on the medial surface of the mandible (see later) and can be seen through the external mandibular fenestra, has a small lateral exposure as an elongated strip along the ventral margin of the angular. The **articular** forms most of the surface that articulates with the quadrate of the skull to form the jaw joint (the surangular contributes a small anterolateral contribution). It has a small exposure posterior to the surangular and will be seen more clearly in medial view. The region of the mandible posterior to the articular surface is the **retroarticular process**, which is much smaller than in the alligator.

In medial view (Figure 8.24b) the dentary contributes mainly to the anterior third of the mandible. Dorsal to

it and forming the medial border of the tooth row is the thin, narrow **supradentary**. This bone continues posteriorly, covered by the **splenial** and prearticular, to fuse with the small and triangular **coronoid**. The general consensus among theropod specialists is that the supradentary is probably not a distinct bone and thus that it is an anterior extension of the coronoid. The splenial is an approximately arrowhead-shaped bone. It encloses the **anterior mylohyoid foramen** and helps form the medial wall of the **Meckelian fossa**, the large space into which adductor musculature descends. Its notched posterior margin forms the anterior margin of the **internal mandibular fenestra**. The anterior tip of the splenial contacts the dentary anteriorly. Its anterodorsal portion overlaps the supradentary before contacting the prearticular. Posteroventrally the splenial helps form the ventral margin of the mandible and contacts the angular. The prearticular is a dorsally-open crescentic or boomerang-shaped bone. Its dorsal edge forms the ventral margin of the Meckelian fossa, and its central portion meets the angular ventrally. Its anterodorsally oriented arm meets the coronoid and overlaps the supradentary. The anterior margin of this arm helps form the internal mandibular fenestra. The posterodorsally oriented arm of the prearticular articulates with the articular. Between the articular surface and retroarticular process, the small **foramen aereum** (not illustrated) pierces the dorsal surface of the articular and communicates with the internal cavity of the quadrate. The triangular coronoid is wedged between the prearticular and surangular, and forms the anterodorsal margin of the Meckelian fossa. The surangular, already noted in lateral view, also has a broad exposure in medial view. This surface forms the lateral wall of the Meckelian fossa. The dorsal edge of the surangular arches medially over the fossa, forming its dorsal margin.

Key Terms: Dinosaur Skull and Mandible

- angular
- anterior mylohyoid foramend
- antorbital fenestra
- antorbital fossa
- articular
- basioccipital
- basisphenoid
- choanae
- coronoid
- dentary
- ectopterygoid
- epipterygoid
- exoccipital
- external mandibular fenestra
- foramen aereum
- foramen magnum
- frontal
- infratemporal fenestra
- internal mandibular fenestra
- jugal
- lacrimal
- lacrimal pneumatic recess
- laterosphenoid
- mandibular condyle
- maxilla
- maxillary fenestra
- Meckelian fossa
- naris
- nasal
- occipital condyle
- opisthotic
- orbit
- palatine
- paraoccipital process
- parietal
- posterior surangular foramen
- postorbital
- posttemporal fenestra
- prearticular
- prefrontal
- premaxilla
- prootic
- pterygoid
- quadrate
- quadrate foramen
- quadratojugal
- retroarticular process
- rostrum of parasphenoid
- sagittal crest
- splenial
- squamosal
- supradentary
- supraoccipital
- supratemporal fenestrae
- surangular
- surangular foramen
- teeth
- transverse nuchal crest
- vomer

9

The Pigeon

INTRODUCTION

The pigeon, *Columba livia*, belongs to Aves or birds, a group within Dinosauria (see Figure 1.15). Although birds have long been recognized as the flying, feathered vertebrates, recent fossil discoveries—among the more exciting paleontological developments of the last few decades—have revealed that feathers first evolved in several groups of theropod dinosaurs. Feathers are therefore not unique to birds and can no longer be used to diagnose them as a group. Nonetheless, birds are the only living vertebrates with feathers. The terms Aves and birds have long been used to refer to the same group of amniotes, but they are not precisely synonymous. Aves is considered the group that includes the ancestor of living birds and all of its descendants. However, the fossil remains of much more primitive birds have long been known to science.

Indeed, birds have a long fossil record, beginning with what has traditionally been considered the earliest bird, the Late Jurassic *Archaeopteryx lithographica*. The group including *Archaeopteryx* and all other birds is termed Avialae and is recognized by, among other features, development of the forelimbs into wings. For most of the century and a half since *Archaeopteryx* was first discovered in 1861, it has provided essentially the only fossil evidence of the possible transitional features that might have occurred in the evolution of birds. As such, it has become an iconic fossil, pivotal for understanding the origin of birds (Foth and Rauhut, 2017). However, the recent discovery of several Late Jurassic feathered dinosaur fossils, particularly from China, has revealed that the events surrounding the origin of birds was less straightforward, at least with respect to which taxon we might designate as the "first" bird (an arbitrary designation in any event, as explained in Chapter 1). Such discoveries are welcome, of course, as the more information available, the better our understanding becomes.

Several years ago, *Archaeopteryx* was "knocked off its perch" as the earliest and most basal bird, with Xu et al. (2011) reporting that the discovery of *Xiaotingia* and their

phylogenetic analysis shifted *Archaeopteryx* from Avialae (and thus the traditional group recognized as birds) to the closely related Deinonychosauria. More recent analyses, including newly described fossil taxa such as *Aurornis* by Godefroit et al. (2013) and *Chongmingia* by Wang et al. (2016), have reestablished *Archaeopteryx* as belonging to Avialae, although it may no longer be the earliest or most basal taxon of this clade.

Among more derived birds, the fossil record documents several substantial radiations during the Jurassic and Cretaceous Periods, of which many became extinct. We may mention as examples Confuciusornithidae (a group of toothless, flying birds that still had mainly separate digits in the manus and a pair of long ornamental tail feathers in, presumably, males), Hesperornithiformes (a group of flightless, diving birds with the forelimb represented only by a splint-like humerus), and Ichthyornithiformes (a group of ancient flying birds). These groups had all become extinct by the end of the Cretaceous.

Modern birds, or Aves, began their radiation during the Cretaceous. They are among the most numerous of vertebrate groups, having diversified to include at least 9700 living species (and molecular studies suggest that there might be twice as many species as currently recognized). Birds are characterized by the complete loss of teeth and a very large sternal **carina** (see later). They are subdivided into Palaeognathae and Neognathae, which are differentiated primarily on palatal features. The paleognaths include two lineages, the flightless ratites (e.g., ostriches, emus, rheas, cassowaries) and the tinamous.

Ratites and tinamous comprise some 54 living species, so most of present-day avian diversity is represented by the neognaths. Twelve main lineages are recognized, but their interrelationships are largely unresolved. Among these may be mentioned the sister groups Anseriformes (ducks, geese, swans) and Galliformes (chickens, grouse, pheasants, turkeys), which together apparently form the outgroup to other neognaths. Procellariiformes (albatrosses, petrels, and shearwaters), Pelicaniformes (pelicans, gannets, cormorants), Sphenisciformes (penguins),

The Dissection of Vertebrates, Third Edition
https://doi.org/10.1016/B978-0-12-410460-0.00009-7

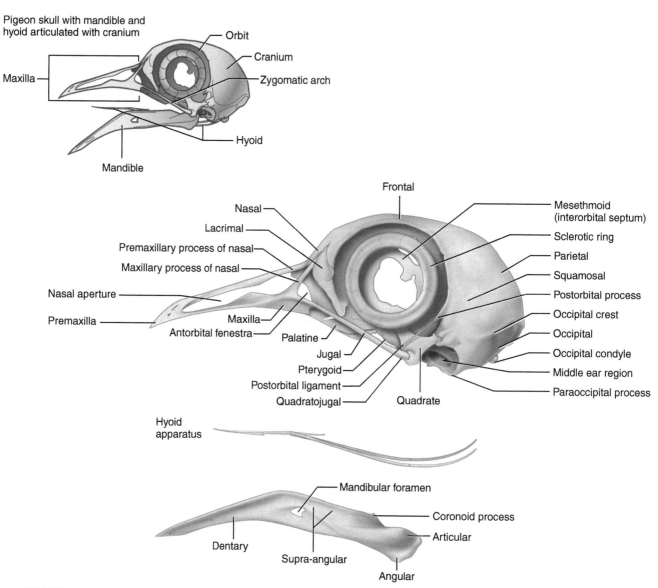

Pigeon skull with mandible and hyoid articulated with cranium

Maxilla

Orbit

Cranium

Zygomatic arch

Hyoid

Mandible

Frontal

Nasal

Lacrimal

Premaxillary process of nasal

Maxillary process of nasal

Nasal aperture

Premaxilla

Maxilla

Antorbital fenestra

Palatine

Jugal

Pterygoid

Postorbital ligament

Quadratojugal

Quadrate

Mesethmoid (interorbital septum)

Sclerotic ring

Parietal

Squamosal

Postorbital process

Occipital crest

Occipital

Occipital condyle

Middle ear region

Paraoccipital process

Hyoid apparatus

Mandibular foramen

Coronoid process

Articular

Dentary

Supra-angular

Angular

FIGURE 9.1 Skull of the pigeon in left lateral view. Inset figure shows elements in position, and large illustration indicates detail.

Gaviformes (loons), and Podicipediformes (grebes) are part of another lineage, as are Falconiformes (eagles, falcons, ospreys) and Strigiformes (owls). Passeriformes, the great group of songbirds, together with Piciformes (barbets, toucans, woodpeckers) and Coraciiformes (hoopoes, hornbills, kingfishers) represent yet another lineage. The pigeon is included with doves in Columbiformes, which is part of the lineage including Charadriiformes (auks, curlews, snipes, terns, gulls, puffins), Ardeidae (bitterns, egrets, herons), and Gruiformes (cranes, coots, rails, bustards).

SECTION I: SKELETON

The structure of birds, most of which can fly, has been highly modified to meet the demands of flight. The skeleton has a number of such modifications, most obviously

perhaps in the forelimbs, which are adapted to form a bony frame for the wings (Figure 9.3). Also, however, the skeleton has undergone changes to reduce weight, clearly an advantage in a flying animal, and mobility, which gives the wings a rigid support. Several bones of many birds are pneumatic—hollow and containing extensions of the respiratory system's air sacs, described in more detail later (Figure 9.4; also see O'Connor, 2004).

Skull, Mandible, and Hyoid Apparatus

The bones of the skull (Figures 9.1 and 9.2) are extensively fused to provide strength while minimizing weight. As the sutures are often obliterated, it is difficult to distinguish many of the bones in an adult individual. Attempting to identify each bone is, at this level of study, impractical, and it is more useful to examine various regions of the skull.

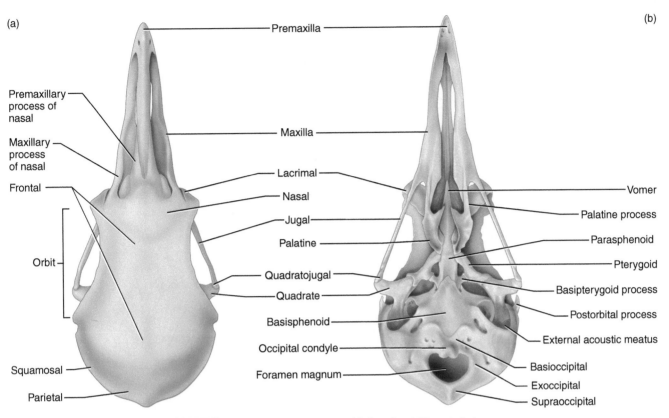

(a)

Premaxilla

Premaxillary process of nasal

Maxillary process of nasal

Frontal

Orbit

Squamosal

Parietal

Maxilla

Lacrimal

Nasal

Jugal

Palatine

Quadratojugal

Quadrate

Basisphenoid

Occipital condyle

Foramen magnum

(b)

Vomer

Palatine process

Parasphenoid

Pterygoid

Basipterygoid process

Postorbital process

External acoustic meatus

Basioccipital

Exoccipital

Supraoccipital

FIGURE 9.2   Skull of the pigeon in (a) dorsal and (b) ventral views.

Among the more notable features of the skull are the large **cranium**, enclosing the enlarged brain of birds, and **orbits**, which house the large eyes (Figures 9.1 and 9.2a). The latter reflect the importance of vision in birds. The skull is a modified diapsid type (see Chapter 1); in modern birds the upper temporal arch has been lost, so the supratemporal and infratemporal fenestrae have merged to form a single opening. Further, the latter has merged with the orbit. The slender, rod-like **zygomatic arch** (jugal bar; see later) forms the ventral margin of the opening. This pattern is convergent on that of mammals, but has, of course, been independently acquired.

Examine the orbits. The often incomplete, bony septum between the orbits is formed by the **mesethmoid**. A circle of small bones, the **sclerotic ring**, supports the eyeball. The **postorbital process** lies at the posteroventral margin of the orbit. Posterior to the orbits, the cranium is composed of the paired **frontals** that cover most of the roof of the skull; the **parietals**, smaller rectangular bones; the **squamosals** on the lateral wall of the cranium (Figures 9.1 and 9.2a); and, posteroventrally, the **occipital** (Figure 9.1), which is formed from several bones (Figure 9.2b) fused into a single unit. These cranial bones can usually be distinguished, although the sutures between them are often indistinguishable.

The **quadrate** lies toward the ventral part of the orbit and is mobile in birds. It has a prominent orbital process, as well as an otic process, that contacts the cranium.

Just ventral to this contact is a large, oval depression, the **auditory meatus**, which is bounded posteroventrally by the **paroccipital** process and leads into the **middle ear**. The bony region, pierced by various foramina, visible within the meatus contains the inner ear. The quadrate has a ventral process that articulates with the **mandible** to form the jaw joint.

The **foramen magnum** is the large opening in the occipital. At its anterior edge is the single **occipital condyle**. It is formed largely by the **basioccipital**, which forms the ventral part of the occipital and extends along the basicranium anterior to the occipital condyle (Figure 9.2b). The paired **exoccipitals** form the lateral portions of the occipital and contribute to the occipital condyle (Baumel and Witmer, 1993), whereas the **supraoccipital** forms the dorsal part of the occipital. The **basisphenoid** lies anterior to the basioccipital and has a tapered anterior end. The **parasphenoid** lies anterior to this tapered end of the basisphenoid. The **basipterygoid processes**, one on either side, extend anterolaterally from the posterior end of the parasphenoid and contact the **pterygoid**, each of which articulates anteriorly with a **palatine**.

The much smaller **antorbital fenestra** (Figure 9.1; a synapomorphy of archosaurs, as noted in Chapters 1 and 8) lies anterior to the orbit; its posterior margin is formed by the **lacrimal**. Anteriorly, a complex of bones form the upper jaw and support the **bill** (Figures 9.1–9.3). This

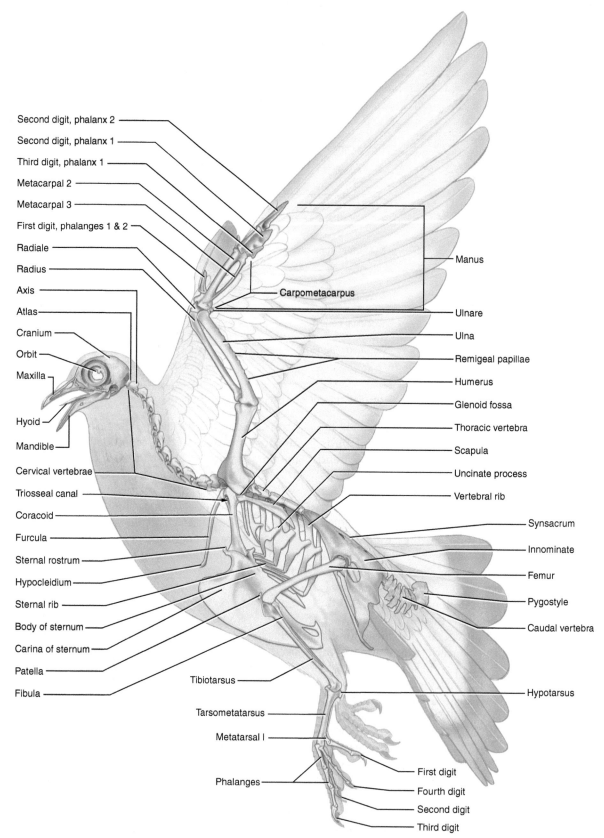

Second digit, phalanx 2
Second digit, phalanx 1
Third digit, phalanx 1
Metacarpal 2
Metacarpal 3
First digit, phalanges 1 & 2
Radiale
Radius
Axis
Atlas
Cranium
Orbit
Maxilla
Hyoid
Mandible
Cervical vertebrae
Triosseal canal
Coracoid
Furcula
Sternal rostrum
Hypocleidium
Sternal rib
Body of sternum
Carina of sternum
Patella
Fibula
Tibiotarsus
Tarsometatarsus
Metatarsal I
Phalanges

Carpometacarpus

Manus
Ulnare
Ulna
Remigeal papillae
Humerus
Glenoid fossa
Thoracic vertebra
Scapula
Uncinate process
Vertebral rib
Synsacrum
Innominate
Femur
Pygostyle
Caudal vertebra
Hypotarsus
First digit
Fourth digit
Second digit
Third digit

FIGURE 9.3   Skeleton of the pigeon in left lateral view, superimposed on body outline.

complex is the **maxilla**[1] (see, for example, Baumel and Witmer, 1993; Zusi, 1993), and it includes the **nasals**, **premaxillae**, and **maxillae** (in this sense, as paired bony elements). From its dorsal contact with the frontal, each nasal sends out two slender processes, one contacting the maxilla, the other contacting the premaxilla (Figure 9.2a). The latter is a large element forming most of the dorsal and ventral portions of the anterior part of the upper jaw. Near its dorsal articulation with the nasal, the premaxillae can, in life, bend or flex slightly dorsally in many birds (see later). The maxilla forms most of the posterior part of the upper jaw ventrally and contacts the palatine posteriorly. Finally, note again the zygomatic arch, a slender bridge between the maxilla anteriorly and the quadrate posteriorly. It is formed almost entirely by the jugal; a small posterior contribution is made by the **quadratojugal**.

The mandible or lower jaw (Figure 9.1) is composed of several bones, but they are difficult to distinguish. The **dentary** forms nearly the anterior half of the mandible. Its posterior end coincides roughly with the **mandibular foramen**. Most of the lateral surface posterior to the foramen is formed by the **supra-angular**. The **angular** is a smaller bone ventral to the supra-angular. The **articular** is a small element that articulates with the quadrate of the skull. Two other bones are present but are on the medial surface of the mandible. The small protuberance, the **coronoid process**, along the posterodorsal margin of the mandible serves for attachment of the tendon for part of the jaw-closing musculature.

The **hyoid apparatus** (Figure 9.1) is a slender, elongated, Y-shaped structure that supports the tongue. It is formed by various elements, but they are difficult to distinguish and are not considered separately.

The capacity of some skull elements (in addition to the lower jaw) of birds to move relative to each other was noted briefly earlier (and for other reptiles in Chapter 8). Such ability is termed **kinesis** and several types (such as prokinesis, amphikinesis, and rhynchokinesis; see Gussekloo et al., 2001 for further details) may occur in modern birds, allowing them to move the upper jaw or part of it. Indeed, as noted by Zusi (1993), the manipulation of objects by the bill involves the motion of both upper and lower jaws. The movement of the entire upper jaw is termed **prokinesis**. The mechanism allowing this movement includes the quadrates, pterygoids, palatines, and zygomatic arches, as well as muscles and ligaments. Prokinesis is initiated by rotation of the quadrates, which imparts motion to the zygomatic arches, palatal complex (including the pterygoids and palatines), and upper jaw. The palatal

complex is divided functionally at the joint between the pterygoids and palatines, where these bones can slide against the parasphenoid (see Figure 9.2b). The nasofrontal and palatomaxillary joints are also involved. These are thin, flexible joints that function as hinges about which rotation may occur. Elevation of the upper jaw is induced when muscles pull the quadrate anterodorsally. This results in anterior displacement of the zygomatic arch and pterygoid (and hence the palatine). In turn, these elements press against the upper jaw and rotate it about the nasofrontal (or prokinetic) joint. Posteroventral rotation of the quadrate reverses these motions to depress the upper jaw.

Postcranial Skeleton

Vertebrae

The vertebral column (Figure 9.3) normally includes 14 **cervical**, 5 **thoracic**, 6 **lumbar**, 2 **sacral**, and 15 **caudal vertebrae**. The column of the pigeon, as with most birds, is notable for its degree of fusion, with the exception of the cervical vertebrae. The atlas and axis are followed by 12 other cervical vertebrae, all fully mobile. The forelimbs of most birds are so specialized for flight that they are unavailable for other functions such as grooming or obtaining food. The long, mobile, S-shaped neck allows some of the functions usually performed by the forelimbs in other vertebrates. The last two cervical vertebrae bear small, floating **ribs** that do not attach to the **sternum** (see later). Nearly all the other cervical vertebrae also bear ribs, but they are fused to the vertebrae.

There are five thoracic vertebrae, the first four of which are fused together. Each thoracic vertebra bears a pair of ribs. The fifth thoracic vertebra is fused to the

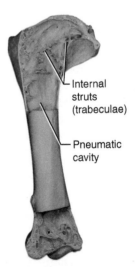

FIGURE 9.4 Humerus of the pigeon with proximal and distal ends sectioned to reveal internal structure.

---

[1]The use of upper mandible (rather than maxilla) for the upper jaw and lower mandible for the lower jaw is common in bird guides, such as those of Sibley (2002) and Alderfer and Dunn (2007).

first of the six lumbar vertebrae, which are all fused together and incorporated into the **synsacrum** (see later), as are the two sacral vertebrae. Of the fifteen caudal vertebrae, the first five are incorporated into the synsacrum, the following six are free and mobile, and the last four are fused together to form the **pygostyle** (Figures 9.5 and 9.6). The extensive fusion of sections of the vertebral column posterior to the neck renders the trunk rather rigid. This inflexibility may be a feature that helps to reduce weight, as it decreases the need for extensive musculature to maintain a streamlined and rigid posture during flight.

Ribs

There are several types of ribs (Figure 9.3). As noted above, independent ribs are associated with the last two cervical vertebrae. These ribs do not reach the sternum and are often referred to as floating ribs (or bicipital cervical ribs). The five ribs associated with the thoracic vertebrae are each formed from a robust, dorsal element, a **vertebral rib**, and a more slender, ventral element, a **sternal rib**. The vertebral rib, as its name implies, extends from the vertebra and articulates with the sternal rib, which in turn articulates with the sternum. In some specimens the last vertebral rib may articulate with the sternal rib of the preceding rib. Note the prominent, posteriorly projecting **uncinate processes**, which overlap the subsequent vertebral rib and help brace the ribs.

Sternum

The **body** of the **sternum** (Figures 9.3 and 9.12) is a large, curved plate of bone. Its most prominent feature is an extremely large vertical plate, the **carina** or keel. The body articulates anteriorly with the **coracoids** (see later) and more posteriorly with the sternal ribs. The large **caudolateral process** projects posteriorly from behind the articular surfaces for the sternal ribs and helps form the **sternal notch**.

Pectoral Girdle and Forelimb

The pectoral girdle of birds is modified to produce a rigid and stable brace for the requirements of flight. These modifications are so pronounced that they usually are retained even in birds that are secondarily flightless. For example, the **scapula**, **coracoid**, and **furcula** are tightly bound near the shoulder joint, and the coracoid is a stout, elongated element bracing the forelimbs against the sternum.

The scapula (Figures 9.3 and 9.12) is an elongated, blade-like bone that narrows anteriorly. Its anterior end forms part of the **glenoid fossa**, which articulates with the **humerus**. The coracoid (Figures 9.3, 9.10, and 9.12) is a stout bone. Its posterodorsal surface articulates with the scapula and completes the glenoid fossa.

Dorsomedially, it articulates with the furcula (see later). The coracoid widens ventrally and articulates with the body of the sternum, thus acting as a strut between the sternum and shoulder joint. The **triosseal canal** is a bony passage formed in many birds by three bones (Mayr, 2017; hence its name)—usually the scapula, coracoid, and furcula—that serves as the passage for the tendon of the **supracoracoideus** muscle (see later). In the pigeon, however, the canal is formed only by the coracoid (Baumel and Witmer, 1993). It is an oval opening located near the dorsal end of the coracoid, just ventral to its articulation with the furcula.

The furcula (or wishbone) is a V-shaped element, with its two arms, or rami, united ventrally. It is variably expanded here, though only slightly so in the pigeon, to form the **hypocleidium** (Baumel and Witmer, 1993). In some birds, such as the pigeon, a ligament (the hypocleidal ligament; see Sreeranjini et al., 2015) extends between the hypocleidium and anterior end of the sternal carina; the ligament is often preserved in mounted pigeon specimens. In other bird species, the hypocleidium may contact the carina directly (Baumel and Witmer, 1993). In several groups of birds, including some parrots, owls, and pigeons, the furcula is not fully ossified, so their two rami are not united ventrally (Bock, 2013). The furcula is commonly considered to represent fused clavicles, which (when present) are separate paired skeletal elements (as described for the cat). The hypocleidium has also been considered to represent a third element, the interclavicle (a midventral element in the pectoral girdle of several groups of tetrapods) contributing to the furcula. However, Vickaryous and Hall (2010) found no evidence that the hypocleidium develops separately from the rest of the furcula and thus no support for its homology with the interclavicle. In contrast to the commonly held view that the furcula is homologous with the clavicles, these authors suggested that homology of the furcula with the interclavicle is just as plausible.

The forelimb (Figure 9.3) consists of a stout proximal humerus, followed by the longer **radius** and **ulna**. The radius is the straighter, more slender bone. The ulna is bowed and has a short **olecranon process** proximally. Its posterior margin has several **remigeal papillae**, knob-like markings for the attachment of flight feathers. The manus consists, as usual, of carpals, metacarpals, and phalanges, but these have been highly modified in birds. Two **carpals** remain unfused: the **ulnare**, a slender bone that articulates with the ulna, and the **radiale**, which articulates with the radius. These carpals are followed by the **carpometacarpus**, an elongated element composed of several carpals, proximally, and three metacarpals fused together.

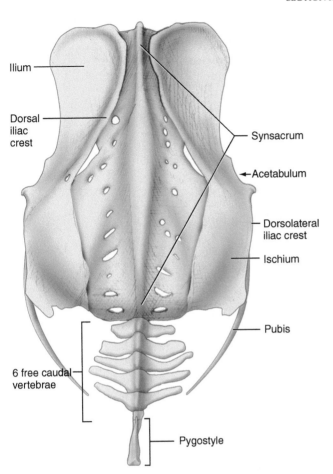

FIGURE 9.5 Posterior part of vertebral column and pelvis of the pigeon in dorsal view.

There has been considerable debate over the homology of the digits of birds, with some researchers considering them homologous with digits 1, 2, and 3 (as is typical of theropod dinosaurs) and other researchers considering them homologous with digits 2, 3, and 4. Recent morphological and developmental analyses suggest that the digits of birds represent digits 1, 2, and 3. The two elongated structures of the manus are **metacarpal II** (robust and nearly straight) and **metacarpal III** (slender and bowed). Although it is difficult to discern, **metacarpal I** (the alular metacarpal) is fused into the proximal end of the carpometacarpus along with the carpals. The phalanges are also highly modified. Those of the alular digit are represented by the short, triangular fused element, including **phalanges 1** and **2**, at the proximal end of the carpometacarpus. **Phalanges 1** and **2** of digit 2 extend distally from metacarpal II. A small triangular element at the articulation between the carpometacarpus and phalanx one of digit 2 is **phalanx 1** of digit 3.

Pelvic Girdle and Hind Limb

The pelvis (Figures 9.3, 9.5, and 9.6) includes left and right **innominate** bones, each of which is formed by an **ilium, ischium**, and **pubis**. The ilium is the largest element, forming the dorsal half of the pelvis. It has two distinct regions, an anterior concave region and a posterior convex region (in dorsal view). The posterior part of the **dorsal iliac crest** separates these regions, just dorsal to the **acetabulum**, the depression that receives the head of the femur (see later). The crest continues posteriorly as

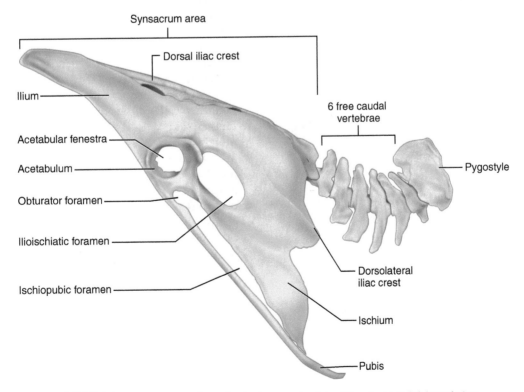

FIGURE 9.6 Posterior part of vertebral column and pelvis of the pigeon in left lateral view.

the **dorsolateral iliac crest**. The ischium lies ventral to the posterior part of the ilium. The separation between these bones is marked approximately by the **ilioischiadic foramen**. The acetabulum is directly anterior to it. Note that the floor of the acetabulum has an opening, the **acetabular fenestra**. The curved pubis is a very slender and elongated bone extending posteroventrally from the acetabulum, along the ventral margin of the ischium. Two other openings may be noted. The **obturator foramen** lies anteriorly between the pubis and ischium, whereas the long and narrow **ischiopubic fenestra** lies more posteriorly.

A notable feature of birds is that the pelvis is solidly and extensively fused to the synsacrum, the posterior series of fused and expanded vertebrae (see earlier). The whole unit, pelvis and synsacrum together, form a rigid platform for muscles of the hind limb and tail, and is part of the system that helps a bird to maintain a stable, streamlined posture during flight.

The hind limb (Figure 9.3) includes several bones. The **femur** is the proximal bone. It is followed by the longer **tibiotarsus**, which is formed by fusion of the tibia and several tarsals. Articulating anteriorly between the femur and tibiotarsus is the small **patella**. A slender, splint-like **fibula** lies along the lateral surface of the tibiotarsus. Several tarsals and the second, third, and fourth **metatarsals** fuse together to form the **tarsometatarsus**, which bears a protuberance, the **hypotarsus**, on its posteroproximal surface. **Metatarsal I** articulates with the posterodistal surface of the tarsometatarsus. There are four digits in the **pes**, each ending in a claw. The first, or **hallux**, has two **phalanges** and is oriented posteriorly; the three remaining digits point anteriorly. This arrangement, the most common among birds, represents the anisodactyl condition. Other arrangements include the zygodactyl (digits 1 and 4 are reversed), heterodactyl (digits 1 and 2 are reversed), syndactyl (digits 2 and 3 are fused together for much of their length), and pamprodactyl (digits 1 and 4 pivot between facing anteriorly and posteriorly) conditions. Digit 2 has three phalanges, digit 3 has four phalanges, and digit 4 has five phalanges.

Key Terms: Skeleton

acetabular fenestra
acetabulum
angular
antorbital fenestra
articular
auditory meatus
basioccipital
basipterygoid processes
basisphenoid
body of **sternum**
bill (beak)
carina (keel)

carpals
carpometacarpus
caudal vertebrae
caudolateral process
cervical vertebrae
coracoid
coronoid process
dentary
dorsal iliac crest
dorsolateral iliac crest
exoccipitals
femur
fibula
foramen magnum
frontals
furcula
glenoid fossa
hallux
humerus
hypocleidium (hypocleideum; furcular apophysis)
hyoid apparatus
hypotarsus
ilioischiadic foramen
ilium
innominate (coxal bone; os coxae)
ischiopubic fenestra
ischium
jugal
kinesis
lacrimal
lumbar vertebrae
mandible
mandibular foramen
maxilla (as a complex forming the upper jaw)
maxillae (as individual bony elements)
mesethmoid
metacarpal I (alular metacarpal)
metacarpal II (major metacarpal)
metacarpal III (minor metacarpal)
metatarsal I
metatarsals
middle ear
nasals
obturator foramen
occipital
occipital condyle
olecranon process
orbits
palatine
parasphenoid
parietals
paroccipital
patella
pes
phalanges

postorbital process
premaxilla
prokinesis
pterygoid
pubis
pygostyle
quadrate
quadratojugal
radiale
radius
remigeal papillae
ribs
sacral vertebrae
scapula
sclerotic ring
squamosals
sternal notch
sternal rib
sternum
supra-angular
supracoracoideus muscle
supraoccipital
synsacrum
tarsals
tarsometatarsus
thoracic vertebrae
tibiotarsus
triosseal canal
ulna
ulnare
uncinate process
vertebral rib
zygomatic arch

SECTION II: EXTERNAL ANATOMY

The **bill** (or beak) is the most prominent feature of the head (Figure 9.7). It includes the maxilla and mandible, which are covered by a horny sheath or **rhamphotheca**. Open the mouth and note the absence of teeth. The **nares** (sing., **naris**) pierce the maxilla. Immediately posterior to them is the **operculum**, a soft swelling of the integument. The eyes are large, as would be expected in a vertebrate that depends largely on vision. The **external acoustic meatus**, leading to the tympanic membrane of the ear, lies posterior and just ventral to the level of the eye, but is concealed by **feathers**.

The most obvious feature of birds is the presence of feathers (Figure 9.7), which in most birds serve primarily for flight and temperature control and will be examined in more detail later. The **collum** (neck) appears shorter than it really is (see Section I) due to the covering of feathers. Its surfaces are designated by particular terms—for example, jugulum refers to the midventral

surface of the neck. The **forelimbs** are strongly modified to form wings. Spread the wings and note the prominence of the feathers. By contrast, the distal part of the hind limb, including the pes, is covered by horny scales.

The general topographic regions of a bird may be recognized as similar to those of most other vertebrates— for example, the back may be referred to as the dorsum. However, a terminology of more precise designations may be used as well and is especially useful in identification— indeed, Proctor and Lynch (1993: 48) noted that even "a casual birdwatcher must learn the basic topography (external anatomy) of birds to understand field guide descriptions of plumages." Several terms are described here, but the interested reader may refer to Clark (1993) and Proctor and Lynch (1993) for further detail on such regions. The **dorsum trunci** designates the dorsal surface of the trunk, with **dorsum** (interscapular region, back) restricted to the dorsal region of the thorax (approximately the surface between the wings). The dorsal region posterior to it is termed the **pyga** (rump; the region dorsal to the pelvic bones), which includes the base of the **uropygium** (see later), the fleshy part of the **cauda** (tail). The latter includes the dorsal and ventral surfaces, as well as most of the uropygium, and associated feathers. The ventral surface includes the **pectus** (breast), followed by the **venter** (abdomen, belly), with the boundary between them coinciding approximately with the posterior border of the sternum and last rib. The flank or **ilia** (as distinct from the ilium, a bone of the pelvic girdle, noted earlier) is the surface between the posterior part of the venter and pyga. Spread the feathers on the dorsal surface of the uropygium to reveal the **uropygial gland** (Figure 9.11). The gland produces an oily secretion used in preening that protects the feathers. Spread the ventral tail feathers and note the slit-like opening of the **cloaca** (Figure 9.14).

There are several types of feathers in birds, but only the larger **contour feathers** are described here. Contour feathers are those that are typically thought of as feathers, such as the flight feathers on the wings and tail, as well as smaller feathers that cover much of the body. Examine a contour feather (Figure 9.8) from the wing of the pigeon. Note the central **shaft** or quill. The **calamus** is the hollow, basal part of the quill, and the **rachis** is the part bearing **barbs**, the tiny, parallel structures that branch from the rachis. Numerous **barbules** branch out from each barb, and each barbule bears tiny **hooklets** that interlock with hooklets of adjacent barbules to help maintain the shape and structure of a feather. Magnification is required to observe barbules and hooklets. The barbs are arranged in two **vanes**. The vanes are usually symmetrical around the rachis in most contour feathers covering the body, but are asymmetric in most of the flight and, to a lesser degree, tail feathers. In flight feathers, the vane facing the leading edge of the wing is narrower than at the trailing edge.

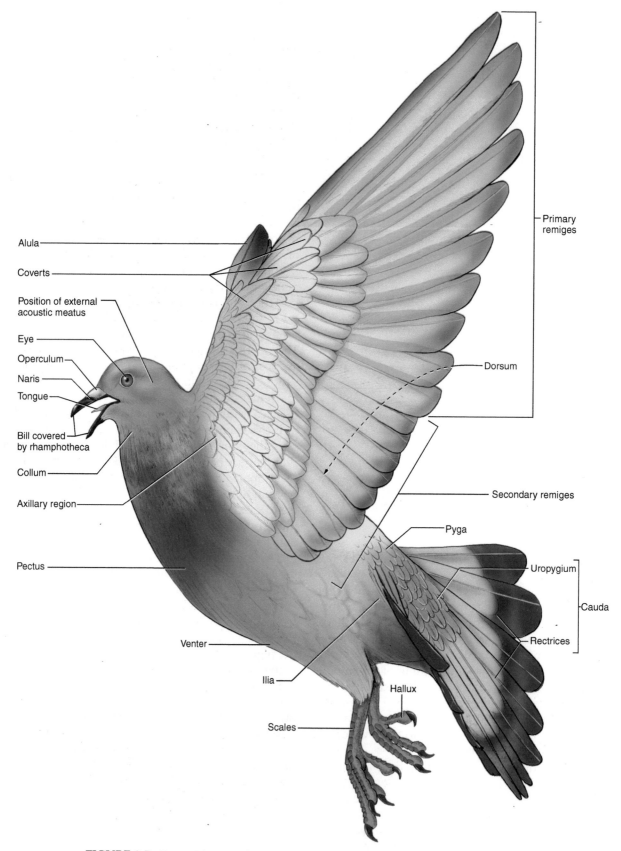

FIGURE 9.7 External features of the pigeon, with forelimb (wing) abducted, in left lateral view.

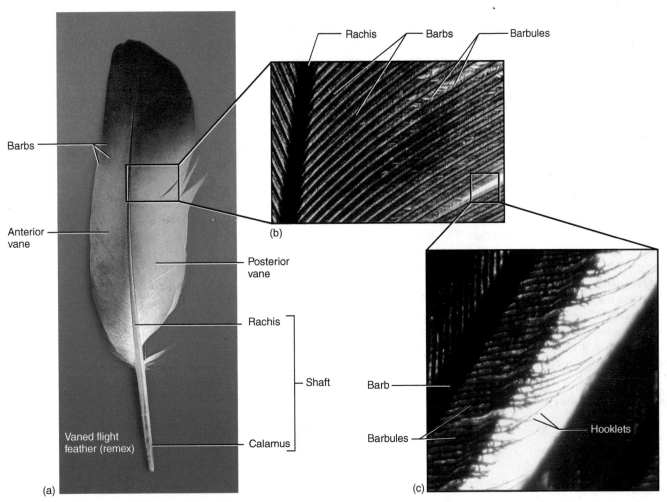

FIGURE 9.8   Feather of the pigeon (a), with blowups (b) and (c) showing successively finer detail of structures.

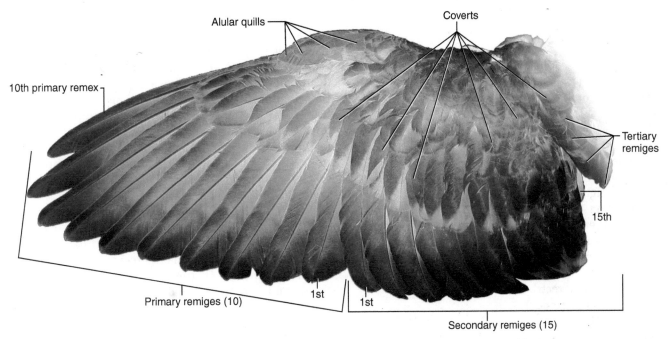

FIGURE 9.9   Extended left wing of the pigeon in dorsal view, showing arrangement of feathers.

THE DISSECTION OF VERTEBRATES

Feathers are designated by their position on the body, and there are numerous different types. Here, only the more typical ones important in flight on the tail (Figure 9.7) and wing (Figure 9.9) are distinguished. The **remiges** (sing., **remex**) are the large feathers of the wing. On the trailing edge of the wing, the **primary remiges** (or simply, primaries) are the feathers attaching to the distal end of the wing, the carpometacarpus and phalanges, and are numbered from 1 to 10 in proximal to distal order. Those attaching to the ulna are the **secondary remiges**, of which there are usually 15, numbered in distal to proximal order. The **tertiary remiges**, usually four in number, lie proximal to the most proximal secondaries. The **coverts**, which vary in size, cover nearly all of the remainder of the wing. The exception is a small patch of (usually four) **alular quills**, which arise from the alular phalanx and together form the **alula** or bastard wing. Though relatively small, the alula is an extremely important flight structure that helps in the avoidance of stalling during low-velocity flight. The feathers of the tail are termed **rectrices** (sing., **rectrix**).

Key Terms: External Anatomy

alula (bastard wing)
alular quills
barbs
barbules
bill (beak)
calamus
cauda (tail)
cloaca
collum (neck)
contour feathers
coverts
dorsum (interscapular region, back)
dorsum trunci
external acoustic meatus
eyes
feathers
forelimbs
hooklets
ilia (flank)
nares (sing., **naris**)
operculum
pectus (breast)
primary remiges
pyga (rump)
rachis
rectrices (sing., **rectrix**)
remiges (sing., **remex**)
rhamphotheca
scales
secondary remiges

shaft
tertiary remiges
uropygial gland
uropygium
vanes
venter (vent, abdomen, belly)

SECTION III: MUSCULATURE

As the flight and wing musculature are among the more interesting features of the pigeon, they are the focus of this dissection. To study the musculature, the skin must be removed. Begin by brushing aside the feathers from the midventral surface of the thoracic region. Note that there are no feathers attached at the midline. Make a longitudinal incision through the skin, but work carefully, as it is quite thin. Extend the incision from the cloaca anteriorly to the anterior part of the breast, and then spread the skin, separating it from the underlying tissue using a blunt probe. In most places the skin will come off readily, but in some spots muscular slips attach to the skin and must be cut through. You will reveal the **pectoralis** (Figure 9.10), the largest and most superficial of the flight muscles. At the anterior part of the pectoralis lies the thin-walled **crop**, a sac-like specialization of the **esophagus** (see later) that is used to store food. Carefully continue the incision anteriorly, being careful not to injure the crop, to just below the bill. Skin the neck, but do not damage the vessels.

Several folds of skin form the wing surface. The large fold between the shoulder and carpus, forming the leading edge of the wing, is the **propatagium**. Several delicate muscles and tendons lie within it. A large **postpatagium** projects posteriorly from the ulna. With the wing outstretched, gently push back on the leading edge of the propatagium. The tension you feel is due to the **long tendon** (see later). Begin skinning the wing ventrally, near the central part of the brachium, and work your way anteriorly. As you approach the leading edge, be watchful of the tendon, a thin, whitish strand, and uncover it. Proceed to skin the rest of the wing up to the carpus. Then continue skinning the dorsal surface of the wing and body.

The dominance of the pectoralis can now be appreciated. It arises from the sternum and, being the main depressor of the wing, inserts mainly on the humerus. The shoulder is covered mainly by several parts of the **deltoideus** muscle (Figures 9.9 and 9.11). Anteriorly, the deltoid is represented by the **pars propatagialis**, a single hypertrophied belly consisting of **long** and **short heads**. The long head is the smaller, anterior, sheet-like portion. The more posterior short head is thicker. The long head is just anterior to a small, triangular slip of

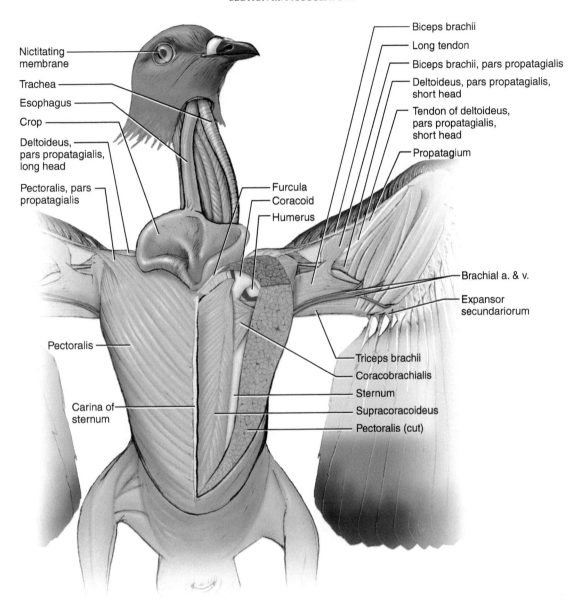

FIGURE 9.10   Pectoral and wing musculature of the pigeon in ventral view. Left-side pectoralis muscle cut away to show underlying structures.

the pectoralis that inserts on the common belly of the pars propatagialis. The tendon of the long head, the long tendon noted earlier, extends laterally toward the carpus and divides into two separate tendons, but these may be difficult to discern. Distally, the short head forms a tendinous sheet that mainly spreads to cover the muscles on the antebrachium. The **biceps brachii** muscle covers the anteromedial part of the brachium. The **pars propatagialis** of the biceps brachii is a slip of the biceps and arises from its anterior edge. Anteriorly, the pars propatagialis fuses with the tendon of the long tendon. Its distal end gives rise to a tendon that angles distally to fuse with one of the divisions of the long tendon. The wing muscles described in this paragraph, except the biceps brachii, serve mainly to alter tension of the wing surface.

They are commonly referred to as tensors and play a significant role in flight aerodynamics.

Several other wing muscles may be noted. Just posterior to the short head of the pars propatagialis of the deltoideus is the **deltoideus major**, which pulls the humerus medially and posteriorly. The triceps brachii has three portions, but only two of them, the **scapulotriceps** and **humerotriceps**, can be readily identified. The third, the coracotriceps, is a small muscle arising from the tendon of the **expansor secundariorum**. This tendon extends from the axilla, or armpit, along the posterior edge of the triceps brachii. Carefully dissect the margin of the postpatagium to find the small and delicate muscular portion, which fans onto the quills of the secondary remiges. The expansor secundariorum acts to spread the secondaries.

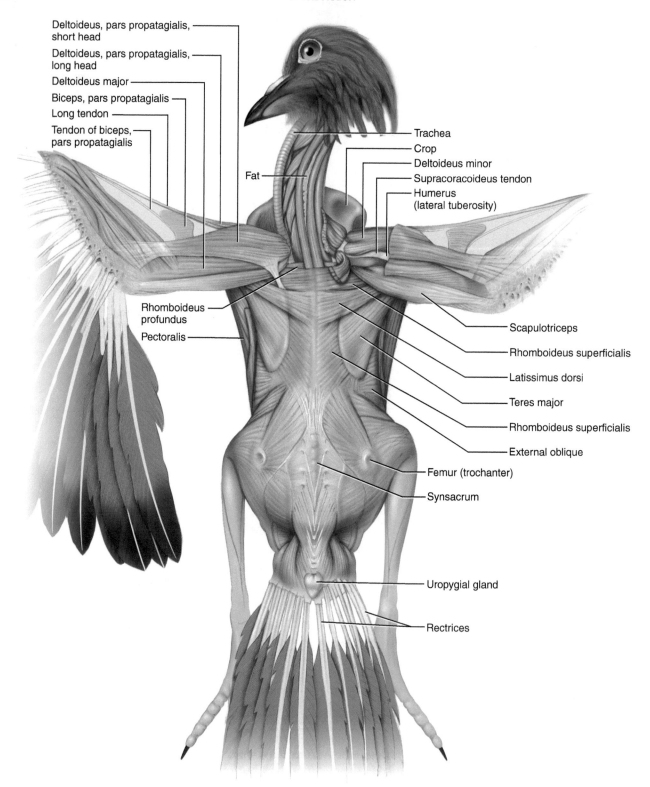

Deltoideus, pars propatagialis, short head

Deltoideus, pars propatagialis, long head

Deltoideus major

Biceps, pars propatagialis

Long tendon

Tendon of biceps, pars propatagialis

Fat

Trachea

Crop

Deltoideus minor

Supracoracoideus tendon

Humerus (lateral tuberosity)

Rhomboideus profundus

Pectoralis

Scapulotriceps

Rhomboideus superficialis

Latissimus dorsi

Teres major

Rhomboideus superficialis

External oblique

Femur (trochanter)

Synsacrum

Uropygial gland

Rectrices

FIGURE 9.11   Musculature of the pigeon in dorsal view.

The main elevator of the wing is the **supracoracoideus**, which lies deep to and is covered entirely by the pectoralis. To expose it, make a cut through the pectoralis near its center and at right angles to the fiber direction, but do so only a few millimeters at a time. Spread the incision as you cut. This will help avoid damaging the underlying supracoracoideus. Once you have cut through the pectoralis and identified the supracoracoideus, cut out a large portion of the pectoralis to expose the supracoracoideus (Figures 9.9 and 9.12). Its fibers

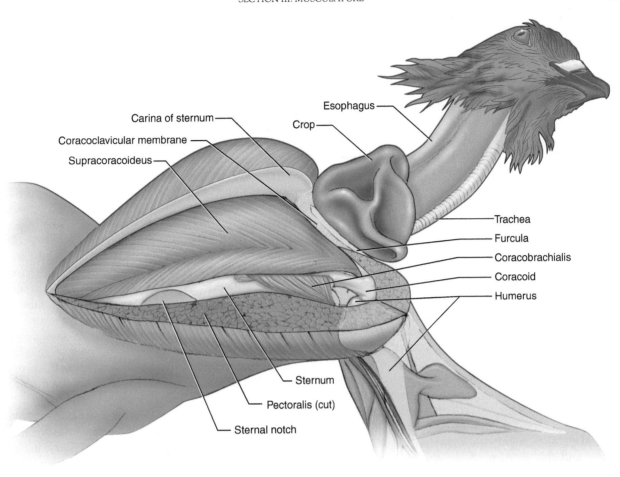

(a)

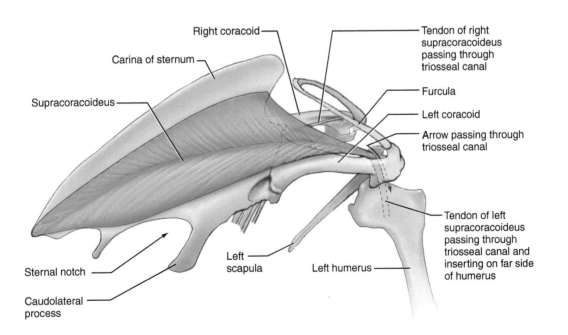

(b)

FIGURE 9.12 Pectoral musculature of the pigeon in left ventrolateral view: (a) left-side pectoralis muscle cut away; and (b) sternum and appendicular skeleton isolated to show supracoracoideus muscle and the course of its tendon through the triosseal canal.

converge toward the middle of the muscle and extend to a stout anterolaterally directed tendon that passes through the triosseal canal onto the dorsal surface of the humerus. This acts as a pulley system, through which the wing is elevated mainly by a muscle, the supracoracoideus, that lies ventral to the wing. This arrangement enhances stability for flight by maintaining the center of gravity below the wings. To see the tendon of the supracoracoideus (Figures 9.9–9.12), cut the deltoideus major and the pars propatagialis of the deltoideus near their center and reflect them. You will uncover the **deltoideus minor**, which contributes to elevation of the wing. The tendon passes just posterior to the posterior margin of the deltoideus minor.

Other musculature that may be noted without much further effort includes the **rhomboideus superficialis** and **rhomboideus profundus**, which extend between vertebral spines and the scapula. The superficialis covers all but the anteriormost portion of the profundus. The **latissimus dorsi** is a relatively small muscle extending laterally from the middorsal line. These muscles form part of the complex of muscles that help support and stabilize the shoulder. Although there are several other back muscles, an interesting condition in birds is the relative lack or smaller size of the back musculature as compared, for example, with that of the cat. Strength and stability of the trunk is not compromised, however, as the skeletal elements are firmly fused together (see earlier), and this in turn allows the musculature to be reduced to minimize weight.

Key Terms: Musculature

biceps brachii
crop
deltoideus
deltoideus major
deltoideus minor
esophagus
expansor secundariorum
humerotriceps
latissimus dorsi
long tendon of long head, pars propatagialis of deltoideus
pars propatagialis of biceps brachii (biceps slip, biceps propatagialis, patagial accessory)
pars propatagialis, long and short heads of deltoideus
pectoralis
postpatagium
propatagium
rhomboideus profundus
rhomboideus superficialis
scapulotriceps
supracoracoideus

SECTION IV: BODY CAVITY, VISCERA, AND VESSELS

The sternum and pectoral musculature must be removed to examine the structures within the body cavity, including viscera and blood vessels. Unfortunately, removal of the sternum and muscles as next described often obliterates much of the extensive respiratory system, the **air sacs** that form auxiliary air pathways in the avian respiratory system. This system was noted earlier in connection with the continuation of this system into several bones of the body. In birds the respiratory system is arranged to produce a continuous stream of air through the lungs, which is made possible by the presence of the air sacs. By contrast, the system in other air-breathing vertebrates, in which air passes into and out of the lungs through the same pathway, results in a residual volume of air in the lungs. The avian system is considerably more efficient and allows the high metabolic levels required for sustained flight. The system is also involved in cooling. Although some of the air sacs can be seen when the body cavity is exposed, the system is best seen and appreciated in a prepared specimen. If a preparation is available, use Figure 9.13 to identify the major air sacs (see also Duncker, 1971, 2004).

Return to the task of exposing the body cavity. Clear the connective tissue from the lateral part of the pectoral muscles, just posterior to the axillary regions, but do not damage any vessels. Make a longitudinal incision through the abdominal muscles, just to one side of the midventral line, and using stout scissors, follow anteriorly along the lateral margin of one of the pectoral muscles, cutting through the ribs as you do so. Lift the sternum as much as possible and clear the connecting tissue until you see the vessels that pass to the sternum so that you avoid damaging them. Using a sharp scalpel, cut through the anterior part of the pectorals, as shown in Figure 9.14. You will also need to cut (with stout scissors) through the furcula and coracoid. Repeat this procedure for the pectorals on the other side. You will then be able to lift the sternum with the pectoral musculature attached to it. Remove the sternum by clearing the connective tissue between it and the deeper structures. Cut through the vessels, the **pectoral arteries** and **veins** (Figure 9.14), extending to the breast as close to the sternum as possible.

Note the **heart** (Figure 9.14) lying anteriorly on the midline. Its four main chambers include the **left** and **right atria** (sing., **atrium**) and **left** and **right ventricles** (a very small **sinus venosus** is also present but will not be identified). The **lungs**, right and left, are tucked laterally to the heart. Probe to find them, and partially remove the serosa covering them to see their spongy texture. Posterior to the heart are the lobes of the **liver**; note that the right lobe is considerably larger than the left.

9

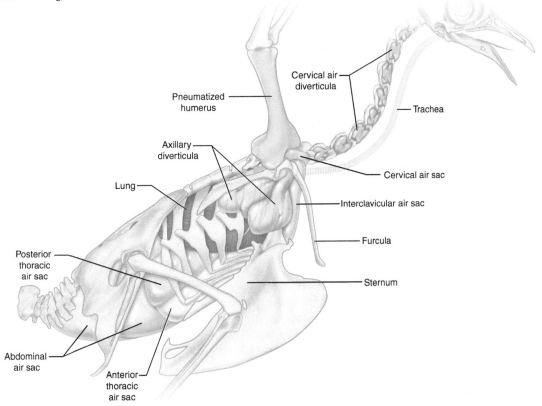

(a) The air sac system of the pigeon,
at maximal filling, in lateral view

Cervical air
diverticula

Pneumatized
humerus

Trachea

Axillary
diverticula

Cervical air sac

Lung

Interclavicular air sac

Furcula

Posterior
thoracic
air sac

Sternum

Abdominal
air sac

Anterior
thoracic
air sac

FIGURE 9.13   The air sac system of the pigeon in
(a) left lateral and (b) ventral views.

Tucked deep to these lobes, on the left side, is the **gizzard**, the very muscular, distal portion of the complex, two-part **stomach** of birds. The glandular **proventriculus**, the anterior part of the stomach (though it is actually a modification of the distal part of the esophagus), continues posteriorly from the **esophagus** and lies deep to the heart, as does the posterior part of the **trachea**. These structures will be seen shortly (Figure 9.15). The highly coiled **small intestine** is relatively long, in contrast to the short and straight **large intestine** (Figures 9.16 and 9.17). The division between these structures is marked by the presence of a pair of small diverticula, the **colic ceca**.

Return to the heart (Figure 9.14). The **ascending aorta** arches anteriorly from the left ventricle and gives off the paired **brachiocephalic arteries** before curving to the right and posteriorly as the **aortic arch**. Each brachiocephalic soon divides into a smaller **common carotid artery** and a larger **subclavian artery** (Figures 9.14 and 9.15). The common carotid is a short artery that extends anteriorly from the base of the neck, where it divides into several arteries, such as the vertebral trunk and superficial cervical artery (which are not illustrated here), as well as its most apparent branch, the **internal carotid artery**. The **external carotid artery** (not illustrated here) in the

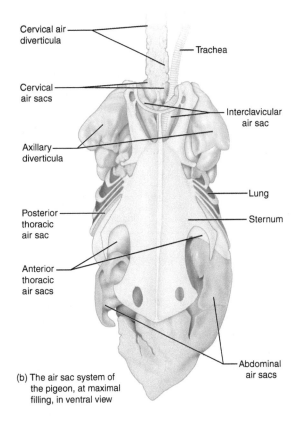

Cervical air
diverticula

Trachea

Cervical
air sacs

Interclavicular
air sac

Axillary
diverticula

Lung

Posterior
thoracic
air sac

Sternum

Anterior
thoracic
air sacs

Abdominal
air sacs

(b) The air sac system of
the pigeon, at maximal
filling, in ventral view

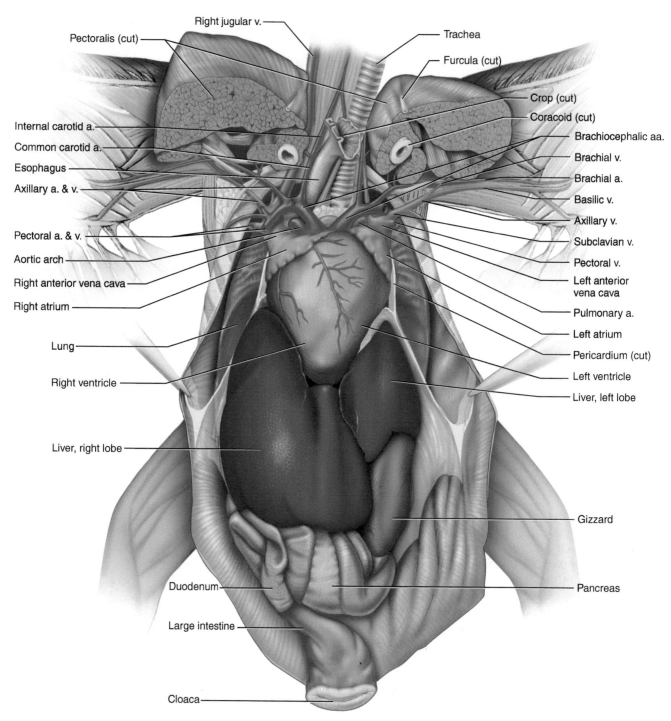

FIGURE 9.14   Trunk of the pigeon in ventral view. Pectoral musculature and sternum removed to reveal heart, viscera, and vessels.

adult is a branch of the anterior end of the internal carotid artery (but not in the fetus; see Baumel, 1993); note that this is not the condition described for the cat, in which the principal branches of the common carotid artery are the internal and external carotid arteries. In the pigeon the internal carotid artery extends anteriorly along the midventral surface of the neck, close beside the internal carotid from the other side of the body. These vessels converge and continue together, deep to the musculature, toward the base of the neck. Each gives off several branches, including the external carotid artery, as noted earlier. The vertebral trunk, one of the branches of the common carotid, gives rise to several, mainly anastomosing, vessels that ascend the neck dorsal to the internal carotid, but it is not necessary to attempt to trace these (for further details of the pattern of vessels, see Baumel,

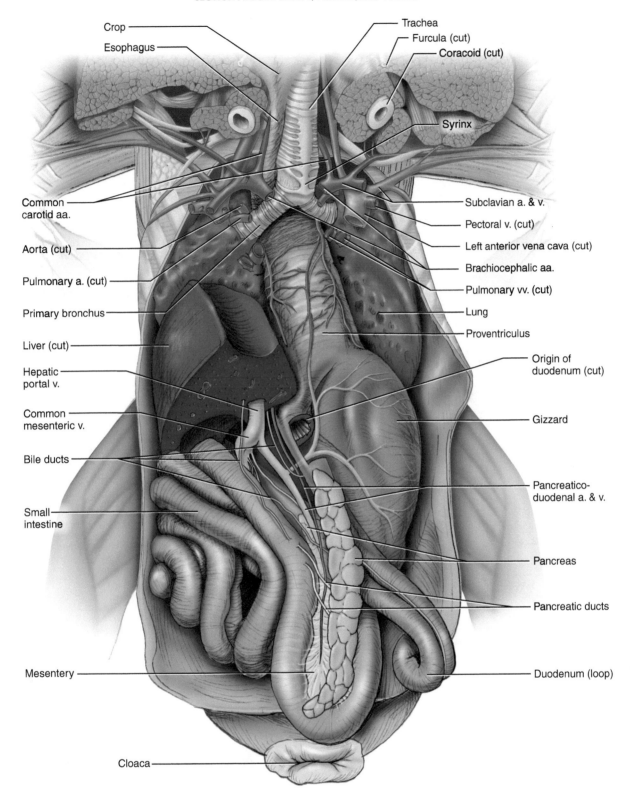

FIGURE 9.15 Trunk of the pigeon in ventral view. Pectoral musculature, sternum, and heart removed to reveal viscera and vessels.

1993). The subclavian artery extends laterally for a short distance (after the origin of the common carotid) before subdividing into the **axillary artery** and pectoral artery, noted already. The latter is a large vessel that quickly branches into several smaller arteries to supply the extensive pectoral musculature. The axillary artery gives off several branches before continuing into the arm as the **brachial artery**.

Most regions supplied by these arteries are drained by veins that ultimately enter the **right** or **left anterior**

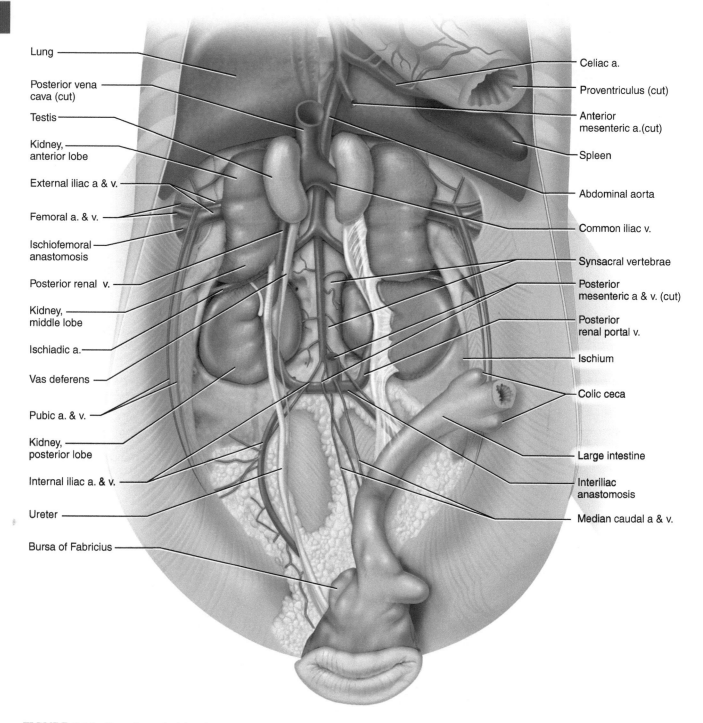

FIGURE 9.16 Posterior end of the pleuroperitoneal cavity of the male pigeon in ventral view. Gizzard and small intestine removed to show urogenital structures, viscera, and vessels.

vena cava, each of which is formed by the confluence of three large vessels, the **jugular**, **subclavian**, and **pectoral** veins. The jugular lies along the lateral surface of the neck, draining the head and neck. The right jugular is usually larger than the left. The subclavian vein is a short segment that mainly receives the **axillary vein**, which in turn receives the **basilic** vein from the arm. It is the basilic and not the **brachial vein** that follows the brachial artery. The brachial vein is represented by an anterior branch of the axillary that divides into two narrow vessels along the anterior margin of the brachial muscle. The pectoral vein, which divides further into branches much as those of the pectoral artery, usually enters the anterior vena

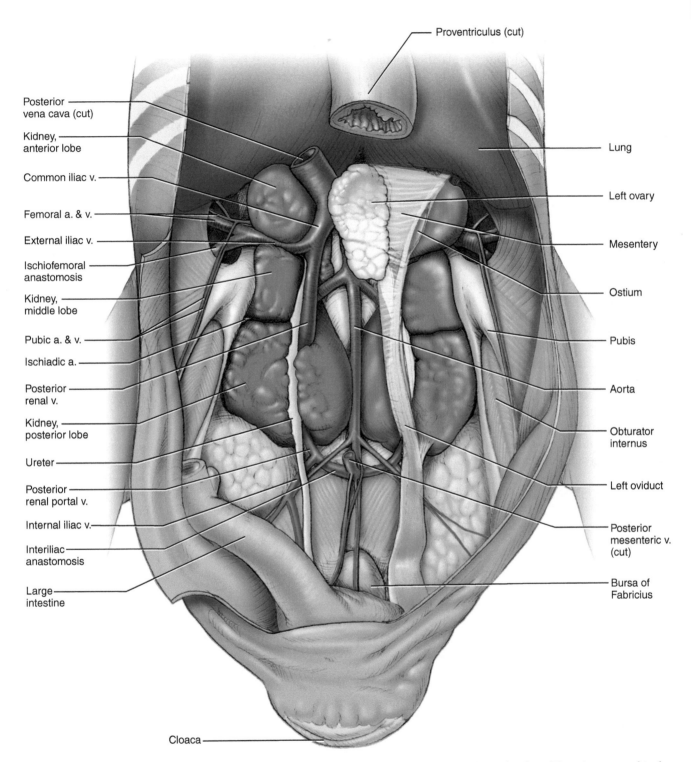

Proventriculus (cut)

Posterior
vena cava (cut)

Kidney,
anterior lobe

Common iliac v.

Femoral a. & v.

External iliac v.

Ischiofemoral
anastomosis

Kidney,
middle lobe

Pubic a. & v.

Ischiadic a.

Posterior
renal v.

Kidney,
posterior lobe

Ureter

Posterior
renal portal v.

Internal iliac v.

Interiliac
anastomosis

Large
intestine

Cloaca

Lung

Left ovary

Mesentery

Ostium

Pubis

Aorta

Obturator
internus

Left oviduct

Posterior
mesenteric v.
(cut)

Bursa of
Fabricius

FIGURE 9.17   Posterior end of the pleuroperitoneal cavity of the female pigeon in ventral view. Gizzard and small intestine removed to show urogenital structures, viscera, and vessels.

cava but enters the subclavian in some specimens. The right anterior vena cava proceeds almost directly posteriorly to enter the sinus venosus, but the left anterior vena cava turns right and crosses the heart to enter the sinus venosus. The **posterior vena cava** (Figures 9.16–9.18) is a large vessel that drains the posterior part of the body. It can be found by lifting the lateral margin of the right atrium. Its branches will be followed shortly.

The **pulmonary trunk** leaves the right ventricle and almost immediately splits into **left** and **right pulmonary arteries** (Figures 9.14 and 9.15) to the lungs. The left pulmonary artery is easily apparent and can be found passing dorsal to the left brachiocephalic artery. There are two **right** and two **left pulmonary veins** that return blood from the lungs. These vessels enter the left atrium separately. They are not easy to find, but will be seen when the heart is removed (Figure 9.15).

Lift the heart and note the great vessels, just described, associated with it. Cut each vessel, and remove the heart. This will expose the lungs and allow you to follow the trachea and esophagus posteriorly (Figure 9.15). The trachea is held open by cartilaginous rings. More posteriorly it bifurcates into left and right **bronchi**, which extend into the lungs. The **syrinx** is the sound-producing organ located at the base of the trachea. Examine the cut vessels, and identify the pulmonary arteries and pulmonary veins. The latter exit the lungs just posterior to the entrance of the bronchi. There are two main veins on each side but it may be difficult to discern them. The pulmonary arteries enter the lungs somewhat anterior to the bronchi.

Remove the left lobe of the liver, being careful not to injure vessels external to it. Once it has been removed, you will be able to follow the esophagus to the stomach. In birds the stomach is a complex, two-part organ. The more proximal proventriculus secretes strong hydrochloric acid and digestive enzymes that begin the chemical digestion of food. The thicker, muscular gizzard performs most of the mechanical breakdown of food.

Next, examine the intestines. Note that they are bound by mesentery. Several vessels will be seen either supplying or draining the viscera. For the time being, identify the large **hepatic portal vein** (Figure 9.15) that extends between the viscera and the right lobe of the liver. Lift the lobe to note the entrance of the vein into the liver. Cut the portion (and only that portion) of the right lobe that is posterior to the entrance of the hepatic portal vein. You will now be able to follow the intestinal tract more easily. The **duodenum**, the first part of the **small intestine**, arises from the junction of the proventriculus and gizzard. It makes a long loop, consisting of ascending and descending portions that lie close together. Between them is the narrow, elongated **pancreas**. Follow the remaining part of the small intestine. It passes into the

short and straight large intestine, which continues posteriorly into the cloaca. A pair of small **colic ceca** (sing., **cecum**) marks the division between the small and large intestines.

Return to the duodenum. Carefully dissect the connective tissue between its ascending and descending portions and the pancreas. You will reveal several structures, including the bile ducts and pancreatic ducts (Figure 9.15). There are three of the latter, but only two are readily apparent. Also, you will reveal a large vein, the **pancreaticoduodenal vein**, accompanied by the smaller **pancreaticoduodenal artery**, associated with the pancreas.

Carefully uncoil the intestines to observe that the **posterior mesenteric vein**, a branch of the hepatic portal system, extends posteriorly and eventually joins the venous system of the posterior abdominal cavity (see Figures 9.16–9.18). Once the hepatic portal system has been studied, find the point where the posterior vena cava, which returns blood from most of the body posterior to the heart, passes through the left lobe of the liver. This occurs dorsal to the entrance of the hepatic portal vein. Cut through the vena cava and remove the remainder of the liver. Cut through the distal end of the proventriculus, just dorsal to its union with the gizzard. Find the **celiac** and **anterior mesenteric arteries** (Figure 9.16) as they emerge from the aorta, just to the right of the distal end of the proventriculus. The ovoid, dark **spleen** lies in this region as well and can vary in size. Follow the celiac artery, noting its main branches. Cut through them and the anterior mesenteric artery. Also cut through the posterior end of the small intestine, just anterior to the colic ceca, and the posterior mesenteric vein. Then remove the digestive tract.

Clear away connective tissue to expose the urogenital structures and the vessels lying on the dorsal wall of the abdominal cavity. Each **kidney** is superficially subdivided into three lobes by the vessels that pass through it. Males (Figure 9.16) possess a pair of approximately bean-shaped **testes**. Each lies on the anterior lobe of a kidney. In nonbreeding males, the testes are much smaller than in breeding males and may be difficult to identify. The **ductus deferens** carries semen from the testis to the cloaca. The ductus deferens leaves, slightly expanded, the dorsomedial side of the testis and extends posteriorly to the cloaca as a relatively straight, narrow tube. It is slightly wider and may be convoluted in breeding males. It passes for most of its length along the medial side of the **ureter**, the wider tube that carries urine from the kidney to the cloaca. The ureter emerges from between the anterior and middle lobes of the kidney. Near the cloaca, the ductus deferens crosses the surface of the ureter and extends to the cloaca, lateral to the ureter.

In females (Figure 9.17) only the left-side reproductive organs are present, those of the right side having degenerated soon after their initial formation. The **ovary** lies on the anterior end of the left kidney. Its morphology and size varies in accordance with the breeding season. The ovary contains numerous spherical **follicles** in various stages of development. Mature ova pass from the ovary and enter the oviduct through its anterior opening, the **ostium**. The **oviduct** is a relatively straight tubular structure (Figure 9.17) that becomes large and convoluted. It conducts the ovum to the cloaca and is subdivided into glandular portions that perform specific roles (such as secreting the shell membrane or the shell itself), but these portions are not identifiable grossly. The left and right ureters of the female are in the same position as in the male. A vestigial posterior portion of the right oviduct may be present along the posteromedial end of the right ureter.

Examine the region where the ureters and genital ducts enter the cloaca and note the **bursa of Fabricius** (Figures 9.16 and 9.17). The bursa is a dorsal outpocketing of the cloaca and has an important role in the immune system, particularly in young individuals, in which it is relatively large.

Return to the posterior vena cava and follow it posteriorly (Figures 9.16–9.18). It is formed at the level of

the anterior part of the kidneys by the confluence of the **right** and **left common iliac veins**. An **anterior renal vein**, draining the anterior lobe of the kidney, enters each common iliac. The common iliac turns laterally between the anterior and medial lobes of the kidney. As it does so, it receives the **posterior renal vein**, a large vessel that extends posteriorly on the kidney with the ureter and ductus deferens in the male and the oviduct (left side only) in the female. The posterior renal vein arises from several branches in the posterior lobe of the kidney and receives a branch from the middle lobe of the kidney.

The **posterior renal portal vein** enters the posterior lobe of the kidney. To follow its path, as well as that of many other veins in this region, kidney tissue must be removed (Figure 9.18). Begin by exposing the posterior portions of the posterior renal vein. Then follow the posterior renal portal vein as it passes anteriorly through the kidney, deep to the posterior renal vein. Between the posterior and middle lobes of the kidney, the renal portal receives the **ischiadic vein**, which is the main vein of the hind limb. The paired **ischiadic arteries** lie ventral to the ischiadic veins and are easily seen. Trace the origin of one from the descending aorta. The artery supplies

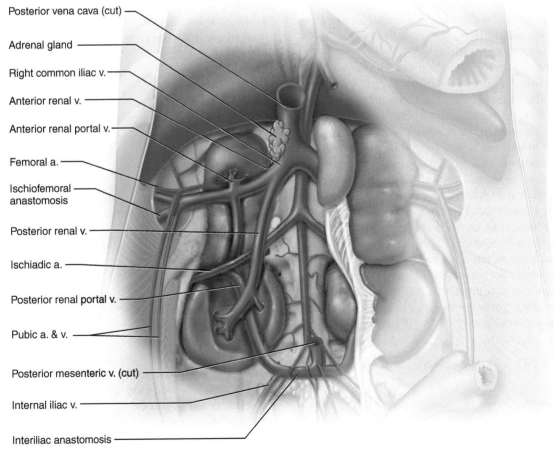

Posterior vena cava (cut)

Adrenal gland

Right common iliac v.

Anterior renal v.

Anterior renal portal v.

Femoral a.

Ischiofemoral anastomosis

Posterior renal v.

Ischiadic a.

Posterior renal portal v.

Pubic a. & v.

Posterior mesenteric v. (cut)

Internal iliac v.

Interiliac anastomosis

FIGURE 9.18   Posterior end of the pleuroperitoneal cavity of the (male) pigeon in ventral view. The right-side kidney has been dissected to reveal the pattern of renal and renal portal veins.

the middle and posterior lobes of the kidney. Follow the ischiadic vein laterally, but do not damage the superficial and narrow **pubic vein** and **artery** lying on the musculature along the pubic bone. The ischiadic vein, accompanied by the ischiadic artery, passes through the ilioischiadic foramen (Figure 9.6) as it enters the abdominal cavity from the hind limb.

Continue to follow the posterior renal portal vein. Between the anterior and middle lobes of the kidney, the renal portal vein unites with the external iliac vein to form the common iliac vein. The **anterior renal portal vein** enters almost directly opposite the entrance of the posterior renal portal vein. Follow the external iliac vein laterally to its tributary, the **femoral vein**, which drains most of the anterior part of the thigh. The main vein of the hind limb is the ischiadic vein, but most of the blood is diverted to the femoral vein by the **ischiofemoral anastomosis**, a large vessel between the ischiadic and femoral veins. It appears as a large branch, larger indeed than the femoral, passing posteriorly deep to musculature. You may follow it by cutting through the musculature. Lastly, note the pubic vein, draining the lateral abdominal wall, near the origin of the external iliac. The **external iliac artery** lies deep to the external iliac vein. Carefully probe to find it. It gives rise to the **femoral artery**, supplying the hind limb, and the pubic artery, supplying the lateral abdominal wall. Trace the external iliac artery to its origin from the descending aorta. Farther anteriorly, note the **anterior renal artery**, which supplies the anterior and middle lobes of the kidney.

Examine the posterior end of the abdominal cavity. The posterior renal portal vein is formed by the confluence of the narrower **internal iliac vein** and the wider, transversely oriented **interiliac anastomosis**. The posterior mesenteric vein, which was cut during removal of the intestines, arises from the middle of the interiliac anastomosis and passes anteriorly as part of the hepatic portal system. A small branch accompanies the **posterior mesenteric artery** to the posterior end of the large intestine. The narrow **median caudal vein** enters the interiliac anastomosis opposite the origin of the posterior mesenteric vein. The internal iliac vein drains the posterolateral region of the abdominal cavity. This vessel is accompanied by the **internal iliac artery**. Beyond the origin of the internal iliac arteries, the descending aorta continues posteriorly as the **median caudal artery**.

Key Terms: Body Cavity, Viscera, and Vessels

air sacs
anterior mesenteric artery
anterior renal artery
anterior renal portal vein
anterior renal vein

anterior vena cava, right and **left**
aortic arch
ascending aorta
atrium (pl., atria), right and **left**
axillary artery
axillary vein
basilic vein
brachial artery
brachial vein
brachiocephalic artery, right and **left**
bronchi
bursa of Fabricius (bursa Fabricii)
celiac artery
colic ceca
common carotid artery
common iliac vein, right and **left**
ductus deferens
duodenum
esophagus
external carotid artery
external iliac artery
femoral artery
femoral vein
follicles
gizzard
heart
hepatic portal vein
interiliac anastomosis
internal carotid artery
internal iliac artery
internal iliac vein
ischiadic artery
ischiadic vein
ischiofemoral anastomosis
jugular vein
kidney
large intestine
liver
lungs
median caudal artery
median caudal vein
ostium
ovary
oviduct
pancreas
pancreaticoduodenal artery
pancreaticoduodenal vein
pectoral artery
pectoral vein
posterior mesenteric artery
posterior mesenteric vein
posterior renal portal vein
posterior renal vein
posterior vena cava
proventriculus

pubic artery
pubic vein
pulmonary artery, right and left
pulmonary trunk
pulmonary vein, right and left
sinus venosus
small intestine
spleen

stomach
subclavian artery
subclavian vein
syrinx
testes
trachea
ureter
ventricle, right and left

9

References

Ahlberg, P. E., & Clack, J. A. (2006). A firm step from water to land. *Nature, 440*(6), 747–749.

Alderfer, J., & Dunn, J. L. (2007). *National geographic birding essentials: All the tools, techniques, and tips you need to begin and become a better birder.* Washington: National Geographic, 224.

Baron, M. G., Norman, D. B., & Barrett, P. M. (2017a). A new hypothesis of dinosaur relationships and early dinosaur evolution. *Nature, 543,* 501–506. https://doi.org/10.1038/nature21700.

Baron, M. G., Norman, D. B., & Barrett, P. M. (2017b). Baron et al. reply. *Nature, 551*(7678), E4–E5. https://doi.org/10.1038/nature24012.

Baskin, D. G., & Detmers, P. A. (1976). Electron microscopic study on the gill bars of amphioxus (*Branchiostoma californiense*) with special reference to neurociliary control. *Cell and Tissue Research, 166*(2), 167–178.

Baumel, J. J. (1993). Systema cardiovasculare. In J. J. Baumel, A. S. King, J. E. Breazile, H. E. Evans, & J. C. Vanden Berge (Eds.), *Handbook of avian anatomy: Nomina anatomica avium* (2nd ed.) (pp. 407–475). Cambridge: Publications of the Nuttall Ornithological Club 23 (pp. 779).

Baumel, J. J., & Witmer, L. M. (1993). Osteologia. In J. J. Baumel, A. S. King, J. E. Breazile, H. E. Evans, & J. C. Vanden Berge (Eds.), *Handbook of avian anatomy: Nomina anatomica avium* (2nd ed.) (pp. 45–132). Cambridge: Publications of the Nuttall Ornithological Club 23 (pp. 779).

Benton, M. (2015). *Vertebrate palaeontology* (4th ed.). Chichester: Wiley Blackwell, 468.

Besoluk, K., & Tipirdamaz, S. (2001). Comparative macroanatomic investigations of the venous drainage of the heart in Akkaraman sheep and Angora goats. *Anatomia, Histologia, Embryologia, 30,* 249–252.

Bever, G. S., Lyson, T. R., Field, D. J., & Bhullar, B.-A. S. (2016). The amniote temporal roof and the diapsid origin of the turtle skull. *Zoology, 119*(6), 471–473.

Biasutto, S. N., Caussa, L. I., & Cirado del Rio, L. E. (2006). Teaching anatomy: Cadavers vs. computers? *Annals of Anatomy, 188,* 187–190.

Bock, W. J. (2013). The furcula and the evolution of avian flight. *Paleontological Journal, 47*(11), 1236–1244.

Brazeau, M. D. (2009). The braincase and jaws of a Devonian 'acanthodian' and modern gnathostome origins. *Nature, 457,* 305–308.

Brazeau, M. D., & Friedman, M. (2014). The characters of Palaeozoic jawed vertebrates. *Zoological Journal of the Linnean Society, 170,* 779–821.

Brazeau, M. D., & Friedman, M. (2015). The origin and early phylogenetic history of jawed vertebrates. *Nature, 520,* 490–497.

Brochu, C. A. (1999). Phylogenetics, taxonomy, and historical biogeography of Alligatoroidea. In T. Rowe, C. A. Brochu, & K. Kishi (Eds.), *Cranial morphology of Alligator mississippiensis and phylogeny of alligatoroidea Journal of Vertebrate Paleontology: Vol. 19.* (pp. 9–100) Supplement to Number 2 (Society of Vertebrate Paleontology Memoir 6).

Brochu, C. A. (2002). Osteology of *Tyrannosaurus rex*: Insights from a nearly complete skeleton and high-resolution computed tomographic analysis of the skull. Society of Vertebrate Paleontology Memoir 7 *Supplement to Journal of Vertebrate Paleontology, 22*(2), 1–138.

Brunetti, R., Gissi, C., Pennati, R., Caicci, F., Gasparini, F., & Manni, L. (2015). Morphological evidence that the molecularly determined *Ciona intestinalis* type A and type B are different species: *Ciona robusta* and *Ciona intestinalis. Journal of Zoological Systematics and Evolutionary Research, 53,* 186–193.

Brusca, R. C., Moore, W., & Shuster, S. M. (2016). *Invertebrates* (3rd ed.). Sunderland: Sinauer Associates Inc. 1104 p.

Budd, G. E., & Olsson, L. (2007). Editorial: A renaissance for evolutionary morphology. *Acta Zoologica, 88,* 1.

de Buffrénil, V., Clarac, F., Fau, M., Martin, S., Martin, B., Pellé, E., et al. (2015). Differentiation and growth of bone ornamentation in vertebrates: A comparative histological study among the crocodylomorpha. *Journal Morphology, 276,* 425–445.

Burighel, P., & Cloney, R. A. (1997). Urochordata: Ascidiacea. In F. W. Harrison, & E. E. Ruppert (Eds.), *Microscopic anatomy of invertebrates Hemichordata, Chaetognatha, and the invertebrate chordates: Vol. 15.* (pp. 221–347). New York: Wiley-Liss.

Burrow, C., Hu, Y., & Young, G. (2016). Placoderms and the evolutionary origin of teeth: A comment on rücklin & donoghue (2015). *Biology Letters, 12.* 20160159. https://doi.org/10.1098/rsbl.2016.0159.

Carr, T. D. (1999). Craniofacial ontogeny in tyrannosauridae (dinosauria, coelurosauria). *Journal of Vertebrate Paleontology, 19*(3), 497–520.

Carrier, J. C., Pratt, H. L., Jr., & Castro, J. I. (2004). Reproductive biology of elasmobranchs. In J. C. Carrier, J. A. Musick, & M. R. Heithus (Eds.), *Biology of sharks and their relatives* (pp. 269–286). Boca Raton, FL: CRC Press.

Charan, N. B., Thompson, W. H., & Carvalho, P. (2007). Functional anatomy of bronchial veins. *Pulmonary Pharmacology and Therapeutics, 20,* 100–103.

Charan, N. B., Turk, M. G., & Dhand, R. (1984). Gross and subgross anatomy of bronchial circulation in sheep. *Journal of Applied Physiology, 57*(3), 658–664.

Chiasson, R. B. (1962). *Laboratory anatomy of the alligator.* Dubuque: W. C. Brown Company. 56 p.

Chiasson, R. B. (1966). *Laboratory anatomy of the perch.* Dubuque: Wm. C. Brown Publishers. 53 p.

Chiasson, R. (1984). *Laboratory anatomy of the pigeon* (3rd ed.). Dubuque: Wm. C. Brown Publishers. 104 p.

Chiasson, R. B., & Radke, W. J. (1993). *Laboratory anatomy of the vertebrates.* Dubuque: Wm. C. Brown Publishers. 224 p.

Chiba, S., Sasaki, A., Nakayama, A., Takamura, K., & Satoh, N. (2004). Development of *Ciona intestinalis* juveniles (through 2nd ascidian stage). *Zoological Science, 21*(3), 285–298.

Clack, J. A. (2012). *Gaining ground The origin and evolution of tetrapods* (2nd ed.). Bloomington: Indiana University Press. 523 p.

Clark, G. A., Jr. (1993). Anatomia topographica externa. In J. J. Baumel, A. S. King, J. E. Breazile, H. E. Evans, & J. C. Vanden Berge (Eds.), *Handbook of avian anatomy: Nomina anatomica avium* (2nd ed.) (pp. 7–17). Cambridge Massachusetts: Publications of the Nuttall Ornithological Club, 23. 779.

Clarke, J. T., Lloyd, G. T., & Friedman, M. (2016). Little evidence for enhanced phenotypic evolution in early teleosts relative to their living fossil sister group. *Proceedings of the National Academy of Sciences United States America, 113*(41), 11531–11536.

Coates, M. (2017). Plenty of fish in the sea. *Nature, 549,* 167–169.

Conrad, J. L. (2008). Phylogeny and systematics of Squamata (Reptilia) based on morphology. *Bulletin of the American Museum of Natural History, 310,* 1–182.

Conrath, C. L., & Musick, J. A. (2012). Reproductive biology of elasmobranchs. In J. C. Carrier, J. A. Musick, & M. R. Heithus (Eds.), *Biology of sharks and their relatives* (2nd ed.) (pp. 291–311). Boca Raton, FL: CRC Press.

Cooper, G., & Schiller, A. L. (1975). *Anatomy of the Guinea pig.* Cambridge: Harvard University Press. 417 p.

Crawford, N. G., Faircloth, B. C., McCormack, J. E., Brumfield, R. T., Winker, K., & Glenn, T. C. (2012). More than 1000 ultraconserved elements provide evidence that turtles are the sister group of archosaurs. *Biology Letters, 8,* 783–786.

Crawford, N. G., Parham, J. F., Sellas, A. B., Faircloth, B. C., Glenn, T. C., Papenfuss, T. J., et al. (2015). A phylogenomic analysis of turtles. *Molecular Phylogenetics and Evolution, 83,* 250–257.

Crouch, J. E. (1969). *Text-atlas of cat anatomy.* Philadelphia: Lea & Febiger. 399 p.

Cundall, D. (1987). Functional morphology. In R. A. Seigel, J. T. Collins, & S. S. Novak (Eds.), *Snakes: Ecology and evolutionary biology* (pp. 106–140). New York: MacMillan.

Cunha, G. R., Risbridger, G., Wang, H., Place, N. J., Grumbach, M., Cunha, T. J., et al. (2014). Development of the external genitalia: Perspectives from the spotted hyena (*Crocuta crocuta*). *Differentiation, 87,* 4–22.

Currie, P. H. (2003). Cranial anatomy of tyrannosaurid dinosaurs from the late cretaceous of alberta, Canada. *Acta Palaeontologica Polonica, 48*(2), 191–226.

Da Silva, L. F. B., Tavares, M., & Soares-Gomes, A. (2008). Population structure of the lancelet *Branchiostoma caribaeum* (cephalochordata: Branchiostomidae) in the baía de Guanabara, rio de Janeiro, southeastern Brazil. *Revista Brasileira de Zoologia, 25,* 617–623.

Daeschler, E. B., Shubin, N. H., & Jenkins, F. A. (2006). A Devonian tetrapod-like fish and the evolution of the tetrapod body plan. *Nature, 440*(6), 757–763.

Darwin, C. (1859). *The origin of species by means of natural selection, or the preservation of favoured races in the struggle for life.* London: John Murray, 432.

Davis, S. P., Finarelli, J. A., & Coates, M. I. (2012). *Acanthodes* and shark-like conditions in the last common ancestor of modern gnathostomes. *Nature, 486,* 247–250.

Delsuc, F., Brinkmann, H., Chourrout, D., & Philippe, H. (2006). Tunicates and not cephalochordates are the closest living relatives of vertebrates. *Nature, 439,* 965–968.

Delsuc, F., Philippe, H., Tsagkogeorga, G., Simion, P., Tilak, M.-K., Turon, X., et al. (2018). A phylogenomic framework and timescale for comparative studies of tunicates. *BMC Biology, 16,* 39. https://doi.org/10.1186/s12915-018-0499-2.

Delsuc, F., Tsagkogeorga, G., Lartillot, N., & Philippe, H. (2008). Additional molecular support for the new chordate phylogeny. *Genesis, 46,* 592–604.

Di Fiore, A., Nesca, A., & D'Armiento, M. (2012). Filogenesi della tiroide: Dalle alghe all'uomo. *Clinica Terapeutica, 163*(2), e73–76.

Diogo, R., Kelly, R. G., Christiaen, L., Levine, M., Ziermann, J. M., Molnar, J. L., et al. (2015). A new heart for a new head in vertebrate cardiopharyngeal evolution. *Nature, 520,* 466–473.

Donaghue, P. C. J., & Rücklin, M. (2016). The ins and outs of the evolutionary origin of teeth. *Evolution and Development, 18*(1), 19–30.

Duelmann, W. E., & Trueb, L. (1994). *Biology of Amphibians.* Baltimore: The Johns Hopkins University Press. 670 p.

Duncker, H.-R. (1971). The lung air sac system of birds. *Advances in Anatomy, Embryology, and Cell Biology, 45*(6), 1–171.

Duncker, H.-R. (2004). Vertebrate lungs: Structure, topography and mechanics: A comparative perspective of the progressive integration of respiratory system, locomotor apparatus and ontogenetic development. *Respiratory Physiology and Neurobiology, 144,* 111–124.

Durán, A. C., Fernández, B., Grimes, A. C., Rodríguez, C., Arqué, J. M., & Sans-Coma, V. (2008). Chondrichthyans have a bulbus arteriosus at the arterial pole of the heart: Morphological and evolutionary implications. *Journal of Anatomy, 213,* 597–606.

Ebert, D. A., & Winton, M. V. (2010). Chondrichthyans of high latitude seas. In J. C. Carrier, J. A. Musick, & M. R. Heithus (Eds.), *Biodiversity, adaptive physiology, and conservation* (pp. 115–158). Boca Raton, FL: CRC Press.

Ecker, A. (1971). *The anatomy of the frog.* Translated by Haslam. Vaals: G. A. Asher & Company N. V. 449 p.

Elbroch, M. (2006). *Animal skulls: A guide to north American species.* Mechanicsburg: Stackpole Books. 726 p.

Evans, S. (2003). *The deep scaly project, Python molurus (On-line).* Digital Morphology. http://digimorph.org/specimens/Python_molurus/.

Fairman, J. E. (1999). *Prosauropod and iguanid jaw musculature: A study on the evolution of form and function* (Unpublished Master of Arts Thesis). Johns Hopkins University. 103 p.

Fishbeck, D. W., & Sebastiani, A. (2015). *Comparative anatomy, manual of vertebrate dissection* (3rd ed.). Englewood: Morton Publishing, 566.

Foth, C., & Rauhut, O. W. M. (2017). Re-evaluation of the Haarlem *Archaeopteryx* and the radiation of maniraptoran theropod dinosaurs. *BMC Evolutionary Biology, 17,* 236. https://doi.org/10.1186/s12862-017-1076-y.

Frank, E. T. B. (1934). *The anatomy of the salamander.* London: Oxford University Press. 381 p.

Frazzetta, T. H. (1959). Studies on the morphology and function of the skull in the Boidae (Serpentes). Part I. Cranial differences between *Python sebae* and *Epicrates cenchris*. *Bulletin of the Museum of Comparative Zoology, 119*(8), 451–472.

Frazzetta, T. H. (1966). Studies on the morphology and function of the skull in the Boidae (Serpentes), Part II. Morphology and function of the jaw apparatus in *Python sebae* and *Python molurus*. *Journal of Morphology, 118,* 217–296.

Friedman, M., & Brazeau, M. D. (2013). A jaw-dropping fish. *Nature, 502,* 175–177.

Frink, R. J., & Merrick, B. (1974). The sheep heart: Coronary and conduction system anatomy with special reference to the presence of an os cordis. *The Anatomical Record, 179,* 189–200.

Gaffney, E. S. (1972). An illustrated glossary of turtle skull nomenclature. *American Museum Novitates, 2486,* 1–33.

Gaffney, E. S. (1979). Comparative cranial morphology of recent and fossil turtles. *Bulletin of the American Museum of Natural History, 164*(2), 65–376.

Gans, C., & Parsons, T. S. (1964). *A photographic atlas of shark anatomy.* Chicago: The University of Chicago Press. 106 p.

Garcia-Fernàndez, J. (2006). Amphioxus: A peaceful anchovy fillet to illuminate chordate evolution (II). *International Journal of Biological Sciences, 2*(3), 93–94.

Gee, H. (2018). Across the bridge: Understanding the origin of the vertebrates. (pp. 288). Chicago: University of Chicago Press

George, J. C., & Berger, A. J. (1966). *Avian myology.* New York: Academic Press. 500 p.

Getty, R. (1975). General heart and blood vessels. In R. Getty (Ed.), *Sisson and grossman's the anatomy of the domestic animals* (5th ed.) (pp. 164–175). Philadelphia: Saunders.

Ghonimi, W., Balah, A., Bareedy, M. H., & Abuel-Atta, A. A. (2014). Os cordis of the mature dromedary camel heart (*Camelus dromedaries*) with special emphasis to the cartilago cordis. *Veterinary Science and Technology, 5*(4), 1–7.

Giles, S., Xu, G.-H., Near, T. J., & Friedman, M. (2017). Early members of 'living fossil' lineage imply later origin of modern ray-finned fishes. *Nature, 549,* 265–268. https://doi.org/10.1038/nature23654.

Giribet, G. (2018). Phylogenomics resolves the evolutionary chronicle of our squirting closest relatives. *BMC Biology, 16,* 49. https://doi.org/10.1186/s12915-018-0517-4.

Gilbert, S. G. (1973a). *Pictorial anatomy of the dogfish.* Seattle: University of Washington Press. 59 p.

Gilbert, S. G. (1973b). *Pictorial anatomy of the Necturus.* Seattle: University of Washington Press. 47 p.

Gilbert, S. G. (1974). *Pictorial anatomy of the frog.* Seattle: University of Washington Press. 63 p.

Gilbert, S. G. (1989). *Pictorial anatomy of the cat* (Revised edition). Toronto: University of Toronto Press. 120 p.

Gilbert, S. G. (1991). *Atlas of general anatomy* (2nd ed.). Minneapolis: Burgess Publishing. 104 p.

Gilbert, S. G. (2000). *Outline of cat anatomy with reference to the human.* Seattle: University of Washington Press. 90 p.

Godefroit, P., Cau, A., Dong-Yu, H., Escuillié, F., Wu, W., & Dyke, G. (2013). A Jurassic avialan dinosaur from China resolves the early phylogenetic history of birds. *Nature, 498*(7454), 359–362.

Goloboff, P. A., Catalano, S. A., Mirande, J. M., Szumik, C. A., Arias, J. S., Källersjö, M., et al. (2009). Phylogenetic analysis of 73060 taxa corroborates major eukaryotic groups. *Cladistics, 25,* 211–230.

Gopalakrishnan, G., Blevins, W. E., & Van Alstine, W. G. (2007). Osteocartilaginous metaplasia in the right atrial myocardium of healthy adult sheep. *Journal of Veterinary Diagnostic Investigation, 19,* 518–524.

Gosselin-Ildari, A. (2004). *AmphibiaTree – multiple institutions, Rana catesbeiana (On-line).* Digital Morphology. http://digimorph.org/specimens/Rana_catesbeiana/.

Grigg, G., & Gans, C. (1993). Morphology and physiology of the crocodylia. In C. G. Glas, G. J. B. Ross, & P. L. Beesley (Eds.), *Fauna of Australia, vol 2A Amphibia and reptilia* (pp. 1–439). Canberra: Australian Government Publishing Service.

Gussekloo, W. S., Vosselman, M. G., & Bout, R. G. (2001). Three-dimensional kinematics of skeletal elements in avian prokinetic and rhynchokinetic skulls determined by Roentgen stereophotogrammetry. *Journal of Experimental Biology, 204,* 1735–1744.

Heimberg, A. M., Cowper-Sal·lari, R., Sémon, M., Donoghue, P. C. J., & Peterson, K. J. (2010). MicroRNAs reveal the interrelationships of hagfish, lampreys, and gnathostomes and the nature of the ancestral vertebrate. *Proceedings of the National Academy of Sciences United States America, 107,* 19379–19383.

Hermanson, J. W., de Lahunta, A., & Evans, H. E. (2019). Miller and Evans' anatomy of the dog. (pp. 981). St. Louis: Elsevier.

Hill, A. J., & Iaizzo, P. A. (2009). Comparative cardiac anatomy. In P. A. Iaizzo (Ed.), *Handbook of cardiac anatomy, physiology, and devices* (2nd ed.) (pp. 87–108). New York: Springer.

Holland, N. D., & Holland, L. Z. (2010). Laboratory spawning and development of the Bahama lancelet, *Asymmetron lucayanum* (Cephalochordata): Fertilization through feeding larvae. *Biological Bulletin, 219,* 132–141.

Holland, L. Z. (2016). Tunicates. *Current Biology, 26,* R141–R156.

Holtz, T. R. (2004). Tyrannosauroidea. In D. B. Weishampel, P. Dodson, & H. Osmólska (Eds.), *The dinosauria* (2nd ed.) (pp. 111–136). Berkeley: University of California Press. 861 p.

Homberger, D. G., & Walker, W. F., Jr. (2003). *Vertebrate dissection* (9th ed.). Belmont, CA: Brooks Cole. 379 p.

Hoshino, Z., & Tokioka, T. (1967). An unusually robust *Ciona* from the northeastern coast of Honsyu Island, Japan. *Publications of the Seto Marine Biological Laboratory, 15*(4), 275–290.

Huber, D. R., Soares, M. C., & Carvalho, M. R. (2011). Cartilaginous fishes cranial muscles. In A. P. Farrell (Ed.), *Encyclopedia of fish physiology: From genome to environment* (Vol. 1) (pp. 449–462). San Diego: Academic Press.

Hurum, J. H., & Sabath, K. (2003). Giant theropod dinosaurs from Asia and North America: Skulls of *Tarbosaurus bataar* and *Tyrannosaurus rex* compared. *Acta Palaeontologica Polonica, 48*(2), 169–190.

Iordansky, N. N. (1973). The skull of the crocodilia. In C. Gans, & T. S. Parsons (Eds.), *Biology of the reptilia, volume 4, morphology D* (pp. 201–262). London: Academic Press. 539 p.

Iordansky, N. N. (2011). Cranial kinesis in lizards (lacertilia): Origin, biomechanics, and evolution. *Biology Bulletin, 38*(9), 852–861.

Jandzik, D., Garnett, A. T., Square, T. A., Cattell, M. V., Yu, J.-K., & Medeiros, D. M. (2015). Evolution of the new vertebrate head by co-option of an ancient chordate skeletal tissue. *Nature, 518,* 534–537.

Janvier, P. (2010). microRNAs revive old views about jawless vertebrate divergence and evolution. *Proceedings of the National Academy of Sciences United States America, 107*(45), 19137–19138. https://doi.org/10.1073/pnas.1014583107.

Janvier, P. (2011). Comparative anatomy: All vertebrates do have jaws. *Current Biology, 21*(17), R661–R663. https://doi.org/10.1016/j.cub.2011.07.014.

Janvier, P. (2015). Facts and fancies about early fossil chordates and vertebrates. *Nature, 520,* 483–489. https://doi.org/10.1038/nature14437.

Jenner, R. A. (2015). Response to stach. *BioScience, 65*(2), 119–120.

Joyce, W. G. (2015). The origin of turtles: A paleontological perspective. *Journal of Experimental Zoology (Molecular and Developmental Evolution), 324B,* 181–193.

Kaji, T., Reimer, J. D., Morov, A. R., Kuratani, S., & Yasu, K. (2016). Amphioxus mouth after dorso-ventral inversion. *Zoological Letters, 2,* 2. https://doi.org/10.1186/s40851-016-0038-3.

Kaneto, S., & Wada, H. (2011). Regeneration of amphioxus oral cirri and its skeletal rods: Implications for the origin of the vertebrate skeleton. *Journal of Experimental Zoology B: Molecular and Developmental Evolution, 316*(6), 409–417.

Kardong, K. V. (2018). *Vertebrates. Comparative anatomy, function, evolution* (8th ed.). New York: McGraw-Hill Higher Education, 790.

King, B., Qiao, T., Lee, M. S. Y., Zhu, M., & Long, J. A. (2017). Bayesian morphological clock methods resurrect placoderm monophyly and reveal rapid early evolution in jawed vertebrates. *Systematic Biology, 66*(4), 499–516.

King, A. S., & McLelland, J. (Eds.). (1985). *Form and function in birds* (Vol. 3). London: Academic Press. 522 p.

Kocot, K. M., Tassia, M. G., Halanych, K. M., & Swalla, B. J. (2018). Phylogenetics offers resolution of major tunicate relationships. *Molecular Phylogenetics and Evolution, 121,* 166–173.

Kon, T., Nohara, M., Nishida, M., Sterrer, W., & Nishikawa, T. (2006). Hidden ancient diversification in the circumtropical lancelet *Asymmetron lucayanum* complex. *Marine Biology, 149,* 875–883.

Konietzko-Meier, D., Shelton, C. D., & Sander, P. M. (2016). The discrepancy between morphological and microanatomical patterns of anamniotic stegocephalian postcrania from the Early Permian Briar Creek Bonebed (Texas). *Comptes Rendus Paleovol, 15,* 103–114.

Konno, K., Kaizu, M., Hotta, K., Horie, T., Sasakura, Y., Ikeo, K., et al. (2010). Distribution and structural diversity of cilia in tadpole larvae of the ascidian *Ciona intestinalis. Developmental Biology, 337,* 42–62.

Konrad, M. W. (2016). Blood circulation in the ascidian tunicate *Corella inflata* (Corellidae). *PeerJ, 4,* e2771. https://doi.org/10.7717/peerj.2771.

Konrad, M. W. (2018). A quantitative study of blood circulation in the developing adult ascidian tunicate *Ciona savignyi* (Cionidae). *PeerJ Preprints, 6,* e26556v1. https://doi.org/10.7287/peerj.preprints.26556v1.

Lacalli, T. (2013). Looking into eye evolution: Amphioxus points the way. *Pigment Cell and Melanoma Research, 26*(2), 162–164.

Lacalli, T., & Stach, T. (2016). Acrania (cephalochordata). In S. H. Schmidt-Rhaesa, & G. Purschke (Eds.), *Structure and evolution of invertebrate nervous systems* (pp. 719–734). Oxford: Oxford University Press.

Langer, M. C., Ezcurra, M. D., Rauhut, O. W. M., Benton, M. J., Knoll, F., McPhee, B. W., et al. (2017). Untangling the dinosaur family tree. *Nature, 551*(7678), E1–E3. https://doi.org/10.1038/nature24011.

Laurin, M. (2010). *How vertebrates left the water.* Berkeley: University of California Press. 199 p.

Laurin, M., & Anderson, J. S. (2004). Meaning of the name tetrapoda in the scientific literature: An exchange. *Systematic Biology, 53,* 68–80.

Laurin, M., & Soler-Gijòn, R. (2006). The oldest known stegocephalian (Sarcopterygii: Temnospondyli) from Spain. *Journal of Vertebrate Paleontology, 26,* 284–299.

Laurin, M., & Soler-Gijòn, R. (2010). Osmotic tolerance and habitat of early stegocephalians: Indirect evidence from parsimony, taphonomy, paleobiogeography, physiology and morphology. In M. Vecoli, & G. Clément (Eds.), *The terrestrialization process: Modelling complex interactions at the biosphere-geosphere interface* (pp. 151–179). London: The Geological Society of London.

Li, C., Wu, X.-C., Rieppel, O., Wang, L.-T., & Zhao, L.-J. (2008). An ancestral turtle from the Late Triassic of southwestern China. *Nature, 456*(27), 497–501.

Liem, K. F., Bemis, W. E., Walker, W. F., Jr., & Grande, L. (2001). *Functional anatomy of the vertebrates. An evolutionary perspective* (3rd ed.). Fort Worth: Harcourt College Publishers. 703 p.

Lowe, C. J., Clarke, D. N., Medeiros, D. M., Rokhsar, D. S., & Gerhart, J. (2015). The deuterostome context of chordate origins. *Nature, 520*, 456–465.

Lyson, T. R., Bever, G. S., Bhullar, B.-A. S., Joyce, W. G., & Gauthier, J. A. (2010). Transitional fossils and the origin of turtles. *Biology Letters, 6*, 830–833.

Lyson, T. R., Sperling, E. A., Heimberg, A. M., Gauthier, J. A., King, B. L., & Peterson, K. J. (2012). MicroRNAs support a turtle + lizard clade. *Biology Letters, 8*(1), 104–107.

Mabragaña, E., Figueroa, D. E., Scenna, L. B., Díaz de Astarloa, J. M., Colonello, J. H., & Delpiani, G. (2011). Chondrichthyan egg cases from the south-west Atlantic Ocean. *Journal of Fish Biology, 79*, 1261–1290.

Mackie, G. O., & Burighel, P. (2005). The nervous system in adult tunicates: Current research directions. *Canadian Journal of Zoology, 83*(1), 151–183.

Maisey, J. G., Miller, R., Pradel, A., Denton, J. S. S., Bronson, A., & Janvier, P. (2017). Pectoral morphology in *doliodus*: Bridging the 'acanthodian'-chondrichthyan divide. *American Museum Novitates, 3875*, 1–15.

Maklad, A., Reed, C., Johnson, N. S., & Fritzch, B. (2014). Anatomy of the lamprey ear: Morphological evidence for occurrence of horizontal semicircular ducts in the labyrinth of *Petromyzon marinus*. *Journal of Anatomy, 224*(4), 432–446.

Manni, L., & Pennati, R. (2016). Tunicata. In A. Schmidt-Rhaesa, S. Harzsch, & G. Purschke (Eds.), *Structure and evolution of invertebrate nervous systems* (pp. 699–718). Oxford: Oxford University Press.

Marinelli, W., & Strenger, A. (1954). *Vergleichende Anatomie und Morphologie der Wirbeltiere*. Wien: Verlag Franz Deuticke. 308 p.

Marjanović, D., & Laurin, M. (2013). The origin(s) of extant amphibians: A review with emphasis on the "lepospondyl hypothesis". *Geodiversitas, 35*, 207–272.

Mason, M. J., Lin, C. C., & Narins, P. M. (2003). Sex differences in the middle ear of the bullfrog (*Rana catesbeiana*). *Brain, Behavior and Evolution, 61*, 91–101.

Mason, M. J., & Narins, P. M. (2002). Vibrometric studies of the middle ear of the bullfrog *Rana catesbeiana* II. The operculum. *Journal of Experimental Biology, 205*, 3167–3176.

May, N. D. S. (1964). *The anatomy of the sheep heart* (2nd ed.). Brisbane: University of Queensland Press, 369.

Mayr, G. (2017). *Avian evolution: The fossil record of birds and it paleobiological significance*. Chichester: Wiley Blackwell. 293 p.

Metzger, K. (2002). Cranial kinesis in lepidosaurs: Skulls in motion. In P. Aerts, K. D'Août, A. Herrel, & R. Van Damme (Eds.), *Topics in functional and ecological vertebrate morphology: A tribute to frits de Vree* (pp. 15–46). Maastricht: Shaker Publishing. 372 p.

Millar, R. H. (1953). Ciona*Liverpool marine biology committee memoirs on typical British marine plants and animals* (Vol. 35) 1–123.

Miyashita, T. (2015). Fishing for jaws in early vertebrate evolution: A new hypothesis of mandibular confinement. *Biological Reviews*. https://doi.org/10.111/brv.12187.

Miyashita, T., & Diogo, R. (2016). Evolution of serial patterns in the vertebrate pharyngeal apparatus and paired appendages via assimilation of dissimilar units. *Frontiers in Ecology and Evolution, 4*, 71. https://doi.org/10.3389/fevo.2016.00071.

Mo, J.-Y., Xu, X., & Evans, S. E. (2010). The evolution of the lepidosaurian lower temporal bar: New perspectives from the late cretaceous of South China. *Proceedings of the Royal Society B, 277*(1679), 331–336.

Modesto, S. P., Scott, D. M., MacDougall, M. J., Sues, H.-D., Evans, D. C., & Reisz, R. R. (2015). The oldest parareptile and the early diversification of reptiles. *Proceedings of the Royal Society B, 282*, 20141912. https://doi.org/10.1098/rspb.2014.1912.

Modesto, S. P., Scott, D. M., & Reisz, R. R. (2009). A new parareptile with temporal fenestration from the Middle Permian of South Africa. *Canadian Journal of Earth Sciences, 46*, 9–20.

Mondéjar-Fernández, J. (2018). On cosmine: Its origins, biology and implications for sarcopterygian interrelationships. *Cybium, 42*(1), 41–65.

Motani, R., Jiang, D.-Y., Chen, G.-B., Tintori, A., Rieppel, O., Ji, C., et al. (2015). A basal ichthyosauriform with a short snout from the Lower Triassic of China. *Nature, 517*, 485–488.

Motta, P. J., & Huber, D. R. (2012). Prey capture behaviour and feeding mechanisms. In J. C. Carrier, J. A. Musick, & M. R. Heithus (Eds.), *Biology of sharks and their relatives* (2nd ed.) (pp. 153–209). Boca Raton, FL: CRC Press.

Motta, P. J., & Wilga, C. D. (1999). Anatomy of the feeding apparatus of the nurse shark, *Ginglystoma cirratum*. *Journal of Morphology, 241*, 33–60.

Motta, P. J., & Wilga, C. D. (2001). Advances in the study of feeding behaviors, mechanisms, and mechanics of sharks. *Environmental Biology of Fishes, 60*, 131–156.

Murdock, D. J. E., Dong, X. P., Repetski, J. E., Marone, F., Stampanoni, M., & Donoghue, P. C. J. (2013). The origin of conodonts and of vertebrate mineralized skeletons. *Nature, 502*, 546–549.

Near, T. J. (2009). Conflict and resolution between phylogenies inferred from molecular and phenotypic data sets for hagfish, lampreys and gnathostomes. *Journal of Experimental Zoology (Molecular and Developmental Evolution), 312B*, 749–761.

Neenan, J. M., Klein, N., & Scheyer, T. M. (2013). European origin of placodont marine reptiles and the evolution of crushing dentition in Placodontia. *Nature Communications, 4*, 1621. https://doi.org/10.1038/ncomms2633.

Nicholas, D., Holland, N. D., & Holland, L. Z. (2010). Laboratory spawning and development of the bahama lancelet, *Asymmetron lucayanum* (cephalochordata): Fertilization through feeding larvae. *Biological Bulletin, 219*(2), 132–141.

Nielsen, C. (2015). Evolution of deuterostomy – and origin of the chordates. *Biological Reviews*, 1–10. https://doi.org/10.1111/brv.12229.

Nielsen, S. E., Bone, Q., Bond, P., & Harper, G. (2007). On particle filtration by amphioxus (*Branchiostoma lanceolatum*). *Journal of the Marine Biological Association of the United Kingdom, 87*, 983–989.

O'Connor, P. M. (2004). Pulmonary pneumaticity in the postcranial skeleton of extant aves: A case study examining anseriformes. *Journal of Morphology, 261*, 141–161.

Oelrich, T. M. (1956). *The anatomy of the head of Ctenosaura pectinata (Iguanidae)* (Vol. 94). Miscellaneous Publications, Museum of Zoology, University of Michigan. 122 p.

Oldham, J. C., & Smith, H. M. (1975). *Laboratory anatomy of the iguana*. Dubuque: W. C. Brown Company. 106 p.

Olori, J. (2004). *Chelydra serpentina*. Digital Morphology. http://digimorph.org/specimens/Chelydra_serpentina/.

Ota, K. G., Fujimoto, S., Oisi, Y., & Kuratani, S. (2011). Identification of vertebra-like elements and their possible differentiation from sclerotomes in the hagfish. *Nature Communications, 2*, 273. https://doi.org/10.1038/ncomms1355.

Ota, K. G., Fujimoto, S., Oisi, Y., & Kuratani, S. (2013). Late development of hagfish vertebral elements. *Journal of Experimental Zoology (Molecular and Developmental Evolution), 320B*, 129–139.

Pennisi, E. (2016). Fossil fishes challenge 'urban legend' of evolution. *Science, 353*(6307), 1483.

Pierce, S. E., Hutchinson, J. R., & Clack, J. A. (2013). Historical perspectives on the evolution of tetrapodomorph movement. *Integrative and Comparative Biology*, 1–15. https://doi.org/10.1093/icb/ict022.

Poss, S. G., & Boschung, H. T. (1996). Lancelets (cephalochordata: Branchiostomatidae): How many species are valid? *Israel Journal of Zoology, 42*, S13–S66.

Pough, F. H., & Janis, C. M. (2018). *Vertebrate life* (10th ed.). New York: Sinauer Associates/Oxford University Press. 552 p.

Proctor, N. S., & Lynch, P. J. (1993). *Manual of ornithology. Avian structure & function*. New Haven: Yale University Press. 340 p.

Qiao, T., King, B., Long, J. A., Ahlberg, P. E., & Zhu, M. (2016). Early gnathostome phylogeny revisited: Multiple method consensus. *PLoS One*, *11*(9), e0163157. https://doi.org/10.1371/journal.pone.0163157.

Reisz, R. R., & Head, J. J. (2008). Turtle origins out to sea. *Nature*, *456*(27), 450–451.

Rieppel, O. (1990). The structure and development of the jaw adductor musculature in the turtle *Chelydra serpentina*. *Zoological Journal of the Linnean Society*, *98*, 27–62.

Rieppel, O. (1993). Patterns of diversity in the reptilian skull. In J. Hanken, & B. K. Hall (Eds.), *The skull volume 2, patterns of structural and systematic diversity* (pp. 344–390). Chicago: The University of Chicago Press. 566 p.

Rieppel, O., & de Braga, M. (1996). Turtles as diapsid reptiles. *Nature*, *384*, 453–455.

Rieppel, O., & Zaher, H. (2001). The development of the skull in *Acrochordus granulatus* (Schneider) (Reptilia: Serpentes), with special consideration of the otico-occipital complex. *Journal of Morphology*, *249*, 252–266.

Romer, A. S. (1997). *Osteology of the reptiles*. Malabar: Krieger Publishing Company. 772 p.

Romer, A. S., & Parson, T. S. (1986). *The vertebrate body* (6th ed.). Philadelphia: Saunders College Publishing. 679 p.

Roseth, C. J., & Saltarelli, W. A. (2014). Human cadavers vs. multimedia simulation: A study of student learning in anatomy. *Anatomical Sciences Education*, *7*(5), 331–339.

Ross, C. (2015). Educational paradigm change to dissect to prosect or to game (simulation) that is the question. *College Quarterly*, *18*(1), 1–4.

Rouse, G. W., Wilson, N. G., Carvajal, J. I., & Vrijenhoel, R. C. (2016). New deep-sea species of *Xenoturbella* and the position of Xenacoelomorpha. *Nature*, *530*, 94–97. https://doi.org/10.1038/nature16545.

Rowe, T., Brochu, C., Colbert, M., Kishi, K., & Merck, J. (2003). *Alligator mississippiensis (On-line)*. Digital Morphology. http://digimorph.org/specimens/Alligator_mississippiensis/adult/.

Rücklin, M., & Donoghue, P. C. J. (2015). *Romundina* and the evolutionary origin of teeth. *Biology Letters*, *11*, 20150326. https://doi.org/10.1098/rsbl.2015.0326.

Rücklin, M., & Donoghue, P. C. J. (2016). Reply to 'placoderms and the evolutionary origin of teeth': Burrow et al. (2016). *Biology Letters*, *12*, 20160526. https://doi.org/10.1098/rsbl.2016.0526.

Ruppert, E. E. (1997). Cephalochordata (Acrania). In F. W. Harrison & E. E. Ruppert (Eds.), Microscopic anatomy of invertebrates: Volume 15, Hemichordata, Chaetognatha, and the invertebrate chordates (pp. 349–504). Wiley-Liss: New York.

Ruppert, E. E. (2005). Key characters uniting hemichordates and chordates: Homologies or homoplasies? *Canadian Journal of Zoology*, *83*, 8–23.

Ruppert, E. E., Nash, T. R., & Smith, A. J. (2000). The size range of suspended particles trapped and ingested by the filter-feeding lancelet *Branchiostoma floridae* (Cephalochordata: Acrania). *Journal of the Marine Biological Association of the United Kingdom*, *80*, 329–332.

Satoh, N. (2016). *Chordate origins and evolution. The molecular evolutionary road to vertebrates*. Amsterdam: Academic Press. 206 p.

Scanlon, J. D. (2005). Cranial morphology of the Plio-Pleistocene giant madtsoiid snake *Wonambi naracoortensis*. *Acta Palaeontologica Polonica*, *50*(1), 139–180.

Schlosser, G. (2015). Vertebrate cranial placodes as evolutionary innovations–the ancestor's tale. *Current Topics in Developmental Biology*, *111*, 235–300.

Schoch, R. R. (2014). *Amphibian evolution, the life of early land vertebrates*. Chichester: Wiley Blackwell. 276 p.

Schoch, R. R., & Sues, H.-D. (2015). A Middle Triassic stem-turtle and the evolution of the turtle body plan. *Nature*, *523*, 584–587. https://doi.org/10.1080/14772019.2017.1354936.

Schoch, R. R., & Sues, H.-D. (2017). Osteology of the Middle Triassic stem-turtle *Pappochelys rosinae* and the early evolution of the turtle skeleton. *Journal of Systematic Palaeontology*. https://doi.org/10.1080/14772019.2017.1354936.

Shaffer, H. B., McCartney-Melstad, E., Near, T. J., Mount, G. G., & Spinks, P. Q. (2017). Phylogenomic analyses of 539 highly informative loci dates a fully resolved time tree for the major clades of living turtles (Testudines). *Molecular Phylogenetics and Evolution*, *115*, 7–15.

Sheikh, A. H., Barry, D. S., Gutierrez, H., Cryan, J. F., & O'Keeffe, G. W. (2016). Cadaveric anatomy in the future of medical education: What is the surgeons view? *Anatomical Sciences Education*, *9*, 203–208.

Shubin, N. H., Daeschler, E. B., & Jenkins, F. A. (2006). The pectoral fin of *Tiktaalik roseae* and the origin of the tetrapod limb. *Nature*, *440*(6), 764–771.

Shubin, N. H., Daeschler, E. B., & Jenkins, F. A., Jr. (2014). Pelvic girdle and fin of *Tiktaalik roseae*. *Proceedings of the National Academy of Sciences United States America*, *111*(3), 893–899.

Schultze, H. -P. (2016). Scales, enamel, cosmine, ganoine, and early osteichthyans. *Comptes Rendus Palevol*, *15*, 83–102.

Sibaja-Cordero, J. A., Troncoso, J. S., & Cortés, J. (2012). The lancelet *Asymmetron lucayanum* complex in cocos island national park, Pacific Costa Rica. *Pacific Science*, *66*(4), 523–528.

Sibley, D. A. (2002). *Sibley's birding basics: How to identify birds, using the clues in feathers, habitats, behaviors, and sounds*. New York: Alfred A Knopf, 168.

Soukup, V., & Kozmik, Z. (2016). Zoology: A new mouth for amphioxus. *Current Biology*, *26*, R355–R376. https://doi.org/10.1016/j.cub.2016.03.016.

Sreeranjini, A. R., Ashok, N., Indu, V. R., Lucy, K. M., Maya, S., & Chungath, J. J. (2015). Gross anatomical features of the sternum of green-winged macaw (*Ara chloroptera*). *Indian Journal of Animal Research*, *49*(6), 860–862.

Stach, T. (2008). Chordate phylogeny and evolution: A not so simple three-taxon problem. *Journal of Zoology*, *276*, 117–141.

Stach, T. (2014). Deuterostome phylogeny – a morphological perspective. In W. Wägele, & T. Bartholomaeus (Eds.), *Deep metazoan phylogeny: The backbone of the tree of life: New insights form analyses of molecules, morphology, and theory of data analysis* (pp. 425–457). Berlin: Walter de Gruyter GmbH. 736 p.

Stach, T. (2015). Science behind, around, and after the tree. *BioScience*, *65*(2), 118–119.

Stokes, M. D., & Holland, N. D. (1998). The lancelet. *American Scientist*, *86*, 552–560.

Stuart, R. R. (1947). *The anatomy of the bull frog* (3rd ed.). Chicago: Denoyer-Geppert Company. 30 p.

Sugand, K., Abrahams, P., & Khurana, A. (2010). The anatomy of anatomy: A review for its modernization. *Anatomical Sciences Education*, *3*, 83–93.

Thomson, R. C., Plachetzki, D. C., Mahler, D. L., & Moore, B. R. (2014). A critical appraisal of the use of microRNA data in phylogenetics. *Proceedings of the National Academy of Sciences United States America*, *111*(35), E3659–E3668. https://doi.org/10.1073/pnas.1407207111.

Trueb, L. (1973). Bones, frogs, and evolution. In J. L. Vial (Ed.), *Evolutionary biology of the Anurans, contemporary research on major problems* (pp. 65–132). Columbia: University of Missouri Press. 470 p.

Trueb, L. (1993). Patterns of cranial diversity among the Lissamphibia. In J. Hanken, & B. K. Hall (Eds.), *The skull volume 2, patterns of structural and systematic diversity* (pp. 255–343). Chicago: The University of Chicago Press. 566 p.

Tsuji, L. A., & Müller, J. (2009). Assembling the history of the parareptilia: Phylogeny, diversification, and a new definition of the clade. *Fossil Record*, *12*(1), 71–81.

Turner, S., Burrow, C. J., Schultze, H.-P., Blieck, A., Reif, W.-E., Rexroad, C. B., et al. (2010). False teeth: Conodont-vertebrate phylogenetic relationships revisited. *Geodiversitas, 32*(4), 545–594.

Van Wyk, J., & Rennie, C. O. (2015). Learning anatomy through dissection: Perceptions of a diverse medical student cohort. *International Journal of Morphology, 33*(1), 89–95.

Vargas, J. A., & Harlan, K. D. (2010). On *Branchiostoma californiense* (cephalochordata) from the gulf of nicoya estuary, Costa Rica. *Revista de Biología Tropical, 58*(4), 1143–1148.

Vickaryous, M. K., & Hall, B. K. (2010). Comparative development of the crocodylian interclavicle and avian furcula, with comments on the homology of dermal elements in the pectoral apparatus. *Journal of Experimental Zoology (Molecular and Developmental Evolution), 314B,* 196–207.

Wake, M. H. (1979). *Hyman's comparative vertebrate anatomy* (3rd ed.). Chicago: The University of Chicago Press.

Wang, M., Wang, X., & Zhou, Z. (2016). A new basal bird from China with implications for morphological diversity in early birds. *Scientific Reports, 6,* 19700. https://doi.org/10.1038/srep19700.

Wang, W., Zhong, J., & Wang, Y.-Q. (2010). Comparative genomic analysis reveals the evolutionary conservation of Pax gene family. *Genes and Genetic Systems, 85*(3), 193–206.

Werneburg, I., & Sánchez-Villagra, M. R. (2009). Timing of organogenesis support basal position of turtles in the amniote tree of life. *BMC Evolutionary Biology, 9,* 82. https://doi.org/10.1186/1471-2148-9-82.

White, T. D., & Folkens, P. A. (2005). *The human bone manual.* Amsterdam: Elsevier Academic Press. 464 p.

Wicht, H., & Lacalli, T. C. (2005). The nervous system of amphioxus: Structure, development, and evolutionary significance. *Canadian Journal of Zoology, 83*(1), 122–150.

Wilga, C. D., & Motta, P. J. (1998). Conservation and variation in the feeding mechanism of the spiny dogfish *Squalus acanthias. Journal of Experimental Biology, 201,* 1345–1358.

Wischnitzer, S., & Wischnitzer, E. (2006). *Atlas and dissection guide for comparative anatomy* (6th ed.). New York: W. H. Freeman and Company. 368 p.

Witmer, L. (1997). The evolution of the antorbital cavity of archosaurs: A study in soft-tissue reconstruction in the fossil record with an analysis of the function of pneumaticity. Society of Vertebrate Paleontology Memoir 3 *Supplement to Journal of Vertebrate Paleontology, 17*(1), 1–73.

Wyneken, J. (2001). *The anatomy of sea turtles.* U.S. Department of Commerce NOAA Technical Memorandum NMFS-SEFSC-470. 172 p.

Xu, X., You, H., Du, K., & Han, F. (2011). An *Archaeopteryx*-like theropod from China and the origin of Avialae. *Nature, 475,* 460–465. https://doi.org/10.1038/nature10288.

Zhong, J., Zhang, J., Mukwaya, E., & Wang, Y. (2009). Revaluation of deuterstome phylogeny and evolutionary relationships among chordate subphyla using mitogenome data. *Journal of Genetics and Genomics, 36,* 151–160.

Zhu, M., Ahlberg, P. E., Zhaohui, P., Zhu, Y., Qiao, T., Zhao, W., et al. (2016). A Silurian maxillate placoderm illuminates jaw evolution. *Science, 354*(6310), 334–336.

Zhu, M., Yu, X., Choo, B., Wang, J., & Jia, L. (2012). An antiarch placoderm shows that pelvic girdles arose at the root of jawed vertebrates. *Biology Letters, 8,* 453–456.

Zhu, M., Xiaobo, Y., Ahlberg, P. E., Choo, B., Lu, J., Qiao, T., et al. (2013). A Silurian placoderm with osteichthyan-like marginal jaw bones. *Nature, 502,* 188–193.

Zug, G. R., Vitt, L. J., & Caldwell, J. P. (2001). *Herpetology: An introductory biology of Amphibians and reptiles* (2nd ed.). San Diego: Academic Press. 630 p.

Zusi, R. L. (1993). Patterns of diversity in the avian skull. In J. Hanken, & B. K. Hall (Eds.), *The skull volume 2, patterns of structural and systematic diversity* (pp. 391–437). Chicago: The University of Chicago Press. 566 p.

Index

Note: 'Page numbers followed by "f" indicate figures, "t" indicate tables'.